U0287083

中国科学院科学出版基金资助出版

国外数学名著系列（影印版） 30

Riem annian Geom etry

Second Edition

黎 曼 几 何

（第二版）

Peter Petersen

科学出版社
北京

图字：01-2006-7388

Peter Petersen：Riemannian Geometry(Second Edition)
ⓒ 2006 Springer Science+Business Media，LLC

This reprint has been authorized by Springer-Verlag(Berlin/Heidelberg/New York) for sale in the People's Republic of China only and not for export therefrom．

图书在版编目(CIP)数据

黎曼几何：第2版＝Riemannian Geometry(Second Edition)/(美)彼得森(Petersen，P.)著．—影印版．—北京：科学出版社，2007
（国外数学名著系列）

ISBN 978-7-03-018294-4

Ⅰ.黎… Ⅱ.彼… Ⅲ.黎曼几何 英文 Ⅳ.O186.12

中国版本图书馆CIP数据核字(2006)第153196号

责任编辑：范庆奎/责任印制：吴兆东/封面设计：黄华斌

科学出版社 出版
北京东黄城根北街16号
邮政编码：100717
http://www.sciencep.com

北京中科印刷有限公司印刷
科学出版社发行 各地新华书店经销
＊

2007年1月第 一 版　　开本：720×1000 1/16
2024年11月第八次印刷　　印张：26 1/2
字数：494 000

定价：148.00元
(如有印装质量问题，我社负责调换)

《国外数学名著系列》(影印版)序

要使我国的数学事业更好地发展起来,需要数学家淡泊名利并付出更艰苦地努力。另一方面,我们也要从客观上为数学家创造更有利的发展数学事业的外部环境,这主要是加强对数学事业的支持与投资力度,使数学家有较好的工作与生活条件,其中也包括改善与加强数学的出版工作。

从出版方面来讲,除了较好较快地出版我们自己的成果外,引进国外的先进出版物无疑也是十分重要与必不可少的。从数学来说,施普林格(Springer)出版社至今仍然是世界上最具权威的出版社。科学出版社影印一批他们出版的好的新书,使我国广大数学家能以较低的价格购买,特别是在边远地区工作的数学家能普遍见到这些书,无疑是对推动我国数学的科研与教学十分有益的事。

这次科学出版社购买了版权,一次影印了23本施普林格出版社出版的数学书,就是一件好事,也是值得继续做下去的事情。大体上分一下,这23本书中,包括基础数学书5本,应用数学书6本与计算数学书12本,其中有些书也具有交叉性质。这些书都是很新的,2000年以后出版的占绝大部分,共计16本,其余的也是1990年以后出版的。这些书可以使读者较快地了解数学某方面的前沿,例如基础数学中的数论、代数与拓扑三本,都是由该领域大数学家编著的"数学百科全书"的分册。对从事这方面研究的数学家了解该领域的前沿与全貌很有帮助。按照学科的特点,基础数学类的书以"经典"为主,应用和计算数学类的书以"前沿"为主。这些书的作者多数是国际知名的大数学家,例如《拓扑学》一书的作者诺维科夫是俄罗斯科学院的院士,曾获"菲尔兹奖"和"沃尔夫数学奖"。这些大数学家的著作无疑将会对我国的科研人员起到非常好的指导作用。

当然,23本书只能涵盖数学的一部分,所以,这项工作还应该继续做下去。更进一步,有些读者面较广的好书还应该翻译成中文出版,使之有更大的读者群。

总之,我对科学出版社影印施普林格出版社的部分数学著作这一举措表示热烈的支持,并盼望这一工作取得更大的成绩。

王 元

2005 年 12 月 3 日

To my wife, Laura

Preface

This book is meant to be an introduction to Riemannian geometry. The reader is assumed to have some knowledge of standard manifold theory, including basic theory of tensors, forms, and Lie groups. At times we shall also assume familiarity with algebraic topology and de Rham cohomology. Specifically, we recommend that the reader is familiar with texts like [14], [63], or [87, vol. 1]. For the readers who have only learned a minimum of tensor analysis we have an appendix which covers Lie derivatives, forms, Stokes' theorem, Čech cohomology, and de Rham cohomology. The reader should also have a nodding acquaintance with ordinary differential equations. For this, a text like [67]is more than sufficient.

Most of the material usually taught in basic Riemannian geometry, as well as several more advanced topics, is presented in this text. Several theorems from chapters 7 to 11 appear for the first time in textbook form. This is particularly surprising as we have included essentially only the material students of Riemannian geometry must know.

The approach we have taken sometimes deviates from the standard path. Aside from the usual variational approach (added in the second edition) we have also developed a more elementary approach that simply uses standard calculus together with some techniques from differential equations. Our motivation for this treatment has been that examples become a natural and integral part of the text rather than a separate item that is sometimes minimized. Another desirable by-product has been that one actually gets the feeling that gradients, Hessians, Laplacians, curvatures, and many other things are actually computable.

We emphasize throughout the text the importance of using the correct type of coordinates depending on the theoretical situation at hand. First, we develop a substitute for the second variation formula by using adapted frames or coordinates. This is the approach mentioned above that can be used as an alternative to variational calculus. These are coordinates naturally associated to a distance function. If, for example we use the function that measures the distance to a point, then the adapted coordinates are nothing but polar coordinates. Next, we have exponential coordinates, which are of fundamental importance in showing that distance functions are smooth. Then distance coordinates are used first to show that distance-preserving maps are smooth, and then later to give good coordinate systems in which the metric is sufficiently controlled so that one can prove, say, Cheeger's finiteness theorem. Finally, we have harmonic coordinates. These coordinates have some magical properties. One, in particular, is that in such coordinates the Ricci curvature is essentially the Laplacian of the metric.

From a more physical viewpoint, the reader will get the idea that we are also using the Hamilton-Jacobi equations instead of only relying on the Euler-Lagrange

equations to develop Riemannian geometry (see [5]for an explanation of these matters). It is simply a matter of taste which path one wishes to follow, but surprisingly,
the Hamilton-Jacobi approach has never been tried systematically in Riemannian
geometry.

The book can be divided into five imaginary parts

Part I: Tensor geometry, consisting of chapters 1-4.

Part II: Classical geodesic geometry, consisting of chapters 5 and 6.

Part III: Geometry à la Bochner and Cartan, consisting of chapters 7 and 8.

Part IV: Comparison geometry, consisting of chapters 9-11.

Appendix: De Rham cohomology.

Chapters 1-8 give a pretty complete picture of some of the most classical results
in Riemannian geometry, while chapters 9-11 explain some of the more recent developments in Riemannian geometry. The individual chapters contain the following
material:

Chapter 1: Riemannian manifolds, isometries, immersions, and submersions are
defined. Homogeneous spaces and covering maps are also briefly mentioned. We
have a discussion on various types of warped products, leading to an elementary
account of why the Hopf fibration is also a Riemannian submersion.

Chapter 2: Many of the tensor constructions one needs on Riemannian manifolds are developed. First the Riemannian connection is defined, and it is shown
how one can use the connection to define the classical notions of Hessian, Laplacian,
and divergence on Riemannian manifolds. We proceed to define all of the important
curvature concepts and discuss a few simple properties. Aside from these important
tensor concepts, we also develop several important formulas that relate curvature
and the underlying metric. These formulas are to some extent our replacement
for the second variation formula. The chapter ends with a short section where
such tensor operations as contractions, type changes, and inner products are briefly
discussed.

Chapter 3: First, we indicate some general situations where it is possible to
diagonalize the curvature operator and Ricci tensor. The rest of the chapter is
devoted to calculating curvatures in several concrete situations such as: spheres,
product spheres, warped products, and doubly warped products. This is used to
exhibit some interesting examples that are Ricci flat and scalar flat. In particular,
we explain how the Riemannian analogue of the Schwarzschild metric can be constructed. Several different models of hyperbolic spaces are mentioned. We have a
section on Lie groups. Here two important examples of left-invariant metrics are
discussed as well the general formulas for the curvatures of bi-invariant metrics.
Finally, we explain how submersions can be used to create new examples. We
have paid detailed attention to the complex projective space. There are also some
general comments on how submersions can be constructed using isometric group
actions.

Chapter 4: Here we concentrate on the special case where the Riemannian manifold is a hypersurface in Euclidean space. In this situation, one gets some special
relations between curvatures. We give examples of simple Riemannian manifolds
that cannot be represented as hypersurface metrics. Finally we give a brief introduction to the global Gauss-Bonnet theorem and its generalization to higher
dimensions.

Chapter 5: This chapter further develops the foundational topics for Riemannian manifolds. These include, the first variation formula, geodesics, Riemannian

manifolds as metric spaces, exponential maps, geodesic completeness versus metric completeness, and maximal domains on which the exponential map is an embedding. The chapter ends with the classification of simply connected space forms and metric characterizations of Riemannian isometries and submersions.

Chapter 6: We cover two more foundational techniques: parallel translation and the second variation formula. Some of the classical results we prove here are: The Hadamard-Cartan theorem, Cartan's center of mass construction in nonpositive curvature and why it shows that the fundamental group of such spaces are torsion free, Preissmann's theorem, Bonnet's diameter estimate, and Synge's lemma. We have supplied two proofs for some of the results dealing with non-positive curvature in order that people can see the difference between using the variational (or Euler-Lagrange) method and the Hamilton-Jacobi method. At the end of the chapter we explain some of the ingredients needed for the classical quarter pinched sphere theorem as well as Berger's proof of this theorem. Sphere theorems will also be revisited in chapter 11.

Chapter 7: Many of the classical and more recent results that arise from the Bochner technique are explained. We start with Killing fields and harmonic 1-forms as Bochner did, and finally, discuss some generalizations to harmonic p-forms. For the more advanced audience we have developed the language of Clifford multiplication for the study p-forms, as we feel that it is an important way of treating this material. The last section contains some more exotic, but important, situations where the Bochner technique is applied to the curvature tensor. These last two sections can easily be skipped in a more elementary course. The Bochner technique gives many nice bounds on the topology of closed manifolds with nonnegative curvature. In the spirit of comparison geometry, we show how Betti numbers of nonnegatively curved spaces are bounded by the prototypical compact flat manifold: the torus.

The importance of the Bochner technique in Riemannian geometry cannot be sufficiently emphasized. It seems that time and again, when people least expect it, new important developments come out of this simple philosophy.

While perhaps only marginally related to the Bochner technique we have also added a discussion on how the presence of Killing fields in positive sectional curvature can lead to topological restrictions. This is a rather new area in Riemannian geometry that has only been developed in the last 15 years.

Chapter 8: Part of the theory of symmetric spaces and holonomy is developed. The standard representations of symmetric spaces as homogeneous spaces and via Lie algebras are explained. We prove Cartan's existence theorem for isometries. We explain how one can compute curvatures in general and make some concrete calculations on several of the Grassmann manifolds including complex projective space. Having done this, we define holonomy for general manifolds, and discuss the de Rham decomposition theorem and several corollaries of it. The above examples are used to give an idea of how one can classify symmetric spaces. Also, we show in the same spirit why symmetric spaces of (non)compact type have (nonpositive) nonnegative curvature operator. Finally, we present a brief overview of how holonomy and symmetric spaces are related with the classification of holonomy groups. This is used in a grand synthesis, with all that has been learned up to this point, to give Gallot and Meyer's classification of compact manifolds with nonnegative curvature operator.

Chapter 9: Manifolds with lower Ricci curvature bounds are investigated in further detail. First, we discuss volume comparison and its uses for Cheng's maximal diameter theorem. Then we investigate some interesting relationships between Ricci curvature and fundamental groups. The strong maximum principle for continuous functions is developed. This result is first used in a warm-up exercise to give a simple proof of Cheng's maximal diameter theorem. We then proceed to prove the Cheeger-Gromoll splitting theorem and discuss its consequences for manifolds with nonnegative Ricci curvature.

Chapter 10: Convergence theory is the main focus of this chapter. First, we introduce the weakest form of convergence: Gromov-Hausdorff convergence. This concept is often useful in many contexts as a way of getting a weak form of convergence. The real object is then to figure out what weak convergence implies, given some stronger side conditions. There is a section which breezes through Hölder spaces, Schauder's elliptic estimates and harmonic coordinates. To facilitate the treatment of the stronger convergence ideas, we have introduced a norm concept for Riemannian manifolds. We hope that these norms will make the subject a little more digestible. The main idea of this chapter is to prove the Cheeger-Gromov convergence theorem, which is called the Convergence Theorem of Riemannian Geometry, and Anderson's generalizations of this theorem to manifolds with bounded Ricci curvature.

Chapter 11: In this chapter we prove some of the more general finiteness theorems that do not fall into the philosophy developed in chapter 10. To begin, we discuss generalized critical point theory and Toponogov's theorem. These two techniques are used throughout the chapter to prove all of the important theorems. First, we probe the mysteries of sphere theorems. These results, while often unappreciated by a larger audience, have been instrumental in developing most of the new ideas in the subject. Comparison theory, injectivity radius estimates, and Toponogov's theorem were first used in a highly nontrivial way to prove the classical quarter pinched sphere theorem of Rauch, Berger, and Klingenberg. Critical point theory was invented by Grove and Shiohama to prove the diameter sphere theorem. After the sphere theorems, we go through some of the major results of comparison geometry: Gromov's Betti number estimate, The Soul theorem of Cheeger and Gromoll, and The Grove-Petersen homotopy finiteness theorem.

Appendix A: Here, some of the important facts about forms and tensors are collected. Since Lie derivatives are used rather heavily at times we have included an initial section on this. Stokes' theorem is proved, and we give a very short and streamlined introduction to Čech and de Rham cohomology. The exposition starts with the assumption that we only work with manifolds that can be covered by finitely many charts where all possible intersections are contractible. This makes it very easy to prove all of the major results, as one can simply use the Poincaré and Meyer-Vietoris lemmas together with induction on the number of charts in the covering.

At the end of each chapter, we give a list of books and papers that cover and often expand on the material in the chapter. We have whenever possible attempted to refer just to books and survey articles. The reader is then invited to go from those sources back to the original papers. For more recent works, we also give journal references if the corresponding books or surveys do not cover all aspects of the original paper. One particularly exhaustive treatment of Riemannian Geometry for the reader who is interested in learning more is [11]. Other valuable texts that expand or complement much of the material covered here are [70], [87]and [90]. There is also a historical survey by Berger (see [10]) that complements this text very well.

A first course should definitely cover chapters 2, 5, and 6 together with whatever one feels is necessary from chapters 1, 3, and 4. Note that chapter 4 is really a world unto itself and is not used in a serious way later in the text. A more advanced course could consist of going through either part III or IV as defined earlier. These parts do not depend in a serious way on each other. One can probably not cover the entire book in two semesters, but one can cover parts I, II, and III or alternatively I, II, and IV depending on one's inclination. It should also be noted that, if one ignores the section on Killing fields in chapter 7, then this material can actually be covered without having been through chapters 5 and 6. Each of the chapters ends with a collection of exercises. These exercises are designed both to reinforce the material covered and to establish some simple results that will be needed later. The reader should at least read and think about all of the exercises, if not actually solve all of them.

There are several people I would like to thank. First and foremost are those students who suffered through my various pedagogical experiments with the teaching of Riemannian geometry. Special thanks go to Marcel Berger, Hao Fang, Semion Shteingold, Chad Sprouse, Marc Troyanov, Gerard Walschap, Nik Weaver, Fred Wilhelm and Hung-Hsi Wu for their constructive criticism of parts of the book. For the second edition I'd also like to thank Edward Fan, Ilkka Holopainen, Geoffrey Mess, Yanir Rubinstein, and Burkhard Wilking for making me aware of typos and other deficiencies in the first edition. I would especially like to thank Joseph Borzellino for his very careful reading of this text, and Peter Blomgren for writing the programs that generated Figures 2.1 and 2.2. Finally I would like to thank Robert Greene, Karsten Grove, and Gregory Kallo for all the discussions on geometry we have had over the years.

The author was supported in part by NSF grants DMS 0204177 and DMS 9971045.

Contents

CHAPTER 1

Riemannian Metrics

In this chapter we shall introduce the category (i.e., sets and maps) that we wish to work with. Without discussing any theory we present many examples of Riemannian manifolds and Riemannian maps. All of these examples will form the foundation for future investigations into constructions of Riemannian manifolds with various interesting properties.

The abstract definition of a Riemannian manifold used today dates back only to the 1930s as it wasn't really until Whitney's work in 1936 that mathematicians obtained a clear understanding of what abstract manifolds were, other than just being submanifolds of Euclidean space. Riemann himself defined Riemannian metrics only on domains in Euclidean space. Riemannian manifolds where then objects that locally looked like a general metric on a domain in Euclidean space, rather than manifolds with an inner product on each tangent space. Before Riemann, Gauss and others only worked with 2-dimensional geometry. The invention of Riemannian geometry is quite curious. The story goes that Gauss was on Riemann's defense committee for his Habilitation (super doctorate). In those days, the candidate was asked to submit three topics in advance, with the implicit understanding that the committee would ask to hear about the first topic (the actual thesis was on Fourier series and the Riemann integral.) Riemann's third topic was "On the Hypotheses which lie at the Foundations of Geometry." Clearly he was hoping that the committee would select from the first two topics, which were on material he had already worked on. Gauss, however, always being in an inquisitive mood, decided he wanted to hear whether Riemann had anything to say about the subject on which he, Gauss, was the reigning expert. So, much to Riemann's dismay, he had to go home and invent Riemannian geometry to satisfy Gauss's curiosity. No doubt Gauss was suitably impressed, a very rare occurrence indeed for him.

From Riemann's work it appears that he worked with changing metrics mostly by multiplying them by a function (conformal change). By conformally changing the standard Euclidean metric he was able to construct all three constant-curvature geometries in one fell swoop for the first time ever. Soon after Riemann's discoveries it was realized that in polar coordinates one can change the metric in a different way, now referred to as a warped product. This also yields in a unified way all constant curvature geometries. Of course, Gauss already knew about polar coordinate representations on surfaces, and rotationally symmetric metrics were studied even earlier. But these examples are much simpler than the higher-dimensional analogues. Throughout this book we shall emphasize the importance of these special warped products and polar coordinates. It is not far to go from warped products to doubly warped products, which will also be defined in this chapter, but they don't seem to have attracted much attention until Schwarzschild discovered a vacuum space-time that wasn't flat. Since then, doubly warped products have been at the heart of many examples and counterexamples in Riemannian geometry.

Another important way of finding Riemannian metrics is by using left-invariant metrics on Lie groups. This leads us to, among other things, the Hopf fibration and Berger spheres. Both of these are of fundamental importance and are also at the core of a large number of examples in Riemannian geometry. These will also be defined here and studied further throughout the book.

1. Riemannian Manifolds and Maps

A *Riemannian manifold* (M, g) consists of a C^∞-manifold M and a Euclidean inner product g_p or $g|_p$ on each of the tangent spaces T_pM of M. In addition we assume that g_p varies smoothly. This means that for any two smooth vector fields X, Y the inner product $g_p(X|_p, Y|_p)$ should be a smooth function of p. The subscript p will be suppressed when it is not needed. Thus we might write $g(X, Y)$ with the understanding that this is to be evaluated at each p where X and Y are defined. When we wish to associate the metric with M we also denote it as g_M. Often we shall also need M to be connected, and thus we make the assumption throughout the book that we work only with connected manifolds.

All inner product spaces of the same dimension are isometric; therefore all tangent spaces T_pM on a Riemannian manifold (M, g) are isometric to the n-dimensional Euclidean space $\mathbb{R}^n$ endowed with its canonical inner product. Hence, all Riemannian manifolds have the same infinitesimal structure not only as manifolds but also as Riemannian manifolds.

EXAMPLE 1. *The simplest and most fundamental Riemannian manifold is Euclidean space* $(\mathbb{R}^n, \text{can})$. *The canonical Riemannian structure "can" is defined by identifying the tangent bundle* $\mathbb{R}^n \times \mathbb{R}^n \simeq T\mathbb{R}^n$ *via the map:* $(p, v) \to$ *velocity of the curve* $t \to p + tv$ *at* $t = 0$. *The standard inner product on* $\mathbb{R}^n$ *is then defined by*

$$g((p, v), (p, w)) = v \cdot w.$$

A *Riemannian isometry* between Riemannian manifolds (M, g_M) and (N, g_N) is a diffeomorphism $F : M \to N$ such that $F^* g_N = g_M$, i.e.,

$$g_N(DF(v), DF(w)) = g_M(v, w)$$

for all tangent vectors $v, w \in T_pM$ and all $p \in M$. In this case F^{-1} is a Riemannian isometry as well.

EXAMPLE 2. *Whenever we have a finite-dimensional vector space* V *with an inner product, we can construct a Riemannian manifold by declaring, as with Euclidean space, that*

$$g((p, v), (p, w)) = v \cdot w.$$

If we have two such Riemannian manifolds (V, g_V) *and* (W, g_W) *of the same dimension, then they are isometric. The Riemannian isometry* $F : V \to W$ *is simply the linear isometry between the two spaces. Thus* $(\mathbb{R}^n, \text{can})$ *is not only the only n-dimensional inner product space, but also the only Riemannian manifold of this simple type.*

Suppose that we have an immersion (or embedding) $F : M \to N$, and that (N, g_N) is a Riemannian manifold. We can then construct a Riemannian metric on

Figure 1.1

M by pulling back g_N to $g_M = F^* g_N$ on M, in other words,

$$g_M(v, w) = g_N(DF(v), DF(w)).$$

Notice that this defines an inner product as $DF(v) = 0$ implies $v = 0$.

A *Riemannian immersion* (or *Riemannian embedding*) is thus an immersion (or embedding) $F : M \to N$ such that $g_M = F^* g_N$. Riemannian immersions are also called *isometric immersions*, but as we shall see below they are almost never distance preserving.

EXAMPLE 3. *We now come to the second most important example. Define*

$$S^n(r) = \{x \in \mathbb{R}^{n+1} : |x| = r\}.$$

This is the Euclidean sphere of radius r. The metric induced from the embedding $S^n(r) \hookrightarrow \mathbb{R}^{n+1}$ is the canonical metric on $S^n(r)$. The unit sphere, or standard sphere, is $S^n = S^n(1) \subset \mathbb{R}^{n+1}$ with the induced metric. In Figure 1.1 is a picture of the unit sphere in $\mathbb{R}^3$ shown with latitudes and longitudes.

EXAMPLE 4. *If $k < n$ there are, of course, several linear isometric immersions $(\mathbb{R}^k, \text{can}) \to (\mathbb{R}^n, \text{can})$. Those are, however, not the only isometric immersions. In fact, any curve $\gamma : \mathbb{R} \to \mathbb{R}^2$ with unit speed, i.e., $|\dot{\gamma}(t)| = 1$ for all $t \in \mathbb{R}$, is an example of an isometric immersion. One could, for example, take*

$$t \to (\cos t, \ \sin t)$$

as an immersion, and

$$t \to \left(\log\left(t + \sqrt{1 + t^2} \right), \sqrt{1 + t^2} \right)$$

as an embedding. A map of the form:

$$F \ : \ \mathbb{R}^k \to \mathbb{R}^{k+1}$$
$$F(x^1, \ldots, x^k) \ = \ (\gamma(x^1), x^2, \ldots, x^k),$$

(where γ fills up the first two entries) will then give an isometric immersion (or embedding) that is not linear. This is counterintuitive in the beginning, but serves to illustrate the difference between a Riemannian immersion and a distance-preserving map. In Figure 1.2 there are two pictures, one of the cylinder, the other of the isometric embedding of $\mathbb{R}^2$ into $\mathbb{R}^3$ just described.

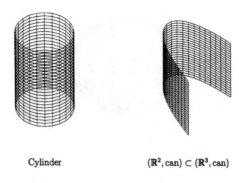

Cylinder $(\mathbb{R}^2, \mathrm{can}) \subset (\mathbb{R}^3, \mathrm{can})$

Figure 1.2

There is also the dual concept of a *Riemannian submersion* $F : (M, g_M) \rightarrow (N, g_N)$. This is a submersion $F : M \rightarrow N$ such that for each $p \in M$, $DF : \ker^{\perp}(DF) \rightarrow T_{F(p)}N$ is a linear isometry. In other words, if $v, w \in T_p M$ are perpendicular to the kernel of $DF : T_p M \rightarrow T_{F(p)}N$, then

$$g_M(v, w) = g_N(DF(v), DF(w)).$$

This is also equivalent to saying that the adjoint $(DF_p)^* : T_{F(p)}N \rightarrow T_p M$ preserves inner products of vectors. Thus the notion is dual to that of a Riemannian immersion.

EXAMPLE 5. *Orthogonal projections* $(\mathbb{R}^n, \mathrm{can}) \rightarrow (\mathbb{R}^k, \mathrm{can})$ *where* $k < n$ *are examples of Riemannian submersions.*

EXAMPLE 6. *A much less trivial example is the* Hopf fibration $S^3(1) \rightarrow S^2(\frac{1}{2})$. *As observed by F. Wilhelm this map can be written as*

$$(z, w) \rightarrow \left(\frac{1}{2} \left(|w|^2 - |z|^2 \right), z\bar{w} \right)$$

if we think of $S^3(1) \subset \mathbb{C}^2$ *and* $S^2(\frac{1}{2}) \subset \mathbb{R} \oplus \mathbb{C}$. *Note that the fiber containing* (z, w) *consists of the points* $(e^{i\theta}z, e^{i\theta}w)$ *and hence* $i(z, w)$ *is tangent to the fiber. Therefore, the tangent vectors that are perpendicular to those points are of the form* $\lambda(-\bar{w}, \bar{z})$, $\lambda \in \mathbb{C}$. *We can check what happens to these tangent vectors by computing*

$$\left(\frac{1}{2} \left(|w + \lambda\bar{z}|^2 - |z - \lambda\bar{w}|^2 \right), (z - \lambda\bar{w}) \overline{(w + \lambda\bar{z})} \right)$$

and then isolating the first order term in λ. *This term is*

$$\left(2\mathrm{Re}\left(\bar{\lambda}zw \right), -\lambda\bar{w}^2 + \bar{\lambda}z^2 \right)$$

and has length $|\lambda|$. *As this is also the length of* $\lambda(-\bar{w}, \bar{z})$ *we have shown that the map is a Riemannian submersion. Below we will examine this example more closely.*

Finally we should mention a very important generalization of Riemannian manifolds. A *semi-* or *pseudo-Riemannian* manifold consists of a manifold and a smoothly varying symmetric bilinear form g on each tangent space. We assume in addition that g is nondegenerate, i.e., for each nonzero $v \in T_p M$ there exists $w \in T_p M$ such that $g(v, w) \neq 0$. This is clearly a generalization of a Riemannian metric where we have the more restrictive assumption that $g(v, v) > 0$ for all nonzero v. Each tangent space admits a splitting $T_p M = P \oplus N$ such that g is positive definite on P and negative definite on N. These subspaces are not unique but it is easy to show that their dimensions are. Continuity of g shows that nearby tangent spaces must have a similar splitting where the subspaces have the same dimension. Thus we can define the *index* of a connected semi-Riemannian manifold as the dimension of the subspace N on which g is negative definite.

EXAMPLE 7. *Let* $n = n_1 + n_2$ *and* $\mathbb{R}^{n_1,n_2} = \mathbb{R}^{n_1} \times \mathbb{R}^{n_2}$. *We can then write vectors in* $\mathbb{R}^{n_1,n_2}$ *as* $v = v_1 + v_2$, *where* $v_1 \in \mathbb{R}^{n_1}$ *and* $v_2 \in \mathbb{R}^{n_2}$. *A natural semi-Riemannian metric of index* n_1 *is defined by*

$$g\left((p,v),(p,w)\right) = -v_1 \cdot w_1 + v_2 \cdot w_2.$$

When $n_1 = 1$ *or* $n_2 = 1$ *this coincides with one or the other version of Minkowski space. We shall use this space in chapter 3.*

Much of the tensor analysis that we shall do on Riemannian manifolds can be carried over to semi-Riemannian manifolds without further ado. It is only when we start using norms of vectors that things won't work out in a similar fashion.

2. Groups and Riemannian Manifolds

We shall study groups of Riemannian isometries on Riemannian manifolds and see how this can be useful in constructing new Riemannian manifolds.

2.1. Isometry Groups. For a Riemannian manifold (M,g) let $\mathrm{Iso}(M) = \mathrm{Iso}(M,g)$ denote the group of Riemannian isometries $F : (M,g) \to (M,g)$ and $\mathrm{Iso}_p(M,g)$ the *isotropy (sub)group* at p, i.e., those $F \in \mathrm{Iso}(M,g)$ with $F(p) = p$. A Riemannian manifold is said to be *homogeneous* if its isometry group acts *transitively*, i.e., for each pair of points $p,q \in M$ there is an $F \in \mathrm{Iso}(M,g)$ such that $F(p) = q$.

EXAMPLE 8.

$$
\begin{aligned}
\mathrm{Iso}(\mathbb{R}^n, \mathrm{can}) &= \mathbb{R}^n \rtimes O(n) \\
&= \{F : \mathbb{R}^n \to \mathbb{R}^n : F(x) = v + Ox, \ v \in \mathbb{R}^n \text{ and } O \in O(n)\}.
\end{aligned}
$$

(Here $H \rtimes G$ *is the semidirect product, with* G *acting on* H *in some way.) The translational part* v *and rotational part* O *are uniquely determined. It is clear that these maps indeed are isometries. To see the converse first observe that* $G(x) = F(x) - F(0)$ *is also a Riemannian isometry. Using that it is a Riemannian isometry, we observe that at* $x = 0$ *the differential* $DG_0 \in O(n)$. *Thus,* G *and* DG_0 *are isometries on Euclidean space, both of which preserve the origin and have the same differential there. It is then a general uniqueness result for Riemannian isometries that* $G = DG_0$ *(see chapter 5). In the exercises to chapter 2 there is a more elementary proof which only works for Euclidean space.*

The isotropy group Iso_p *is apparently always isomorphic to* $O(n)$, *so we see that* $\mathbb{R}^n \simeq \mathrm{Iso}/\mathrm{Iso}_p$ *for any* $p \in \mathbb{R}^n$. *This is in fact always true for homogeneous spaces.*

EXAMPLE 9. *On the sphere*

$$\mathrm{Iso}(S^n(r), \mathrm{can}) = O(n+1) = \mathrm{Iso}_0(\mathbb{R}^{n+1}, \mathrm{can}).$$

It is again clear that $O(n+1) \subset \mathrm{Iso}(S^n(r), \mathrm{can})$. Conversely, if $F \in \mathrm{Iso}(S^n(r), \mathrm{can})$ extend it to

$$\tilde{F} \; : \; \mathbb{R}^{n+1} \to \mathbb{R}^{n+1}$$
$$\tilde{F}(x) \; = \; |x| \cdot r^{-1} \cdot F\left(x \cdot |x|^{-1} \cdot r\right),$$
$$\tilde{F}(0) \; = \; 0.$$

Then check that

$$\tilde{F} \in \mathrm{Iso}_0(\mathbb{R}^{n+1}, \mathrm{can}) = O(n+1).$$

This time the isotropy groups are isomorphic to $O(n)$, that is, those elements of $O(n+1)$ fixing a 1-dimensional linear subspace of $\mathbb{R}^{n+1}$. In particular, $O(n+1)/O(n) \simeq S^n$.

2.2. Lie Groups. More generally, consider a Lie group G. The tangent space can be trivialized

$$TG \simeq G \times T_e G$$

by using left (or right) translations on G. Therefore, any inner product on $T_e G$ induces a *left-invariant* Riemannian metric on G i.e., left translations are Riemannian isometries. It is obviously also true that any Riemannian metric on G for which all left translations are Riemannian isometries is of this form. In contrast to $\mathbb{R}^n$, not all of these Riemannian metrics need be isometric to each other. A Lie group might therefore not come with a canonical metric.

If H is a closed subgroup of G, then we know that G/H is a manifold. If we endow G with a metric such that *right translation* by elements in H act by isometries, then there is a unique Riemannian metric on G/H making the projection $G \to G/H$ into a Riemannian submersion. If in addition the metric is also left invariant then G acts by isometries on G/H (on the left) thus making G/H into a homogeneous space.

We shall investigate the next two examples further in chapter 3.

EXAMPLE 10. *The idea of taking the quotient of a Lie group by a subgroup can be generalized. Consider $S^{2n+1}(1) \subset \mathbb{C}^{n+1}$. $S^1 = \{\lambda \in \mathbb{C} : |\lambda| = 1\}$ acts by complex scalar multiplication on both S^{2n+1} and $\mathbb{C}^{n+1}$; furthermore this action is by isometries. We know that the quotient $S^{2n+1}/S^1 = \mathbb{C}P^n$, and since the action of S^1 is by isometries, we induce a metric on $\mathbb{C}P^n$ such that $S^{2n+1} \to \mathbb{C}P^n$ is a Riemannian submersion. This metric is called the Fubini-Study metric. When $n = 1$, this turns into the Hopf fibration $S^3(1) \to \mathbb{C}P^1 = S^2(\frac{1}{2})$.*

EXAMPLE 11. *One of the most important nontrivial Lie groups is $SU(2)$, which is defined as*

$$SU(2) \; = \; \left\{ A \in M_{2\times 2}(\mathbb{C}) : \det A = 1, A^* = A^{-1} \right\}$$
$$= \; \left\{ \begin{bmatrix} z & w \\ -\bar{w} & \bar{z} \end{bmatrix} : |z|^2 + |w|^2 = 1 \right\}$$
$$= \; S^3(1).$$

The Lie algebra $\mathfrak{su}(2)$ of $SU(2)$ is

$$\mathfrak{su}(2) = \left\{ \begin{bmatrix} i\alpha & \beta + i\gamma \\ -\beta + i\gamma & -i\alpha \end{bmatrix} : \alpha, \beta, \gamma \in \mathbb{R} \right\}$$

and is spanned by

$$X_1 = \begin{bmatrix} i & 0 \\ 0 & -i \end{bmatrix}, X_2 = \begin{bmatrix} 0 & 1 \\ -1 & 0 \end{bmatrix}, X_3 = \begin{bmatrix} 0 & i \\ i & 0 \end{bmatrix}.$$

We can think of these matrices as left-invariant vector fields on $SU(2)$. If we declare them to be orthonormal, then we get a left-invariant metric on $SU(2)$, which as we shall later see is $S^3(1)$. If instead we declare the vectors to be orthogonal, X_1 to have length ε, and the other two to be unit vectors, we get a very important 1-parameter family of metrics g_ε on $SU(2) = S^3$. These distorted spheres are called Berger spheres. Note that scalar multiplication on $S^3 \subset \mathbb{C}^2$ corresponds to multiplication on the left by the matrices

$$\begin{bmatrix} e^{i\theta} & 0 \\ 0 & e^{-i\theta} \end{bmatrix}$$

as

$$\begin{bmatrix} e^{i\theta} & 0 \\ 0 & e^{-i\theta} \end{bmatrix} \begin{bmatrix} z & w \\ -\bar{w} & \bar{z} \end{bmatrix} = \begin{bmatrix} e^{i\theta}z & e^{i\theta}w \\ -e^{-i\theta}\bar{w} & e^{-i\theta}\bar{z} \end{bmatrix}$$

Thus X_1 is exactly tangent to the orbits of the Hopf circle action. The Berger spheres are therefore obtained from the canonical metric by multiplying the metric along the Hopf fiber by ε^2.

2.3. Covering Maps. Discrete groups also commonly occur in geometry, often as deck transformations or covering groups. Suppose that $F : M \to N$ is a covering map. Then F is, in particular, both an immersion and a submersion. Thus, any Riemannian metric on N induces a Riemannian metric on M, making F into an isometric immersion, also called a *Riemannian covering*. Since $\dim M = \dim N$, F must, in fact, be a *local isometry*, i.e., for every $p \in M$ there is a neighborhood $U \ni p$ in M such that $F|_U : U \to F(U)$ is a Riemannian isometry. Notice that the pullback metric on M has considerable symmetry. For if $q \in V \subset N$ is evenly covered by $\{U_p\}_{p \in F^{-1}(q)}$, then all the sets V and U_p are isometric to each other. In fact, if F is a normal covering, i.e., there is a group Γ of deck transformations acting on M such that:

$$F^{-1}(p) = \{g(q) : F(q) = p \text{ and } g \in \Gamma\},$$

then Γ acts by isometries on the pullback metric. This can be used in the opposite direction. Namely, if $N = M/\Gamma$ and M is a Riemannian manifold, where Γ acts by isometries, then there is a unique Riemannian metric on N such that the quotient map is a local isometry.

EXAMPLE 12. *If we fix a basis v_1, v_2 for $\mathbb{R}^2$, then $\mathbb{Z}^2$ acts by isometries through the translations*

$$(n, m) \to (x \to x + nv_1 + mv_2).$$

The orbit of the origin looks like a lattice. The quotient is a torus T^2 with some metric on it. Note that T^2 is itself an Abelian Lie group and that these metrics are invariant with respect to the Lie group multiplication. These metrics will depend on v_1 and v_2 so they need not be isometric to each other.

EXAMPLE 13. *The involution $-I$ on $S^n(1) \subset \mathbb{R}^{n+1}$ is an isometry and induces a Riemannian covering $S^n \to \mathbb{R}P^n$.*

3. Local Representations of Metrics

3.1. Einstein Summation Convention. We shall often use the index and summation convention introduced by Einstein. Given a vector space V, such as the tangent space of a manifold, we use subscripts for vectors in V. Thus a basis of V is denoted by $v_1, \ldots, v_n$. Given a vector $v \in V$ we can then write it as a linear combination of these basis vectors as follows

$$v = \sum_i \alpha^i v_i = \alpha^i v_i = \begin{bmatrix} v_1 & \cdots & v_n \end{bmatrix} \begin{bmatrix} \alpha^1 \\ \vdots \\ \alpha^n \end{bmatrix}.$$

Here we use superscripts on the coefficients and then automatically sum over indices that are repeated as both sub- and superscripts. If we define a dual basis v^i for the dual space $V^* = \mathrm{Hom}\,(V, \mathbb{R})$ as follows:

$$v^i\,(v_j) = \delta^i_j,$$

then the coefficients can also be computed via

$$\alpha^i = v^i\,(v).$$

It is therefore convenient to use superscripts for dual bases in V^*. The matrix representation $\left(\alpha^j_i\right)$ of a linear map $L : V \to V$ is found by solving

$$L\,(v_i) \quad = \quad \alpha^j_i v_j,$$

$$\begin{bmatrix} L\,(v_1) & \cdots & L\,(v_n) \end{bmatrix} = \begin{bmatrix} v_1 & \cdots & v_n \end{bmatrix} \begin{bmatrix} \alpha^1_1 & \cdots & \alpha^1_n \\ \vdots & \ddots & \vdots \\ \alpha^n_1 & \cdots & \alpha^n_n \end{bmatrix}$$

In other words

$$\alpha^j_i = v^j\,(L\,(v_i)).$$

As already indicated subscripts refer to the column number and superscripts to the row number.

When the objects under consideration are defined on manifolds, the conventions carry over as follows. Cartesian coordinates on $\mathbb{R}^n$ and coordinates on a manifold have superscripts $\left(x^i\right)$, as they are the coefficients of the vector corresponding to this point. Coordinate vector fields therefore look like

$$\partial_i = \frac{\partial}{\partial x^i},$$

and consequently they have subscripts. This is natural, as they form a basis for the tangent space. The dual 1-forms

$$dx^i$$

satisfy

$$dx^j\,(\partial_i) = \delta^j_i$$

and therefore form the natural dual basis for the cotangent space.

Einstein notation is not only useful when one doesn't want to write summation symbols, it also shows when certain coordinate- (or basis-) dependent definitions are invariant under change of coordinates. Examples occur throughout the book.

For now, let us just consider a very simple situation, namely, the velocity field of a curve $c : I \to \mathbb{R}^n$. In coordinates, the curve is written

$$
\begin{aligned}
c(t) &= \left(x^i(t) \right) \\
&= x^i(t) e_i,
\end{aligned}
$$

if e_i is the standard basis for $\mathbb{R}^n$. The velocity field is now defined as the vector

$$
\dot{c}(t) = \left(\dot{x}^i(t) \right).
$$

Using the coordinate vector fields this can also be written as

$$
\dot{c}(t) = \frac{dx^i}{dt} \frac{\partial}{\partial x^i} = \dot{x}^i(t) \partial_i.
$$

In a coordinate system on a general manifold we could then try to use this as our definition for the velocity field of a curve. In this case we must show that it gives the same answer in different coordinates. This is simply because the chain rule tells us that

$$
\dot{x}^i(t) = dx^i \left(\dot{c}(t) \right),
$$

and then observing that, we have used the above definition for finding the components of a vector in a given basis.

Generally speaking, we shall, when it is convenient, use Einstein notation. When giving coordinate-dependent definitions we shall be careful that they are given in a form where they obviously conform to this philosophy and are consequently easily seen to be invariantly defined.

3.2. Coordinate Representations. On a manifold M we can multiply 1-forms to get bilinear forms:

$$
\theta_1 \cdot \theta_2(v, w) = \theta_1(v) \cdot \theta_2(w).
$$

Note that $\theta_1 \cdot \theta_2 \neq \theta_2 \cdot \theta_1$. Given coordinates $x(p) = (x^1, \ldots, x^n)$ on an open set U of M, we can thus construct bilinear forms $dx^i \cdot dx^j$. If in addition M has a Riemannian metric g, then we can write

$$
g = g(\partial_i, \partial_j) dx^i \cdot dx^j
$$

because

$$
\begin{aligned}
g(v, w) &= g(dx^i(v)\partial_i,\ dx^j(w)\partial_j) \\
&= g(\partial_i, \partial_j) dx^i(v) \cdot dx^j(w).
\end{aligned}
$$

The functions $g(\partial_i, \partial_j)$ are denoted by g_{ij}. This gives us a representation of g in local coordinates as a positive definite symmetric matrix with entries parametrized over U. Initially one might think that this gives us a way of concretely describing Riemannian metrics. That, however, is a bit optimistic. Just think about how many manifolds you know with a good covering of coordinate charts together with corresponding transition functions. On the other hand, coordinate representations are often a good theoretical tool for doing abstract calculations.

EXAMPLE 14. *The canonical metric on $\mathbb{R}^n$ in the identity chart is*

$$
g = \delta_{ij} dx^i dx^j = \sum_{i=1}^{n} \left(dx^i \right)^2.
$$

EXAMPLE 15. *On $\mathbb{R}^2 - \{half\ line\}$ we also have polar coordinates (r, θ). In these coordinates the canonical metric looks like*

$$g = dr^2 + r^2 d\theta^2.$$

In other words,

$$g_{rr} = 1, \ g_{r\theta} = g_{\theta r} = 0, \ g_{\theta\theta} = r^2.$$

To see this recall that

$$\begin{aligned} x^1 &= r \cos\theta, \\ x^2 &= r \sin\theta. \end{aligned}$$

Thus,

$$\begin{aligned} dx^1 &= \cos\theta dr - r \sin\theta d\theta, \\ dx^2 &= \sin\theta dr + r \cos\theta d\theta, \end{aligned}$$

which gives

$$\begin{aligned} g &= (dx^1)^2 + (dx^2)^2 \\ &= (\cos\theta dr - r \sin\theta d\theta)^2 + (\sin\theta dr + r \cos\theta d\theta)^2 \\ &= (\cos^2\theta + \sin^2\theta)dr^2 + (r \cos\theta\sin\theta - r \cos\theta\sin\theta)dr d\theta \\ &\quad + (r \cos\theta\sin\theta - r \cos\theta\sin\theta)d\theta dr + (r^2 \sin^2\theta)d\theta^2 + (r^2 \cos^2\theta)d\theta^2 \\ &= dr^2 + r^2 d\theta^2 \end{aligned}$$

3.3. Frame Representations. A similar way of representing the metric is by choosing a *frame* $X_1, \ldots, X_n$ on an open set U of M, i.e., n linearly independent vector fields on U, where $n = \dim M$. If $\sigma^1, \ldots, \sigma^n$ is the coframe, i.e., the 1-forms such that $\sigma^i(X_j) = \delta^i_j$, then the metric can be written as

$$g = g_{ij}\sigma^i\sigma^j,$$

where $g_{ij} = g(X_i, X_j)$.

EXAMPLE 16. *Any left-invariant metric on a Lie group G can be written as*

$$(\sigma^1)^2 + \cdots + (\sigma^n)^2$$

using a coframing dual to left-invariant vector fields $X_1, \ldots, X_n$ forming an orthonormal basis for T_eG. If instead we just begin with a framing of left-invariant vector fields $X_1, \ldots, X_n$ and dual coframing $\sigma^1, \ldots, \sigma^n$, then any left-invariant metric g depends only on its values on T_eG and can therefore be written $g = g_{ij}\sigma^i\sigma^j$, where g_{ij} is a positive definite symmetric matrix with real-valued entries. The Berger sphere can, for example, be written

$$g_\varepsilon = \varepsilon^2(\sigma^1)^2 + (\sigma^2)^2 + (\sigma^3)^2,$$

where $\sigma^i(X_j) = \delta^i_j$.

EXAMPLE 17. *A surface of revolution consists of a curve*

$$\gamma(t) = (r(t), z(t)) : I \to \mathbb{R}^2,$$

where $I \subset \mathbb{R}$ is open and $r(t) > 0$ for all t. By rotating this curve around the z-axis, we get a surface that can be represented as

$$(t, \theta) \to f(t, \theta) = (r(t)\cos\theta, r(t)\sin\theta, z(t)).$$

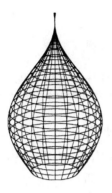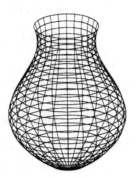

Figure 1.3

This is a cylindrical coordinate representation, and we have a natural frame $\partial_t, \partial_\theta$ on all of the surface with dual coframe $dt, d\theta$. We wish to write down the induced metric $dx^2 + dy^2 + dz^2$ from $\mathbb{R}^3$ in this frame. Observe that

$$
\begin{aligned}
dx &= \dot{r}\cos(\theta)\, dt - r\sin(\theta)\, d\theta, \\
dy &= \dot{r}\sin(\theta)\, dt + r\cos(\theta)\, d\theta, \\
dz &= \dot{z}\, dt.
\end{aligned}
$$

so

$$
\begin{aligned}
dx^2 + dy^2 + dz^2 &= \left(\dot{r}\cos(\theta)\, dt - r\sin(\theta)\, d\theta\right)^2 \\
&\quad + \left(\dot{r}\sin(\theta)\, dt + r\cos(\theta)\, d\theta\right)^2 + \left(\dot{z}\, dt\right)^2 \\
&= \left(\dot{r}^2 + \dot{z}^2\right) dt^2 + r^2 d\theta^2.
\end{aligned}
$$

Thus
$$
g = (\dot{r}^2 + \dot{z}^2)dt^2 + r^2 d\theta^2.
$$

If the curve is parametrized by arc length, we have the simpler formula:
$$
g = dt^2 + r^2 d\theta^2.
$$

This is reminiscent of our polar coordinate description of $\mathbb{R}^2$. In Figure 1.3 there are two pictures of surfaces of revolution. The first shows that when $r = 0$ the metric looks pinched and therefore destroys the manifold. In the second, r starts out being zero, but this time the metric appears smooth, as r has vertical tangent to begin with.

EXAMPLE 18. On $I \times S^1$ we also have the frame $\partial_t, \partial_\theta$ with coframe $dt, d\theta$. Metrics of the form
$$
g = \eta^2(t)dt^2 + \varphi^2(t)d\theta^2
$$
are called rotationally symmetric since η and φ do not depend on θ. We can, by change of coordinates on I, generally assume that $\eta = 1$. Note that not all rotationally symmetric metrics come from surfaces of revolution. For if $dt^2 + r^2 d\theta^2$ is a surface of revolution, then $\dot{z}^2 + \dot{r}^2 = 1$. Whence $|\dot{r}| \le 1$.

EXAMPLE 19. $S^2(r) \subset \mathbb{R}^3$ is a surface of revolution. Just revolve
$$
t \to (r\sin(tr^{-1}), r\cos(tr^{-1}))
$$

around the z-axis. The metric looks like

$$dt^2 + r^2 \sin^2\left(\frac{t}{r}\right) d\theta^2.$$

Note that $r\sin(tr^{-1}) \to t$ *as* $r \to \infty$, *so very large spheres look like Euclidean space. By changing* r *to* ir, *we arrive at some interesting rotationally symmetric metrics:*

$$dt^2 + r^2 \sinh^2(\frac{t}{r}) d\theta^2,$$

that are not surfaces of revolution. If we let $\mathrm{sn}_k(t)$ *denote the unique solution to*

$$\begin{aligned}
\ddot{x}(t) + k \cdot x(t) &= 0, \\
x(0) &= 0, \\
\dot{x}(0) &= 1,
\end{aligned}$$

then we have a 1-parameter family

$$dt^2 + \mathrm{sn}_k^2(t) d\theta^2$$

of rotationally symmetric metrics. (The notation sn_k *will be used throughout the text, it should not be confused with Jacobi's elliptic function* $\mathrm{sn}(k, u)$.) *When* $k = 0$, *this is* $\mathbb{R}^2$; *when* $k > 0$, *we get* $S^2\left(\frac{1}{\sqrt{k}}\right)$; *and when* $k < 0$, *we arrive at the* hyperbolic *(from* $\sinh$) *metrics from above.*

3.4. Polar Versus Cartesian Coordinates. In the rotationally symmetric examples we haven't discussed what happens when $\varphi(t) = 0$. In the revolution case, the curve clearly needs to have a vertical tangent in order to look smooth. To be specific, assume that we have $dt^2 + \varphi^2(t) d\theta^2$, $\varphi : [0, b) \to [0, \infty)$, where $\varphi(0) = 0$ and $\varphi(t) > 0$ for $t > 0$. All other situations can be translated or reflected into this position. We assume that φ is smooth, so we can rewrite it as $\varphi(t) = t\psi(t)$ for some smooth $\psi(t) > 0$ for $t > 0$. Now introduce "Cartesian coordinates"

$$\begin{aligned}
x &= t\cos\theta, \\
y &= t\sin\theta
\end{aligned}$$

near $t = 0$. Then $t^2 = x^2 + y^2$ and

$$\begin{aligned}
\left[\begin{array}{c} dt \\ d\theta \end{array}\right] &= \left[\begin{array}{cc} \cos(\theta) & \sin(\theta) \\ -t^{-1}\sin(\theta) & t^{-1}\cos(\theta) \end{array}\right] \left[\begin{array}{c} dx \\ dy \end{array}\right] \\
&= \left[\begin{array}{cc} t^{-1}x & t^{-1}y \\ -t^{-2}y & t^{-2}x \end{array}\right] \left[\begin{array}{c} dx \\ dy \end{array}\right].
\end{aligned}$$

Thus,

$$\begin{aligned}
dt^2 + \varphi^2(t) d\theta^2 &= dt^2 + t^2\psi^2(t) d\theta^2 \\
&= t^{-2}(xdx + ydy)^2 + (t^2)\psi^2(t)t^{-4}(-ydx + xdy)^2 \\
&= t^{-2}x^2 dx^2 + t^{-2}xydxdy + t^{-2}xydydx \\
&\quad + t^{-2}y^2 dy^2 + t^{-2}\psi^2(t)(xdy - ydx)^2 \\
&= t^{-2}(x^2 + \psi^2(t)y^2)dx^2 + t^{-2}(xy - xy\psi^2(t))dxdy \\
&\quad + t^{-2}(xy - xy\psi^2(t))dydx + t^{-2}(\psi^2(t)x^2 + y^2)dy^2,
\end{aligned}$$

whence

$$g_{xx} = \frac{(x^2 + \psi^2(t)y^2)}{x^2 + y^2} = 1 + \frac{\psi^2(t) - 1}{t^2} \cdot y^2,$$

$$g_{xy} = g_{yx} = \frac{1 - \psi^2(t)}{t^2} \cdot xy,$$

$$g_{yy} = \frac{(\psi^2(t)x^2 + y^2)}{x^2 + y^2} = 1 + \frac{\psi^2(t) - 1}{t^2} \cdot x^2,$$

and we need to check for smoothness of the functions at $(x, y) = 0$ (or $t = 0$). For this we must obviously check that the function

$$\frac{\psi^2(t) - 1}{t^2}$$

is smooth at $t = 0$. First, it is clearly necessary that $\psi(0) = 1$; this is the vertical tangent condition. Second, if ψ is given by a power series we see that it must further satisfy: $\dot{\psi}(0) = \psi^{(3)}(0) = \cdots = 0$. With a little more work these conditions can be seen to be sufficient when ψ is merely smooth. If we translate back to φ, we get that the metric is smooth at $t = 0$ iff $\varphi^{(\text{even})}(0) = 0$ and $\dot{\varphi}(0) = 1$.

These conditions are all satisfied by the metrics $dt^2 + \text{sn}_k^2(t)d\theta^2$, where $t \in [0, \infty)$ when $k \le 0$ and $t \in [0, \frac{\pi}{\sqrt{k}}]$ for $k > 0$. Note that in this case $\text{sn}_k(t)$ is real analytic.

4. Doubly Warped Products

4.1. Doubly Warped Products in General. We can more generally consider metrics on $I \times S^{n-1}$ of the type $dt^2 + \varphi^2(t)ds_{n-1}^2$, where ds_{n-1}^2 is the canonical metric on $S^{n-1}(1) \subset \mathbb{R}^n$. Even more general are metrics of the type:

$$dt^2 + \varphi^2(t)ds_p^2 + \psi^2(t)ds_q^2$$

on $I \times S^p \times S^q$. The first type are again called *rotationally symmetric*, while those of the second type are a special class of *doubly warped products*. As for smoothness, when $\varphi(t) = 0$ we can easily check that the situation for rotationally symmetric metrics is identical to what happened in the previous section. For the doubly warped product observe that nondegeneracy of the metric implies that φ and ψ cannot both be zero at the same time. However, we have the following lemmas:

PROPOSITION 1. *If $\varphi : (0, b) \to (0, \infty)$ is smooth and $\varphi(0) = 0$, then we get a smooth metric at $t = 0$ iff*

$$\varphi^{(\text{even})}(0) = 0,$$
$$\dot{\varphi}(0) = 1,$$

and

$$\psi(0) > 0,$$
$$\psi^{(\text{odd})}(0) = 0.$$

The topology near $t = 0$ in this case is $\mathbb{R}^{p+1} \times S^q$.

PROPOSITION 2. *If $\varphi : (0, b) \to (0, \infty)$ is smooth and $\varphi(b) = 0$, then we get a smooth metric at $t = b$ iff*

$$\varphi^{(\text{even})}(b) = 0,$$
$$\dot{\varphi}(b) = -1,$$

and

$$\psi(b) > 0,$$
$$\psi^{(odd)}(b) = 0.$$

The topology near $t = b$ in this case is again $\mathbb{R}^{p+1} \times S^q$.

By adjusting and possibly changing the roles of these function we can get three different types of topologies.

- $\varphi, \psi : [0, \infty) \to [0, \infty)$ are both positive on all of $(0, \infty)$. Then we have a smooth metric on $\mathbb{R}^{p+1} \times S^q$ if φ, ψ satisfy the first proposition.
- $\varphi, \psi : [0, b] \to [0, \infty)$ are both positive on $(0, b)$ and satisfy both propositions. Then we get a smooth metric on $S^{p+1} \times S^q$.
- $\varphi, \psi : [0, b] \to [0, \infty)$ as in the second type but the roles of ψ and φ are interchanged at $t = b$. Then we get a smooth metric on S^{p+q+1}!

4.2. Spheres as Warped Products. First let us show how the standard sphere can be written as a rotationally symmetric metric in all dimensions. The metrics $dr^2 + sn_k^2(r)ds_{n-1}^2$ are analogous to the surfaces from the last section. So when $k = 0$ we get $(\mathbb{R}^n, can)$, and when $k = 1$ we get $(S^n(1), can)$. To see the last statement observe that we have a map

$$F : (0, \pi) \times \mathbb{R}^n \to \mathbb{R} \times \mathbb{R}^n,$$
$$F(r, z) = (t, x) = (\cos(r), \sin(r) \cdot z),$$

which reduces to a map

$$G : (0, \pi) \times S^{n-1} \to \mathbb{R} \times \mathbb{R}^n,$$
$$G(r, z) = (\cos(r), \sin(r) \cdot z).$$

Thus, G really maps into the unit sphere in $\mathbb{R}^{n+1}$. To see that G is a Riemannian isometry we just compute the canonical metric on $\mathbb{R} \times \mathbb{R}^n$ using the coordinates $(\cos(r), \sin(r) \cdot z)$. To do the calculation we use that

$$1 = \left(z^1\right)^2 + \cdots + \left(z^n\right)^2,$$
$$0 = d\left(\left(z^1\right)^2 + \cdots + \left(z^n\right)^2\right) = 2\left(z^1 dz^1 + \cdots + z^n dz^n\right)$$

$$\begin{aligned}
can &= dt^2 + \sum \delta_{ij} dx^i dx^j \\
&= (d\cos(r))^2 + \sum \delta_{ij} d\left(\sin(r) z^i\right) d\left(\sin(r) z^j\right) \\
&= \sin^2(r) dr^2 + \sum \delta_{ij} \left(z^i \cos(r) dr + \sin(r) dz^i\right)\left(z^j \cos(r) dr + \sin(r) dz^j\right) \\
&= \sin^2(r) dr^2 + \sum \delta_{ij} z^i z^j \cos^2(r) dr^2 + \sum \delta_{ij} z^i \cos(r) \sin(r) dr dz^j \\
&\quad + \sum \delta_{ij} z^j \cos(r) \sin(r) dz^i dr + \sum \delta_{ij} \sin^2(r) dz^i dz^j \\
&= \sin^2(r) dr^2 + \cos^2(r) dr^2 \sum \delta_{ij} z^i z^j + \sin^2(r) \sum \delta_{ij} dz^i dz^j \\
&\quad + \cos(r) \sin(r) dr \sum z^i dz^i + \cos(r) \sin(r) \left(\sum z^i dz^i\right) dr \\
&= dr^2 + \sin^2(r) \left(\left(dz^1\right)^2 + \cdots + \left(dz^n\right)^2\right)
\end{aligned}$$

The claim now follows from the fact that $(dz^1)^2 + \cdots + (dz^n)^2$ restricted to S^{n-1} is exactly the canonical metric ds^2_{n-1}.

The metrics

$$dt^2 + \sin^2(t)ds^2_p + \cos^2(t)ds^2_q, \ t \in [0, \tfrac{\pi}{2}],$$

are also $(S^{p+q+1}(1), \mathrm{can})$. Namely, we have $S^p \subset \mathbb{R}^{p+1}$ and $S^q \subset \mathbb{R}^{q+1}$, so we can map

$$\begin{aligned}(0, \tfrac{\pi}{2}) \times S^p \times S^q \ &\to \ \mathbb{R}^{p+1} \times \mathbb{R}^{q+1}, \\ (t, x, y) \ &\to \ (x \cdot \sin(t), y \cdot \cos(t)),\end{aligned}$$

where $x \in \mathbb{R}^{p+1}$, $y \in \mathbb{R}^{q+1}$ have $|x| = |y| = 1$. These embeddings clearly map into the unit sphere. The computations that the map is a Riemannian isometry are similar to the above calculations.

4.3. The Hopf Fibration. With all this in mind, let us revisit the Hopf fibration $S^3(1) \to S^2\left(\tfrac{1}{2}\right)$ and show that it is a Riemannian submersion between the spaces indicated. On $S^3(1)$, write the metric as

$$dt^2 + \sin^2(t)d\theta^2_1 + \cos^2(t)d\theta^2_2, \ t \in \left[0, \frac{\pi}{2}\right],$$

and use complex coordinates

$$(t, e^{i\theta_1}, e^{i\theta_2}) \to (\sin(t)e^{i\theta_1}, \cos(t)e^{i\theta_2})$$

to describe the isometric embedding

$$\left(0, \frac{\pi}{2}\right) \times S^1 \times S^1 \hookrightarrow S^3(1) \subset \mathbb{C}^2.$$

Since the Hopf fibers come from complex scalar multiplication, we see that they are of the form

$$\theta \to (t, e^{i(\theta_1+\theta)}, e^{i(\theta_2+\theta)}).$$

On $S^2\left(\tfrac{1}{2}\right)$ use the metric

$$dr^2 + \frac{\sin^2(2r)}{4}d\theta^2, \ r \in \left[0, \frac{\pi}{2}\right],$$

with coordinates

$$(r, e^{i\theta}) \to \left(\frac{1}{2}\cos(2r), \frac{1}{2}\sin(2r)e^{i\theta}\right).$$

The Hopf fibration in these coordinates looks like

$$(t, e^{i\theta_1}, e^{i\theta_2}) \to (t, e^{i(\theta_1-\theta_2)}).$$

This conforms with Wilhelm's map defined earlier if we observe that

$$(\sin(t)e^{i\theta_1}, \cos(t)e^{i\theta_2})$$

is supposed to be mapped to

$$\left(\frac{1}{2}\left(\cos^2 t - \sin^2 t\right), \sin(t)\cos(t)\, e^{i(\theta_1-\theta_2)}\right) = \left(\frac{1}{2}\cos(2t), \frac{1}{2}\sin(2t)\, e^{i(\theta_1-\theta_2)}\right)$$

On $S^3(1)$ we have an orthogonal framing

$$\left\{\partial_{\theta_1} + \partial_{\theta_2}, \ \partial_t, \ \frac{\cos^2(t)\partial_{\theta_1} - \sin^2(t)\partial_{\theta_2}}{\cos(t)\sin(t)}\right\},$$

where the first vector is tangent to the Hopf fiber and the two other vectors have unit length. On $S^2\left(\frac{1}{2}\right)$

$$\{\partial_r, \frac{2}{\sin(2r)}\partial_\theta\}$$

is an orthonormal frame. The Hopf map clearly maps

$$\partial_t \quad \to \quad \partial_r,$$

$$\frac{\cos^2(t)\partial_{\theta_1} - \sin^2(t)\partial_{\theta_2}}{\cos(t)\sin(t)} \quad \to \quad \frac{\cos^2(r)\partial_\theta + \sin^2(r)\partial_\theta}{\cos(r)\sin(r)} = \frac{2}{\sin(2r)}\cdot\partial_\theta,$$

thus showing that it is an isometry on vectors perpendicular to the fiber.

Notice also that the map

$$(t, e^{i\theta_1}, e^{i\theta_2}) \to (\cos(t)e^{i\theta_1}, \sin(t)e^{i\theta_2}) \to \begin{pmatrix} \cos(t)e^{i\theta_1} & \sin(t)e^{i\theta_2} \\ -\sin(t)e^{-i\theta_2} & \cos(t)e^{-i\theta_1} \end{pmatrix}$$

gives us the promised isometry from $S^3(1)$ to $SU(2)$, where $SU(2)$ has the left-invariant metric described earlier.

The map

$$I \times S^1 \times S^1 \quad \to \quad I \times S^1$$
$$(t, e^{i\theta_1}, e^{i\theta_2}) \quad \to \quad (t, e^{i(\theta_1-\theta_2)})$$

is in fact always a Riemannian submersion when the domain is endowed with the doubly warped product metric

$$dt^2 + \varphi^2(t)d\theta_1^2 + \psi^2(t)d\theta_2^2$$

and the target has the rotationally symmetric metric

$$dr^2 + \frac{(\varphi(t)\cdot\psi(t))^2}{\varphi^2(t) + \psi^2(t)}d\theta^2.$$

This submersion can be generalized to higher dimensions as follows: On $I \times S^{2n+1} \times S^1$ consider the doubly warped product metric

$$dt^2 + \varphi^2(t)ds_{2n+1}^2 + \psi^2(t)d\theta^2.$$

The unit circle acts by complex scalar multiplication on both S^{2n+1} and S^1 and consequently induces a free isometric action on this space (if $\lambda \in S^1$ and $(z, w) \in S^{2n+1} \times S^1$, then $\lambda \cdot (z, w) = (\lambda z, \lambda w)$.) The quotient map

$$I \times S^{2n+1} \times S^1 \to I \times \left((S^{2n+1} \times S^1)/S^1\right)$$

can be made into a Riemannian submersion by choosing an appropriate metric on the quotient space. To find this metric, we split the canonical metric

$$ds_{2n+1}^2 = h + g,$$

where h corresponds to the metric along the Hopf fiber and g is the orthogonal component. In other words, if $pr : T_pS^{2n+1} \to T_pS^{2n+1}$ is the orthogonal projection (with respect to ds_{2n+1}^2) whose image is the distribution generated by the Hopf action, then

$$h(v, w) = ds_{2n+1}^2(pr(v), pr(w))$$

and

$$g(v, w) = ds_{2n+1}^2(v - pr(v), w - pr(w)).$$

We can then redefine

$$dt^2 + \varphi^2(t)ds_{2n+1}^2 + \psi^2(t)d\theta^2 = dt^2 + \varphi^2(t)g + \varphi^2(t)h + \psi^2(t)d\theta^2.$$

Now notice that $\left(S^{2n+1} \times S^1\right)/S^1 = S^{2n+1}$ and that the S^1 only collapses the Hopf fiber while leaving the orthogonal component to the Hopf fiber unchanged. In analogy with the above example, we therefore get that the metric on $I \times S^{2n+1}$ can be written

$$dt^2 + \varphi^2(t)g + \frac{(\varphi(t) \cdot \psi(t))^2}{\varphi^2(t) + \psi^2(t)}h.$$

In the case where $n = 0$ we recapture the previous case, as g doesn't appear. When $n = 1$, the decomposition: $ds_3^2 = h + g$ can also be written

$$\begin{aligned} ds_3^2 &= (\sigma^1)^2 + (\sigma^2)^2 + (\sigma^3)^2, \\ h &= (\sigma^1)^2, \\ g &= (\sigma^2)^2 + (\sigma^3)^2, \end{aligned}$$

where $\{\sigma^1, \sigma^2, \sigma^3\}$ is the coframing coming from the identification $S^3 \simeq SU(2)$. The Riemannian submersion in this case can therefore be written

$$\left(I \times S^3 \times S^1, \ dt^2 + \varphi^2(t)\left[(\sigma^1)^2 + (\sigma^2)^2 + (\sigma^3)^2\right] + \psi^2(t)d\theta^2\right)$$
$$\downarrow$$
$$\left(I \times S^3, \ dt^2 + \varphi^2(t)[(\sigma^2)^2 + (\sigma^3)^2] + \frac{(\varphi(t)\cdot\psi(t))^2}{\varphi^2(t)+\psi^2(t)}(\sigma^1)^2\right).$$

If we let $\varphi = \sin(t)$, $\psi = \cos(t)$, and $t \in I = \left[0, \frac{\pi}{2}\right]$, then we get the generalized Hopf fibration $S^{2n+3} \to \mathbb{C}P^{n+1}$ defined by

$$\left(0, \frac{\pi}{2}\right) \times \left(S^{2n+1} \times S^1\right) \to \left(0, \frac{\pi}{2}\right) \times \left(\left(S^{2n+1} \times S^1\right)/S^1\right)$$

as a Riemannian submersion, and the Fubini-Study metric on $\mathbb{C}P^{n+1}$ can be represented as

$$dt^2 + \sin^2(t)(g + \cos^2(t)h).$$

5. Exercises

(1) On product manifolds $M \times N$ one has special product metrics $g = g_M + g_N$, where g_M, g_N are metrics on M, N respectively.
   (a) Show that $(\mathbb{R}^n, \mathrm{can}) = \left(\mathbb{R}, dt^2\right) \times \cdots \times \left(\mathbb{R}, dt^2\right)$.
   (b) Show that the flat square torus

$$T^2 = \mathbb{R}^2/\mathbb{Z}^2 = \left(S^1, \left(\frac{1}{2\pi}\right)^2 d\theta^2\right) \times \left(S^1, \left(\frac{1}{2\pi}\right)^2 d\theta^2\right).$$

   (c) Show that

$$F(\theta_1, \theta_2) = \frac{1}{2}(\cos\theta_1, \sin\theta_1, \cos\theta_2, \sin\theta_2)$$

   is a Riemannian embedding: $T^2 \to \mathbb{R}^4$.
(2) Suppose we have an isometric group action G on (M, g) such that the quotient space M/G is a manifold and the quotient map a submersion. Show that there is a unique Riemannian metric on the quotient making the quotient map a Riemannian submersion.

(3) Construct paper models of the Riemannian manifolds $\left(\mathbb{R}^2, dt^2 + a^2 t^2 d\theta^2\right)$. If $a = 1$, this is of course the Euclidean plane, and when $a < 1$, they look like cones. What do they look like when $a > 1$?

(4) Suppose φ and ψ are positive on $(0, \infty)$ and consider the Riemannian submersion

$$\left((0, \infty) \times S^3 \times S^1, \; dt^2 + \varphi^2(t)\left[(\sigma^1)^2 + (\sigma^2)^2 + (\sigma^3)^2\right] + \psi^2(t) d\theta^2\right)$$
$$\downarrow$$
$$\left((0, \infty) \times S^3, \; dt^2 + \varphi^2(t)[(\sigma^2)^2 + (\sigma^3)^2] + \frac{(\varphi(t) \cdot \psi(t))^2}{\varphi^2(t) + \psi^2(t)}(\sigma^1)^2\right).$$

Define $f = \varphi$ and $h = \frac{(\varphi(t) \cdot \psi(t))^2}{\varphi^2(t) + \psi^2(t)}$ and assume that

$$\begin{aligned} f(0) &> 0, \\ f^{(\text{odd})}(0) &= 0, \end{aligned}$$

and

$$\begin{aligned} h(0) &= 0, \\ h'(0) &= k, \\ h^{(\text{even})}(0) &= 0, \end{aligned}$$

where k is a positive integer. Show that the above construction yields a smooth metric on the vector bundle over S^2 with Euler number $\pm k$. Hint: Away from the zero section this vector bundle is $(0, \infty) \times S^3/\mathbb{Z}_k$, where $S^3/\mathbb{Z}_k$ is the quotient of S^3 by the cyclic group of order k acting on the Hopf fiber. You should use the submersion description and then realize this vector bundle as a submersion of $S^3 \times \mathbb{R}^2$. When $k = 2$, this becomes the tangent bundle to S^2. When $k = 1$, it looks like $\mathbb{C}P^2 - \{\text{point}\}$.

(5) Let G be a compact Lie group

(a) Show that G admits a bi-invariant metric, i.e., both right and left translations are isometries. Hint: Fix a left invariant metric g_L and a volume form $\omega = \sigma^1 \wedge \cdots \wedge \sigma^1$ where σ^i are left invariant 1-forms. Then define g as the average over right translations:

$$g(v, w) = \frac{1}{\int \omega} \int g_L(DR_x(v), DR_x(w)) \, \omega.$$

(b) Show that the *inner automorphism* $\text{Ad}_h(x) = hxh^{-1}$ is a Riemannian isometry. Conclude that its differential at $x = e$ denoted by the same letters

$$\text{Ad}_h : \mathfrak{g} \to \mathfrak{g}$$

is a linear isometry with respect to g.

(c) Use this to show that the adjoint action

$$\begin{aligned} \text{ad}_U &: \quad \mathfrak{g} \to \mathfrak{g}, \\ \text{ad}_U(X) &= [U, X] \end{aligned}$$

is skew-symmetric, i.e.,

$$g([U, X], Y) = -g(X, [U, Y]).$$

Hint: It is shown in the appendix that $U \to \text{ad}_U$ is the differential of $h \to \text{Ad}_h$. (See also chapter 3).

(6) Let V be an n-dimensional vector space with a symmetric nondegenerate bilinear form g of index p.

(a) Show that there exists a basis $e_1, ..., e_n$ such that $g(e_i, e_j) = 0$ if $i \neq j$, $g(e_i, e_i) = -1$ if $i = 1, ..., p$ and $g(e_i, e_i) = 1$ if $i = p + 1, ..., n$. Thus V is isometric to $\mathbb{R}^{p,q}$.

(b) Show that for any v we have the expansion

$$v = \sum_{i=1}^{n} \frac{g(v, e_i)}{g(e_i, e_i)} e_i$$

$$= -\sum_{i=1}^{p} g(v, e_i) e_i + \sum_{i=p+1}^{n} g(v, e_i) e_i.$$

(c) Let $L : V \to V$ be a linear operator. Show that

$$\operatorname{tr}(L) = \sum_{i=1}^{n} \frac{g(L(e_i), e_i)}{g(e_i, e_i)}.$$

CHAPTER 2

Curvature

With the comforting feeling that there are indeed a variety of Riemannian manifolds out there, we shall now immerse ourselves in the theory. In this chapter we confine ourselves to infinitesimal considerations. The most important and often also least understood object of Riemannian geometry is the connection and its function as covariant differentiation. We shall give a motivation of this concept that depends on exterior and Lie derivatives (The basic definitions and properties of Lie derivatives are recaptured in the appendix). It is hoped that this makes the concept a little less of a *deus ex machina*. Covariant differentiation, in turn, gives us nice formulae for exterior derivatives, Lie derivatives, divergence and much more (see also the appendix). It is also important in the development of curvature which is the central theme of Riemannian geometry. The idea of a Riemannian metric having curvature, while intuitively appealing and natural, is for most people the stumbling block for further progress into the realm of geometry.

In the third section of the chapter we shall study what we call the fundamental equations of Riemannian geometry. These equations relate curvature to the Hessian of certain geometrically defined functions (Riemannian submersions onto intervals). These formulae hold all the information that we shall need when computing curvatures in new examples and also for studying Riemannian geometry in the abstract.

Surprisingly, the idea of a connection postdates Riemann's introduction of the curvature tensor. Riemann discovered the Riemannian curvature tensor as a second-order term in the Taylor expansion of a Riemannian metric at a point, where coordinates are chosen such that the zeroth-order term is the Euclidean metric and the first-order term is zero. Lipschitz, Killing, and Christoffel introduced the connection in various ways as an intermediate step in computing the curvature. Also, they found it was a natural invariant for what is called the equivalence problem in Riemannian geometry. This problem, which seems rather odd nowadays (although it really is important), comes out of the problem one faces when writing the same metric in two different coordinates. Namely, how is one to know that they are the same or equivalent. The idea is to find invariants of the metric that can be computed in coordinates and then try to show that two metrics are equivalent if their invariant expressions are equal. After this early work by the above-mentioned German mathematicians, an Italian school around Levi-Civita, Ricci, Bianchi et al. began systematically to study Riemannian metrics and tensor analysis. They eventually defined parallel translation and through that clarified the use of the connection. Hence the name Levi-Civita connection for the Riemannian connection. Most of their work was still local in nature and mainly centered on developing tensor analysis as a tool for describing physical phenomena such as stress, torque, and divergence. At the beginning of the twentieth century Minkowski started developing the geometry of space-time with the hope of using it for Einstein's new special relativity theory. It was this work that eventually enabled Einstein to give a geometric formulation of general relativity theory. Since then, tensor calculus, connections, and curvature have become an indispensable language for many theoretical physicists.

Much of what we do in this chapter carries over to the semi-Riemannian setting. The connection and curvature tensor are generalized without changes. But the formulas for divergence and Ricci curvature do require some modifications. The thing to watch for is that the trace of an operator has a slightly different formula in this setting (see exercises to chapter 1).

1. Connections

1.1. Directional Differentiation. First we shall introduce some important notation. There are many ways of denoting the *directional derivative* of a function on a manifold. Given a function $f : M \to \mathbb{R}$ and a vector field Y on M we will use the following ways of writing the directional derivative of f in the direction of Y

$$\nabla_Y f = D_Y f = L_Y f = df(Y) = Y(f).$$

If we have a function $f : M \to \mathbb{R}$ on a manifold, then the differential $df :$ $TM \to \mathbb{R}$ measures the change in the function. In local coordinates, $df = \partial_i(f)dx^i$. If, in addition, M is equipped with a Riemannian metric g, then we also have the *gradient* of f, denoted by $\operatorname{grad} f = \nabla f$, defined as the vector field satisfying $g(v, \nabla f) = df(v)$ for all $v \in TM$. In local coordinates this reads, $\nabla f = g^{ij}\partial_i(f)\partial_j$, where g^{ij} is the inverse of the matrix g_{ij} (see also the next section). Defined in this way, the gradient clearly depends on the metric. But is there a way of defining a gradient vector field of a function without using Riemannian metrics? The answer is no and can be understood as follows. On $\mathbb{R}^n$ the gradient is defined as

$$\nabla f = \delta^{ij}\partial_i(f)\partial_j = \sum_{i=1}^{n} \partial_i(f)\partial_i.$$

But this formula depends on the fact that we used Cartesian coordinates. If instead we had used polar coordinates on $\mathbb{R}^2$, say, then we mostly have that

$$\begin{aligned} \nabla f &= \partial_x(f)\partial_x + \partial_y(f)\partial_y \\ &\neq \partial_r(f)\partial_r + \partial_\theta(f)\partial_\theta, \end{aligned}$$

One rule of thumb for items that are invariantly defined is that they should satisfy the Einstein summation convention, where one sums over identical super- and subscripts. Thus, $df = \partial_i(f)dx^i$ is invariantly defined, while $\nabla f = \partial_i(f)\partial_i$ is not. The metric $g = g_{ij}dx^idx^j$ and gradient $\nabla f = g^{ij}\partial_i(f)\partial_j$ are invariant expressions that also depend on our choice of metric.

1.2. Covariant Differentiation. We now come to the question of attaching a meaning to the change of a vector field. In $\mathbb{R}^n$ we can use the standard Cartesian coordinate vector fields to write $X = a^i\partial_i$. If we think of the coordinate vector fields as being constant, then it is natural to define the *covariant derivative* of X in the direction of Y as

$$\nabla_Y X = \left(\nabla_Y a^i\right)\partial_i = d\left(a^i\right)(Y)\partial_i.$$

Thus we measure the change in X by measuring how the coefficients change. Therefore, a vector field with constant coefficients does not change. This formula clearly depends on the fact that we used Cartesian coordinates and is not invariant under change of coordinates. If we take the coordinate vector fields

$$\begin{aligned}
\partial_r &= \frac{1}{r}\left(x\partial_x + y\partial_y\right)\\
\partial_\theta &= -y\partial_x + x\partial_y
\end{aligned}$$

that come from polar coordinates in $\mathbb{R}^2$, then we see that they are not constant.

In order to better understand what is happening we need to find a coordinate independent definition of this change. This is done most easily by splitting the problem of defining the change in a vector field X into two problems.

First, we can measure the change in X by asking whether or not X is a gradient field. If $i_X g = \theta_X$ is the 1-form dual to X, i.e., $(i_X g)(Y) = g(X, Y)$, then we know that X is locally the gradient of a function if and only if $d\theta_X = 0$. In general, the 2-form $d\theta_X$ therefore measures the extend to which X is a gradient field.

Second, we can measure how a vector field X changes the metric via the Lie derivative $L_X g$. This is a symmetric $(0, 2)$-tensor as opposed to the skew-symmetric $(0, 2)$-tensor $d\theta_X$. If F^t is the local flow for X, then we see that $L_X g = 0$ if and only if F^t are isometries (see also chapter 7). If this happens then we say that X is a *Killing field*. Lie derivatives will be used heavily below. The results we use are standard from manifold theory and are all explained in the appendix.

In case $X = \nabla f$ is a gradient field the expression $L_{\nabla f} g$ is essentially the Hessian of f. We can prove this in $\mathbb{R}^n$ were we already know what the Hessian should be. Let

$$\begin{aligned}
X &= \nabla f = a^i \partial_i,\\
a^i &= \partial_i f,
\end{aligned}$$

then

$$\begin{aligned}
L_X\left(\delta_{ij}dx^i dx^j\right) &= \left(L_X\delta_{ij}\right) + \delta_{ij}L_X\left(dx^i\right)dx^j + \delta_{ij}dx^i L_X\left(dx^j\right)\\
&= 0 + \delta_{ij}\left(dL_X\left(x^i\right)\right)dx^j + \delta_{ij}dx^i\left(dL_X\left(x^j\right)\right)\\
&= \delta_{ij}\left(da^i\right)dx^j + \delta_{ij}dx^i da^j\\
&= \delta_{ij}\left(\partial_k a^i\right)dx^k dx^j + \delta_{ij}dx^i\left(\partial_k a^j\right)dx^k\\
&= \partial_k a^i dx^k dx^i + \partial_k a^i dx^i dx^k\\
&= \left(\partial_k a^i + \partial_i a^k\right)dx^i dx^k\\
&= \left(\partial_k\partial_i f + \partial_i\partial_k f\right)dx^i dx^k\\
&= 2\left(\partial_i\partial_k f\right)dx^i dx^k\\
&= 2\mathrm{Hess}f.
\end{aligned}$$

From this calculation we can also quickly see what the Killing fields on $\mathbb{R}^n$ should be. If $X = a^i \partial_i$, then X is a Killing field iff $\partial_k a^i + \partial_i a^k = 0$. This shows that

$$
\begin{aligned}
\partial_j \partial_k a^i &= -\partial_j \partial_i a^k \\
&= -\partial_i \partial_j a^k \\
&= \partial_i \partial_k a^j \\
&= \partial_k \partial_i a^j \\
&= -\partial_k \partial_j a^i \\
&= -\partial_j \partial_k a^i.
\end{aligned}
$$

Thus we have $\partial_j \partial_k a^i = 0$ and hence

$$
a^i = \alpha^i_j x^j + \beta^i
$$

with the extra conditions that

$$
\alpha^i_j = \partial_j a^i = -\partial_i a^j = -\alpha^j_i.
$$

The angular field ∂_θ is therefore a Killing field. This also follows from the fact that the corresponding flow is matrix multiplication by the orthogonal matrix

$$
\begin{bmatrix} \cos(t) & -\sin(t) \\ \sin(t) & \cos(t) \end{bmatrix}.
$$

More generally one can show that the flow of the Killing field X is

$$
\begin{aligned}
F^t(x) &= \exp(At) x + t\beta, \\
A &= [\alpha^i_j], \\
\beta &= [\beta^i].
\end{aligned}
$$

In this way we see that a vector field on $\mathbb{R}^n$ is constant iff it is a Killing field that is also a gradient field.

The important observation we can make on $\mathbb{R}^n$ is that

PROPOSITION 3. *The covariant derivative in $\mathbb{R}^n$ is given by the implicit formula:*

$$
2g(\nabla_Y X, Z) = (L_X g)(Y, Z) + (d\theta_X)(Y, Z).
$$

PROOF. Since both sides are tensorial in Y and Z it suffices to check the formula on the Cartesian coordinate vector fields. Write $X = a^i \partial_i$ and calculate the right hand side

$$
\begin{aligned}
(L_X g)(\partial_k, \partial_l) + (d\theta_X)(\partial_k, \partial_l) &= D_X \delta_{kl} - g(L_X \partial_k, \partial_l) - g(\partial_k, L_X \partial_l) \\
&\quad + \partial_k g(X, \partial_l) - \partial_l g(X, \partial_k) - g(X, [\partial_k, \partial_l]) \\
&= -g(L_{a^i \partial_i} \partial_k, \partial_l) - g(\partial_k, L_{a^j \partial_j} \partial_l) \\
&\quad + \partial_k a^l - \partial_l a^k \\
&= -g(-(\partial_k a^i)\partial_i, \partial_l) - g(\partial_k, -(\partial_l a^j)\partial_j) \\
&\quad + \partial_k a^l - \partial_l a^k \\
&= +\partial_k a^l + \partial_l a^k + \partial_k a^l - \partial_l a^k \\
&= 2\partial_k a^l \\
&= 2g((\partial_k a^i)\partial_i, \partial_l) \\
&= 2g(\nabla_{\partial_k} X, \partial_l).
\end{aligned}
$$

$\square$

Since the right hand side in the formula for $\nabla_Y X$ makes sense on any Riemannian manifold we can use this to give an implicit definition of the *covariant derivative* of X in the direction of Y. This covariant derivative turns out to be uniquely determined by the following properties.

THEOREM 1. (The Fundamental Theorem of Riemannian Geometry) *The assignment* $X \to \nabla X$ *on* (M, g) *is uniquely defined by the following properties:*

(1) $Y \to \nabla_Y X$ *is a* $(1, 1)$-*tensor:*

$$\nabla_{\alpha v + \beta w} X = \alpha \nabla_v X + \beta \nabla_w X.$$

(2) $X \to \nabla_Y X$ *is a derivation:*

$$\begin{aligned} \nabla_Y (X_1 + X_2) &= \nabla_Y X_1 + \nabla_Y X_2, \\ \nabla_Y (fX) &= (D_Y f) X + f \nabla_Y X \end{aligned}$$

for functions $f : \mathbb{R}^n \to \mathbb{R}$.

(3) Covariant differentiation is torsion free:

$$\nabla_X Y - \nabla_Y X = [X, Y].$$

(4) Covariant differentiation is metric:

$$D_Z g (X, Y) = g (\nabla_Z X, Y) + g (X, \nabla_Z Y).$$

PROOF. We have already established (1) by using that

$$(L_X g) (Y, Z) + (d\theta_X) (Y, Z)$$

is tensorial in Y and Z. This also shows that the expression is linear in X. To check the derivation rule we observe that

$$\begin{aligned} L_{fX} g + d\theta_{fX} &= f L_X g + df \cdot \theta_X + \theta_X \cdot df + d (f \theta_X) \\ &= f L_X g + df \cdot \theta_X + \theta_X \cdot df + df \wedge \theta_X + f d\theta_X \\ &= f (L_X g + d\theta_X) + df \cdot \theta_X + \theta_X \cdot df + df \cdot \theta_X - \theta_X \cdot df \\ &= f (L_X g + d\theta_X) + 2 df \cdot \theta_X. \end{aligned}$$

Thus

$$\begin{aligned} 2g (\nabla_Y (fX), Z) &= f 2g (\nabla_Y X, Z) + 2 df (Y) g (X, Z) \\ &= 2g (f \nabla_Y X + df (Y) X, Z) \\ &= 2g (f \nabla_Y X + (D_Y f) X, Z) \end{aligned}$$

To establish the next two claims it is convenient to do the following expansion also known as *Koszul's formula*.

$$\begin{aligned} 2g (\nabla_Y X, Z) &= (L_X g) (Y, Z) + (d\theta_X) (Y, Z) \\ &= D_X g (Y, Z) - g ([X, Y], Z) - g (Y, [X, Z]) \\ &\quad + D_Y \theta_X (Z) - D_Z \theta_X (Y) - \theta_X ([X, Y]) \\ &= D_X g (Y, Z) - g ([X, Y], Z) - g (Y, [X, Z]) \\ &\quad + D_Y g (X, Z) - D_Z g (X, Y) - g (X, [Y, Z]) \\ &= D_X g (Y, Z) + D_Y g (Z, X) - D_Z g (X, Y) \\ &\quad - g ([X, Y], Z) - g ([Y, Z], X) + g ([Z, X], Y). \end{aligned}$$

We then see that (3) follows from

$$
\begin{aligned}
2g\left(\nabla_X Y - \nabla_Y X, Z\right) &= D_Y g\left(X, Z\right) + D_X g\left(Z, Y\right) - D_Z g\left(Y, X\right) \\
&\quad -g\left([Y, X], Z\right) - g\left([X, Z], Y\right) + g\left([Z, Y], X\right) \\
&\quad -D_X g\left(Y, Z\right) - D_Y g\left(Z, X\right) + D_Z g\left(X, Y\right) \\
&\quad +g\left([X, Y], Z\right) + g\left([Y, Z], X\right) - g\left([Z, X], Y\right) \\
&= 2g\left([X, Y], Z\right).
\end{aligned}
$$

And (4) from

$$
\begin{aligned}
2g\left(\nabla_Z X, Y\right) + 2g\left(X, \nabla_Z Y\right) &= D_X g\left(Z, Y\right) + D_Z g\left(Y, X\right) - D_Y g\left(X, Z\right) \\
&\quad -g\left([X, Z], Y\right) - g\left([Z, Y], X\right) + g\left([Y, X], Z\right) \\
&\quad +D_Y g\left(Z, X\right) + D_Z g\left(X, Y\right) - D_X g\left(Y, Z\right) \\
&\quad -g\left([Y, Z], X\right) - g\left([Z, X], Y\right) + g\left([X, Y], Z\right) \\
&= 2D_Z g\left(X, Y\right).
\end{aligned}
$$

Conversely, if we have a covariant derivative $\bar{\nabla}_Y X$ with these four properties, then

$$
\begin{aligned}
2g\left(\nabla_Y X, Z\right) &= \left(L_X g\right)\left(Y, Z\right) + \left(d\theta_X\right)\left(Y, Z\right) \\
&= D_X g\left(Y, Z\right) + D_Y g\left(Z, X\right) - D_Z g\left(X, Y\right) \\
&\quad -g\left([X, Y], Z\right) - g\left([Y, Z], X\right) + g\left([Z, X], Y\right) \\
&= g\left(\bar{\nabla}_X Y, Z\right) + g\left(Y, \bar{\nabla}_X Z\right) + g\left(\bar{\nabla}_Y Z, X\right) + g\left(Z, \bar{\nabla}_Y X\right) \\
&\quad -g\left(\bar{\nabla}_Z X, Y\right) - g\left(X, \bar{\nabla}_Z Y\right) + g\left(\bar{\nabla}_Z X, Y\right) - g\left(\bar{\nabla}_X Z, Y\right) \\
&\quad -g\left(\bar{\nabla}_X Y, Z\right) + g\left(\bar{\nabla}_Y X, Z\right) - g\left(\bar{\nabla}_Y Z, X\right) + g\left(\bar{\nabla}_Z Y, X\right) \\
&= 2g\left(\bar{\nabla}_Y X, Z\right)
\end{aligned}
$$

showing that $\nabla_Y X = \bar{\nabla}_Y X$. $\square$

Any assignment on a manifold that satisfies (1) and (2) is called an *affine connection*. If (M, g) is a Riemannian manifold and we have a connection which in addition also satisfies (3) and (4), then we call it a *Riemannian connection*. As we just saw, this connection is uniquely defined by these four properties and is given implicitly through the formula

$$
2g\left(\nabla_Y X, Z\right) = \left(L_X g\right)\left(Y, Z\right) + \left(d\theta_X\right)\left(Y, Z\right).
$$

Before proceeding we need to discuss how $\nabla_Y X$ depends on X and Y. Since $\nabla_Y X$ is tensorial in Y, we see that the value of $\nabla_Y X$ at $p \in M$ depends only on $Y|_p$. But in what way does it depend on X? Since $X \to \nabla_Y X$ is a derivation, it is definitely not tensorial in X. Therefore, we can not expect that $\left(\nabla_Y X\right)|_p$ depends only on $X|_p$ and $Y|_p$. The next two lemmas explore how $\left(\nabla_Y X\right)|_p$ depends on X.

LEMMA 1. *Let M be a manifold and ∇ an affine connection on M. If $p \in M$, $v \in T_p M$, and X, Y are vector fields on M such that $X = Y$ in a neighborhood $U \ni p$, then $\nabla_v X = \nabla_v Y$.*

PROOF. Choose $\lambda : M \to \mathbb{R}$ such that $\lambda \equiv 0$ on $M - U$ and $\lambda \equiv 1$ in a neighborhood of p. Then $\lambda X = \lambda Y$ on M. Thus

$$
\nabla_v \lambda X = \lambda(p) \nabla_v X + d\lambda(v) \cdot X(p) = \nabla_v X
$$

since $d\lambda|_p = 0$ and $\lambda(p) = 1$. In particular,

$$\begin{aligned}
\nabla_v X &= \nabla_v \lambda X \\
&= \nabla_v \lambda Y \\
&= \nabla_v Y
\end{aligned}$$

□

For a Riemannian connection we could also have used the Koszul formula to prove this since the right hand side of that formula can be localized. This lemma tells us an important thing. Namely, if a vector field X is defined only on an open subset of M, then ∇X still makes sense on this subset. Therefore, we can use coordinate vector fields or more generally frames to compute ∇ locally.

LEMMA 2. *Let M be a manifold and ∇ an affine connection on M. If X is a vector field on M and $\gamma : I \to M$ a smooth curve with $\dot\gamma(0) = v \in T_pM$, then $\nabla_v X$ depends only on the values of X along γ, i.e., if $X \circ \gamma = Y \circ \gamma$, then $\nabla_{\dot\gamma} X = \nabla_{\dot\gamma} Y$.*

PROOF. Choose a framing $\{Z_1, \ldots, Z_n\}$ in a neighborhood of p and write $Y = \sum \alpha^i \cdot Z_i$, $X = \sum \beta^i Z_i$ on this neighborhood. From the assumption that $X \circ \gamma = Y \circ \gamma$ we get that $\alpha^i \circ \gamma = \beta^i \circ \gamma$. Thus,

$$\begin{aligned}
\nabla_v Y &= \nabla_v \alpha^i Z_i \\
&= \alpha^i(p)\nabla_v Z_i + Z_i(p)d\alpha^i(v) \\
&= \beta^i(p)\nabla_v Z_i + Z_i(p)d\beta^i(v) \\
&= \nabla_v X.
\end{aligned}$$

□

This shows that $\nabla_v X$ makes sense as long as X is prescribed along some curve (or submanifold) that has v as a tangent.

It will occasionally be convenient to use coordinates or orthonormal frames with certain nice properties. We say that a coordinate system is *normal* at p if $g_{ij}|_p = \delta_{ij}$ and $\partial_k g_{ij}|_p = 0$. An orthonormal frame E_i is *normal* at $p \in M$ if $\nabla_v E_i(p) = 0$ for all $i = 1, \ldots, n$ and $v \in T_pM$. It is an easy exercise to show that such coordinates and frames always exist.

1.3. Derivatives of Tensors. The connection, as we shall see, is incredibly useful in generalizing many of the well-known concepts (such as Hessian, Laplacian, divergence) from multivariable calculus to the Riemannian setting.

If S is a $(0,r)$- or $(1,r)$-tensor field, then we can define a *covariant derivative* ∇S that we interpret as a $(0, r+1)$- or $(1, r+1)$-tensor field. (Remember that a vector field X is a $(1,0)$-tensor field and ∇X is a $(1,1)$-tensor field.) The main idea is to make sure that Leibniz' rule holds. So for a $(1,1)$-tensor S we should have

$$\nabla_X (S(Y)) = (\nabla_X S)(Y) + S(\nabla_X Y).$$

Therefore, it seems reasonable to define ∇S as

$$\begin{aligned}
\nabla S(X, Y) &= (\nabla_X S)(Y) \\
&= \nabla_X (S(Y)) - S(\nabla_X Y).
\end{aligned}$$

In other words

$$\nabla_X S = [\nabla_X, S].$$

It is easily checked that $\nabla_X S$ is still tensorial in Y.

More generally, define

$$
\begin{aligned}
\nabla S(X, Y_1, \ldots, Y_r) &= (\nabla_X S)(Y_1, \ldots, Y_r) \\
&= \nabla_X (S(Y_1, \ldots, Y_r)) - \sum_{i=1}^{r} S(Y_1, \ldots, \nabla_X Y_i, \ldots, Y_r).
\end{aligned}
$$

Here ∇_X is interpreted as the directional derivative when applied to a function, while we use it as covariant differentiation on vector fields.

A tensor is said to be *parallel* if $\nabla S \equiv 0$. In $(\mathbb{R}^n, \mathrm{can})$ one can easily see that if a tensor is written in Cartesian coordinates, then it is parallel iff it has constant coefficients. Thus $\nabla X \equiv 0$ for constant vector fields. On a Riemannian manifold (M, g) we always have that $\nabla g \equiv 0$ since

$$
(\nabla g)(X, Y_1, Y_2) = \nabla_X (g(Y_1, Y_2)) - g(\nabla_X Y_1, Y_2) - g(Y_1, \nabla_X Y_2) = 0
$$

from property (4) of the connection.

If $f : M \to \mathbb{R}$ is smooth, then we already have ∇f defined as the vector field satisfying

$$
g(\nabla f, v) = D_v f = df(v).
$$

There is some confusion here, with ∇f now also being defined as df. In any given context it will generally be clear what we mean. The *Hessian* $\mathrm{Hess} f$ is defined as the symmetric $(0, 2)$-tensor $\frac{1}{2} L_{\nabla f} g$. We know that this conforms with our definition on $\mathbb{R}^n$. It can also be defined as a self-adjoint $(1, 1)$-tensor by $S(X) = \nabla_X \nabla f$. These two tensors are naturally related by

$$
\mathrm{Hess} f(X, Y) = g(S(X), Y).
$$

To see this we observe that $d(\theta_{\nabla f}) = 0$ so

$$
\begin{aligned}
2g(S(X), Y) &= 2g(\nabla_X \nabla f, Y) \\
&= (L_{\nabla f} g)(Y, Z) + d(\theta_{\nabla f})(Y, Z) \\
&= 2\mathrm{Hess} f(X, Y).
\end{aligned}
$$

The trace of S is the *Laplacian*, and we will use the notation $\Delta f = \mathrm{tr}(S)$. On $\mathbb{R}^n$ this is also written as $\Delta f = \mathrm{div} \nabla f$. The *divergence* of a vector field, $\mathrm{div} X$, on (M, g) is defined as

$$
\mathrm{div} X = \mathrm{tr}(\nabla X).
$$

In coordinates this is

$$
\mathrm{tr}(\nabla X) = dx^i (\nabla_{\partial_i} X),
$$

and with respect to an orthonormal basis

$$
\mathrm{tr}(\nabla X) = \sum_{i=1}^{n} g(\nabla_{e_i} X, e_i).
$$

Thus, also

$$
\Delta f = \mathrm{tr}(\nabla(\nabla f)) = \mathrm{div}(\nabla f).
$$

In analogy with our definition of $\mathrm{div} X$ we can also define the divergence of a $(1, r)$-tensor S to be the $(0, r)$-tensor

$$
(\mathrm{div} S)(v_1, \ldots, v_r) = \mathrm{tr}(w \to (\nabla_w S)(v_1, \ldots, v_r)).
$$

For a $(\cdot, r)$-tensor field S we define the *second covariant derivative* $\nabla^2 S$ as the $(\cdot, r+2)$-tensor field

$$
\begin{aligned}
\left(\nabla^2_{X_1, X_2} S\right)(Y_1, \ldots, Y_r) &= \left(\nabla_{X_1}(\nabla S)\right)(X_2, Y_1, \ldots, Y_r) \\
&= \left(\nabla_{X_1}(\nabla_{X_2} S)\right)(Y_1, \ldots, Y_r) - \left(\nabla_{\nabla_{X_1} X_2} S\right)(Y_1, \ldots, Y_r).
\end{aligned}
$$

With this we get the $(0, 2)$ version of the Hessian of a function defined as

$$
\begin{aligned}
\nabla^2_{X, Y} f &= \nabla_X \nabla_Y f - \nabla_{\nabla_X Y} f \\
&= \nabla_X g(Y, \nabla f) - g(\nabla_X Y, \nabla f) \\
&= g(Y, \nabla_X \nabla f) \\
&= g(S(X), Y).
\end{aligned}
$$

The second covariant derivative on functions is symmetric in X and Y. For more general tensors, however, this will not be the case. The defect in the second covariant derivative not being symmetric is a central feature in Riemannian geometry and is at the heart of the difference between Euclidean geometry and all other Riemannian geometries.

From the new formula for the Hessian we see that the Laplacian can be written as

$$
\Delta f = \sum_{i=1}^n \nabla^2_{E_i, E_i} f.
$$

2. The Connection in Local Coordinates

In a local coordinate system the metric is written as $g = g_{ij} dx^i dx^j$. So if $X = a^i \partial_i$ and $Y = b^j \partial_j$ are vector fields, then

$$
g(X, Y) = g_{ij} a^i b^j.
$$

We can also compute the dual 1-form θ_X to X by:

$$
\begin{aligned}
\theta_X &= g(X, \cdot) \\
&= g_{ij} dx^i(X) dx^j(\cdot) \\
&= g_{ij} a^i dx^j.
\end{aligned}
$$

The inverse of the matrix $[g_{ij}]$ is denoted $[g^{ij}]$. Thus we have

$$
\delta^i_j = g^{ik} g_{kj}.
$$

The vector field X dual to a 1-form $\omega = \alpha_i dx^i$ is defined implicitly by

$$
g(X, Y) = \omega(Y).
$$

In other words we have

$$
\theta_X = g_{ij} a^i dx^j = \alpha_j dx^j = \omega.
$$

This shows that

$$
g_{ij} a^i = \alpha_j.
$$

In order to isolate a^i we have to multiply by g^{kj} on both sides and also use the symmetry of g_{ij}

$$
\begin{aligned}
g^{kj}\alpha_j &= g^{kj}g_{ij}a^i \\
&= g^{kj}g_{ji}a^i \\
&= \delta^k_i a^i \\
&= a^k.
\end{aligned}
$$

Therefore

$$
\begin{aligned}
X &= a^i\partial_i \\
&= g^{ij}\alpha_j\partial_i.
\end{aligned}
$$

The gradient field of a function is a particularly important example of this construction

$$
\begin{aligned}
\nabla f &= g^{ij}\partial_j f\partial_i, \\
df &= \partial_j f dx^j.
\end{aligned}
$$

We now go on to find a formula for $\nabla_Y X$ in local coordinates

$$
\begin{aligned}
\nabla_Y X &= \nabla_{b^i\partial_i} a^j\partial_j \\
&= b^i\nabla_{\partial_i} a^j\partial_j \\
&= b^i\partial_i\left(a^j\right)\partial_j + b^i a^j\nabla_{\partial_i}\partial_j \\
&= b^i\partial_i\left(a^j\right)\partial_j + b^i a^j\Gamma^k_{ij}\partial_k
\end{aligned}
$$

where we simply expanded the term $\nabla_{\partial_i}\partial_j$ in local coordinates. The first part of this formula is what we expect to get when using Cartesian coordinates in $\mathbb{R}^n$. The second part is the correction term coming from having a more general coordinate system and also a non-Euclidean metric. Our next goal is to find a formula for Γ^k_{ij} in terms of the metric. To this end we can simply use our defining implicit formula for the connection keeping in mind that there are no Lie bracket terms. On the left hand side we have

$$
\begin{aligned}
2g\left(\nabla_{\partial_i}\partial_j, \partial_l\right) &= 2g\left(\Gamma^k_{ij}\partial_k, \partial_l\right) \\
&= 2\Gamma^k_{ij}g_{kl},
\end{aligned}
$$

and on the right hand side

$$
\begin{aligned}
\left(L_{\partial_j}g\right)\left(\partial_i, \partial_l\right) + d\theta_{\partial_j}\left(\partial_i, \partial_l\right) &= \partial_j g_{il} + \partial_i\left(\theta_{\partial_j}\left(\partial_l\right)\right) - \partial_l\left(\theta_{\partial_j}\left(\partial_i\right)\right) \\
&= \partial_j g_{il} + \partial_i g_{jl} - \partial_l g_{ji}.
\end{aligned}
$$

Multiplying by g^{lm} on both sides then yields

$$
\begin{aligned}
2\Gamma^m_{ij} &= 2\Gamma^k_{ij}\delta^m_k \\
&= 2\Gamma^k_{ij}g_{kl}g^{lm} \\
&= \left(\partial_j g_{il} + \partial_i g_{jl} - \partial_l g_{ji}\right)g^{lm}.
\end{aligned}
$$

Thus we have the formula

$$\Gamma_{ij}^{k} = \frac{1}{2}g^{lk}\left(\partial_{j}g_{il}+\partial_{i}g_{jl}-\partial_{l}g_{ji}\right)$$

$$= \frac{1}{2}g^{kl}\left(\partial_{j}g_{il}+\partial_{i}g_{jl}-\partial_{l}g_{ji}\right)$$

$$= \frac{1}{2}g^{kl}\Gamma_{ij,k}$$

The symbols

$$\Gamma_{ij,k} = \frac{1}{2}\left(\partial_{j}g_{ik}+\partial_{i}g_{jk}-\partial_{k}g_{ji}\right)$$

$$= g\left(\nabla_{\partial_{i}}\partial_{j},\partial_{k}\right)$$

are called the Christoffel symbols of the first kind, while Γ_{ij}^{k} are the Christoffel symbols of the second kind. Classically the following notation has also been used

$$\left\{\begin{matrix}k\\ij\end{matrix}\right\} = \Gamma_{ij}^{k},$$

$$[ij,k] = \Gamma_{ij,k}$$

so as not to think that these things define a tensor. The reason why they are not tensorial comes from the fact that they may be zero in one coordinate system but not zero in another. A good example of this comes from the plane where the Christoffel symbols are zero in Cartesian coordinates, but not in polar coordinates:

$$\Gamma_{\theta\theta,r} = \frac{1}{2}\left(\partial_{\theta}g_{\theta r}+\partial_{\theta}g_{\theta r}-\partial_{r}g_{\theta\theta}\right)$$

$$= -\frac{1}{2}\partial_{r}\left(r^{2}\right)$$

$$= -r.$$

In fact, it is always possible to find coordinates around a point $p \in M$ such that

$$g_{ij}|_{p} = \delta_{ij},$$

$$\partial_{k}g_{ij}|_{p} = 0.$$

In particular,

$$g_{ij}|_{p} = \delta_{ij},$$

$$\Gamma_{ij}^{k}|_{p} = 0.$$

The covariant derivative is then computed exactly as in Euclidean space

$$\nabla_{Y}X|_{p} = \left(\nabla_{b^{i}\partial_{i}}a^{j}\partial_{j}\right)|_{p}$$

$$= b^{i}\left(p\right)\partial_{i}\left(a^{j}\right)|_{p}\partial_{j}|_{p}.$$

The torsion free property of the connection is equivalent to saying that the Christoffel symbols are symmetric in ij as

$$\Gamma_{ij}^{k}\partial_{k} = \nabla_{\partial_{i}}\partial_{j}$$

$$= \nabla_{\partial_{j}}\partial_{i}$$

$$= \Gamma_{ji}^{k}\partial_{k}.$$

The metric property of the connection becomes

$$\partial_k g_{ij} = g\left(\nabla_{\partial_k}\partial_i, \partial_j\right) + g\left(\partial_i, \nabla_{\partial_k}\partial_j\right)$$
$$= \Gamma_{ki,j} + \Gamma_{kj,i}.$$

This shows that the Christoffel symbols completely determine the derivatives of the metric.

Just as the metric could be used to give a formula for the gradient in local coordinates we can use the Christoffel symbols to get a local coordinate formula for the Hessian of a function. This is done as follows

$$
\begin{aligned}
2\mathrm{Hess} f\left(\partial_i, \partial_j\right) &= \left(L_{\nabla f} g\right)\left(\partial_i, \partial_j\right) \\
&= D_{\nabla f} g_{ij} - g\left(L_{\nabla f}\partial_i, \partial_j\right) - g\left(\partial_i, L_{\nabla f}\partial_j\right) \\
&= g^{kl}\left(\partial_k f\right)\left(\partial_l g_{ij}\right) \\
&\quad + g\left(L_{\partial_i}\left(g^{kl}\left(\partial_k f\right)\partial_l\right), \partial_j\right) \\
&\quad + g\left(\partial_i, L_{\partial_j}\left(g^{kl}\left(\partial_k f\right)\partial_l\right)\right) \\
&= \left(\partial_k f\right) g^{kl}\left(\partial_l g_{ij}\right) \\
&\quad + \partial_i\left(g^{kl}\left(\partial_k f\right)\right) g_{lj} \\
&\quad + \partial_j\left(g^{kl}\left(\partial_k f\right)\right) g_{il} \\
&= \left(\partial_k f\right) g^{kl}\left(\partial_l g_{ij}\right) \\
&\quad + \left(\partial_i\partial_k f\right) g^{kl} g_{lj} + \left(\partial_j\partial_k f\right) g^{kl} g_{il} \\
&\quad + \left(\partial_i g^{kl}\right)\left(\partial_k f\right) g_{lj} + \left(\partial_j g^{kl}\right)\left(\partial_k f\right) g_{il} \\
&= 2\partial_i\partial_j f \\
&\quad + \left(\partial_k f\right)\left(\left(\partial_i g^{kl}\right) g_{lj} + \left(\partial_j g^{kl}\right) g_{il} + g^{kl}\left(\partial_l g_{ij}\right)\right)
\end{aligned}
$$

To compute $\partial_i g^{jk}$ we note that

$$
\begin{aligned}
0 &= \partial_i \delta_l^j \\
&= \partial_i\left(g^{jk} g_{kl}\right) \\
&= \left(\partial_i g^{jk}\right) g_{kl} + g^{jk}\left(\partial_i g_{kl}\right)
\end{aligned}
$$

Thus we have

$$
\begin{aligned}
2\mathrm{Hess} f\left(\partial_i, \partial_j\right) &= 2\partial_i\partial_j f \\
&\quad + \left(\partial_k f\right)\left(\left(\partial_i g^{kl}\right) g_{lj} + \left(\partial_j g^{kl}\right) g_{il} + g^{kl}\left(\partial_l g_{ij}\right)\right) \\
&= 2\partial_i\partial_j f \\
&\quad + \left(\partial_k f\right)\left(-g^{kl}\partial_i g_{lj} - g^{kl}\partial_j g_{li} + g^{kl}\left(\partial_l g_{ij}\right)\right) \\
&= 2\partial_i\partial_j f - g^{kl}\left(\partial_i g_{lj} + \partial_j g_{li} - \partial_l g_{ij}\right)\partial_k f \\
&= 2\left(\partial_i\partial_j f - \Gamma_{ij}^k \partial_k f\right).
\end{aligned}
$$

3. Curvature

Having now developed the idea of covariant derivatives and explained their relation to the classical concepts of gradient, Hessian, and Laplacian, one might hope that somehow these concepts carry over to tensors. As we have seen, this is true with one important exception, namely, the most important tensor for us, the Riemannian metric g. This tensor is parallel and therefore has no gradient, etc.

Instead, we think of the connection itself as a sort of gradient of the metric. The next question then is, what should the Laplacian and Hessian be? The answer is, curvature.

Any connection on a manifold gives rise to a *curvature tensor*. This operator measures in some sense how far away the connection is from being our standard connection on $\mathbb{R}^n$, which we assume is our canonical curvature-free, or flat, space. If we are on a Riemannian manifold, then it is possible to take traces of this curvature operator to obtain various averaged curvatures.

3.1. The Curvature Tensor. We shall work exclusively in the Riemannian setting. So let (M, g) be a Riemannian manifold and ∇ the Riemannian connection. The curvature tensor is a $(1, 3)$-tensor defined by

$$
\begin{aligned}
R(X, Y)Z &= \nabla^2_{X,Y} Z - \nabla^2_{Y,X} Z \\
&= \nabla_X \nabla_Y Z - \nabla_Y \nabla_X - \nabla_{[X,Y]} Z \\
&= [\nabla_X, \nabla_Y] Z - \nabla_{[X,Y]} Z.
\end{aligned}
$$

on vector fields X, Y, Z. Of course, it needs to be proved that this is indeed a tensor. Since both of the second covariant derivatives are tensorial in X and Y, we need only check that R is tensorial in Z. This is easily done:

$$
\begin{aligned}
R(X, Y) fZ &= \nabla^2_{X,Y} (fZ) - \nabla^2_{Y,X} (fZ) \\
&= f\nabla^2_{X,Y} (Z) - f\nabla^2_{Y,X} (Z) \\
&\quad + \left(\nabla^2_{X,Y} f\right) Z - \left(\nabla^2_{Y,X} f\right) Z \\
&\quad + \left(\nabla_Y f\right) \nabla_X Z + \left(\nabla_X f\right) \nabla_Y Z \\
&\quad - \left(\nabla_X f\right) \nabla_Y Z - \left(\nabla_Y f\right) \nabla_X Z \\
&= f\left(\nabla^2_{X,Y} (Z) - \nabla^2_{Y,X} (Z)\right) \\
&= fR(X, Y) Z.
\end{aligned}
$$

Notice that X, Y appear skew-symmetrically in $R(X, Y)Z$, while Z plays its own role on top of the line, hence the unusual notation. One could also write $R_{X,Y} Z$. Using the metric g we can change this to a $(0, 4)$-tensor as follows:

$$
R(X, Y, Z, W) = g(R(X, Y)Z, W).
$$

The variables are now treated on a more equal footing, which is also justified by the next proposition.

PROPOSITION 4. *The Riemannian curvature tensor $R(X, Y, Z, W)$ satisfies the following properties:*

(1) R is skew-symmetric in the first two and last two entries:

$$
R(X, Y, Z, W) = -R(Y, X, Z, W) = R(Y, X, W, Z).
$$

(2) R is symmetric between the first two and last two entries:

$$
R(X, Y, Z, W) = R(Z, W, X, Y).
$$

(3) R satisfies a cyclic permutation property called Bianchi's first identity:

$$
R(X, Y)Z + R(Z, X)Y + R(Y, Z)X = 0.
$$

(4) ∇R satisfies a cyclic permutation property called Bianchi's second identity:

$$
(\nabla_Z R)(X, Y) W + (\nabla_X R)(Y, Z) W + (\nabla_Y R)(Z, X) W = 0.
$$

PROOF. The first part of (1) has already been established. For part two of (1) observe that $[X, Y]$ is the unique vector field defined by

$$D_X D_Y f - D_Y D_X f - D_{[X,Y]} f = 0.$$

In other words, $R(X, Y)f = 0$. This is the idea behind the calculations that follow:

$$
\begin{aligned}
g(R(X,Y)Z, Z) &= g(\nabla_X \nabla_Y Z, Z) - g(\nabla_Y \nabla_X Z, Z) - g(\nabla_{[X,Y]} Z, Z) \\
&= D_X g(\nabla_Y Z, Z) - g(\nabla_Y Z, \nabla_X Z) \\
&\quad - D_Y g(\nabla_X Z, Z) + g(\nabla_X Z, \nabla_Y Z) - \frac{1}{2} D_{[X,Y]} g(Z, Z) \\
&= \frac{1}{2} D_X D_Y g(Z, Z) - \frac{1}{2} D_Y D_X g(Z, Z) - \frac{1}{2} D_{[X,Y]} g(Z, Z) \\
&= 0.
\end{aligned}
$$

Now (1) follows by polarizing the identity $R(X, Y, Z, Z) = 0$ in Z.

Part (3) is proved using the torsion free property of the connection. We introduce some special notation. Let T be any mapping with 3 vector field variables and values that can be added. Summing over cyclic permutations of the variables gives us a new map

$$\mathfrak{S} T(X, Y, Z) = T(X, Y, Z) + T(Z, X, Y) + T(Y, Z, X)$$

that is invariant under cyclic permutations. Note that T doesn't have to be a tensor. As an example we can use $T(X, Y, Z) = [X, [Y, Z]]$ and observe that the Jacobi identity for vector fields says:

$$\mathfrak{S}[X, [Y, Z]] = 0.$$

For the proof of (3) we have

$$
\begin{aligned}
\mathfrak{S} R(X,Y)Z &= \mathfrak{S} \nabla_X \nabla_Y Z - \mathfrak{S} \nabla_Y \nabla_X Z - \mathfrak{S} \nabla_{[X,Y]} Z \\
&= \mathfrak{S} \nabla_Z \nabla_X Y - \mathfrak{S} \nabla_Z \nabla_Y X - \mathfrak{S} \nabla_{[X,Y]} Z \\
&= \mathfrak{S} \nabla_Z (\nabla_X Y - \nabla_Y X) - \mathfrak{S} \nabla_{[X,Y]} Z \\
&= \mathfrak{S}[X, [Y, Z]] \\
&= 0.
\end{aligned}
$$

Part (2) is a combinatorial consequence of (1) and (3):

$$
\begin{aligned}
R(X, Y, Z, W) &= -R(Z, X, Y, W) - R(Y, Z, X, W) \\
&= R(Z, X, W, Y) + R(Y, Z, W, X) \\
&= -R(W, Z, X, Y) - R(X, W, Z, Y) \\
&\quad - R(W, Y, Z, X) - R(Z, W, Y, X) \\
&= 2R(Z, W, X, Y) + R(X, W, Y, Z) + R(W, Y, X, Z) \\
&= 2R(Z, W, X, Y) - R(Y, X, W, Z) \\
&= 2R(Z, W, X, Y) - R(X, Y, Z, W),
\end{aligned}
$$

which implies $2R(X, Y, Z, W) = 2R(Z, W, X, Y)$.

Now for part (4). We use again the cyclic sum notation and in addition that

$$R(X, Y)Z = [\nabla_X, \nabla_Y]Z - \nabla_{[X,Y]} Z,$$

$$
\begin{aligned}
(\nabla_Z R)(X,Y)W &= \nabla_Z (R(X,Y)W) - R(\nabla_Z X,Y)W \\
&\quad - R(X,\nabla_Z Y)W - R(X,Y)\nabla_Z W \\
&= [\nabla_Z, R(X,Y)]W - R(\nabla_Z X,Y)W - R(X,\nabla_Z Y)W.
\end{aligned}
$$

Keeping in mind that we only do cyclic sums over X,Y,Z and that we have the Jacobi identity for operators:

$$
\mathfrak{S}\,[\nabla_X,[\nabla_Y,\nabla_Z]] = 0
$$

we obtain

$$
\begin{aligned}
\mathfrak{S}\,(\nabla_X R)(Y,Z)W &= \mathfrak{S}\,[\nabla_X, R(Y,Z)]W - \mathfrak{S}\,R(\nabla_X Y, Z)W - \mathfrak{S}\,R(Y,\nabla_X Z)W \\
&= \mathfrak{S}\,[\nabla_X,[\nabla_Y,\nabla_Z]]W - \mathfrak{S}\,[\nabla_X, \nabla_{[Y,Z]}]W \\
&\quad - \mathfrak{S}\,R(\nabla_X Y, Z)W - \mathfrak{S}\,R(Y,\nabla_X Z)W \\
&= -\mathfrak{S}\,[\nabla_X, \nabla_{[Y,Z]}]W - \mathfrak{S}\,R(\nabla_X Y, Z)W + \mathfrak{S}\,R(\nabla_Y X, Z)W \\
&= -\mathfrak{S}\,[\nabla_X, \nabla_{[Y,Z]}]W - \mathfrak{S}\,R([X,Y], Z)W \\
&= -\mathfrak{S}\,[\nabla_X, \nabla_{[Y,Z]}]W - \mathfrak{S}\,[\nabla_{[X,Y]}, \nabla_Z]W + \mathfrak{S}\,\nabla_{[[X,Y],Z]}W \\
&= \mathfrak{S}\,[\nabla_{[X,Y]}, \nabla_Z]W - \mathfrak{S}\,[\nabla_{[X,Y]}, \nabla_Z]W \\
&= 0.
\end{aligned}
$$

$\square$

Notice that part (1) is related to the fact that ∇ is metric, i.e.,

$$
d(g(X,Y)) = g(\nabla X, Y) + g(X, \nabla Y),
$$

while part (3) follows from ∇ being torsion free, i.e.,

$$
\nabla_X Y - \nabla_Y X = [X,Y].
$$

EXAMPLE 20. $(\mathbb{R}^n, \text{can})$ *has* $R \equiv 0$ *since* $\nabla_{\partial_i}\partial_j = 0$ *for the standard Cartesian coordinates.*

More generally for any tensor field S of type $(\cdot, r)$ we can define the curvature as the new $(\cdot, r)$ tensor field

$$
R(X,Y)S = \nabla^2_{X,Y}S - \nabla^2_{Y,X}S.
$$

Again one needs to check that this is indeed a tensor. This is done in the same way we checked that $R(X,Y)Z$ was tensorial in Z. Clearly, $R(X,Y)S$ is also tensorial and skew symmetric in X and Y.

From the curvature tensor R we can derive several different curvature concepts.

3.2. The Curvature Operator. First recall that we have the space $\Lambda^2 M$ of bivectors. If e_i is an orthonormal basis for $T_p M$, then the inner product on $\Lambda^2_p M$ is such that the bivectors $e_i \wedge e_j$, $i < j$ will form an orthonormal basis. The inner product that $\Lambda^2 M$ inherits in this way is also denoted by g. Alternatively, we can define the inner product g on $\Lambda^2_p M$ using

$$
\begin{aligned}
g(x \wedge y, v \wedge w) &= g(x,v)\,g(y,w) - g(x,w)\,g(y,v) \\
&= \det \begin{pmatrix} g(x,v) & g(x,w) \\ g(y,v) & g(y,w) \end{pmatrix}
\end{aligned}
$$

and then extend it by linearity to all of $\Lambda_p^2 M$. It is also useful to interpret bivectors as skew symmetric maps. This is done by the formula:

$$(v \wedge w)(x) = g(w,x)v - g(v,x)w.$$

With this definition we have a Bianchi or Jacobi type identity:

$$(x \wedge y)(z) + (y \wedge z)(x) + (z \wedge x)(y) = 0.$$

From the symmetry properties of the curvature tensor we see that R actually defines a symmetric bilinear map

$$R \quad : \quad \Lambda^2 M \times \Lambda^2 M \to \mathbb{R}$$
$$R(X \wedge Y, V \wedge W) \quad = \quad R(X,Y,W,V).$$

Note the reversal of V and W! The relation

$$g(\mathfrak{R}(X \wedge Y), V \wedge W) = R(X \wedge Y, V \wedge W)$$

therefore defines a self-adjoint operator $\mathfrak{R} : \Lambda^2 M \to \Lambda^2 M$. This operator is called the *curvature operator*. It is clearly just a different manifestation of the curvature tensor. The switch between Z and W is related to our definition of the next curvature concept.

3.3. Sectional Curvature. For any $v \in T_p M$ let

$$R_v(w) = R(w,v)v : T_p M \to T_p M$$

be the *directional curvature operator* . This operator is also known as the *tidal force operator*. The latter name accurately describes in physical terms the meaning of the tensor. The above conditions imply that this operator is self-adjoint and that v is always a zero eigenvector. The normalized quadratic form

$$\begin{aligned}
\sec(v,w) \quad &= \quad \frac{g(R_v(w),w)}{g(v,v)g(w,w) - g(v,w)^2} \\
&= \quad \frac{g(R(w,v)v,w)}{g(v \wedge w, v \wedge w)} \\
&= \quad \frac{g(\mathfrak{R}(v \wedge w), v \wedge w)}{(\mathrm{area}\square(v,w))^2}
\end{aligned}$$

is called the *sectional curvature* of (v,w). Since the denominator is the square of the area of the parallelogram $\{tv + sw : 0 \le t, \ s \le 1\}$, we can easily check that $\sec(v,w)$ depends only on the plane $\pi = \mathrm{span}\{v,w\}$. One of the important relationships between directional and sectional curvature is the following observation by Riemann.

PROPOSITION 5. (Riemann, 1854) *The following properties are equivalent:*
(1) $\sec(\pi) = k$ *for all 2-planes in* $T_p M$.
(2) $R(v_1, v_2)v_3 = k(v_1 \wedge v_2)(v_3)$ *for all* $v_1, v_2, v_3 \in T_p M$.
(3) $R_v(w) = k \cdot (w - g(w,v)v) = k \cdot pr_{v\perp}(w)$ *for all* $w \in T_p M$ *and* $|v| = 1$.
(4) $\mathfrak{R}(\omega) = k \cdot \omega$ *for all* $\omega \in \Lambda_p^2 M$.

PROOF. $(2) \Rightarrow (3) \Rightarrow (1)$ are easy. For $(1) \Rightarrow (2)$ we introduce the multilinear maps:

$$\begin{aligned}
R_k(v_1, v_2)v_3 \quad &= \quad k(v_1 \wedge v_2)(v_3), \\
R_k(v_1, v_2, v_3, v_4) \quad &= \quad kg((v_1 \wedge v_2)(v_3), v_4).
\end{aligned}$$

The first observation is that these maps behave exactly like the curvature tensor in that they satisfy properties 1, 2, and 3 of the above proposition. Now consider the map

$$T\left(v_1, v_2, v_3, v_4\right) = R\left(v_1, v_2, v_3, v_4\right) - R_k\left(v_1, v_2, v_3, v_4\right)$$

which also satisfies the same symmetry properties. Moreover, the assumption that $\sec = k$ implies

$$T\left(v, w, w, v\right) = 0$$

for all $v, w \in T_p M$. Using polarization $w = w_1 + w_2$ we get

$$
\begin{aligned}
0 &= T\left(v, w_1 + w_2, w_1 + w_2, v\right) \\
&= T\left(v, w_1, w_2, v\right) + T\left(v, w_2, w_1, v\right) \\
&= 2T\left(v, w_1, w_2, v\right) \\
&= -2T\left(v, w_1, v, w_2\right).
\end{aligned}
$$

Using properties 1 and 2 of the curvature tensor we now see that T is alternating in all four variables. That, however, is in violation of Bianchi's first identity unless $T = 0$, which is exactly what we wish to prove.

To see why (2) $\Rightarrow$ (4), choose an orthonormal basis e_i for $T_p M$; then $e_i \wedge e_j$, $i < j$, is a basis for $\Lambda_p^2 M$. Using (2) we see that

$$
\begin{aligned}
g\left(\Re\left(e_i \wedge e_j\right), e_t \wedge e_s\right) &= R(e_i, e_j, e_s, e_t) \\
&= k \cdot (g(e_j, e_s)g(e_i, e_t) - g(e_i, e_s)g(e_j, e_t)) \\
&= k \cdot g\left(e_i \wedge e_j, e_t \wedge e_s\right).
\end{aligned}
$$

But this implies that

$$\Re\left(e_i \wedge e_j\right) = k \cdot \left(e_i \wedge e_j\right).$$

For (4) $\Rightarrow$ (1) just observe that if $\{v, w\}$ are orthogonal unit vectors, then

$$k = g\left(\Re\left(v \wedge w\right), v \wedge w\right) = \sec\left(v, w\right).$$

$\square$

A Riemannian manifold (M, g) that satisfies either of these four conditions for all $p \in M$ and the same $k \in \mathbb{R}$ for all $p \in M$ is said to have *constant curvature* k. So far we only know that $(\mathbb{R}^n, \mathrm{can})$ has curvature zero. In chapter 3 we shall prove that $dr^2 + \mathrm{sn}_k^2(r)ds_{n-1}^2$ has constant curvature k. When $k > 0$, recall that these represent $\left(S^n\left(\frac{1}{\sqrt{k}}\right), \mathrm{can}\right)$, while when $k < 0$ we still don't have a good picture yet. A whole section in chapter 3 is devoted to these constant negative curvature metrics.

3.4. Ricci Curvature. Our next curvature is the Ricci curvature, which should be thought of as the Laplacian of g.

The *Ricci curvature* Ric is a trace of R. If $e_1, \ldots, e_n \in T_p M$ is an orthonormal basis, then

$$
\begin{aligned}
\mathrm{Ric}(v, w) &= \mathrm{tr}\left(x \to R\left(x, v\right) w\right) \\
&= \sum_{i=1}^{n} g\left(R\left(e_i, v\right) w, e_i\right) \\
&= \sum_{i=1}^{n} g\left(R\left(v, e_i\right) e_i, w\right) \\
&= \sum_{i=1}^{n} g\left(R\left(e_i, w\right) v, e_i\right).
\end{aligned}
$$

Thus Ric is a symmetric bilinear form. It could also be defined as the symmetric $(1,1)$-tensor

$$
\mathrm{Ric}(v) = \sum_{i=1}^{n} R\left(v, e_i\right) e_i.
$$

We adopt the language that $\mathrm{Ric} \geq k$ if all eigenvalues of $\mathrm{Ric}(v)$ are $\geq k$. In $(0,2)$ language this means more precisely that $\mathrm{Ric}\left(v, v\right) \geq k g\left(v, v\right)$ for all v. If (M, g) satisfies $\mathrm{Ric}(v) = k \cdot v$, or equivalently $\mathrm{Ric}(v, w) = k \cdot g(v, w)$, then (M, g) is said to be an *Einstein manifold* with *Einstein constant* k. If (M, g) has constant curvature k, then (M, g) is also Einstein with Einstein constant $(n-1)k$.

In chapter 3 we shall exhibit several interesting Einstein metrics that do not have constant curvature. Three basic types are

(1) $(S^n(1) \times S^n(1), \, ds_n^2 + ds_n^2)$ with Einstein constant $n - 1$.

(2) The Fubini-Study metric on $\mathbb{C}P^n$ with Einstein constant $2n + 2$.

(3) The Schwarzschild metric on $\mathbb{R}^2 \times S^2$, which is a doubly warped product metric: $dr^2 + \varphi^2(r)d\theta^2 + \psi^2(r)ds_2^2$ with Einstein constant 0.

If $v \in T_p M$ is a unit vector and we complete it to an orthonormal basis $\{v, e_2, \ldots, e_n\}$ for $T_p M$, then

$$
\mathrm{Ric}\left(v, v\right) = g\left(R\left(v, v\right) v, v\right) + \sum_{i=2}^{n} g\left(R\left(e_i, v\right) v, e_i\right) = \sum_{i=2}^{n} \sec\left(v, e_i\right).
$$

Thus, when $n = 2$, there is no difference from an informational point of view in knowing R or Ric. This is actually also true in dimension $n = 3$, because if $\{e_1, e_2, e_3\}$ is an orthonormal basis for $T_p M$, then

$$
\begin{aligned}
\sec\left(e_1, e_2\right) + \sec\left(e_1, e_3\right) &= \mathrm{Ric}\left(e_1, e_1\right), \\
\sec\left(e_1, e_2\right) + \sec\left(e_2, e_3\right) &= \mathrm{Ric}\left(e_2, e_2\right), \\
\sec\left(e_1, e_3\right) + \sec\left(e_2, e_3\right) &= \mathrm{Ric}\left(e_3, e_3\right).
\end{aligned}
$$

In other words:

$$
\begin{bmatrix} 1 & 0 & 1 \\ 1 & 1 & 0 \\ 0 & 1 & 1 \end{bmatrix} \begin{bmatrix} \sec\left(e_1, e_2\right) \\ \sec\left(e_2, e_3\right) \\ \sec\left(e_1, e_3\right) \end{bmatrix} = \begin{bmatrix} \mathrm{Ric}\left(e_1, e_1\right) \\ \mathrm{Ric}\left(e_2, e_2\right) \\ \mathrm{Ric}\left(e_3, e_3\right) \end{bmatrix}.
$$

Here, the matrix has $\det = 2$, therefore any sectional curvature can be computed from Ric. In particular, we see that (M^3, g) is Einstein iff (M^3, g) has constant sectional curvature. The search for Einstein metrics should therefore begin in dimension 4.

3.5. Scalar Curvature. The last curvature quantity we wish to mention is the *scalar curvature*:

$$\text{scal} = \text{tr}\,(\text{Ric}) = 2 \cdot \text{tr}\mathfrak{R}.$$

Notice that scal depends only on $p \in M$ and is therefore a function, $\text{scal} : M \to \mathbb{R}$. In an orthonormal basis $e_1, \ldots, e_n$ for $T_p M$ we have

$$
\begin{aligned}
\text{scal} &= \text{tr}\,(\text{Ric}) \\
&= \sum_{j=1}^{n} g\,(\text{Ric}\,(e_j), e_j) \\
&= \sum_{j=1}^{n}\sum_{i=1}^{n} g\,(R\,(e_i, e_j)\,e_j, e_i) \\
&= \sum_{i,j=1}^{n} g\,(\mathfrak{R}\,(e_i \wedge e_j), e_i \wedge e_j) \\
&= 2\sum_{i<j} g\,(\mathfrak{R}\,(e_i \wedge e_j), e_i \wedge e_j) \\
&= 2\text{tr}\mathfrak{R} \\
&= 2\sum_{i<j} \sec\,(e_i, e_j)\,.
\end{aligned}
$$

When $n = 2$ we see that $\text{scal}(p) = 2 \cdot \sec(T_p M)$. In chapter 3 we shall show that when $n = 3$ there are metrics with constant scalar curvature that are not Einstein. When $n \geq 3$ there is also another interesting phenomenon occurring related to scalar curvature.

LEMMA 3. (Schur, 1886) *Suppose that a Riemannian manifold (M, g) of dimension $n \geq 3$ satisfies one of the following two conditions:*
a) $\sec(\pi) = f(p)$ *for all 2-planes* $\pi \subset T_p M$, $p \in M$.
b) $\text{Ric}(v) = (n-1) \cdot f(p) \cdot v$ *for all* $v \in T_p M$, $p \in M$.
Then in either case f must be constant. In other words, the metric has constant curvature or is Einstein, respectively.

PROOF. It clearly suffices to show (b), as the conditions for (a) imply that (b) holds. To show (b) we need the important identity:

$$d\text{scal} = 2\text{div}\,(\text{Ric})\,.$$

Let us see how this implies (b). First we have

$$
\begin{aligned}
d\text{scal} &= d\text{tr}\,(\text{Ric}) \\
&= d\,(n \cdot (n-1) \cdot f) \\
&= n \cdot (n-1) \cdot df.
\end{aligned}
$$

On the other hand

$$
\begin{aligned}
2\mathrm{div}\,(\mathrm{Ric})\,(v) &= 2\sum g\left(\left(\nabla_{e_i}\mathrm{Ric}\right)(v),e_i\right) \\
&= 2\sum g\left(\left(\nabla_{e_i}\left((n-1)\,f\cdot I\right)\right)(v),e_i\right) \\
&= 2\sum g\left((n-1)\left(\nabla_{e_i}f\right)v,e_i\right)+2\sum g\left((n-1)\,f\left(\nabla_{e_i}I\right)(v),e_i\right) \\
&= 2\left(n-1\right)g\left(v,\sum\left(\nabla_{e_i}f\right)e_i\right) \\
&= 2\left(n-1\right)g\left(v,\nabla f\right) \\
&= 2\left(n-1\right)df\left(v\right).
\end{aligned}
$$

Thus, we have shown that $n\cdot df = 2\cdot df$, but this is impossible unless $n=2$ or $df\equiv 0$ (i.e., f is constant). $\qquad\square$

PROPOSITION 6.

$$
dtr\,(\mathrm{Ric}) = 2\mathrm{div}\,(\mathrm{Ric})\,.
$$

PROOF. The identity is proved by a long and uninspired calculation that uses the second Bianchi identity. Choose a normal orthonormal frame E_i at $p\in M$, i.e., $\nabla E_i|_p = 0$, and let W be a vector field such that $\nabla W|_p = 0$. Using the second Bianchi identity

$$
\begin{aligned}
(dtr\,(\mathrm{Ric}))\,(W)\,(p) &= D_W\sum g\left(\mathrm{Ric}\,(E_i),E_i\right) \\
&= D_W\sum g\left(R\,(E_i,E_j)\,E_j,E_i\right) \\
&= \sum g\left(\nabla_W\left(R\,(E_i,E_j)\,E_j\right),E_i\right) \\
&= \sum g\left(\left(\nabla_W R\right)(E_i,E_j)\,E_j,E_i\right) \\
&= -\sum g\left(\left(\nabla_{E_j}R\right)(W,E_i)\,E_j,E_i\right) \\
&\quad -\sum g\left(\left(\nabla_{E_i}R\right)(E_j,W)\,E_j,E_i\right) \\
&= -\sum\left(\nabla_{E_j}R\right)(W,E_i,E_j,E_i)-\sum\left(\nabla_{E_i}R\right)(E_j,W,E_j,E_i) \\
&= \sum\left(\nabla_{E_j}R\right)(E_j,E_i,E_i,W)+\sum\left(\nabla_{E_i}R\right)(E_i,E_j,E_j,W) \\
&= 2\sum\left(\nabla_{E_j}R\right)(E_j,E_i,E_i,W) \\
&= 2\sum\nabla_{E_j}\left(R\,(E_j,E_i,E_i,W)\right) \\
&= 2\sum\nabla_{E_j}g\left(\mathrm{Ric}\,(E_j),W\right) \\
&= 2\sum\nabla_{E_j}g\left(\mathrm{Ric}\,(W),E_j\right) \\
&= 2\sum g\left(\nabla_{E_j}\left(\mathrm{Ric}\,(W)\right),E_j\right) \\
&= 2\sum g\left(\left(\nabla_{E_j}\mathrm{Ric}\right)(W),E_j\right) \\
&= 2\mathrm{div}\,(\mathrm{Ric})\,(W)\,(p)\,.
\end{aligned}
$$

$\qquad\square$

COROLLARY 1. *An $n\,(>2)$-dimensional Riemannian manifold (M,g) is Einstein iff*

$$\text{Ric} = \frac{\text{scal}}{n}g.$$

3.6. Curvature in Local Coordinates. As with the connection it is sometimes convenient to know what the curvature tensor looks like in local coordinates. We first observe that if $X = \alpha^i \partial_i$, $Y = \beta^j \partial_j$, $Z = \gamma^k \partial_k$, then we can write

$$
\begin{aligned}
R(X,Y)Z &= \alpha^i \beta^j \gamma^k R_{ijk}^l \partial_l, \\
R_{ijk}^l \partial_l &= R(\partial_i, \partial_j)\partial_k.
\end{aligned}
$$

Using the definition of R we see that

$$
\begin{aligned}
R_{ijk}^l \partial_l &= R(\partial_i, \partial_j)\partial_k \\
&= \nabla_{\partial_i}\nabla_{\partial_j}\partial_k - \nabla_{\partial_j}\nabla_{\partial_i}\partial_k \\
&= \nabla_{\partial_i}\left(\Gamma_{jk}^s \partial_s\right) - \nabla_{\partial_j}\left(\Gamma_{ik}^t \partial_t\right) \\
&= \partial_i\left(\Gamma_{jk}^s\right)\partial_s + \Gamma_{jk}^s \nabla_{\partial_i}\partial_s \\
&\quad -\partial_j\left(\Gamma_{ik}^t\right)\partial_t - \Gamma_{ik}^t \nabla_{\partial_j}\partial_t \\
&= \partial_i\left(\Gamma_{jk}^l\right)\partial_l - \partial_j\left(\Gamma_{ik}^l\right)\partial_l \\
&\quad +\Gamma_{jk}^s\Gamma_{is}^l\partial_l - \Gamma_{ik}^t\Gamma_{jt}^l\partial_l \\
&= \left(\partial_i\Gamma_{jk}^l - \partial_j\Gamma_{ik}^l + \Gamma_{jk}^s\Gamma_{is}^l - \Gamma_{ik}^s\Gamma_{js}^l\right)\partial_l.
\end{aligned}
$$

So

$$R_{ijk}^l = \partial_i\Gamma_{jk}^l - \partial_j\Gamma_{ik}^l + \Gamma_{jk}^s\Gamma_{is}^l - \Gamma_{ik}^s\Gamma_{js}^l.$$

This coordinate expression can also be used, in conjunction with the properties of the Christoffel symbols, to prove all of the symmetry properties of the curvature tensor. The formula clearly simplifies if we are at a point p where $\Gamma_{ij}^k|_p = 0$

$$R_{ijk}^l|_p = \partial_i\Gamma_{jk}^l|_p - \partial_j\Gamma_{ik}^l|_p.$$

If we use the formulas for the Christoffel symbols we can evidently get an expression for R_{ijk}^l that depends on the metric g_{ij} and its first two derivatives.

4. The Fundamental Curvature Equations

In this section we are going to study how curvature comes up naturally in the investigation of certain types of functions. This will lead us to various formulae that make it possible to calculate the curvature tensor on all of the rotationally symmetric and doubly warped product metrics from chapter 1. With this information we can then exhibit the above mentioned examples. This will be accomplished in the next chapter.

4.1. Distance Functions. The functions we wish to look into are *distance functions*. As we don't have a concept of distance yet, we will say that $r : U \to \mathbb{R}$, where $U \subset (M,g)$ is open, is a *distance function* if $|\nabla r| \equiv 1$ on U. Distance functions are therefore simply solutions to the *Hamilton-Jacobi equation*

$$|\nabla r|^2 = 1.$$

This is a nonlinear first-order PDE and can be solved by the method of characteristics see e.g. [5]. For now we shall assume that solutions exist and investigate their

Figure 2.1

properties. Later, when we have developed the theory of geodesics, we shall show the existence of such functions and also show that their name is appropriate.

EXAMPLE 21. *On* $(\mathbb{R}^n, \text{can})$ *define* $r(x) = |x - y|$. *Then* r *is smooth on* $\mathbb{R}^n - \{y\}$ *and has* $|\nabla r| \equiv 1$. *If we have two different points* $\{y, z\}$, *then*

$$r(x) = d(x, \{y, z\}) = \min\{d(x, y),\ d(x, z)\}$$

is smooth away from $\{y, z\}$ *and the hyperplane* $\{x \in \mathbb{R}^n : |x - y| = |x - z|\}$ *equidistant from* y *and* z.

EXAMPLE 22. *More generally if* $M \subset \mathbb{R}^n$ *is a submanifold, then it can be shown that*

$$r(x) = d(x, M) = \inf\{d(x, y) : y \in M\}$$

is a distance function on some open set $U \subset \mathbb{R}^n$. *If* M *is an orientable hypersurface, then we can see this as follows. Since* M *is orientable, we can choose a unit normal vector field* N *on* M. *Now "coordinatize"* $\mathbb{R}^n$ *as* $x = tN + y$, *where* $t \in \mathbb{R}$, $y \in M$. *In some neighborhood* U *of* M *these "coordinates" are actually well-defined. In other words, there is some function* $\varepsilon(y) : M \to (0, \infty)$ *such that any point in*

$$U = \{tN + y : y \in M,\ |t| < \varepsilon(y)\}$$

has unique coordinates (t, y). *We can now define* $r(x) = t$ *on* U *or* $r(x) = d(x, M) = |t|$ *on* $U - M$. *Both functions will then define distance functions on their respective domains. Here* r *is usually referred to as the* signed *distance to* M, *while* f *is just the regular distance. Figure 2.1 shows some pictures of the level sets of a distance function together with the orthogonal trajectories that form the integral curves for the gradient of the distance function.*

EXAMPLE 23. *On* $I \times M$, *where* $I \subset \mathbb{R}$, *is an interval we have metrics of the form* $dr^2 + g_r$, *where* dr^2 *is the standard metric on* I *and* g_r *is a metric on* $\{r\} \times M$ *that depends on* r. *In this case the projection* $I \times M \to I$ *is a distance function. Special cases of this situation are rotationally symmetric metrics, doubly warped products, and our submersion metrics on* $I \times S^{2n-1}$.

LEMMA 4. *Given* $r : U \to I \subset \mathbb{R}$, *then* r *is a distance function iff* r *is a Riemannian submersion.*

Proof. In general, we have $dr(v) = g(\nabla r, v)$, so $Dr(v) = dr(v)\partial_t = 0$ iff $v \perp \nabla r$. Thus, v is perpendicular to the kernel of Dr iff it is proportional to ∇r. For such $v = \alpha \nabla r$ we have that

$$Dr(v) = \alpha Dr(\nabla r) = \alpha g(\nabla r, \nabla r)\partial_t.$$

Now ∂_t has length 1 in I, so

$$\begin{aligned} |v| &= |\alpha||\nabla r|, \\ |Dr(v)| &= |\alpha||\nabla r|^2. \end{aligned}$$

Thus, r is a Riemannian submersion iff $|\nabla r| = 1$ □

Before continuing we need some simplifying notation. A distance function $r : U \to \mathbb{R}$ is fixed and $U \subset (M, g)$ is an open subset of a Riemannian manifold. The gradient ∇r will usually be denoted by $\partial_r = \nabla r$. The ∂_r notation comes from our warped product metrics $dr^2 + g_r$. The level sets for r are denoted $U_r = \{x \in U : r(x) = r\}$, and the induced metric on U_r is g_r. In this spirit ∇^r, R^r are the Riemannian connection and curvature on (U_r, g_r). The $(1,1)$ version of the Hessian of r is denoted by $S(\cdot) = \nabla.\partial_r$, i.e., $\mathrm{Hess}\, r(X, Y) = g(S(X), Y)$. S stands for second derivative or *shape operator* or *second fundamental form*, depending on the point of view of the observer. The last two terms are more or less synonymous and refer to the shape of (U_r, g_r) in $(U, g) \subset (M, g)$. The idea is that $S = \nabla \partial_r$ measures how the induced metric on U_r changes by computing how the unit normal to U_r changes.

Example 24. *Let $M \subset \mathbb{R}^n$ be an orientable hypersurface, N the unit normal, and S the shape operator defined by $S(v) = \nabla_v N$ for $v \in TM$. If $S \equiv 0$ on M then N must be a constant vector field on M, and hence M is an open subset of the hyperplane*

$$H = \{x + p \in \mathbb{R}^n : x \cdot N_p = 0\},$$

where $p \in M$ is fixed. As an explicit example of this, recall our isometric immersion or embedding $(\mathbb{R}^{n-1}, \mathrm{can}) \to (\mathbb{R}^n, \mathrm{can})$ from chapter 1 defined by

$$(x^1, \ldots, x^{n-1}) \to (\gamma(x^1), x^2, \ldots, x^{n-1}),$$

where γ is a unit speed curve $\gamma : \mathbb{R} \to \mathbb{R}^2$. In this case,

$$N = (N(x^1), 0, \ldots, 0)$$

is a unit normal, where $N(x^1)$ is the unit normal to γ in $\mathbb{R}^2$. We can write this as

$$N = (-\dot{\gamma}^2(x^1), \dot{\gamma}^1(x^1), 0, \ldots, 0)$$

in Cartesian coordinates. So

$$\begin{aligned} \nabla N &= -d(\dot{\gamma}^2)\partial_1 + d(\dot{\gamma}^1)\partial_2 \\ &= -\ddot{\gamma}^2 dx^1 \partial_1 + \ddot{\gamma}^1 dx^1 \partial_2 \\ &= (-\ddot{\gamma}^2 \partial_1 + \ddot{\gamma}^1 \partial_2) dx^1. \end{aligned}$$

Thus, $S \equiv 0$ iff $\ddot{\gamma}^1 = \ddot{\gamma}^2 = 0$ iff γ is a straight line iff M is an open subset of a hyperplane. The shape operator therefore really captures the idea that the hypersurface bends in $\mathbb{R}^n$, even though $\mathbb{R}^{n-1}$ cannot be seen to bend inside itself.

We have seen here the difference between *extrinsic* and *intrinsic* geometry. Intrinsic geometry is everything we can do on a Riemannian manifold (M, g) that does not depend on how (M, g) might be isometrically immersed in some other Riemannian manifold. Extrinsic geometry is the study of how an isometric immersion $(M, g) \to (N, g_N)$ bends (M, g) inside (N, g_N). Thus, the curvature tensor on (M, g) measures how the space bends intrinsically, while the shape operator measures extrinsic bending.

4.2. Curvature Equations. We are now ready to establish our first fundamental equation.

THEOREM 2. (The Radial Curvature Equation) *If $U \subset (M, g)$ is an open set and $r : U \to \mathbb{R}$ a distance function, then*

$$\nabla_{\partial_r} S + S^2 = -R_{\partial_r}.$$

PROOF. We proceed by straightforward computation. If X is a vector field on U, then

$$
\begin{aligned}
(\nabla_{\partial_r} S)(X) + S^2(X) &= \nabla_{\partial_r}(S(X)) - S(\nabla_{\partial_r} X) + S(S(X)) \\
&= \nabla_{\partial_r} \nabla_X \partial_r - \nabla_{\nabla_{\partial_r} X} \partial_r + \nabla_{\nabla_X \partial_r} \partial_r \\
&= \nabla_{\partial_r} \nabla_X \partial_r - \nabla_{\nabla_{\partial_r} X - \nabla_X \partial_r} \partial_r \\
&= \nabla_{\partial_r} \nabla_X \partial_r - \nabla_{[\partial_r, X]} \partial_r.
\end{aligned}
$$

In order for this to equal $-R(X, \partial_r)\partial_r$ we only need to check what happened to $-\nabla_X \nabla_{\partial_r} \partial_r$. However, as $\partial_r = \nabla r$ is unit, we see that for any vector field Y on U:

$$
\begin{aligned}
g(\nabla_{\partial_r} \partial_r, Y) &= \operatorname{Hess} r (\partial_r, Y) \\
&= \operatorname{Hess} r (Y, \partial_r) \\
&= g(\nabla_Y \partial_r, \partial_r) \\
&= \frac{1}{2} D_Y g(\partial_r, \partial_r) \\
&= \frac{1}{2} D_Y 1 = 0.
\end{aligned}
$$

In particular, $\nabla_{\partial_r} \partial_r = S(\partial_r) = 0$ on all of U. $\square$

This result tells us two things: First, that ∂_r is always a zero eigenvector for S and secondly how certain "radial curvatures" relate to the Hessian of r. The Hessian of a generic function cannot, of course, exhibit such predictable behavior (namely, being a solution to a PDE). It is only geometrically relevant functions that behave so nicely.

The second and third fundamental equations are also known as the *Gauss equations* and *Codazzi-Mainardi equations*, respectively. They will be proved simultaneously but stated separately. For a vector we use the notation for decomposing it into normal and tangential components to U_r:

$$
\begin{aligned}
v &= \tan v + \operatorname{nor} v \\
&= v - g(v, \partial_r)\partial_r + g(v, \partial_r)\partial_r.
\end{aligned}
$$

THEOREM 3. (The Tangential Curvature Equation)

$$
\begin{aligned}
\tan R(X, Y)Z &= R^r(X, Y)Z - (S(X) \wedge S(Y))(Z), \\
g(R(X, Y)Z, W) &= g_r(R^r(X, Y)Z, W) - \operatorname{II}(Y, Z)\operatorname{II}(X, W) + \operatorname{II}(X, Z)\operatorname{II}(Y, W).
\end{aligned}
$$

Here X, Y, Z, W are tangent to the level sets U_r and

$$\mathrm{II}\,(U,V) = \mathrm{Hess}r\,(U,V) = g\,(S\,(U)\,,V)$$

is the classical second fundamental form.

THEOREM 4. (The Normal or Mixed Curvature Equation)

$$
\begin{aligned}
g\,(R(X,Y)Z, \partial_r) &= g(-(\nabla_X S)(Y) + (\nabla_Y S)(X), Z) \\
&= -(\nabla_X \mathrm{II})\,(Y, Z) + (\nabla_Y \mathrm{II})\,(X, Z)\,.
\end{aligned}
$$

where X, Y, Z are tangent to the level sets U_r.

PROOF. The proofs hinge on the important fact that if X, Y are vector fields that are tangent to the level sets U_r, then:

$$
\begin{aligned}
\nabla_X^r Y &= \tan(\nabla_X Y) \\
&= \nabla_X Y - g\,(\nabla_X Y, \partial_r)\,\partial_r \\
&= \nabla_X Y + g(S(X), Y)\partial_r \\
&= \nabla_X Y + \mathrm{II}(X, Y)\partial_r
\end{aligned}
$$

Here the first equality is a consequence of the uniqueness of the Riemannian connection on (U_r, g_r). One can check either that $\tan(\nabla_X Y)$ satisfies properties 1-4 of a Riemannian connection or alternatively that it satisfies the Koszul formula. The latter task is almost immediate. The second and fourth equality are obvious. The third follows as $Y \perp \partial_r$ implies

$$
\begin{aligned}
0 &= \nabla_X g(Y, \partial_r) \\
&= g(\nabla_X Y, \partial_r) + g(Y, S(X)),
\end{aligned}
$$

whence

$$g(S(X), Y) = -g(\nabla_X Y, \partial_r).$$

Both of the curvature equations are now verified by calculating $R(X,Y)Z$ using

$$\nabla_X Y = \nabla_X^r Y - g(S(X), Y) \cdot \partial_r.$$

$$
\begin{aligned}
R(X,Y)Z &= \nabla_X \nabla_Y Z - \nabla_Y \nabla_X Z - \nabla_{[X,Y]} Z \\
&= \nabla_X (\nabla_Y^r Z - g(S(Y), Z) \cdot \partial_r) - \nabla_Y (\nabla_X^r Z - g(S(X), Z) \cdot \partial_r) \\
&\quad - \nabla_{[X,Y]}^r Z + g(S([X,Y]), Z) \cdot \partial_r \\
&= \nabla_X \nabla_Y^r Z - \nabla_Y \nabla_X^r Z - \nabla_{[X,Y]}^r Z \\
&\quad - \nabla_X (g(S(Y), Z) \cdot \partial_r) + \nabla_Y (g(S(X), Z) \cdot \partial_r) + g(S([X,Y]), Z) \cdot \partial_r \\
&= R^r(X,Y)Z - g(S(X), \nabla_Y Z) \cdot \partial_r + g(S(Y), \nabla_X Z) \cdot \partial_r \\
&\quad - g(S(Y), \nabla_X Z) \cdot \partial_r + g(S(X), \nabla_Y Z) \cdot \partial_r \\
&\quad - g(\nabla_X S(Y), Z) \cdot \partial_r + g(\nabla_Y S(X), Z) \cdot \partial_r + g(S([X,Y]), Z) \cdot \partial_r \\
&\quad - g(S(Y), Z)S(X) + g(S(X), Z)S(Y) \\
&= R^r(X,Y)Z - (S(X) \wedge S(Y))\,(Z) \\
&\quad + g(-(\nabla_X S)(Y) + (\nabla_Y S)(X), Z) \cdot \partial_r
\end{aligned}
$$

This establishes the first part of each formula. The second parts follow from using the definitions of the involved concepts. □

The three fundamental equations give us a way of computing curvature tensors by induction on dimension. More precisely, if we know how to do computations on U_r and also how to compute S, then we can compute anything on U. We shall clarify and exploit this philosophy in subsequent chapters.

Here we confine ourselves to some low dimensional observations. Recall that the three curvature quantities sec, Ric, and scal obeyed some special relationships in dimensions 2 and 3. Curiously enough this also manifests itself in our three fundamental equations.

If M has dimension 1, then there aren't too many distance functions. Our equations don't even seem to apply here since the level sets are points. This is related to the fact that $R \equiv 0$ on all 1 dimensional spaces.

If M has dimension 2, then any distance function $r : U \subset M \to \mathbb{R}$ has 1-dimensional level sets. Thus $R^r \equiv 0$ and the three vectors X, Y and Z are proportional. Our equations therefore reduce to the single equation:

$$\nabla_{\partial_r} S + S^2 = -R_{\partial_r}.$$

Actually, since $S(\partial_r) = 0$, we know that S depends only on its value on a unit vector $v \in TU_r$ thus $S(v) = \alpha v$, where $\alpha = \mathrm{tr} S = \Delta r$. The radial curvature equation can therefore be reduced to:

$$\partial_r(\Delta r) + (\Delta r)^2 = -\sec(T_p M).$$

To be even more concrete, we have that g_r on U_r can be written: $g_r = \varphi^2(r, \theta) d\theta^2$; so

$$g = dr^2 + \varphi^2(r, \theta) d\theta^2,$$

and since

$$
\begin{aligned}
\varphi \partial_r \varphi &= \frac{1}{2} \partial_r g\left(\partial_\theta, \partial_\theta\right) \\
&= g\left(\nabla_{\partial_r} \partial_\theta, \partial_\theta\right) \\
&= g\left(S\left(\partial_\theta\right), \partial_\theta\right) \\
&= \alpha\left|\partial_\theta\right|^2 \\
&= \alpha \varphi^2,
\end{aligned}
$$

we have

$$\mathrm{tr} S = \frac{\partial_r \varphi}{\varphi},$$

implying

$$-\sec(T_p M) = \frac{\partial_r^2 \varphi}{\varphi}.$$

When M has dimension 3, the level sets of r are 2-dimensional. The radial curvature equation therefore doesn't reduce, but in the other two equations we have that one of the three vectors X, Y, Z is a linear combination of the other two. We might as well assume that $X \perp Y$ and $Z = X$ or Y. So, if $\{X, Y, \partial_r\}$ represents an orthonormal framing, then the complete curvature tensor depends on the quantities: $g(R(X, \partial_r)\partial_r, Y)$, $g(R(X, \partial_r)\partial_r, X)$, $g(R(Y, \partial_r)\partial_r, Y)$, $g(R(X, Y)Y, X)$, $g(R(X, Y)Y, \partial_r)$, $g(R(Y, X)X, \partial_r)$. The first three quantities can be computed from the radial curvature equation, the fourth from the tangential curvature equation, and the last two from the mixed curvature equation.

In the special case where $M^3 = \mathbb{R}^3$, $R = 0$, the tangential curvature equation is particularly interesting:

$$
\begin{aligned}
\sec(T_p U_r) &= R^r(X,Y,Y,X) \\
&= g(S(X),X)g(S(Y),Y) - g(S(X),Y)g(S(X),Y) \\
&= \det S
\end{aligned}
$$

This was *Gauss's wonderful observation!* namely, that the extrinsic quantity $\det S$ for U_r is actually the intrinsic quantity, $\sec(T_p U_r)$.

Finally, in dimension 4 everything reaches its most general level. We can start with an orthonormal framing $\{X, Y, Z, \partial_r\}$, and there will be twenty curvature quantities to compute.

5. The Equations of Riemannian Geometry

In this section we shall investigate the connection between the metric tensor and curvature. This is done by using the radial curvature equation together with some new formulae. Having established these fundamental equations, we shall introduce some useful vector fields that make it possible to see how the curvature influences the metric in some unexpected ways.

Recall from the end of the last section that we arrived at a very nice formula for the relationship between the metric and curvature on a surface, namely, if $g = dr^2 + \varphi^2(r, \theta)d\theta^2$, then $\partial_r^2 \varphi = -\sec \cdot \varphi$. This formula can be used not only to compute curvatures from knowledge of the metric, but also in reverse to conclude things about the metric from the curvature. This relationship, which is classical for surfaces, will be generalized in this section to manifolds of any dimension and then extensively used throughout the entire text as a universal tool for understanding the relationship between the metric and curvature.

5.1. The Coordinate-Free Equations. We need to introduce an ad hoc concept for Hessians and symmetric bilinear forms on Riemannian manifolds. If $B(X,Y)$ is a symmetric $(0,2)$-tensor and $L(X)$ the corresponding self-adjoint $(1,1)$-tensor defined via

$$
g(L(X),Y) = B(X,Y),
$$

then the *square* of B is the symmetric bilinear form corresponding to L^2

$$
B^2(X,Y) = g(L^2(X),Y) = g(L(X),L(Y)).
$$

Note that this symmetric bilinear form is always nonnegative, i.e., $B^2(X,X) \geq 0$ for all X.

PROPOSITION 7. *If we have a smooth distance function* $r : (U,g) \to \mathbb{R}$ *and denote* $\nabla r = \partial_r$, *then*

(1) $L_{\partial_r} g = 2\text{Hess}r$,
(2) $(\nabla_{\partial_r} \text{Hess}r)(X,Y) + \text{Hess}^2 r(X,Y) = -R(X,\partial_r,\partial_r,Y)$,
(3) $(L_{\partial_r} \text{Hess}r)(X,Y) - \text{Hess}^2 r(X,Y) = -R(X,\partial_r,\partial_r,Y)$.

PROOF. (1) is simply the definition of the Hessian.

To prove (2) and (3) we use that $\nabla_{\partial_r} \partial_r = 0$ and perform virtually the same calculations that were used for the radial curvature equation. Keep in mind that

$\nabla_X \partial_r = S(X)$ is the self-adjoint operator corresponding to Hess$r$.

$$
\begin{aligned}
(\nabla_{\partial_r} \mathrm{Hess}r)(X,Y) &= \partial_r \mathrm{Hess}r(X,Y) - \mathrm{Hess}r(\nabla_{\partial_r}X,Y) - \mathrm{Hess}r(X,\nabla_{\partial_r}Y) \\
&= \partial_r g(\nabla_X \partial_r, Y) - g\left(\nabla_{\nabla_{\partial_r}X}\partial_r, Y\right) - g(\nabla_X \partial_r, \nabla_{\partial_r}Y) \\
&= g(\nabla_{\partial_r}\nabla_X \partial_r, Y) - g\left(\nabla_{\nabla_{\partial_r}X}\partial_r, Y\right) \\
&\quad + g(\nabla_X \partial_r, \nabla_{\partial_r}Y) - g(\nabla_X \partial_r, \nabla_{\partial_r}Y) \\
&= g(R(\partial_r, X)\partial_r, Y) - g(\nabla_{\nabla_X \partial_r}\partial_r, Y) \\
&= -R(X, \partial_r, \partial_r, Y) - g(\nabla_Y \partial_r, \nabla_X \partial_r) \\
&= -R(X, \partial_r, \partial_r, Y) - \mathrm{Hess}^2 r(X,Y).
\end{aligned}
$$

$$
\begin{aligned}
(L_{\partial_r} \mathrm{Hess}r)(X,Y) &= \partial_r \mathrm{Hess}r(X,Y) - \mathrm{Hess}r([\partial_r, X], Y) - \mathrm{Hess}r(X, [\partial_r, Y]) \\
&= \partial_r g(\nabla_X \partial_r, Y) - g\left(\nabla_{[\partial_r, X]}\partial_r, Y\right) - g(\nabla_X \partial_r, [\partial_r, Y]) \\
&= g(\nabla_{\partial_r}\nabla_X \partial_r, Y) - g\left(\nabla_{[\partial_r, X]}\partial_r, Y\right) \\
&\quad + g(\nabla_X \partial_r, \nabla_{\partial_r}Y) - g(\nabla_X \partial_r, \nabla_{\partial_r}Y - \nabla_Y \partial_r) \\
&= g(R(\partial_r, X)\partial_r, Y) + g(\nabla_X \partial_r, \nabla_Y \partial_r) \\
&= -R(X, \partial_r, \partial_r, Y) + \mathrm{Hess}^2 r(X,Y).
\end{aligned}
$$

$\square$

The first equation shows how the Hessian controls the metric. The second and third equations give us control over the Hessian if we have information about the curvature. These two equations are different in a very subtle way. The third equation is at the moment the easiest to work with as it only uses Lie derivatives and hence can be put in a nice form in an appropriate coordinate system. The second equation is ultimately more useful, but requires that we find a way of making it easier to interpret.

In the next two sections we shall see how appropriate choices for vector fields can give us a better understanding of these fundamental equations.

5.2. Jacobi Fields. A *Jacobi field* for a smooth distance function r is a smooth vector field J that does not depend on r, i.e., it satisfies the *Jacobi equation*

$$L_{\partial_r} J = 0.$$

This is a first order linear PDE, which can be solved by the *method of characteristics*. To see how this is done we locally select a coordinate system $(r, x^2, ..., x^n)$ where r is the first coordinate. Then $J = a^r \partial_r + a^i \partial_i$ and the Jacobi equation becomes:

$$
\begin{aligned}
0 &= L_{\partial_r} J \\
&= L_{\partial_r}\left(a^r \partial_r + a^i \partial_i\right) \\
&= \partial_r(a^r)\partial_r + \partial_r(a^i)\partial_i.
\end{aligned}
$$

Thus the coefficients a^r, a^i have to be independent of r as already indicated. What is more, we can construct such Jacobi fields knowing the values on a hypersurface $H \subset M$ where $(x^2, ..., x^n)|_H$ is a coordinate system. In this case ∂_r is transverse to H and so we can solve the equations by declaring that a^r, a^i are constant along the integral curves for ∂_r. Note that the coordinate vector fields are themselves Jacobi fields. Jacobi fields satisfy a more general second order equation, also known as the *Jacobi Equation*:

$$\nabla_{\partial_r}\nabla_{\partial_r} J = -R(J, \partial_r)\partial_r,$$

since

$$\begin{aligned}
-R\left(J,\partial_r\right)\partial_r &= R\left(\partial_r, J\right)\partial_r \\
&= \nabla_{\partial_r}\nabla_J\partial_r - \nabla_J\nabla_{\partial_r}\partial_r - \nabla_{[\partial_r,J]}\partial_r \\
&= \nabla_{\partial_r}\nabla_J\partial_r \\
&= \nabla_{\partial_r}\nabla_{\partial_r}J.
\end{aligned}$$

This is a second order equation and must therefore have more solutions than the above first order equation. This equation will be studied further in chapter 3 for rotationally symmetric metrics and for general Riemannian manifolds in chapter 6.

If we evaluate equations (1) and (3) on Jacobi fields we obtain

(1) $\partial_r\left(g\left(J_1, J_2\right)\right) = 2\mathrm{Hess}r\left(J_1, J_2\right),$

(3) $\partial_r\left(\mathrm{Hess}r\left(J_1, J_2\right)\right) - \mathrm{Hess}^2 r\left(J_1, J_2\right) = -R\left(J_1, \partial_r, \partial_r, J_2\right).$

As we now only have directional derivatives we have a much simpler version of the fundamental equations. Therefore, there is a much better chance of predicting how g and Hess$r$ change depending on our knowledge of Hess$r$ and R respectively.

This can be reduced a bit further if we take a product neighborhood $\Omega = (a,b) \times H \subset M$ such that $r\left(t, z\right) = t$. On this product the metric has the form

$$g = dr^2 + g_r$$

where g_r is a one parameter family of metrics on H. If J is a vector field on H, then there is a unique extension to a Jacobi field on $\Omega = (a, b) \times H$. First observe that

$$\begin{aligned}
\mathrm{Hess}r\left(\partial_r, J\right) &= g\left(\nabla_{\partial_r}\partial_r, J\right) = 0, \\
g_r\left(\partial_r, J\right) &= 0.
\end{aligned}$$

Thus we only need to consider the restrictions of g and Hess$r$ to H. By doing this we obtain

$$\partial_r g = \partial_r g_r = 2\mathrm{Hess}r$$

The fundamental equations can therefore be written as

(1) $\partial_r g_r = 2\mathrm{Hess}r,$

(3) $\partial_r\mathrm{Hess}r - \mathrm{Hess}^2 r = -R\left(\cdot, \partial_r, \partial_r, \cdot\right).$

There is a sticky point that is hidden in (3). Namely, how to extract information from R and pass it on to the Hessian. As we usually make assumptions about the sectional curvature we should try to rewrite this term. This can be done as follows:

$$\begin{aligned}
R\left(X, \partial_r, \partial_r, X\right) &= \sec\left(X, \partial_r\right)\left(g\left(X, X\right)g\left(\partial_r, \partial_r\right) - \left(g\left(X, \partial_r\right)\right)^2\right) \\
&= \sec\left(X, \partial_r\right)g\left(X - g\left(X, \partial_r\right)\partial_r, X - g\left(X, \partial_r\right)\partial_r\right) \\
&= \sec\left(X, \partial_r\right)g_r\left(X, X\right).
\end{aligned}$$

So if we evaluate (3) on a Jacobi field J we obtain

$$\partial_r\left(\mathrm{Hess}r\left(J, J\right)\right) - \mathrm{Hess}^2 r\left(J, J\right) = -\sec\left(J, \partial_r\right)g_r\left(J, J\right).$$

This means that (1) and (3) are coupled as we have not eliminated the metric from (3). The next subsection shows how we can deal with this by evaluating on different vector fields.

Nevertheless, we have reduced (1) and (3) to a set of ODEs where r is the independent variable along the integral curve for ∂_r through p.

5.3. Parallel Fields. A *parallel field* for a smooth distance function is a vector field X such that:

$$\nabla_{\partial_r} X = 0.$$

This is, like the Jacobi equation, a first order linear PDE and can be solved in a similar manner. There is, however, one crucial difference: Parallel fields are almost never Jacobi fields.

If we evaluate g on a pair of parallel fields we see that

$$\partial_r g\left(X, Y\right) = g\left(\nabla_{\partial_r} X, Y\right) + g\left(X, \nabla_{\partial_r} Y\right) = 0.$$

This means that (1) is not simplified by using parallel fields. The second equation, on the other hand, now looks like

$$\partial_r \left(\operatorname{Hess} r\left(X, Y\right)\right) + \operatorname{Hess}^2 r\left(X, Y\right) = -R\left(X, \partial_r, \partial_r, Y\right).$$

If we rewrite this in terms of sectional curvature we obtain as above

$$\partial_r \left(\operatorname{Hess} r\left(X, X\right)\right) + \operatorname{Hess}^2 r\left(X, X\right) = -\sec\left(X, \partial_r\right) g_r\left(X, X\right).$$

But this time we know that $g_r\left(X, X\right)$ is constant in r as X is parallel. We can even assume that $g\left(X, \partial_r\right) = 0$ and $g\left(X, X\right) = 1$ by first projecting X onto H and then scaling it. Therefore, (2) takes the form

$$\partial_r \left(\operatorname{Hess} r\left(X, X\right)\right) + \operatorname{Hess}^2 r\left(X, X\right) = -\sec\left(X, \partial_r\right)$$

on unit parallel fields that are orthogonal to ∂_r. In this way we really have decoupled the equation for the Hessian from the metric. This allows us to glean information about the Hessian from information about sectional curvature. Equation (1), when rewritten using Jacobi fields, then gives us information about the metric from the information we just obtained about the Hessian using parallel fields.

5.4. Conjugate Points. In general, we might think of the curvatures R_{∂_r} as being given. They could be constant or merely satisfy some inequality. We then wish to investigate how the curvature influences the metric. Equation (1) is linear. Thus the metric can't blow up in finite time unless the Hessian also blows up. However, if we assume that the curvature is bounded, then equation (2) tells us that, if the Hessian blows up, then it must be decreasing in r, hence it can only go to $-\infty$. Going back to (1), we then conclude that the only degeneration which can occur along an integral curve for ∂_r, is that the metric stops being positive definite. We say that the distance function r develops a *conjugate, or focal, point* along this integral curve if this occurs. Below we have some pictures of how conjugate points can develop. Note that as the metric itself is Euclidean, these singularities exist only in the coordinates, not in the metric.

Figure 2.2

It is worthwhile investigating equations (2) and (3) a little further. If we rewrite them as

$$(2) \quad (\nabla_{\partial_r}\text{Hess}r)(X,X) = -R(X,\partial_r,\partial_r,X) - \text{Hess}^2 r(X,X),$$
$$(3) \quad (L_{\partial_r}\text{Hess}r)(X,X) = -R(X,\partial_r,\partial_r,X) + \text{Hess}^2 r(X,X),$$

then we can think of the curvatures as representing fixed *external forces*, while $\text{Hess}^2 r$ describes an *internal reaction (or interaction)*. The reaction term is always of a fixed sign and, it will try to force Hess$r$ blow up in finite time. If, for instance sec ≤ 0, then $L_{\partial_r}\text{Hess}r$ is positive. Therefore, if Hess$r$ is positive at some point, then it will stay positive. On the other hand, if sec ≥ 0, then $\nabla_{\partial_r}\text{Hess}r$ is negative, forcing Hess$r$ to stay nonpositive if it is nonpositive at a point.

In chapters 6, 7, 9, and 11 we shall study and exploit this in much greater detail.

6. Some Tensor Concepts

In this section we shall collect together some notational baggage that is needed from time to time.

6.1. Type Change. The inner product structure on the tangent spaces to a Riemannian manifold makes it possible to view tensors in different ways. We saw this with the Hessian and the Ricci tensor. This is nothing but the elementary observation that a bilinear map can be interpreted as a linear map when one has an inner product present.

If, in general, we have an (s,t)-tensor T, we view it as a section in the bundle

$$\underbrace{TM \otimes \cdots \otimes TM}_{s \text{ times}} \otimes \underbrace{T^*M \otimes \cdots \otimes T^*M}_{t \text{ times}}$$

Then given a Riemannian metric g on M, we can make it into an $(s-k, t+k)$-tensor for any $k \in \mathbb{Z}$ such that both $s-k$ and $t+k$ are nonnegative. Abstractly, this is done as follows: On a Riemannian manifold TM is naturally isomorphic to T^*M; the isomorphism is given by sending $v \in TM$ to the linear map $(w \to g(v,w)) \in T^*M$. Using this isomorphism we can therefore replace TM by T^*M or vice versa and thus change the type of the tensor.

At a more concrete level what happens is this: We select a frame $E_1, \ldots, E_n$ and construct the coframe $\sigma^1, \ldots, \sigma^n$. The vectors and covectors (in T^*M) can be written as

$$\begin{aligned} v &= v^i E_i = \sigma^i \left(v \right) E_i, \\ \omega &= \alpha_j \sigma^j = \omega \left(E_j \right) \sigma^j. \end{aligned}$$

The tensor T can now be written as

$$T = T^{i_1 \cdots i_s}_{j_1 \cdots j_t} E_{i_1} \otimes \cdots \otimes E_{i_s} \otimes \sigma^{j_1} \otimes \cdots \otimes \sigma^{j_t}.$$

Now we need to know how we can change E_i into a covector and σ^j into a vector. As before, the dual to E_i is the covector $w \to g \left(E_i, w \right)$, which can be written as

$$g \left(E_i, w \right) = g \left(E_i, E_j \right) \sigma^j \left(w \right) = g_{ij} \sigma^j \left(w \right).$$

Conversely, we have to find the vector v corresponding to the covector σ^j. The defining property is

$$g \left(v, w \right) = \sigma^j \left(w \right).$$

Thus, we have

$$g \left(v, E_i \right) = \delta^j_i.$$

If we write $v = v^k E_k$, this gives

$$g_{ki} v^k = \delta^j_i.$$

Letting g^{ij} denote the ijth entry in the inverse of (g_{ij}), we therefore have

$$v = v^i E_i = g^{ij} E_i.$$

Thus,

$$\begin{aligned} E_i &\to g_{ij} \sigma^j, \\ \sigma^j &\to g^{ij} E_i. \end{aligned}$$

Note that using Einstein notation properly will help keep track of the correct way of doing things as long as the inverse of g is given with superscript indices. With this formula one can easily change types of tensors by replacing Es with σs and vice versa. Note that if we used coordinate vector fields in our frame, then one really needs to invert the metric, but if we had chosen an orthonormal frame, then one simply moves indices up and down as the metric coefficients satisfy $g_{ij} = \delta_{ij}$.

Let us list some examples:

The Ricci tensor: We write the Ricci tensor as a $(1,1)$-tensor: $\mathrm{Ric} \left(E_i \right) = \mathrm{Ric}^j_i E_j$; thus

$$\mathrm{Ric} = \mathrm{Ric}^i_j \cdot E_i \otimes \sigma^j.$$

As a $(0,2)$-tensor it will look like

$$\mathrm{Ric} = \mathrm{Ric}_{jk} \cdot \sigma^j \otimes \sigma^k = g^i_{ji} \mathrm{Ric}_k \cdot \sigma^j \otimes \sigma^k,$$

while as a $(2,0)$-tensor acting on covectors it will be

$$\mathrm{Ric} = \mathrm{Ric}^{ik} \cdot E_i \otimes E_k = g^{ij} \mathrm{Ric}^k_j \cdot E_i \otimes E_k.$$

The curvature tensor: We start with the $(1,3)$-curvature tensor $R \left(X, Y \right) Z$, which we write as

$$R = R^l_{ijk} \cdot E_l \otimes \sigma^i \otimes \sigma^j \otimes \sigma^k.$$

As a $(0,4)$-tensor we get

$$
\begin{aligned}
R &= R_{ijkl} \cdot \sigma^i \otimes \sigma^j \otimes \sigma^k \otimes \sigma^l \\
&= R^s_{ijk} g_{sl} \cdot \sigma^i \otimes \sigma^j \otimes \sigma^k \otimes \sigma^l,
\end{aligned}
$$

while as a $(2,2)$-tensor we have:

$$
\begin{aligned}
R &= R^{kl}_{ij} \cdot E_k \otimes E_l \otimes \sigma^i \otimes \sigma^j \\
&= R^l_{ijs} g^{sk} \cdot E_k \otimes E_l \otimes \sigma^i \otimes \sigma^j.
\end{aligned}
$$

Here, however, we must watch out, because there are several different ways of doing this. We choose to raise the last index, but we could also have chosen any other index, thus yielding different $(2,2)$-tensors. The way we did it gives essentially the curvature operator.

6.2. Contractions. Contractions are simply traces of tensors. Thus, the contraction of a $(1,1)$-tensor $T = T^i_j \cdot E_i \otimes \sigma^j$ is simply its trace:

$$
C(T) = \mathrm{tr}\, T = T^i_i.
$$

If instead we had a $(0,2)$-tensor T, then we could, using the Riemannian structure, first change it to a $(1,1)$-tensor and then take the trace

$$
\begin{aligned}
C(T) &= C\left(T_{ij} \cdot \sigma^i \otimes \sigma^j\right) \\
&= C\left(T_{ik} g^{kj} \cdot E_k \otimes \sigma^j\right) \\
&= T_{ik} g^{ki}.
\end{aligned}
$$

In this way the Ricci tensor becomes a contraction:

$$
\begin{aligned}
\mathrm{Ric} &= \mathrm{Ric}^i_j \cdot E_i \otimes \sigma^j \\
&= R^{kj}_{ik} \cdot E_i \otimes \sigma^j \\
&= R^j_{iks} g^{sk} \cdot E_i \otimes \sigma^j,
\end{aligned}
$$

or

$$
\begin{aligned}
\mathrm{Ric} &= \mathrm{Ric}_{ij} \cdot \sigma^i \otimes \sigma^j \\
&= g^{kl} R_{iklj} \cdot \sigma^i \otimes \sigma^j,
\end{aligned}
$$

which after type change can be seen to give the same expressions. The scalar curvature can be expressed as:

$$
\begin{aligned}
\mathrm{scal} &= \mathrm{tr}(\mathrm{Ric}) \\
&= \mathrm{Ric}^i_i \\
&= R^i_{iks} g^{sk} \\
&= \mathrm{Ric}_{ik} g^{ki} \\
&= R_{ijkl} g^{jk} g^{il}.
\end{aligned}
$$

Again, it is necessary to be careful to specify over which indices one contracts in order to get the right answer.

Note that the divergence of a $(1,k)$-tensor S is nothing but a contraction of the covariant derivative ∇S of the tensor. Here one contracts against the new variable introduced by the covariant differentiation.

6.3. Norms of Tensors. There are several conventions in Riemannian geometry for how one should measure the norm of a linear map. Essentially, there are two different norms in use, the *operator norm* and the *Euclidean norm*. The former is defined for a linear map $L : V \to W$ between inner product spaces as

$$|L| = \sup_{|v|=1} |Lv|$$

The Euclidean norm, in contrast, is given by

$$|L| = \sqrt{\operatorname{tr}(L^* \circ L)} = \sqrt{\operatorname{tr}(L \circ L^*)},$$

where $L^* : W \to V$ is the adjoint. Despite the fact that we use the same notation for these norms, they are almost never equal. If, for instance, $L : V \to V$ is self adjoint and $\lambda_1 \le \cdots \le \lambda_n$ the eigenvalues of L counted with multiplicities, then the operator norm is: $\max\{|\lambda_1|, |\lambda_n|\}$, while the Euclidean norm is $\sqrt{\lambda_1^2 + \cdots + \lambda_n^2}$. The Euclidean norm also has the advantage of actually coming from an inner product:

$$\langle L_1, L_2 \rangle = \operatorname{tr} L_1 \circ L_2^* = \operatorname{tr} L_2 \circ L_1^*.$$

As a general rule we shall always use the Euclidean norm.

It is worthwhile to see how the Euclidean norm of some simple tensors can be computed on a Riemannian manifold. Note that this computation uses type changes to compute adjoints and contractions to take traces.

Let us start with a $(1,1)$-tensor $T = T_j^i \cdot E_i \otimes \sigma^j$. We think of this as a linear map $TM \to TM$. Then the adjoint is first of all the dual map $T^* : T^*M \to T^*M$, which we then change to $T^* : TM \to TM$. This means that

$$T^* = T_i^j \cdot \sigma^i \otimes E_j,$$

which after type change becomes

$$T^* = T_l^k g^{lj} g_{ki} \cdot E_j \otimes \sigma^i.$$

Finally,

$$|T|^2 = T_j^i T_l^k g^{lj} g_{ki}.$$

If the frame is orthonormal, this takes the simple form of

$$|T|^2 = T_j^i T_i^j.$$

For a $(0,2)$-tensor $T = T_{ij} \cdot \sigma^i \otimes \sigma^j$ we first have to change type and then proceed as above. In the end one gets the nice formula

$$|T|^2 = T_{ij} T^{ij}.$$

6.4. Positional Notation. A final remark is in order. Many of the above notations could be streamlined even further so as to rid ourselves of some of the notational problems we have introduced by the way in which we write tensors in frames. Namely, tensors $TM \to TM$ (section of $TM \otimes T^*M$) and $T^*M \to T^*M$ (section of $T^*M \otimes TM$) seem to be written in the same way, and this causes some confusion when computing their Euclidean norms. That is, the only difference between the two objects $\sigma \otimes E$ and $E \otimes \sigma$ is in the ordering, not in what they actually do. We simply interpret the first as a map $TM \to TM$ and then the second as $T^*M \to T^*M$, but the roles could have been reversed, and both could be interpreted as maps $TM \to TM$. This can indeed cause great confusion.

One way to at least keep the ordering straight when writing tensors out in coordinates is to be even more careful with our indices and how they are written down. Thus, a tensor T that is a section of $T^*M \otimes TM \otimes T^*M$ should really be written as

$$T = T_i{}^j{}_k \cdot \sigma^i \otimes E_j \otimes \sigma^k.$$

Our standard $(1,1)$-tensor (section of $TM \otimes T^*M$) could therefore be written

$$T = T^i{}_j \cdot E_i \otimes \sigma^j,$$

while the adjoint (section of $T^*M \otimes TM$) before type change is

$$\begin{aligned} T^* &= T_k{}^l \cdot \sigma^k \otimes E_l \\ &= T^i{}_j g_{ki} g^{lj} \cdot \sigma^k \otimes E_l. \end{aligned}$$

Thus, we have the nice formula

$$|T|^2 = T^i{}_j T_i{}^j.$$

In the case of the curvature tensor one would normally write

$$R = R^l{}_{ijk} \cdot E_l \otimes \sigma^i \otimes \sigma^j \otimes \sigma^k,$$

and when changing to the $(2,2)$ version we have

$$\begin{aligned} R &= R^{kl}{}_{ij} \cdot E_k \otimes E_l \otimes \sigma^i \otimes \sigma^j \\ &= R^l{}_{ijs} g^{sk} \cdot E_k \otimes E_l \otimes \sigma^i \otimes \sigma^j. \end{aligned}$$

It is then clear how to keep track of the other $(2,2)$ versions by writing

$$R_i{}^{jk}{}_l = R_{ist}{}^u g^{js} g^{kt} g_{lu}.$$

Nice as this notation is, it is not used consistently in the literature, probably due to typesetting problems. It would be convenient to use it, but in most cases one can usually keep track of things anyway. Most of this notation can of course also be avoided by using invariant (coordinate-free) notation, but often it is necessary to do coordinate or frame computations both in abstract and concrete situations.

To this we can add yet another piece of notation that is often seen. Namely, if S is a $(1, k)$-tensor written in a frame as:

$$S = S^i_{j_1 \cdots j_k} \cdot E_i \otimes \sigma^{j_1} \otimes \cdots \otimes \sigma^{j_k},$$

Then the covariant derivative is a $(1, k+1)$-tensor that can be written as

$$\nabla S = S^i_{j_1 \cdots j_k, j_{k+1}} \cdot E_i \otimes \sigma^{j_1} \otimes \cdots \otimes \sigma^{j_k} \otimes \sigma^{j_{k+1}}.$$

The coefficient $S^i_{j_1 \cdots j_k, j_{k+1}}$ can be computed via the formula

$$\begin{aligned} \nabla_{E_{j_{k+1}}} S &= D_{E_{j_{k+1}}} \left(S^i_{j_1 \cdots j_k} \right) \cdot E_i \otimes \sigma^{j_1} \otimes \cdots \otimes \sigma^{j_k} \\ &+ S^i_{j_1 \cdots j_k} \cdot \nabla_{E_{j_{k+1}}} \left(E_i \otimes \sigma^{j_1} \otimes \cdots \otimes \sigma^{j_k} \right), \end{aligned}$$

where one must find the expression for

$$\begin{aligned} \nabla_{E_{j_{k+1}}} \left(E_i \otimes \sigma^{j_1} \otimes \cdots \otimes \sigma^{j_k} \right) &= \left(\nabla_{E_{j_{k+1}}} E_i \right) \otimes \sigma^{j_1} \otimes \cdots \otimes \sigma^{j_k} \\ &+ E_i \otimes \left(\nabla_{E_{j_{k+1}}} \sigma^{j_1} \right) \otimes \cdots \otimes \sigma^{j_k} \\ &\cdots \\ &+ E_i \otimes \sigma^{j_1} \otimes \cdots \otimes \left(\nabla_{E_{j_{k+1}}} \sigma^{j_k} \right) \end{aligned}$$

by writing each of the terms $\left(\nabla_{E_{j_{k+1}}} E_i \right), \left(\nabla_{E_{j_{k+1}}} \sigma^{j_1} \right), \ldots, \left(\nabla_{E_{j_{k+1}}} \sigma^{j_k} \right)$ in terms of the frame and coframe and substitute back into the formula.

7. Further Study

It is still too early to give useful references. In the upcoming chapters we shall mention several other books on geometry that the reader might wish to consult. At this stage we shall only list the authoritative guide [**60**]. Every differential geometer must have a copy of these tomes, but their effective usefulness has probably passed away. In a way, it is the Bourbaki of differential geometry and should be treated as such.

8. Exercises

(1) Show that the connection on Euclidean space is the only affine connection such that $\nabla X = 0$ for all constant vector fields X.

(2) If $F : M \to M$ is a diffeomorphism, then the push-forward of a vector field is defined as

$$(F_* X)\,|_p = DF\left(X|_{F^{-1}(p)}\right).$$

Let F be an isometry on (M, g).
   (a) Show that $F_* (\nabla_X Y) = \nabla_{F_* X} F_* Y$ for all vector fields.
   (b) If $(M, g) = (\mathbb{R}^n, \mathrm{can})$, then isometries are of the form $F(x) = Ox + b$, where $O \in O(n)$ and $b \in \mathbb{R}^n$. Hint: Show that F maps constant vector fields to constant vector fields.

(3) Let G be a Lie group. Show that there is a unique affine connection such that $\nabla X = 0$ for all left invariant vector fields. Show that this connection is torsion free iff the Lie algebra is Abelian.

(4) Show that if X is a vector field of constant length on a Riemannian manifold, then $\nabla_v X$ is always perpendicular to X.

(5) For any $p \in (M, g)$ and orthonormal basis $e_1, \ldots, e_n$ for $T_p M$, show that there is an orthonormal frame $E_1, \ldots, E_n$ in a neighborhood of p such that $E_i = e_i$ and $(\nabla E_i)\,|_p = 0$. Hint: Fix an orthonormal frame $\bar{E}_i$ near $p \in M$ with $\bar{E}_i(p) = e_i$. If we define $E_i = \alpha_i^j \bar{E}_j$, where $\left[\alpha_i^j(x)\right] \in SO(n)$ and $\alpha_i^j(p) = \delta_i^j$, then this will yield the desired frame provided that the $D_{e_k} \alpha_i^j$ are appropriately prescribed.

(6) (Riemann) As in the previous problem, but now show that there are coordinates $x^1, \ldots, x^n$ such that $\partial_i = e_i$ and $\nabla \partial_i = 0$ at p. These conditions imply that the metric coefficients satisfy $g_{ij} = \delta_{ij}$ and $\partial_k g_{ij} = 0$ at p. Such coordinates are called normal coordinates at p. Show that in normal coordinates g viewed as a matrix function of x has the expansion

$$
\begin{aligned}
g &= \sum_{i,j=1}^n g_{ij}\, dx^i dx^j \\
&= \sum_{i=1}^n dx^i dx^i \\
&\quad + \sum_{i<j,k<l} R_{ijkl}\left(x^i dx^j - x^j dx^i\right)\left(x^k dx^l - x^l dx^k\right) + o\left(|x|^2\right),
\end{aligned}
$$

where $R_{ijkl} = g\left(R\left(\partial_i, \partial_j\right)\partial_k, \partial_l\right)(p)$. In dimension 2 this formula reduces to

$$
\begin{aligned}
g &= dx^2 + dy^2 + R_{1212}\left(xdy - ydx\right)^2 + o\left(x^2 + y^2\right) \\
&= dx^2 + dy^2 - \sec\left(p\right)\left(xdy - ydx\right)^2 + o\left(x^2 + y^2\right).
\end{aligned}
$$

(7) Let M be an n-dimensional submanifold of $\mathbb{R}^{n+m}$ with the induced metric and assume that we have a local coordinate system given by a parametrization $x^s\left(u^1, ..., u^n\right)$, $s = 1, ..., n + m$. Show that in these coordinates we have:

(a)
$$
g_{ij} = \sum_{s=1}^{n+m} \frac{\partial x^s}{\partial u^i} \frac{\partial x^s}{\partial u^j}.
$$

(b)
$$
\Gamma_{ij,k} = \sum_{s=1}^{n+m} \frac{\partial x^s}{\partial u^k} \frac{\partial^2 x^s}{\partial u^i \partial u^j}.
$$

(c) R_{ijkl} depends only on the first and second partials of x^s.

(8) Show that $\mathrm{Hess}f = \nabla df$.

(9) Let r be a distance function and $S\left(X\right) = \nabla_X \partial_r$ the $(1,1)$ version of the Hessian. Show that

$$
\begin{aligned}
L_{\partial_r} S &= \nabla_{\partial_r} S, \\
L_{\partial_r} S + S^2 &= -R_{\partial_r}.
\end{aligned}
$$

How do you reconcile this with what happens for the fundamental equations for the $(0,2)$-version of the Hessian?

(10) Let (M, g) be oriented and define the Riemannian volume form $d\mathrm{vol}$ as follows:
$$
d\mathrm{vol}\left(v_1, \ldots, v_n\right) = \det\left(g\left(v_i, e_j\right)\right),
$$
where $e_1, \ldots, e_n$ is a positively oriented orthonormal basis for T_pM.

(a) Show that if $v_1, \ldots, v_n$ is positively oriented, then
$$
d\mathrm{vol}\left(v_1, \ldots, v_n\right) = \sqrt{\det\left(g\left(v_i, v_j\right)\right)}.
$$

(b) Show that the volume form is parallel.

(c) Show that in positively oriented coordinates,
$$
d\mathrm{vol} = \sqrt{\det\left(g_{ij}\right)}dx^1 \wedge \cdots \wedge dx^n.
$$

(d) If X is a vector field, show that
$$
L_X d\mathrm{vol} = \mathrm{div}\left(X\right) d\mathrm{vol}.
$$

(e) Conclude that the Laplacian has the formula
$$
\Delta u = \frac{1}{\sqrt{\det\left(g_{ij}\right)}} \partial_k \left(\sqrt{\det\left(g_{ij}\right)} g^{kl} \partial_l u\right).
$$
Given that the coordinates are normal at p we get as in Euclidean space that
$$
\Delta f\left(p\right) = \sum_{i=1}^{n} \partial_i \partial_i f.
$$

(11) Let (M, g) be a oriented Riemannian manifold with volume form $d\mathrm{vol}$ as above.

(a) If f has compact support, then

$$\int_M \Delta f \cdot d\mathrm{vol} = 0.$$

(b) Show that

$$\mathrm{div}\,(f \cdot X) = g\,(\nabla f, X) + f \cdot \mathrm{div} X.$$

(c) Show that

$$\Delta\,(f_1 \cdot f_2) = (\Delta f_1) \cdot f_2 + 2g\,(\nabla f_1, \nabla f_2) + f_1 \cdot (\Delta f_2)\,.$$

(d) Establish the integration by parts formula for functions with compact support:

$$\int_M f_1 \cdot \Delta f_2 \cdot d\mathrm{vol} = -\int_M g\,(\nabla f_1, \nabla f_2) \cdot d\mathrm{vol}.$$

(e) Conclude that if f is sub- or superharmonic (i.e., $\Delta f \geq 0$ or $\Delta f \leq 0$) then f is constant. (Hint: first show $\Delta f = 0$; then use integration by parts on $f \cdot \Delta f$.) This result is known as the *weak maximum principle*. More generally, one can show that any subharmonic (respectively superharmonic) function that has a global maximum (respectively minimum) must be constant. For this one does not need f to have compact support. This result is usually referred to as the *strong maximum principle*.

(12) A vector field and its corresponding flow is said to be *incompressible* if $\mathrm{div} X = 0$.

(a) Show that X is incompressible iff the local flows it generates are volume preserving (i.e., leave the Riemannian volume form invariant).

(b) Let X be a unit vector field X on $\mathbb{R}^2$. Show that $\nabla X = 0$ if X is incompressible.

(c) Find a unit vector field X on $\mathbb{R}^3$ that is incompressible but where $\nabla X \neq 0$.

(13) Let X be a unit vector field on (M, g) such that $\nabla_X X = 0$.

(a) Show that X is locally the gradient of a distance function iff the orthogonal distribution is integrable.

(b) Show that X is the gradient of a distance function in a neighborhood of $p \in M$ iff the orthogonal distribution has an integral submanifold through p. Hint: It might help to show that $L_X \theta_X = 0$.

(c) Find X with the given conditions so that it is not a gradient field. Hint: Consider S^3.

(14) Given an orthonormal frame $E_1, \ldots, E_n$ on (M, g), define the *structure constants* c_{ij}^k by $[E_i, E_j] = c_{ij}^k E_k$. Then define the Γs and Rs by

$$\begin{aligned}
\nabla_{E_i} E_j &= \Gamma_{ij}^k E_k, \\
R\,(E_i, E_j)\,E_k &= R_{ijk}^l E_l
\end{aligned}$$

and compute them in terms of the cs. Notice that on Lie groups with left-invariant metrics the structure constants can be assumed to be constant. In this case, computations simplify considerably.

(15) There is yet another effective method for computing the connection and curvatures, namely, the *Cartan formalism*. Let (M, g) be a Riemannian manifold. Given a frame $E_1, \ldots, E_n$, the connection can be written

$$\nabla E_i = \omega_i^j E_j,$$

where ω_i^j are 1-forms. Thus,

$$\nabla_v E_i = \omega_i^j(v) E_j.$$

Suppose now that the frame is orthonormal and let ω^i be the dual coframe, i.e., $\omega^i(E_j) = \delta_j^i$. Show that the *connection forms* satisfy

$$\begin{aligned} \omega_i^j &= -\omega_j^i, \\ d\omega^i &= \omega^j \wedge \omega_j^i. \end{aligned}$$

These two equations can, conversely, be used to compute the connection forms given the orthonormal frame. Therefore, if the metric is given by declaring a certain frame to be orthonormal, then this method can be very effective in computing the connection.

If we think of $\left[\omega_i^j \right]$ as a matrix, then it represents a 1-form with values in the skew-symmetric $n \times n$ matrices, or in other words, with values in the Lie algebra $\mathfrak{so}(n)$ for $O(n)$.

The *curvature forms* Ω_i^j are 2-forms with values in $\mathfrak{so}(n)$. They are defined as

$$R(\cdot, \cdot) E_i = \Omega_i^j E_j.$$

Show that they satisfy

$$d\omega_i^j = \omega_i^k \wedge \omega_k^j + \Omega_i^j.$$

When reducing to Riemannian metrics on surfaces we obtain for an orthonormal frame E_1, E_2 with coframe ω^1, ω^2

$$\begin{aligned} d\omega^1 &= \omega^2 \wedge \omega_2^1, \\ d\omega^2 &= -\omega^1 \wedge \omega_2^1, \\ d\omega_2^1 &= \Omega_2^1, \\ \Omega_2^1 &= \sec \cdot dvol. \end{aligned}$$

(16) Show that a Riemannian manifold with parallel Ricci tensor has constant scalar curvature. In chapter 3 it will be shown that the converse is not true, and also that a metric with parallel curvature tensor doesn't have to be Einstein.

(17) Show that if R is the $(1, 3)$-curvature tensor and Ric the $(0, 2)$-Ricci tensor, then

$$(\operatorname{div} R)(X, Y, Z) = (\nabla_X \operatorname{Ric})(Y, Z) - (\nabla_Y \operatorname{Ric})(X, Z).$$

Conclude that $\operatorname{div} R = 0$ if $\nabla \operatorname{Ric} = 0$. Then show that $\operatorname{div} R = 0$ iff the $(1, 1)$ Ricci tensor satisfies:

$$(\nabla_X \operatorname{Ric})(Y) = (\nabla_Y \operatorname{Ric})(X) \quad \text{for all } X, Y.$$

(18) Let G be a Lie group with a bi-invariant metric. Using left-invariant fields establish the following formulas. Hint: First go back to the exercises to chapter 1 and take a peek at chapter 3 where some of these things are proved.

(a) $\nabla_X Y = \frac{1}{2}[X, Y]$.

(b) $R(X, Y)Z = \frac{1}{4}[Z, [X, Y]]$.

(c) $g(R(X, Y)Z, W) = -\frac{1}{4}(g([X, Y], [Z, W]))$. Conclude that the sectional curvatures are nonnegative.

(d) Show that the curvature operator is also nonnegative by showing that:

$$g\left(\mathfrak{R}\left(\sum_{i=1}^{k} X_i \wedge Y_i\right), \left(\sum_{i=1}^{k} X_i \wedge Y_i\right)\right) = \frac{1}{4}\left|\sum_{i=1}^{k}[X_i, Y_i]\right|^2.$$

(e) Show that $\mathrm{Ric}(X, X) = 0$ iff X commutes with all other left-invariant vector fields. Thus G has positive Ricci curvature if the center of G is discrete.

(f) Consider the linear map $\Lambda^2 \mathfrak{g} \to [\mathfrak{g}, \mathfrak{g}]$ that sends $X \wedge Y$ to $[X, Y]$. Show that the sectional curvature is positive iff this map is an isomorphism. Conclude that this can only happen if $n = 3$ and $\mathfrak{g} = \mathfrak{su}(2)$.

(19) It is illustrative to use the Cartan formalism in the above problem and compute all quantities in terms of the structure constants for the Lie algebra. Given that the metric is bi-invariant, it follows that with respect to an orthonormal basis they satisfy

$$c_{ij}^k = -c_{ji}^k = c_{jk}^i.$$

The first equality is skew-symmetry of the Lie bracket, and the second is bi-invariance of the metric.

(20) Suppose we have two Riemannian manifolds (M, g_M) and (N, g_N). Then the product has a natural product metric $(M \times N, g_M + g_N)$. Let X be a vector field on M and Y one on N, show that if we regard these as vector fields on $M \times N$, then $\nabla_X Y = 0$. Conclude that $\sec(X, Y) = 0$. This means that product metrics always have many curvatures that are zero.

(21) Suppose we have two distributions E and F on (M, g), that are orthogonal complements of each other in TM. In addition, assume that the distributions are parallel i.e., if two vector fields X and Y are tangent to, say, E, then $\nabla_X Y$ is also tangent to E.

(a) Show that the distributions are integrable.

(b) Show that around any point in M there is a product neighborhood $U = V_E \times V_F$ such that $(U, g) = (V_E \times V_F, g|_E + g|_F)$, where $g|_E$ and $g|_F$ are the restrictions of g to the two distributions. In other words, M is locally a product metric.

(22) Let X be a parallel vector field on (M, g). Show that X has constant length. Show that X generates parallel distributions, one that contains X and the other that is the orthogonal complement to X. Conclude that locally the metric is a product with an interval $(U, g) = (V \times I, g|_{TV} + dt^2)$.

(23) For 3-dimensional manifolds, show that if the curvature operator in diagonal form looks like

$$\begin{pmatrix} \alpha & 0 & 0 \\ 0 & \beta & 0 \\ 0 & 0 & \gamma \end{pmatrix},$$

then the Ricci curvature has a diagonal form like

$$\begin{pmatrix} \alpha + \beta & 0 & 0 \\ 0 & \beta + \gamma & 0 \\ 0 & 0 & \alpha + \gamma \end{pmatrix}.$$

Moreover, the numbers α, β, γ must be sectional curvatures.

(24) The *Einstein tensor* on a Riemannian manifold is defined as

$$G = \text{Ric} - \frac{\text{scal}}{2} \cdot I.$$

Show that $G = 0$ in dimension 2 and that $\text{div} G = 0$ in higher dimensions. This tensor is supposed to measure the mass/energy distribution. The fact that it is divergence free tells us that energy and momentum are conserved. In a vacuum, one therefore imagines that $G = 0$. Show that this happens in dimensions > 2 iff the metric is Ricci flat.

(25) This exercise will give you a way of finding the curvature tensor from the sectional curvatures. Using the Bianchi identity show that

$$-6R(X,Y,Z,W) = \frac{\partial^2}{\partial s \partial t}\Big|_{s=t=0} \{R(X + sZ, Y + tW, Y + tW, X + sZ)$$
$$- R(X + sW, Y + tZ, Y + tZ, X + sW)\}.$$

(26) Using polarization show that the norm of the curvature operator on $\Lambda^2 T_p M$ is bounded by

$$|\mathfrak{R}|_p \le c(n) |\sec|_p$$

for some constant $c(n)$ depending on dimension, and where $|\sec|_p$ denotes the largest absolute value for any sectional curvature of a plane in $T_p M$.

(27) We can artificially complexify the tangent bundle to a manifold: $T_{\mathbb{C}} M = TM \otimes \mathbb{C}$. If we have a Riemannian structure, we can extend all the accompanying tensors to this realm. The metric tensor, in particular, gets extended as follows:

$$g_{\mathbb{C}}(v_1 + iv_2, w_1 + iw_2) = g(v_1, w_1) - g(v_2, w_2) + i(g(v_1, w_2) + g(v_2, w_1)).$$

This means that a vector can have complex length zero without being trivial. Such vectors are called *isotropic*. Clearly, they must have the form $v_1 + iv_2$, where $|v_1| = |v_2|$ and $g(v_1, v_2) = 0$. More generally, we can have isotropic subspaces, i.e., those subspace on which $g_{\mathbb{C}}$ vanishes. If, for instance, a plane is generated by two isotropic vectors $v_1 + iv_2$ and $w_1 + iw_2$, where v_1, v_2, w_1, w_2 are orthogonal, then the plane is isotropic. Note that one must be in dimension ≥ 4 to have isotropic planes. We now say that the isotropic curvatures are positive, if "sectional" curvatures on isotropic planes are positive. This means that if $v_1 + iv_2$ and $w_1 + iw_2$ span the plane and v_1, v_2, w_1, w_2 are orthogonal, then

$$0 < R(v_1 + iv_2, w_1 + iw_2, w_1 - iw_2, v_1 - iv_2).$$

(a) Show that the expression $R\left(v_1 + iv_2, w_1 + iw_2, w_1 - iw_2, v_1 - iv_2\right)$ is always a real number.

(b) Show that if the original metric is strictly quarter pinched, i.e., all sectional curvatures lie in an open interval of the form $\left(\frac{1}{4}k, k\right)$, then the isotropic curvatures are positive.

(c) Show that if the sum of the two smallest eigenvalues of the original curvature operator is positive, then the isotropic curvatures are positive.

(28) Consider a Riemannian metric (M, g). Now *scale* the metric by multiplying it by a number λ^2. Then we get a new Riemannian manifold $\left(M, \lambda^2 g\right)$. Show that the new connection and $(1,3)$-curvature tensor remain the same, but that sec, scal, and $\mathfrak{R}$ all get multiplied by λ^{-2}.

(29) For a $(1,1)$-tensor T on a Riemannian manifold, show that if E_i is an orthonormal basis, then

$$|T|^2 = \sum |T\left(E_i\right)|^2.$$

(30) If we have two tensors S, T of the same type (r, s), $r = 0, 1$, define the inner product

$$g\left(S, T\right)$$

and show that

$$D_X g\left(S, T\right) = g\left(\nabla_X S, T\right) + g\left(S, \nabla_X T\right).$$

If S is symmetric and T skew-symmetric show that $g\left(S, T\right) = 0$.

(31) Recall that complex manifolds have complex tangent spaces. Thus we can multiply vectors by $\sqrt{-1}$. As a generalization of this we can define an *almost complex* structure. This is a $(1,1)$-tensor J such that $J^2 = -I$. Show that the *Nijenhuis tensor:*

$$N\left(X, Y\right) = [J\left(X\right), J\left(Y\right)] - J\left([J\left(X\right), Y]\right) - J\left([X, J\left(Y\right)]\right) - [X, Y]$$

is indeed a tensor. If J comes from a complex structure then $N = 0$, conversely Newlander&Nirenberg have shown that J comes from a complex structure if $N = 0$.

A *Hermitian structure* on a Riemannian manifold (M, g) is an almost complex structure J such that

$$g\left(J\left(X\right), J\left(Y\right)\right) = g\left(X, Y\right).$$

The *Kähler form* of a Hermitian structure is

$$\omega\left(X, Y\right) = g\left(J\left(X\right), Y\right).$$

Show that ω is a 2-form. Show that $d\omega = 0$ iff $\nabla J = 0$. If the Kähler form is closed, then we call the metric a Kähler metric.

CHAPTER 3

Examples

We are now ready to compute the curvature tensors on the examples we constructed earlier. After computing these quantities in general, we will try to find examples of manifolds with constant sectional, Ricci, and scalar curvature. In particular, we shall look at the standard product metrics on spheres and also construct the Riemannian version of the Schwarzschild metric.

The examples we present here include a selection of important techniques such as: Conformal change, left-invariant metrics, Riemannian submersion constructions etc. We shall not always develop the techniques in full detail. Rather we shall show how they work in some simple, but important, examples.

1. Computational Simplifications

Before we present the examples it will be useful to have some general results that deal with how one finds the range of the various curvatures.

PROPOSITION 8. *Let e_i be an orthonormal basis for T_pM. If $e_i \wedge e_j$ diagonalize the curvature operator*

$$\mathfrak{R}(e_i \wedge e_j) = \lambda_{ij} e_i \wedge e_j,$$

then for any plane π in T_pM we have $\sec(\pi) \in [\min \lambda_{ij}, \max \lambda_{ij}]$.

PROOF. If v, w form an orthonormal basis for π, then we have $\sec(\pi) = g(\mathfrak{R}(v \wedge w), (v \wedge w))$, so the result is immediate. $\quad\square$

PROPOSITION 9. *Let e_i be an orthonormal basis for T_pM and suppose that $R(e_i, e_j) e_k = 0$ if the indices are mutually distinct; then $e_i \wedge e_j$ diagonalize the curvature operator.*

PROOF. If we use

$$
\begin{aligned}
g(\mathfrak{R}(e_i \wedge e_j), (e_k \wedge e_l)) &= -g(R(e_i, e_j) e_k, e_l) \\
&= g(R(e_i, e_j) e_l, e_k),
\end{aligned}
$$

then we see that this expression is 0 when i, j, k are mutually distinct or if i, j, l are mutually distinct. Thus, the expression can only be nonzero when $\{k, l\} = \{i, j\}$. This gives the result. $\quad\square$

We shall see that in all rotationally symmetric and doubly warped products we can find e_i such that $R(e_i, e_j) e_k = 0$. In this case, the curvature operator can then be computed by finding the expressions $R(e_i, e_j, e_j, e_i)$. In general, however, this will not happen.

There is also a more general situation where we can find the range of the Ricci curvatures:

PROPOSITION 10. *Let e_i be an orthonormal basis for T_pM and suppose that*

$$g\left(R\left(e_i, e_j\right)e_k, e_l\right) = 0$$

if three of the indices are mutually distinct, then e_i diagonalize Ric.

PROOF. Recall that

$$g\left(\text{Ric}\left(e_i\right), e_j\right) = \sum_{k=1}^{n} g\left(R\left(e_i, e_k\right)e_k, e_j\right),$$

so if we assume that $i \neq j$, then $g\left(R\left(e_i, e_k\right)e_k, e_j\right) = 0$ unless k is either i or j. However, if $k = i, j$, then the expression is zero from the symmetry properties. Thus, e_i must diagonalize Ric. $\square$

2. Warped Products

So far, all we know about curvature is that $(\mathbb{R}^n, \text{can})$ has curvature that vanishes. Using this, let us figure out what the curvature tensor is on $(S^{n-1}(r), \text{can})$.

2.1. Spheres. On $\mathbb{R}^n$ we have the distance function $r(x) = |x|$ and the polar coordinate representation:

$$g = dr^2 + g_r = dr^2 + r^2 ds_{n-1}^2,$$

where ds_{n-1}^2 is the canonical metric on $S^{n-1}(1)$. The level sets are $U_r = S^{n-1}(r)$ with the usual induced metric $g_r = r^2 ds_{n-1}^2$. The differential is given by

$$dr = \sum \frac{x^i}{r} dx^i,$$

while the gradient is

$$\partial_r = \frac{1}{r} x^i \partial_i.$$

Since ds_{n-1}^2 is independent of r we can compute the Hessian as follows:

$$
\begin{aligned}
2\text{Hess}r &= L_{\partial_r} g \\
&= L_{\partial_r}\left(dr^2\right) + L_{\partial_r}\left(r^2 ds_{n-1}^2\right) \\
&= L_{\partial_r}\left(dr\right)dr + dr L_{\partial_r}\left(dr\right) + \partial_r\left(r^2\right) ds_{n-1}^2 + r^2 L_{\partial_r}\left(ds_{n-1}^2\right) \\
&= \partial_r\left(r^2\right) ds_{n-1}^2 \\
&= 2r ds_{n-1}^2 \\
&= 2\frac{1}{r} g_r.
\end{aligned}
$$

The tangential curvature equation then tells us that

$$R^r(X, Y)Z = r^{-2}(g_r(Y, Z)X - g_r(X, Z)Y),$$

since the curvature on $\mathbb{R}^n$ is zero. In particular, if e_i is any orthonormal basis, we see that $R^r\left(e_i, e_j\right)e_k = 0$ when the indices are mutually distinct. Therefore, $(S^{n-1}(r), \text{can})$ has constant curvature r^{-2}, provided that $n \geq 3$. This justifies our notation that S_k^n is the rotational symmetric metric $dr^2 + \text{sn}_k^2(r)ds_{n-1}^2$ when $k \geq 0$, as these metrics have curvature k in this case. Below we shall see that this is also true when $k < 0$.

2.2. Product Spheres. Let us next compute the curvatures on the product spheres

$$S_a^n \times S_b^m = S^n \left(\frac{1}{\sqrt{a}} \right) \times S^m \left(\frac{1}{\sqrt{b}} \right).$$

We saw that the metric g_r on $S^n(r)$ is $g_r = r^2 ds_n^2$, so we can write

$$S_a^n \times S_b^m = \left(S^n \times S^m, \frac{1}{a} ds_n^2 + \frac{1}{b} ds_m^2 \right).$$

Let Y be a unit vector field on S^n, V a unit vector field on on S^m, and X a unit vector field on either S^n or S^m that is perpendicular to both Y and V. The Koszul formula then shows

$$
\begin{aligned}
2g\left(\nabla_Y X, V\right) &= g\left([Y,X],V\right) + g\left([V,Y],X\right) - g\left([X,V],Y\right) \\
&= g\left([Y,X],V\right) - g\left([X,V],Y\right) \\
&= 0,
\end{aligned}
$$

as $[Y,X]$ is either zero or tangent to S^n and likewise with $[X,V]$. Thus $\nabla_Y X = 0$ if X is tangent to S^m. And $\nabla_Y X$ is tangent to S^n if X is tangent to S^n, showing that $\nabla_Y X$ can be computed on S_a^n. This shows that if X,Y are tangent to S^n and U,V tangent to S^m, then

$$
\begin{aligned}
\Re(X \wedge V) &= 0, \\
\Re(X \wedge Y) &= aX \wedge Y, \\
\Re(U \wedge V) &= bU \wedge V.
\end{aligned}
$$

In particular, all sectional curvatures lie in the interval $[0, \max\{a,b\}]$. From this we see

$$
\begin{aligned}
\mathrm{Ric}(X) &= (n-1)aX, \\
\mathrm{Ric}\,(V) &= (m-1)\,bV, \\
\mathrm{scal} &= n(n-1)a + m(m-1)b.
\end{aligned}
$$

Therefore, we can conclude that $S_a^n \times S_b^m$ always has constant scalar curvature, is an Einstein manifold exactly when $(n-1)a = (m-1)b$ (which requires $n, m \geq 2$ or $n = m = 1$), and has constant sectional curvature only when $n = m = 1$. Note also that the curvature tensor on $S_a^n \times S_b^m$ is always parallel.

2.3. Rotationally Symmetric Metrics. Let us look at what happens for a general rotationally symmetric metric

$$dr^2 + \varphi^2 ds_{n-1}^2.$$

We shall compute all of the relevant terms below and also check that the fundamental equations hold. The metric is of the form $g = dr^2 + g_r$ on $(a,b) \times S^{n-1}$, with $g_r = \varphi^2 ds_{n-1}^2$. As ds_{n-1}^2 does not depend on r we have that

$$
\begin{aligned}
2\mathrm{Hess}r &= L_{\partial_r} g_r \\
&= L_{\partial_r} \left(\varphi^2 ds_{n-1}^2 \right) \\
&= \partial_r \left(\varphi^2 \right) ds_{n-1}^2 + \varphi^2 L_{\partial_r} \left(ds_{n-1}^2 \right) \\
&= 2\varphi \left(\partial_r \varphi \right) ds_{n-1}^2 \\
&= 2 \frac{\partial_r \varphi}{\varphi} g_r.
\end{aligned}
$$

The Lie and covariant derivatives of the Hessian are now computed as follows:

$$
\begin{aligned}
L_{\partial_r} \mathrm{Hess}r &= L_{\partial_r}\left(\frac{\partial_r \varphi}{\varphi} g_r\right) \\
&= \partial_r\left(\frac{\partial_r \varphi}{\varphi}\right) g_r + \frac{\partial_r \varphi}{\varphi} L_{\partial_r}(g_r) \\
&= \frac{(\partial_r^2 \varphi)\,\varphi - (\partial_r \varphi)^2}{\varphi^2} g_r + 2\left(\frac{\partial_r \varphi}{\varphi}\right)^2 g_r \\
&= \frac{\partial_r^2 \varphi}{\varphi} g_r + \left(\frac{\partial_r \varphi}{\varphi}\right)^2 g_r \\
&= \frac{\partial_r^2 \varphi}{\varphi} g_r + \mathrm{Hess}^2 r
\end{aligned}
$$

$$
\begin{aligned}
\nabla_{\partial_r} \mathrm{Hess}r &= \nabla_{\partial_r}\left(\frac{\partial_r \varphi}{\varphi} g_r\right) \\
&= \partial_r\left(\frac{\partial_r \varphi}{\varphi}\right) g_r + \frac{\partial_r \varphi}{\varphi} \nabla_{\partial_r}(g_r) \\
&= \frac{(\partial_r^2 \varphi)\,\varphi - (\partial_r \varphi)^2}{\varphi^2} g_r \\
&= \frac{\partial_r^2 \varphi}{\varphi} g_r - \left(\frac{\partial_r \varphi}{\varphi}\right)^2 g_r \\
&= \frac{\partial_r^2 \varphi}{\varphi} g_r - \mathrm{Hess}^2 r
\end{aligned}
$$

The fundamental equations then show that when restricted to S^{n-1} we have

$$
\mathrm{Hess}r = \frac{\partial_r \varphi}{\varphi} g_r,
$$

$$
R(\cdot, \partial_r, \partial_r, \cdot) = -\frac{\partial_r^2 \varphi}{\varphi} g_r.
$$

This shows that

$$
\nabla_X \partial_r = \begin{cases} \frac{\partial_r \varphi}{\varphi} X & \text{if } X \text{ is tangent to } S^{n-1}, \\ 0 & \text{if } X = \partial_r, \end{cases}
$$

$$
R(X, \partial_r) \partial_r = \begin{cases} -\frac{\partial_r^2 \varphi}{\varphi} X & \text{if } X \text{ is tangent to } S^{n-1}, \\ 0 & \text{if } X = \partial_r. \end{cases}
$$

If we restrict a general vector field X on $(a, b) \times S^{n-1}$ to S^{n-1}, then it has the form $X = f \partial_r + Z$, where $f : S^{n-1} \to \mathbb{R}$ and Z is a smooth vector field on S^{n-1}. Therefore, all Jacobi fields for r are of the form $J = f \partial_r + Z$, where $f : S^{n-1} \to \mathbb{R}$ and Z is a smooth vector field on S^{n-1}.

The parallel fields can be found by using the same initial conditions on S^{n-1}. However, we have to adjust the field as we move away from r_0 in order to keep the field parallel. Thus we consider

$$
X = \lambda(r) f \partial_r + \mu(r) Z
$$

and calculate

$$
\begin{aligned}
\nabla_{\partial_r} X &= (\partial_r \lambda) f \partial_r + (\partial_r \mu) Z + \mu \nabla_{\partial_r} Z \\
&= (\partial_r \lambda) f \partial_r + (\partial_r \mu) Z + \mu \nabla_Z \partial_r \\
&= (\partial_r \lambda) f \partial_r + (\partial_r \mu) Z + \mu \frac{\partial_r \varphi}{\varphi} Z \\
&= (\partial_r \lambda) f \partial_r + \left(\partial_r \mu + \frac{\partial_r \varphi}{\varphi} \mu \right) Z.
\end{aligned}
$$

This shows that

$$
\begin{aligned}
\lambda &\equiv \alpha \in \mathbb{R}, \\
\mu &= \beta \frac{1}{\varphi}, \beta \in \mathbb{R}.
\end{aligned}
$$

After adjusting f and Z any parallel field can then be written as: $X = f \partial_r + \frac{1}{\varphi} Z$. Thus Jacobi fields are only parallel when φ is constant. We can now also solve the more general second order Jacobi equation. We assume again that

$$
J = \lambda(r) f \partial_r + \mu(r) Z
$$

and get

$$
\begin{aligned}
\frac{\partial_r^2 \varphi}{\varphi} \mu Z &= -R(J, \partial_r) \partial_r \\
&= \nabla_{\partial_r} \nabla_{\partial_r} J \\
&= \nabla_{\partial_r} \left((\partial_r \lambda) f \partial_r + \left(\partial_r \mu + \frac{\partial_r \varphi}{\varphi} \mu \right) Z \right) \\
&= (\partial_r^2 \lambda) f \partial_r + \left(\partial_r^2 \mu + \partial_r \left(\frac{\partial_r \varphi}{\varphi} \mu \right) \right) Z + \left(\partial_r \mu + \frac{\partial_r \varphi}{\varphi} \mu \right) \frac{\partial_r \varphi}{\varphi} Z \\
&= (\partial_r^2 \lambda) f \partial_r + \left(\partial_r^2 \mu + 2 \frac{\partial_r \varphi}{\varphi} \partial_r \mu + \frac{\partial_r^2 \varphi}{\varphi} \mu \right) Z
\end{aligned}
$$

Thus

$$
\begin{aligned}
\lambda &= \alpha + \beta r, \alpha, \beta \in \mathbb{R}, \\
\mu &= \gamma \int \frac{1}{\varphi^2} dr + \delta, \gamma, \delta \in \mathbb{R}.
\end{aligned}
$$

After adjusting f and Z we see that all such fields must be of the form:

$$
J = (f_0 + r f_1) \partial_r + Z_0 + \left(\int \frac{1}{\varphi^2} dr \right) Z_1,
$$

$$
\begin{aligned}
f_0, f_1 &: \quad S^{n-1} \to \mathbb{R}, \\
Z_0, Z_1 &: \quad S^{n-1} \to TS^{n-1}.
\end{aligned}
$$

To be even more specific we can let $n = 2$ and ∂_θ be the angular vector field on S^1:

- If $\varphi(r) = r$, then the metric is just the Euclidean metric on $\mathbb{R}^2$. The generalized Jacobi fields look like

$$
(f_0 + r f_1) \partial_r + \left(h_0 + \frac{1}{r} h_1 \right) \partial_\theta,
$$

$$
f_0, f_1, h_0, h_1 : S^1 \to \mathbb{R}.
$$

- If $\varphi(r) = 1$, then the metric is a flat cylinder and the generalized Jacobi fields look like

$$(f_0 + rf_1)\,\partial_r + (h_0 + rh_1)\,\partial_\theta,$$
$$f_0, f_1, h_0, h_1 : S^1 \to \mathbb{R}.$$

We shall now compute the other curvatures on

$$\left(I \times S^{n-1}, dr^2 + \varphi^2(r)ds_{n-1}^2\right)$$

coming from the tangential and mixed curvature equations

$$g\left(R(X,Y)V,W\right) = g_r\left(R^r(X,Y)V,W\right) - \mathrm{II}(Y,V)\mathrm{II}(X,W) + \mathrm{II}(X,V)\mathrm{II}(Y,W),$$
$$g\left(R(X,Y)Z,\partial_r\right) = -\left(\nabla_X\mathrm{II}\right)(Y,Z) + \left(\nabla_Y\mathrm{II}\right)(X,Z).$$

Using that g_r is the metric of curvature $\frac{1}{\varphi^2}$ on the sphere, we get

$$g_r\left(R^r(X,Y)V,W\right) = \frac{1}{\varphi^2}g_r(X \wedge Y, W \wedge V).$$

Combining this with $\mathrm{II} = \mathrm{Hess}r$ we obtain

$$g\left(R(X,Y)V,W\right) = \frac{1 - (\partial_r\varphi)^2}{\varphi^2}g_r(X \wedge Y, W \wedge V).$$

Finally we note that the mixed curvature vanishes as $\frac{\partial_r\varphi}{\varphi}$ depends only on r :

$$\begin{aligned}
\nabla_X\mathrm{II} &= \nabla_X\left(\frac{\partial_r\varphi}{\varphi}g_r\right) \\
&= D_X\left(\frac{\partial_r\varphi}{\varphi}\right)g_r + \frac{\partial_r\varphi}{\varphi}\nabla_X g_r \\
&= 0.
\end{aligned}$$

From this we can conclude

$$\begin{aligned}
\mathfrak{R}(X \wedge \partial_r) &= -\frac{\partial_r^2\varphi}{\varphi}X \wedge \partial_r = -\frac{\ddot{\varphi}}{\varphi}X \wedge \partial_r, \\
\mathfrak{R}(X \wedge Y) &= \frac{1 - (\partial_r\varphi)^2}{\varphi^2}X \wedge Y = \frac{1 - \dot{\varphi}^2}{\varphi^2}X \wedge Y
\end{aligned}$$

In particular, we have diagonalized $\mathfrak{R}$. Hence all sectional curvatures lie between the two values $-\frac{\ddot{\varphi}}{\varphi}$ and $\frac{1-\dot{\varphi}^2}{\varphi^2}$. Furthermore, if we select an orthonormal basis E_i

where $E_1 = \partial_r$, then the Ricci tensor and scalar curvature are

$$\begin{aligned}
\mathrm{Ric}\,(X) &= \sum_{i=1}^{n} R\,(X, E_i)\,E_i \\
&= \sum_{i=1}^{n-1} R\,(X, E_i)\,E_i + R\,(X, \partial_r)\,\partial_r \\
&= \left((n-2)\frac{1-\dot\varphi^2}{\varphi^2} - \frac{\ddot\varphi}{\varphi} \right) X \\
\mathrm{Ric}\,(\partial_r) &= -(n-1)\frac{\ddot\varphi}{\varphi}\partial_r \\
\mathrm{scal} &= -(n-1)\frac{\ddot\varphi}{\varphi} + (n-1)\left((n-2)\frac{1-\dot\varphi^2}{\varphi^2} - \frac{\ddot\varphi}{\varphi} \right) \\
&= -2(n-1)\frac{\ddot\varphi}{\varphi} + (n-1)(n-2)\frac{1-\dot\varphi^2}{\varphi^2}
\end{aligned}$$

Notice that when $n = 2$, we have $\sec = -\frac{\ddot\varphi}{\varphi}$, because there are no tangential curvatures. This makes for quite a difference between 2- and higher-dimensional rotationally symmetric metrics.

Constant curvature: First, we should compute the curvature of:

$$dr^2 + \mathrm{sn}_k^2(r)ds_{n-1}^2 \text{ on } S_k^n.$$

Since $\varphi = \mathrm{sn}_k$ solves $\ddot\varphi + k\varphi = 0$ we see that $\sec(X, \partial_r) = k$. To compute $\sec(X, Y) = \frac{1-\dot\varphi^2}{\varphi^2}$, just recall that $\mathrm{sn}_k(r) = \frac{1}{\sqrt{k}} \sin\left(\sqrt{k}r \right)$ (even when $k < 0$), so

$$\begin{aligned}
\dot\varphi &= \cos(\sqrt{k}\,r), \\
1 - \dot\varphi^2 &= \sin^2(\sqrt{k}\,r) = k\varphi^2.
\end{aligned}$$

Thus, all sectional curvatures are equal to k, just as promised.

Next let us see if we can find any interesting Ricci flat or scalar flat examples.

Ricci flat metrics: A Ricci flat metric must satisfy

$$\begin{aligned}
\frac{\ddot\varphi}{\varphi} &= 0, \\
(n-2)\frac{1-\dot\varphi^2}{\varphi^2} - \frac{\ddot\varphi}{\varphi} &= 0.
\end{aligned}$$

Hence, if $n > 2$, we must have $\ddot\varphi \equiv 0$ and $\dot\varphi^2 \equiv 1$. Thus, $\varphi(r) = a \pm r$. In case $n = 2$ we only need $\ddot\varphi = 0$. In any case, the only Ricci flat rotationally symmetric metrics are, in fact, flat.

Scalar flat metrics: To find scalar flat metrics we need to solve

$$2(n-1)\left[-\frac{\ddot\varphi}{\varphi} + \frac{n-2}{2} \cdot \frac{1-\dot\varphi^2}{\varphi^2} \right] = 0$$

when $n \geq 3$. We rewrite this equation as

$$-\varphi\ddot\varphi + \frac{n-2}{2}(1-\dot\varphi^2) = 0.$$

This is an autonomous second-order equation. The change of variables

$$\dot{\varphi} = G(\varphi),$$
$$\ddot{\varphi} = G'\dot{\varphi} = G'G$$

will yield a first-order equation:

$$-\varphi G'G + \frac{n-2}{2}(1 - G^2) = 0.$$

Using separation of variables, we see that G and φ are related by

$$\dot{\varphi}^2 = G^2 = 1 + C\varphi^{2-n},$$

which after differentiation becomes:

$$2\ddot{\varphi}\dot{\varphi} = (2-n)C\varphi^{1-n}\dot{\varphi}.$$

To analyze the solutions to this equation that are positive and therefore yield Riemannian metrics, we need to study the cases $C > 0$, $C = 0$, $C < 0$ separately. But first, notice that if $C \neq 0$, then we cannot have that $\varphi(a) = 0$, as this would imply $\ddot{\varphi}(a) = \infty$.

$C = 0$: In this case, we have $\ddot{\varphi} \equiv 0$ and $\dot{\varphi}^2(0) = 1$. Thus, $\varphi = a + r$ is the only solution and the metric is the standard Euclidean metric.

$C > 0$: First, observe that from the equation

$$2\ddot{\varphi} = (2-n)C\varphi^{1-n}$$

we get that φ is concave. Thus, if φ is extended to its maximal interval, it must cross the "x-axis," but as pointed out above this means that $\ddot{\varphi}$ becomes undefined, and therefore we don't get any metrics this way.

$C < 0$: This time the solution is convex and doesn't have to cross the "x-axis" as before. Thus we can assume that it is positive wherever defined. We claim that φ must exist for all time. Otherwise, we could find $a \in \mathbb{R}$ where $\varphi(t) \to \infty$ as $t \to a$ from the left or right. But then $\dot{\varphi}^2(t) \to 0$, which is clearly impossible. Next, observe that $\varphi \to \infty$ as $t \to \pm\infty$ as $\dot{\varphi}^2(t)$ doesn't converge to 0. Finally, we can conclude that φ must have a unique positive minimum. Using translational invariance of the solutions, we can assume that this minimum is achieved at $t = 0$.

So assume that $\varphi(0) = \alpha > 0$ and in addition $\dot{\varphi}(0) = 0$. We get the relation

$$0 = \dot{\varphi}^2(0) = 1 + C \cdot \alpha^{2-n},$$

which tells us that $C = -\alpha^{n-2}$. Let $\varphi_\alpha(r)$ denote this solution. Thus, we have a scalar flat rotationally symmetric metric on $\mathbb{R} \times S^{n-1}$. Notice that φ_α is also even, and so $(r, x) \to (-r, -x)$ is an isometry on

$$\left(\mathbb{R} \times S^{n-1}, dr^2 + \varphi_\alpha^2(r)ds_{n-1}^2\right).$$

We therefore get a Riemannian covering map

$$\mathbb{R} \times S^{n-1} \to \tau(\mathbb{R}P^{n-1})$$

and a scalar flat metric on $\tau(\mathbb{R}P^{n-1})$, the tautological line bundle over $\mathbb{R}P^{n-1}$. One can prove that $\varphi_\alpha(r) \geq |r|$ for all $r \in \mathbb{R}$ and that $\varphi_\alpha(r) \cdot |r|^{-1} \to 1$ as $r \to \infty$. Thus

$$\dot{\varphi}^2 = 1 - \alpha^{n-2}\varphi^{2-n} \simeq 1 - \alpha^{n-2}|r|^{2-n},$$
$$2\ddot{\varphi} \simeq (n-2)\alpha^{n-2}|r|^{1-n}$$

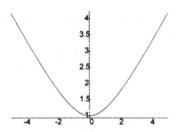

Graph of φ when $n = 3$ and $\alpha = 1$

Figure 3.1

as $r \to \infty$. This means, in particular, that all sectional curvatures are $\simeq |r|^{-n}$ as $r \to \infty$. The rotationally symmetric metric $dr^2 + \varphi_\alpha^2(r)ds_{n-1}^2$ therefore looks very much like $dr^2 + r^2 ds_{n-1}^2$ at ∞. Figure 3.1 is a picture of the warping function when $\alpha = 1$ and $n = 3$.

We shall in chapter 5 be able to show that $\mathbb{R} \times S^{n-1}$, $n \geq 3$, does not admit a (complete) constant curvature metric. Later in chapter 9, we will see that if $\mathbb{R} \times S^{n-1}$ has $\mathrm{Ric} \equiv 0$, then S^{n-1} also has a metric with $\mathrm{Ric} \equiv 0$. When $n = 3$ or 4 this means that S^2 and S^3 have flat metrics, and we shall see in chapter 5 that this is not possible. Thus we have found a manifold with a nice scalar flat metric that does not carry any Ricci flat or constant curvature metrics.

2.4. Doubly Warped Products. We wish to compute the curvatures on

$$\left(I \times S^p \times S^q, dr^2 + \varphi^2(r)ds_p^2 + \psi^2(r)ds_q^2 \right).$$

This time the Hessian looks like

$$\mathrm{Hess}\, r = (\partial_r \varphi)\, \varphi ds_p^2 + (\partial_r \psi)\, \psi ds_q^2.$$

and we see as in the rotationally symmetric case that

$$\nabla_X \mathrm{II} = 0.$$

Thus the mixed curvatures vanish. Let X, Y be tangent to S^p and V, W tangent to S^q. Using our curvature calculations from the rotationally symmetric case and the product sphere case we obtain

$$\Re\left(\partial_r \wedge X \right) = -\frac{\ddot{\varphi}}{\varphi} \partial_r \wedge X,$$

$$\Re\left(\partial_r \wedge V \right) = -\frac{\ddot{\psi}}{\psi} \partial_r \wedge V,$$

$$\Re\left(X \wedge Y \right) = \frac{1 - \dot{\varphi}^2}{\varphi^2} X \wedge Y,$$

$$\Re\left(U \wedge V \right) = \frac{1 - \dot{\psi}^2}{\psi^2} U \wedge V,$$

$$\Re\left(X \wedge V \right) = -\frac{\dot{\varphi}\dot{\psi}}{\varphi\psi} X \wedge V.$$

From this we can see that all sectional curvatures are convex linear combinations of

$$\frac{-\ddot{\varphi}}{\varphi}, \frac{-\ddot{\psi}}{\psi}, \frac{1-\dot{\varphi}^2}{\varphi^2}, \frac{1-\dot{\psi}^2}{\psi^2}, \frac{-\dot{\varphi}\dot{\psi}}{\varphi\psi}.$$

Moreover,

$$\mathrm{Ric}(\partial_r) = \left(-p\frac{\ddot{\varphi}}{\varphi} - q\frac{\ddot{\psi}}{\psi}\right)\partial_r,$$

$$\mathrm{Ric}(X) = \left(\frac{-\ddot{\varphi}}{\varphi} + (p-1)\frac{1-\dot{\varphi}^2}{\varphi^2} - q\cdot\frac{\dot{\varphi}\dot{\psi}}{\varphi\psi}\right)X,$$

$$\mathrm{Ric}(V) = \left(\frac{-\ddot{\psi}}{\psi} + (q-1)\frac{1-\dot{\psi}^2}{\psi^2} - p\cdot\frac{\dot{\varphi}\dot{\psi}}{\varphi\psi}\right)V.$$

2.5. The Schwarzschild Metric. We wish to find a Ricci flat metric on $\mathbb{R}^2 \times S^2$, so let $p = 1$ and $q = 2$ in the above doubly warped product case. This means we have to solve the following three equations simultaneously:

$$\frac{-\ddot{\varphi}}{\varphi} - 2\frac{\ddot{\psi}}{\psi} = 0,$$

$$\frac{-\ddot{\varphi}}{\varphi} - 2\frac{\dot{\varphi}\dot{\psi}}{\varphi\psi} = 0,$$

$$\frac{-\ddot{\psi}}{\psi} + \frac{1-\dot{\psi}^2}{\psi^2} - \frac{\dot{\varphi}\dot{\psi}}{\varphi\psi} = 0.$$

Subtracting the first two equations gives

$$\frac{\ddot{\psi}}{\psi} = \frac{\dot{\varphi}\dot{\psi}}{\varphi\psi}.$$

This is equivalent to $\left(\frac{\dot{\psi}}{\varphi}\right) = \alpha$, for some constant α. Thus, $\dot{\psi} = \alpha\varphi$ and $\ddot{\psi} = \alpha\dot{\varphi}$. Inserting this into the three equations we get

$$-\frac{\ddot{\varphi}}{\varphi} - 2\frac{\alpha\dot{\varphi}}{\psi} = 0,$$

$$-\frac{\ddot{\varphi}}{\varphi} - 2\frac{\alpha\dot{\varphi}}{\psi} = 0,$$

$$-\frac{\alpha\dot{\varphi}}{\psi} + \frac{1-\alpha^2\varphi^2}{\psi^2} - \frac{\alpha\dot{\varphi}}{\psi} = 0,$$

$$\dot{\psi} = \alpha\varphi,$$

which reduces to

$$-\frac{\ddot{\varphi}}{\varphi} - 2\frac{\alpha\dot{\varphi}}{\psi} = 0,$$

$$\frac{1-\alpha^2\varphi^2}{\psi^2} - 2\frac{\alpha\dot{\varphi}}{\psi} = 0,$$

$$\dot{\psi} = \alpha\varphi,$$

which implies

$$\frac{1 - \alpha^2 \varphi^2}{2\alpha\dot{\varphi}} = \psi,$$

$$-\frac{\ddot{\varphi}}{\varphi} - \frac{4\alpha^2 \dot{\varphi}^2}{1 - \alpha^2 \varphi^2} = 0,$$

$$\dot{\psi} = \alpha\varphi,$$

which implies

$$2\psi\ddot{\psi} - (1 - \dot{\psi}^2) = 0,$$

$$-\frac{\ddot{\varphi}}{\varphi} - \frac{4\alpha^2 \dot{\varphi}^2}{1 - \alpha^2 \varphi^2} = 0,$$

$$\dot{\psi} = \alpha\varphi.$$

Now, $\psi = r$ solves the first equation. This means that $\varphi = \frac{1}{\alpha}$, which also solves the second equation. The metric, however, lives on $S^1 \times \mathbb{R}^3$ rather than $\mathbb{R}^2 \times S^2$, and it is the standard flat metric on this space. To get more complicated solutions, assume $\dot{\psi}^2 = G(\psi)$, $2\ddot{\psi} = G'$. Then the first equation becomes

$$\psi G' + G = 1,$$

so

$$G = 1 + C\psi^{-1}, \ C \in \mathbb{R}.$$

Translating back we get

$$\dot{\psi}^2 = 1 + C\psi^{-1},$$

$$2\ddot{\psi} = -C\psi^{-2},$$

$$\dot{\psi} = \alpha\varphi,$$

as the equation

$$-\frac{\ddot{\varphi}}{\varphi} - \frac{4\alpha^2 \dot{\varphi}^2}{1 - \alpha^2 \varphi^2} = 0$$

is now redundant. Also, since we want a metric on $\mathbb{R}^2 \times S^2$, we may assume that $\varphi(0) = 0$, $\dot{\varphi}(0) = 1$, and $\psi(0) = \beta > 0$. This actually gives all the requirements for a smooth metric. First, ψ is forced to be even if it solves the above equation. Consequently, φ is odd. The constants α, β, and C are related through

$$0 = \dot{\psi}^2(0) = 1 + C \cdot \beta^{-1},$$

so

$$C = -\beta,$$
$$2\alpha = 2\alpha\dot{\varphi}(0) = 2\ddot{\psi}(0) = -C\beta^{-2} = \beta^{-1}.$$

For given $\beta > 0$, let the solutions be denoted by φ_β and ψ_β. Since $\psi_\beta(0) = \beta > 0$ and $\ddot{\psi}_\beta = \frac{\beta}{2}\psi_\beta^{-2}$, we have that ψ_β is convex as long as it is positive. We can then prove as in the scalar flat case that ψ is defined for all r and that $\psi(r) \sim |r|$ as $r \to \pm\infty$.

Thus, the metric looks like $S^1 \times \mathbb{R}^3$ at infinity, where the metric on S^1 is multiplied by $(2 \cdot \beta)^2$. Therefore, the Schwarzschild metric is a Ricci flat metric on $\mathbb{R}^2 \times S^2$

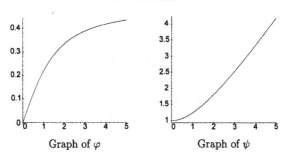

Graph of φ Graph of ψ

Figure 3.2

that at infinity looks approximately like the flat metric on $S^1 \times \mathbb{R}^3$. Both warping functions are sketched below in the case where $\alpha = 1$.

3. Hyperbolic Space

We have a pretty good picture of spheres and Euclidean space as models for constant curvature spaces. We even know what the symmetry groups are. We don't have a similarly good picture for spaces of constant negative curvature. This is partly because these metrics are not hypersurface metrics in Euclidean space. There are, however, several different good models. To explain them we also need to expand our general knowledge a little.

3.1. The Rotationally Symmetric Model. We define H^n to be the rotationally symmetric metric $dr^2 + \sinh^2(r)ds_{n-1}^2$ on $\mathbb{R}^n$ of constant curvature -1. As with all rotationally symmetric metrics, we see that $O(n)$ acts by isometries in a natural way. But it is not clear that H^n is homogeneous from this description as the origin is singled out as being fixed by the $O(n)$ action.

3.2. The Upper Half Plane Model. Let
$$M = \{(x^1, \ldots, x^n) \in \mathbb{R}^n : x^n > 0\}$$
and let
$$g = \left(\frac{1}{x^n}\right)^2 \left((dx^1)^2 + \cdots + (dx^n)^2\right).$$
Thus $\frac{1}{x^n}dx^1, \ldots, \frac{1}{x^n}dx^n$ is an orthonormal coframing on M. This can be used to check that the curvature is $\equiv -1$. Another way is to notice that
$$g = dr^2 + (e^{-r})^2 \left((dx^1)^2 + \cdots + (dx^{n-1})^2\right),$$
where $r = \log(x^n)$, and then use the fundamental equations. In this case the metric is on $\mathbb{R}^n = \mathbb{R} \times \mathbb{R}^{n-1}$. In particular, Iso $(\mathbb{R}^{n-1}) = \mathbb{R}^{n-1} \rtimes O(n-1)$ acts by isometries on M. There is no fixed point for the action and it acts transitively on the hypersurfaces $r = $ constant.

3.3. The Riemann Model. If (M, g) is a Riemannian manifold and φ is positive on M, then we can get a new Riemannian manifold $(M, \varphi^2 g)$. Such a change in metric is called a *conformal change,* and φ^2 is referred to as the *conformal factor.* The upper half plane model is a conformal change of the Euclidean metric. More generally we can ask when
$$\varphi^2 \cdot \left((dx^1)^2 + \cdots + (dx^n)^2\right)$$

Figure 3.3

has constant curvature? Clearly, $\varphi \cdot dx^1, \ldots, \varphi \cdot dx^n$ is an orthonormal coframing, and $\frac{1}{\varphi}\partial_1, \ldots, \frac{1}{\varphi}\partial_n$ is an orthonormal framing. We can use the Koszul formula to compute $\nabla_{\partial_i}\partial_j$ and hence the curvature tensor. This tedious task is done in [87, vols. II and IV]. Using

$$\varphi = \left(1 + \frac{k}{4}r^2\right)^{-1}$$

gives a metric of constant curvature k on $\mathbb{R}^n$ if $k \geq 0$ and on $B(0, -4k^{-1})$ if $k < 0$.

3.4. The Imaginary Unit Sphere Model. Our last model exhibits H^n as a hypersurface in Minkowski space by analogy with $S^n(1) \subset \mathbb{R}^{n+1}$. A discussion of this model can also be found in chapter 8. Minkowski space is the physicists' model for space-time. Topologically, the space is $\mathbb{R}^{n+1}$, but we use a semi-Riemannian metric. If $(x^0, x^1, \ldots, x^n)$ are Cartesian coordinates on $\mathbb{R}^{1,n}$, then we have the indefinite metric:

$$g = -(dx^0)^2 + (dx^1)^2 + \cdots + (dx^n)^2$$

In other words, the framing $\partial_0, \partial_1, \ldots, \partial_n$ consists of orthogonal vectors where $|\partial_0|^2 = -1$ and $|\partial_i|^2 = 1$, $i = 1, \ldots, n$. The zeroth coordinate is singled out as having imaginary norm, this is the physicists' time variable. The distance "spheres" in this space of radius $i \cdot r$ satisfy the equation

$$-(x^0)^2 + (x^1)^2 + \cdots + (x^n)^2 = -r^2.$$

A picture of this set in Minkowski 3-space is given in Figure 3.3.

With this in mind it seems reasonable to study the "distance" function

$$\begin{aligned} r(x) &= \left|-(x^0)^2 + (x^1)^2 + \cdots + (x^n)^2\right|^{1/2} \\ &= \left((x^0)^2 - (x^1)^2 - \cdots - (x^n)^2\right)^{1/2} \end{aligned}$$

on the connected open set

$$U = \left\{x \in \mathbb{R}^{n+1} : -(x^0)^2 + (x^1)^2 + \cdots + (x^n)^2 < 0, x^0 > 0\right\}.$$

The level sets $H(r) \subset U$ are diffeomorphic to $\mathbb{R}^n$ and look like hyperbolae of revolution. Furthermore, if we restrict the Minkowski metric g to these level sets, they induce Riemannian metrics on $H(r)$. This is because

$$dr = \frac{1}{r}(x^0 dx^0 - x^1 dx^1 - \cdots - x^n dx^n),$$

so any tangent vector $v \in TH(r)$ satisfies

$$0 = dr(v) = (-x^0 v^0 + x^1 v^1 + \cdots + x^n v^n) = g(x, v);$$

and therefore, for any such v,

$$
\begin{aligned}
g(v, v) &= -(v^0)^2 + (v^1)^2 + \cdots + (v^n)^2 \\
&= -\frac{(x^1 v^1 + \cdots + x^n v^n)^2}{(x^0)^2} + (v^1)^2 + \cdots + (v^n)^2 \\
&\geq -\frac{((x^1)^2 + \cdots + (x^n)^2)((v^1)^2 + \cdots + (v^n)^2)}{(x^0)^2} + (v^1)^2 + \cdots + (v^n)^2 \\
&= \left(-1 + \frac{r^2}{(x^0)^2}\right)((v^1)^2 + \cdots + (v^n)^2) + (v^1)^2 + \cdots + (v^n)^2 \\
&= \frac{r^2}{(x^0)^2}((v^1)^2 + \cdots + (v^n)^2) \geq 0.
\end{aligned}
$$

This shows that g is positive definite on $H(r)$. Our claim is that $H(r)$ with the induced metric has constant curvature $-r^{-2}$. There are several ways to check this. One way is to observe that

$$
\begin{aligned}
(0, \infty) \times S^{n-1} &\rightarrow \mathbb{R}^{1,n} \\
(t, x) &\rightarrow r(\cosh(t), \sinh(t) \cdot x)
\end{aligned}
$$

defines a Riemannian isometry from $dt^2 + r^2 \sinh^2\left(\frac{t}{r}\right) ds_{n-1}^2$ to $H(r)$, if $x \in S^{n-1} \subset \mathbb{R}^n$ is viewed as a vector in $\mathbb{R}^n$. This also shows that at least two of our models are equal. We could also compute gradients, etc., and use the tangential curvature equation as we did for the sphere. In outline this works out as follows. The Minkowski gradient $\nabla r = \alpha^i \partial_i$ must satisfy

$$
\begin{aligned}
g(\nabla r, v) &= dr(v), \\
-\alpha^0 v^0 + \alpha^1 v^1 + \cdots + \alpha^n v^n &= \frac{1}{r}(x^0 v^0 - x^1 v^1 - \cdots - x^n v^n),
\end{aligned}
$$

or equivalently, $\nabla r = g^{ij} \partial_i(r) \partial_j$, so $\nabla r = \frac{-1}{r} x^i \partial_i$. This is clearly not the same as the Euclidean gradient, but aside from the minus sign it corresponds exactly to the gradient for the distance function in $\mathbb{R}^{n+1}$ that has $S^n(1)$ as level sets. Also,

$$g(\nabla r, \nabla r) = \frac{1}{r^2}\left(-(x^0)^2 + (x^1)^2 + \cdots + (x^n)^2\right) = -1,$$

so we are working with an (imaginary) distance function, which aside from the sign should satisfy all of the fundamental equations we have already established. The Minkowski connection on $\mathbb{R}^{1,n}$ of course satisfies all of the same properties as the Riemannian connection and can in particular be found using the Koszul formula. But since $g(\partial_i, \partial_j)$ is always constant and $[\partial_i, \partial_j] = 0$, we see that $\nabla_{\partial_i} \partial_j = 0$. Hence, we get just the standard Euclidean connection and therefore the curvature tensor $R = 0$ as well. With all this information one can easily compute $\text{Hess}\, r$ and check using the tangential curvature equation that $H(r)$ indeed has constant curvature $-r^{-2}$.

Finally, we should compute the isometry group $\text{Iso}(H(r))$. On $\mathbb{R}^{1,n}$ the linear isometries that preserve the Minkowski metric are denoted by

$$O(1, n) = \left\{L : \mathbb{R}^{1,n} \rightarrow \mathbb{R}^{1,n} : g(Lv, Lv) = g(v, v)\right\}.$$

One can, as in the case of the sphere, see that these are isometries on $H(r)$ as long as we they preserve the condition $x^0 > 0$. The group of those isometries is denoted $O^+(1, n)$. The isotropy group that preserves $(r, 0, \ldots, 0)$ can be identified with $O(n)$ (isometries we get from the metric being rotationally symmetric). One can also easily check that $O^+(1, n)$ acts transitively on $H(r)$.

With all this we now have a fairly complete picture of all the space forms S_k^n, i.e., our models for constant curvature. We shall in chapter 5 prove that in a suitable sense these are the only simply connected Riemannian manifolds of constant curvature.

4. Metrics on Lie Groups

We are going to study some general features of left-invariant metrics and show how things simplify in the bi-invariant situation. There are also two examples of left-invariant metrics. The first represents H^2, and the other is the Berger sphere.

4.1. Generalities on Left-Invariant Metrics.
We construct a metric on a Lie group G by fixing an Euclidean metric $(,)$ on $T_e G$ and then translating it to $T_g M$ using left translation $L_g(x) = gx$. The metric is also denoted (X, Y) on G so as not to confuse it with elements $g \in G$. With this metric, L_g becomes an isometry for all g since

$$
\begin{aligned}
(DL_g)|_h &= \left(DL_{ghh^{-1}}\right)|_h \\
&= \left(D\left(L_{gh} \circ L_{h^{-1}}\right)\right)|_h \\
&= (DL_{gh})|_e \circ (DL_{h^{-1}})|_h \\
&= (DL_{gh})|_e \circ ((DL_h)|_e)^{-1}
\end{aligned}
$$

and we have assumed that $(DL_{gh})|_e$ and $(DL_h)|_e$ are isometries.

We know that left-invariant fields X, i.e., $DL_g(X|_h) = X|_{gh}$ are completely determined by their value at the identity. We can therefore identify $T_e M$ with $\mathfrak{g}$, the space of left-invariant fields. Note that $\mathfrak{g}$ is in a natural way a vector space as addition of left-invariant fields is left-invariant. It is also a Lie algebra as the vector field Lie bracket of two such fields is again left-invariant. In the appendix we show that on matrix groups the Lie bracket is simply the commutator of the matrices in $T_e M$ representing the vector fields.

If $X \in \mathfrak{g}$, then the integral curve through $e \in G$ is denoted by $\exp(tX)$. In case of a matrix group the standard matrix exponential e^{tX} is in fact the integral curve since

$$
\begin{aligned}
\frac{d}{dt}|_{t=t_0}\left(e^{tX}\right) &= \frac{d}{dt}|_{s=0}\left(e^{(t_0+s)X}\right) \\
&= \frac{d}{dt}|_{s=0}\left(e^{t_0 X} e^{sX}\right) \\
&= \frac{d}{dt}|_{s=0}\left(L_{e^{t_0}x} e^{sX}\right) \\
&= D\left(L_{e^{t_0}x}\right)\left(\frac{d}{dt}|_{s=0} e^{sX}\right) \\
&= D\left(L_{e^{t_0}x}\right)(X|_I) \\
&= X|_{e^{t_0}x}.
\end{aligned}
$$

The key property for $t \to \exp(tX)$ to be the integral curve for X is similarly that the derivative at $t = 0$ is $X|_e$ and that $t \to \exp(tX)$ is a homomorphism

$$\exp((t+s)X) = \exp(tX) \exp(sX).$$

The entire flow for X can now be written as follows

$$F^t(x) = x \exp(tX) = L_x \exp(tX) = R_{\exp(sX)}(x).$$

The curious thing is that the flow maps $F^t : G \to G$ don't act by isometries unless the metric is also invariant under right-translations, i.e., the metric is bi-invariant. In particular, the elements of $\mathfrak{g}$ are not in general Killing fields.

We can give a fairly reasonable way of checking that a left-invariant metric is also bi-invariant. The inner automorphism $x \to gxg^{-1}$ is usually denoted $\text{Ad}_g(x) = gxg^{-1}$ on Lie groups and is called the *adjoint action* of G on G. The differential of this action at $e \in G$ is a linear map $\text{Ad}_g : \mathfrak{g} \to \mathfrak{g}$ denoted by the same symbol, and called the *adjoint action* of G on $\mathfrak{g}$. It is in fact a Lie algebra isomorphism. These two adjoint actions are related by

$$\text{Ad}_g(\exp(tX)) = \exp(t\text{Ad}_g(X)).$$

This is quite simple to prove. It only suffices to check that $t \to \text{Ad}_g(\exp(tX))$ is a homomorphism with differential $\text{Ad}_g(X)$ at $t = 0$. The latter follows from the definition of the differential of a map and the former by noting that it is the composition of two homomorphisms $x \to \text{Ad}_g(x)$ and $t \to \exp(tX)$. We can now give our criterion for bi-invariance.

PROPOSITION 11. *A left-invariant metric is bi-invariant if and only if the adjoint action on the Lie algebra is by isometries.*

PROOF. In case the metric is bi-invariant we know that both L_g and $R_{g^{-1}}$ act by isometries. Thus also $\text{Ad}_g = L_g \circ R_{g^{-1}}$ acts by isometries. The differential is therefore a linear isometry on the Lie algebra.

Now assume that $\text{Ad}_g : \mathfrak{g} \to \mathfrak{g}$ is always an isometry. Using that

$$(DR_g)|_h = (DR_{hg})|_e \circ ((DR_h)|_e)^{-1}$$

it clearly suffices to prove that $(DR_g)|_e$ is always an isometry. This follows from

$$\begin{aligned} R_g &= L_g \circ \text{Ad}_{g^{-1}}, \\ (DR_g)|_e &= D(L_g)|_e \circ \text{Ad}_{g^{-1}}. \end{aligned}$$

$\square$

In the next two subsections we shall see how this can be used to check whether metrics are bi-invariant in some specific matrix group examples.

Before giving examples of how to compute the connection and curvatures for left-invariant metrics we present the general and simpler situation of bi-invariant metrics.

PROPOSITION 12. *Let G be a Lie group with a bi-invariant metric $(,)$. If $X, Y, Z, W \in \mathfrak{g}$, then*

$$\nabla_Y X = \frac{1}{2}[Y, X],$$

$$R(X, Y)Z = -\frac{1}{4}[[X, Y], Z],$$

$$R(X, Y, Z, W) = \frac{1}{4}([X, Y], [W, Z]).$$

In particular, the sectional curvature is always nonnegative.

PROOF. We first need to construct the *adjoint action* $\mathrm{ad}_X : \mathfrak{g} \to \mathfrak{g}$ of the Lie algebra on the Lie algebra. If we think of the adjoint action of the Lie group on the Lie algebra as a homomorphism $\mathrm{Ad} : G \to \mathrm{Aut}(\mathfrak{g})$, then $\mathrm{ad} : \mathfrak{g} \to \mathrm{End}(\mathfrak{g})$ is simply the differential $\mathrm{ad} = D(\mathrm{Ad})|_e$. In the section on Lie derivatives in the appendix is it shown that $\mathrm{ad}_X(Y) = [X, Y]$. The bi-invariance of the metric shows that the image $\mathrm{Ad}(G) \subset O(\mathfrak{g})$ lies in the group of orthogonal linear maps on $\mathfrak{g}$. This immediately shows that the image of ad lies in the set of skew-adjoint maps since

$$\begin{aligned}
0 &= \frac{d}{dt}(Y, Z)|_{t=0} \\
&= \frac{d}{dt}\left(\mathrm{Ad}_{\exp(tX)}(Y), \mathrm{Ad}_{\exp(tX)}(Z)\right)|_{t=0} \\
&= (\mathrm{ad}_X Y, Z) + (Y, \mathrm{ad}_X Y).
\end{aligned}$$

Keeping this skew-symmetry in mind we can use the Koszul formula on $X, Y, Z \in \mathfrak{g}$ to see that

$$\begin{aligned}
2(\nabla_Y X, Z) &= D_X(Y, Z) + D_Y(Z, X) - D_Z(X, Y) \\
&\quad - ([X, Y], Z) - ([Y, Z], X) + ([Z, X], Y) \\
&= -([X, Y], Z) - ([Y, Z], X) + ([Z, X], Y) \\
&= -([X, Y], Z) + ([Y, X], Z) + ([X, Y], Z) \\
&= ([Y, X], Z).
\end{aligned}$$

As for the curvature we then have

$$\begin{aligned}
R(X, Y)Z &= \nabla_X \nabla_Y Z - \nabla_Y \nabla_X Z - \nabla_{[X,Y]} Z \\
&= \frac{1}{2}\nabla_X[Y, Z] - \frac{1}{2}\nabla_Y[X, Z] - \frac{1}{2}[[X, Y], Z] \\
&= \frac{1}{4}[X, [Y, Z]] - \frac{1}{4}[Y, [X, Z]] - \frac{1}{2}[[X, Y], Z] \\
&= \frac{1}{4}[X, [Y, Z]] + \frac{1}{4}[Y, [Z, X]] + \frac{1}{4}[Z, [X, Y]] - \frac{1}{4}[[X, Y], Z] \\
&= -\frac{1}{4}[[X, Y], Z],
\end{aligned}$$

and finally

$$
\begin{aligned}
(R(X,Y)Z,W) &= -\frac{1}{4}([[X,Y],Z],W) \\
&= \frac{1}{4}([Z,[X,Y]],W) \\
&= -\frac{1}{4}([Z,W],[X,Y]) \\
&= \frac{1}{4}([X,Y],[W,Z]).
\end{aligned}
$$

$\square$

We note that Lie groups with bi-invariant metrics always have non-negative sectional curvature and with a little more work that the curvature operator is also non-negative.

4.2. Hyperbolic Space as a Lie Group. Let G be the 2-dimensional Lie group

$$
G = \left\{ \begin{bmatrix} \alpha & \beta \\ 0 & 1 \end{bmatrix} : \alpha > 0, \beta \in \mathbb{R} \right\}.
$$

Notice that the first row can be identified with the upper half plane. The Lie algebra of G is

$$
\mathfrak{g} = \left\{ \begin{bmatrix} a & b \\ 0 & 0 \end{bmatrix} : a, b \in \mathbb{R} \right\}.
$$

If we define

$$
X = \begin{bmatrix} 1 & 0 \\ 0 & 0 \end{bmatrix}, Y = \begin{bmatrix} 0 & 1 \\ 0 & 0 \end{bmatrix},
$$

then

$$
[X,Y] = XY - YX = Y.
$$

Now declare $\{X,Y\}$ to be an orthonormal frame on G. Then use the Koszul formula to compute

$$
\nabla_X X = 0, \ \nabla_Y Y = X, \ \nabla_X Y = 0, \ \nabla_Y X = \nabla_X Y - [X,Y] = -Y.
$$

Hence,

$$
R(X,Y)Y = \nabla_X \nabla_Y Y - \nabla_Y \nabla_X Y - \nabla_{[X,Y]} Y = \nabla_X X - 0 - \nabla_Y Y = -X,
$$

which implies that G has constant curvature -1.

We can also compute Ad_g:

$$
\begin{aligned}
\mathrm{Ad}_{\begin{bmatrix} \alpha & \beta \\ 0 & 1 \end{bmatrix}} \begin{bmatrix} a & b \\ 0 & 0 \end{bmatrix} &= \begin{bmatrix} \alpha & \beta \\ 0 & 1 \end{bmatrix} \begin{bmatrix} a & b \\ 0 & 0 \end{bmatrix} \begin{bmatrix} \alpha & \beta \\ 0 & 1 \end{bmatrix}^{-1} \\
&= \begin{bmatrix} a & -a\beta + b\alpha \\ 0 & 0 \end{bmatrix} \\
&= aX + (-a\beta + b\alpha)Y
\end{aligned}
$$

The orthonormal basis

$$
\begin{bmatrix} 1 & 0 \\ 0 & 0 \end{bmatrix}, \begin{bmatrix} 0 & 1 \\ 0 & 0 \end{bmatrix}
$$

is therefore mapped to the basis

$$\begin{bmatrix} 1 & -\beta \\ 0 & 0 \end{bmatrix}, \begin{bmatrix} 0 & \alpha \\ 0 & 0 \end{bmatrix}.$$

This, however, is not an orthonormal basis unless $\beta = 0$ and $\alpha = 1$. The metric is therefore not bi-invariant, nor are the left-invariant fields Killing fields.

This example can be generalized to higher dimensions. Thus, the upper half plane is in a natural way a Lie group with a left invariant metric of constant curvature -1. This is in sharp contrast to the spheres, where only $S^3 = SU(2)$ and $S^1 = SO(2)$ are Lie groups.

4.3. Berger Spheres. On $SU(2)$ we have the left-invariant metric where $\lambda_1^{-1} X_1, \lambda_2^{-1} X_2, \lambda_3^{-1} X_3$ is an orthonormal frame and $[X_i, X_{i+1}] = 2X_{i+2}$ (indices are mod 3), as mentioned in chapter 1. The Koszul formula is:

$$2 \left(\nabla_{X_i} X_j, X_k \right) = \left([X_i, X_j], X_k \right) + \left([X_k, X_i], X_j \right) - \left([X_j, X_k], X_i \right).$$

From this we can quickly see that like with a bi-invariant metric we have:

$$\nabla_{X_i} X_i = 0.$$

We can also see that

$$\nabla_{X_i} X_{i+1} = \left(\frac{\lambda_{i+2}^2 + \lambda_{i+1}^2 - \lambda_i^2}{\lambda_{i+2}^2} \right) X_{i+2},$$

$$\nabla_{X_{i+1}} X_i = [X_{i+1}, X_i] + \nabla_{X_i} X_{i+1}$$

$$= \left(\frac{-\lambda_{i+2}^2 + \lambda_{i+1}^2 - \lambda_i^2}{\lambda_{i+2}^2} \right) X_{i+2}$$

This shows that

$$R(X_i, X_{i+1}) X_{i+2} = \nabla_{X_i} \nabla_{X_{i+1}} X_{i+2}$$
$$- \nabla_{X_{i+1}} \nabla_{X_i} X_{i+2} - \nabla_{[X_i, X_{i+1}]} X_{i+2}$$
$$= 0 - 0 - 0.$$

Thus all curvatures between three distinct vectors vanish.

The special case of the Berger spheres occurs when $\lambda_1 = \varepsilon < 1$, $\lambda_2 = \lambda_3 = 1$. In this case

$$\nabla_{X_1} X_2 = \left(2 - \varepsilon^2 \right) X_3, \ \nabla_{X_2} X_1 = -\varepsilon^2 X_3$$
$$\nabla_{X_2} X_3 = X_1, \ \nabla_{X_3} X_2 = -X_1,$$
$$\nabla_{X_3} X_1 = \varepsilon^2 X_2, \ \nabla_{X_1} X_3 = \left(\varepsilon^2 - 2 \right) X_2.$$

and

$$R(X_1, X_2) X_2 = \varepsilon^2 X_1,$$
$$R(X_3, X_1) X_1 = \varepsilon^4 X_3,$$
$$R(X_2, X_3) X_3 = \left(4 - 3\varepsilon^2 \right) X_2$$

$$\mathfrak{R}(X_1 \wedge X_2) = \varepsilon^2 X_1 \wedge X_2,$$
$$\mathfrak{R}(X_3 \wedge X_1) = \varepsilon^2 X_3 \wedge X_1,$$
$$\mathfrak{R}(X_2 \wedge X_3) = \left(4 - 3\varepsilon^2 \right) X_2 \wedge X_3$$

Thus all sectional curvatures must lie in the interval $\left[\varepsilon^2, 4 - 3\varepsilon^2\right]$. Note that as $\varepsilon \to 0$ the sectional curvature $\sec\left(X_2, X_3\right) \to 4$, which is the curvature of the base space $S^2\left(\frac{1}{2}\right)$ in the Hopf fibration.

We should also consider the adjoint action in this case. The standard orthogonal basis X_1, X_2, X_3 is mapped to

$$\mathrm{Ad}_{\begin{bmatrix} z & w \\ -\bar{w} & \bar{z} \end{bmatrix}} X_1 = \left(|z|^2 - |w|^2\right) X_1 - 2\,\mathrm{Re}\left(wz\right) X_2 - 2\,\mathrm{Im}\left(wz\right) X_3,$$

$$\mathrm{Ad}_{\begin{bmatrix} z & w \\ -\bar{w} & \bar{z} \end{bmatrix}} X_2 = 2i\,\mathrm{Im}\left(z\bar{w}\right) X_1 + \mathrm{Re}\left(w^2 + z^2\right) X_2 + \mathrm{Im}\left(w^2 + z^2\right) X_3,$$

$$\mathrm{Ad}_{\begin{bmatrix} z & w \\ -\bar{w} & \bar{z} \end{bmatrix}} X_3 = 2\,\mathrm{Re}\left(z\bar{w}\right) X_1 + \mathrm{Re}\left(iz^2 - iw^2\right) X_2 + \mathrm{Im}\left(iz^2 - iw^2\right) X_3,$$

If the three vectors X_1, X_2, X_3 have the same length, then we see that the adjoint action is by isometries, otherwise it is not.

5. Riemannian Submersions

In this section we shall develop some formulas for curvatures that relate to Riemannian submersions. The situation is quite similar to that of distance functions, which as we know are Riemannian submersions. In this case, however, we shall try to determine the curvature of the base space from information about the total space. Thus the situation is actually dual to what we have studied so far.

5.1. Riemannian Submersions and Curvatures.
Throughout this section let $F : (M, \bar{g}) \to (N, g)$ be a Riemannian submersion. Like with the metrics we shall use the notation $\bar{p}$ and p as well as $\bar{X}$ and X for points and vector fields that are F-related, i.e., $F\left(\bar{p}\right) = p$ and $DF\left(\bar{X}\right) = X$. The *vertical distribution* consists of the tangent spaces to the preimages $F^{-1}\left(p\right)$ and is therefore given by $\mathcal{V}_{\bar{p}} = \ker DF_{\bar{p}} \subset T_{\bar{p}}M$. The *horizontal distribution* is the orthogonal complement $\mathcal{H}_{\bar{p}} = \left(\mathcal{V}_{\bar{p}}\right)^{\perp} \subset T_{\bar{p}}M$. The fact that F is a Riemannian submersion means that $DF : \mathcal{H}_{\bar{p}} \to T_pN$ is an isometry for all $\bar{p} \in M$. Given a vector field X on N we can always find a unique horizontal vector field $\bar{X}$ on M that is F related to X. We say that $\bar{X}$ is a *basic horizontal* lift of X. Any vector in M can be decomposed into horizontal and vertical parts: $v = v^{\mathcal{V}} + v^{\mathcal{H}}$.

The next proposition gives some important properties for relationships between vertical and basic horizontal vector fields.

PROPOSITION 13. *Let V be a vertical vector field on M and X, Y, Z vector fields on N with basic horizontal lifts $\bar{X}, \bar{Y}, \bar{Z}$.*
(1) $\left[V, \bar{X}\right]$ is vertical,
(2) $\left(L_V \bar{g}\right)\left(\bar{X}, \bar{Y}\right) = D_V \bar{g}\left(\bar{X}, \bar{Y}\right) = 0$,
(3) $\bar{g}\left(\left[\bar{X}, \bar{Y}\right], V\right) = 2\bar{g}\left(\nabla_{\bar{X}} \bar{Y}, V\right) = -2\bar{g}\left(\nabla_V \bar{X}, \bar{Y}\right) = 2\bar{g}\left(\nabla_{\bar{Y}} V, \bar{X}\right)$,
(4) $\nabla_{\bar{X}} \bar{Y} = \overline{\nabla_X Y} + \frac{1}{2}\left[\bar{X}, \bar{Y}\right]^{\mathcal{V}}$.

PROOF. (1): $\bar{X}$ is F related to X and V is F related to the zero vector field on N. Thus

$$DF\left(\left[\bar{X}, V\right]\right) = \left[DF\left(\bar{X}\right), DF\left(V\right)\right] = [X, 0] = 0.$$

(2): We use (1) to see that

$$(L_V \bar{g})(\bar{X}, \bar{Y}) = D_V \bar{g}(\bar{X}, \bar{Y}) - \bar{g}([V, \bar{X}], \bar{Y}) - \bar{g}(\bar{X}, [V, \bar{Y}])$$
$$= D_V \bar{g}(\bar{X}, \bar{Y}).$$

Next we use that F is a Riemannian submersion to conclude that $\bar{g}(\bar{X}, \bar{Y}) = g(X, Y)$. But this implies that the inner product is constant in the direction of the vertical distribution.

(3): Using (1) and (2) the Koszul formula in all cases reduce to

$$2\bar{g}(\nabla_{\bar{X}} \bar{Y}, V) = \bar{g}([\bar{X}, \bar{Y}], V),$$
$$2\bar{g}(\nabla_V \bar{X}, \bar{Y}) = -\bar{g}([\bar{X}, \bar{Y}], V),$$
$$2\bar{g}(\nabla_{\bar{Y}} V, \bar{X}) = \bar{g}([\bar{X}, \bar{Y}], V).$$

This proves the claims.

(4) We have just seen in (3) that $\frac{1}{2}[\bar{X}, \bar{Y}]^V$ is the vertical component of $\nabla_{\bar{X}} \bar{Y}$. We know that $\overline{\nabla_X Y}$ is horizontal so it only remains to be seen that it is the horizontal component of $\nabla_{\bar{X}} \bar{Y}$. The Koszul formula together with F relatedness of the fields and the fact that inner products are the same in M and N show that

$$2\bar{g}(\nabla_{\bar{X}} \bar{Y}, \bar{Z}) = 2g(\nabla_X Y, Z)$$
$$= 2\bar{g}(\overline{\nabla_X Y}, \bar{Z}).$$

□

Note that the map that takes horizontal vector fields X, Y on M to $[X, Y]^V$ measures the extent to which the horizontal distribution is integrable in the sense of Frobenius. It is in fact tensorial as well as skew-symmetric since

$$[X, fY]^V = f[X, Y]^V + (D_X f) Y^V = f[X, Y]^V.$$

Therefore, it defines a map $\mathcal{H} \times \mathcal{H} \to \mathcal{V}$ called the *integrability tensor*.

EXAMPLE 25. *In the case of the Hopf map $S^3(1) \to S^2\left(\frac{1}{2}\right)$ we have that X_1 is vertical and X_2, X_3 are horizontal. However, X_2, X_3 are not basic. Still, we know that $[X_2, X_3] = 2X_1$ so the horizontal distribution cannot be integrable.*

We are now ready to give a formula for the curvature tensor on N in terms of the curvature tensor on M and the integrability tensor.

THEOREM 5. (B. O'Neill and A. Grey) *Let R be the curvature tensor on N and $\bar{R}$ the curvature tensor on M, then*

$$g(R(X, Y)Y, X) = \bar{g}(\bar{R}(\bar{X}, \bar{Y})\bar{Y}, \bar{X}) + \frac{3}{4}\left|[\bar{X}, \bar{Y}]^V\right|^2.$$

PROOF. The proof is a direct calculation using the above properties. We calculate the full curvature tensor so let X, Y, Z, H be vector fields on M with zero Lie brackets. This forces the corresponding Lie brackets $[\bar{X}, \bar{Y}]$, etc. in M to be vertical.

$$\bar{g}\left(\bar{R}\left(\bar{X},\bar{Y}\right)\bar{Z},\bar{H}\right) \;=\; \bar{g}\left(\nabla_{\bar{X}}\nabla_{\bar{Y}}\bar{Z}-\nabla_{\bar{Y}}\nabla_{\bar{X}}\bar{Z}-\nabla_{[\bar{X},\bar{Y}]}\bar{Z},\bar{H}\right)$$

$$=\; \bar{g}\left(\nabla_{\bar{X}}\left(\overline{\nabla_{Y}Z}+\frac{1}{2}\left[\bar{Y},\bar{Z}\right]\right),\bar{H}\right)$$

$$-\bar{g}\left(\nabla_{\bar{Y}}\left(\overline{\nabla_{X}Z}+\frac{1}{2}\left[\bar{X},\bar{Z}\right]\right),\bar{H}\right)$$

$$+\bar{g}\left(\left[\bar{Z},\bar{H}\right],\left[\bar{X},\bar{Y}\right]\right)$$

$$=\; \bar{g}\left(\overline{\nabla_{X}\nabla_{Y}Z}+\frac{1}{2}\left[\bar{X},\overline{\nabla_{Y}Z}\right]^{\mathcal{V}}+\frac{1}{2}\nabla_{\bar{X}}\left[\bar{Y},\bar{Z}\right],\bar{H}\right)$$

$$-\bar{g}\left(\overline{\nabla_{Y}\nabla_{X}Z}+\frac{1}{2}\left[\bar{Y},\overline{\nabla_{X}Z}\right]^{\mathcal{V}}+\frac{1}{2}\nabla_{\bar{Y}}\left[\bar{X},\bar{Z}\right],\bar{H}\right)$$

$$-\frac{1}{2}\bar{g}\left(\left[\bar{X},\bar{Y}\right],\left[\bar{H},\bar{Z}\right]\right)$$

$$=\; g\left(R\left(X,Y\right)Z,H\right)$$

$$-\frac{1}{2}\bar{g}\left(\left[\bar{Y},\bar{Z}\right],\nabla_{\bar{X}}\bar{H}\right)+\frac{1}{2}\bar{g}\left(\left[\bar{X},\bar{Z}\right],\nabla_{\bar{Y}}\bar{H}\right)$$

$$-\frac{1}{2}\bar{g}\left(\left[\bar{X},\bar{Y}\right],\left[\bar{H},\bar{Z}\right]\right)$$

$$=\; g\left(R\left(X,Y\right)Z,H\right)$$

$$-\frac{1}{4}\bar{g}\left(\left[\bar{Y},\bar{Z}\right],\left[\bar{X},\bar{H}\right]\right)+\frac{1}{4}\bar{g}\left(\left[\bar{X},\bar{Z}\right],\left[\bar{Y},\bar{H}\right]\right)$$

$$-\frac{1}{2}\bar{g}\left(\left[\bar{X},\bar{Y}\right],\left[\bar{H},\bar{Z}\right]\right)$$

Letting $X = H$ and $Y = Z$ we get the above formula. $\square$

More generally, one can find formulae for $\bar{R}$ where the variables are various combinations of basic horizontal and vertical fields.

5.2. Riemannian Submersions and Lie Groups.

One can find many examples of manifolds with nonnegative or positive curvature using the previous theorem. In this section we shall explain the terminology in the general setting. The types of examples often come about by having $(M,\bar{g})$ with a free compact group action G by isometries and using $N = M/G$. Examples are:

$$\mathbb{C}P^{n} \;=\; S^{2n+1}/S^{1},$$
$$TS^{n} \;=\; \left(SO\left(n+1\right)\times\mathbb{R}^{n}\right)/SO\left(n\right),$$
$$N \;=\; SU\left(3\right)/T^{2}.$$

The complex projective space will be studied further in the next subsection.

The most important general example of a Riemannian submersion comes about by having an isometric group action by G on M such that the quotient space is a manifold $N = M/G$. Such a submersion is also called *fiber homogeneous* as the group acts transitively on the fibers of the submersion. In this case we have a natural map $F : M \to N$ that takes orbits to point, i.e., $p = \{x \cdot \bar{p} : x \in G\}$ for $\bar{p} \in M$. The vertical space $\mathcal{V}_{\bar{p}}$ then consists of the vectors that are tangent to the action. These directions can be found using the Killing fields generated by G. If

$X \in \mathfrak{g} = T_e G$, then we get a vector $X|_{\bar{p}} \in T_{\bar{p}} M$ by the formula

$$X|_{\bar{p}} = \frac{d}{dt} \left(\exp\left(tX \right) \cdot \bar{p} \right)|_{t=0},$$

This means that the flow for X on M is defined by $F^t(\bar{p}) = \exp\left(tX \right) \cdot \bar{p}$. As the map $\bar{p} \to x \cdot \bar{p}$ is assumed to be an isometry for all $x \in G$ we get that the flow acts by isometries. This means that X is a Killing field. The next observation is that the action preserves the vertical distribution, i.e., $Dx\left(\mathcal{V}_{\bar{p}}\right) = \mathcal{V}_{x \cdot \bar{p}}$. Using the Killing fields this follows from

$$
\begin{aligned}
Dx\left(X|_{\bar{p}}\right) &= Dx\left(\frac{d}{dt}\left(\exp\left(tX\right)\cdot\bar{p}\right)|_{t=0}\right) \\
&= \frac{d}{dt}\left(x\cdot\left(\exp\left(tX\right)\cdot\bar{p}\right)\right)|_{t=0} \\
&= \frac{d}{dt}\left(\left(x\exp\left(tX\right)x^{-1}\right)\cdot x\cdot\bar{p}\right)|_{t=0} \\
&= \left(\left(\mathrm{Ad}_x\left(\exp\left(tX\right)\right)\right)\cdot x\cdot\bar{p}\right)|_{t=0} \\
&= \frac{d}{dt}\left(\left(\exp\left(t\mathrm{Ad}_xX\right)\right)\cdot x\cdot\bar{p}\right)|_{t=0} \\
&= \left(\mathrm{Ad}_x\left(X\right)\right)|_{x\cdot\bar{p}}.
\end{aligned}
$$

Thus $Dx\left(X|_{\bar{p}}\right)$ comes from first conjugating X via the adjoint action in $T_e G$ and then evaluating it at $x \cdot \bar{p}$. Since $\left(\mathrm{Ad}_x\left(X\right)\right)|_{x \cdot \bar{p}} \in \mathcal{V}_{x \cdot \bar{p}}$ we get that Dx maps the vertical spaces to vertical spaces. However, it doesn't preserve the Killing fields in the way one might have hoped for. As Dx is a linear isometry it also preserves the orthogonal complements. These complements are our horizontal spaces $\mathcal{H}_{\bar{p}} = \left(\mathcal{V}_{\bar{p}}\right)^{\perp} \subset T_{\bar{p}} M$. We know that $DF : \mathcal{H}_{\bar{p}} \to T_p N$ is an isomorphism. We have also seen that all of the spaces $\mathcal{H}_{x \cdot \bar{p}}$ are isometric to $\mathcal{H}_{\bar{p}}$ via Dx. We can therefore define the Riemannian metric on $T_p N$ using the isomorphism $DF : \mathcal{H}_{\bar{p}} \to T_p N$. This means that $F : M \to N$ defines a Riemannian submersion.

In the above discussion we did not discuss what conditions to put on the action of G on M in order to ensure that the quotient becomes a nice manifold. If G is compact and acts freely, then this will happen. In the next subsection we consider the special case of complex projective spaces as a quotient of a sphere. There is also a general way of getting new metrics on M it self from having a general isometric group action. This will be considered in the last subsection.

5.3. Complex Projective Space. Recall that $\mathbb{C}P^n = S^{2n+1}/S^1$, where S^1 acts by complex scalar multiplication on $S^{2n+1} \subset \mathbb{C}^{n+1}$. If we write the metric

$$ds_{2n+1}^2 = dr^2 + \sin^2(r)ds_{2n-1}^2 + \cos^2(r)d\theta^2,$$

then we can think of the S^1 action on S^{2n+1} as acting separately on S^{2n-1} and S^1. Then

$$\mathbb{C}P^n = \left[0, \frac{\pi}{2}\right] \times \left(\left(S^{2n-1} \times S^1\right)/S^1\right),$$

and the metric can be written as

$$dr^2 + \sin^2(r)\left(g + \cos^2(r)h\right).$$

If we restrict our attention to the case where $n = 2$ the metric can be written as

$$dr^2 + \sin^2(r)\left(\cos^2(r)(\sigma^1)^2 + (\sigma^2)^2 + (\sigma^3)^2\right).$$

This is a bit different from the warped product metrics we have seen so far. It is certainly still possible to apply the general techniques of distance functions to compute the curvature tensor, however, we shall instead use the Riemannian submersion apparatus that was developed in the previous section. We shall also consider the general case rather than $n = 2$.

The O'Neill formula from the previous section immediately shows that $\mathbb{C}P^n$ has sectional curvature ≥ 1. Let V be the unit vector field on S^{2n+1} that is tangent to the S^1 action. Then $\sqrt{-1}V$ is the unit inward pointing normal vector to $S^{2n+1} \subset \mathbb{C}^{n+1}$. This shows that the horizontal distribution, which is orthogonal to V, is invariant under multiplication by $\sqrt{-1}$. This corresponds to the fact that $\mathbb{C}P^n$ has a complex structure. It also tells us what the integrability tensor for this submersion is. If we let $\bar{X}, \bar{Y}$ be basic horizontal vector fields and denote the canonical Euclidean metric on $\mathbb{C}^{n+1}$ by $\bar{g}$, then

$$
\begin{aligned}
\bar{g}\left(\frac{1}{2}\left[\bar{X}, \bar{Y}\right], V\right) &= \bar{g}\left(\nabla_{\bar{X}}^{S^{2n+1}}\bar{Y}, V\right) \\
&= \bar{g}\left(\nabla_{\bar{X}}^{\mathbb{C}^{n+1}}\bar{Y}, V\right) \\
&= -\bar{g}\left(\bar{Y}, \nabla_{\bar{X}}^{\mathbb{C}^{n+1}}V\right) \\
&= \bar{g}\left(\bar{Y}, \sqrt{-1}\nabla_{\bar{X}}^{\mathbb{C}^{n+1}}\sqrt{-1}V\right) \\
&= \bar{g}\left(\bar{Y}, \sqrt{-1}\bar{X}\right).
\end{aligned}
$$

Thus

$$
\frac{1}{2}\left[\bar{X}, \bar{Y}\right]^V = \bar{g}\left(\bar{Y}, \sqrt{-1}\bar{X}\right)V.
$$

If we let X, Y be orthonormal on $\mathbb{C}P^n$, then the horizontal lifts $\bar{X}, \bar{Y}$ are also orthonormal so

$$
\begin{aligned}
\sec(X, Y) &= 1 + \frac{3}{4}\left|\left[\bar{X}, \bar{Y}\right]^V\right|^2 \\
&= 1 + 3\left|\bar{g}\left(\bar{Y}, \sqrt{-1}\bar{X}\right)\right|^2 \\
&\leq 4,
\end{aligned}
$$

with equality precisely when $\bar{Y} = \pm\sqrt{-1}\bar{X}$.

The proof of the O'Neill formula in fact gave us a formula for the full curvature tensor. One can use that formula on an orthonormal set of vectors of the form $X, \sqrt{-1}X, Y, \sqrt{-1}Y$ to see that the curvature operator is not diagonalized on a decomposable basis of the form $E_i \wedge E_j$ as was the case in the previous examples. In fact it is diagonalized by expressions of the form

$$
\begin{aligned}
&X \wedge \sqrt{-1}X \pm Y \wedge \sqrt{-1}Y, \\
&X \wedge Y \pm \sqrt{-1}X \wedge \sqrt{-1}Y, \\
&X \wedge \sqrt{-1}Y \pm Y \wedge \sqrt{-1}X
\end{aligned}
$$

and has eigenvalues that lie in the interval $[0, 6]$.

We can also see that this metric on $\mathbb{C}P^n$ is Einstein with Einstein constant $2n + 2$. If we fix a unit vector X and an orthonormal basis for the complement

$E_0, ..., E_{2n-2}$ so that the lifts satisfy $\sqrt{-1}X = \bar{E}_0$, then we get that

$$
\begin{aligned}
\text{Ric}\,(X, X) &= \sum_{i=0}^{2n-2} \sec\,(X, E_i) \\
&= \sec\,(X, E_0) + \sum_{i=1}^{2n-2} \sec\,(X, E_i) \\
&= 1 + 3\left|\bar{g}\left(\bar{E}_0, \sqrt{-1}\bar{X}\right)\right|^2 + \sum_{i=1}^{2n-2} \left(1 + 3\left|\bar{g}\left(\bar{E}_i, \sqrt{-1}\bar{X}\right)\right|^2\right) \\
&= 1 + 3\left|\bar{g}\left(\sqrt{-1}\bar{X}, \sqrt{-1}\bar{X}\right)\right|^2 + \sum_{i=1}^{2n-2} \left(1 + 3\,|0|^2\right) \\
&= 1 + 3 + 2n - 2 \\
&= 2n + 2.
\end{aligned}
$$

5.4. Berger-Cheeger Perturbations. The constructions we do here where first considered by Cheeger and where based on a slightly different construction by Berger that is explained at the end.

Fix a Riemannian manifold (M, g) and a Lie group G with a left-invariant metric $(,)$. If G acts by isometries on M, then it also acts by isometries on $M \times G$ if we use the product metric $g + \lambda\,(,)$ and $\lambda > 0$ is a positive scalar. As G acts freely on itself it also acts freely on $M \times G$. The quotient $(M \times G)/G$ is also denoted by $M \times_G G$. Since G acts freely, the natural map $M \to M \times G \to M \times_G G$ is a bijection. Thus the quotient is in a natural way a manifold diffeomorphic to M. The quotient map $Q : M \times G \to M$ is given by $Q\,(p, x) = x^{-1}p$.

As G acts by isometries when using any of the metrics $g + \lambda\,(,)$ we get a submersion metric g_λ on $M = M \times_G G$. We wish to study this perturbed metric's relation to the original metric g. The tangent space $T_p M$ is naturally decomposed into the vectors $\mathcal{V}_p$ that are tangent to the action and the orthogonal complement $\mathcal{H}_p$. Unlike the case where G acts freely on M this decomposition is not necessarily a nicely defined distribution. It might happen that G fixes certain but not all points in M. At points p that are fixed we see that $\mathcal{V}_p = \{0\}$. At other point $\mathcal{V}_p \neq \{0\}$. The nomenclature is, however, not inappropriate. If we select $X \in T_e G$ and let X be the corresponding Killing field on M, then $(X|_p, X) \in T_p M \times T_e G$ is a vertical direction for this action at $(p, e) \in M \times G$. Therefore, $\mathcal{V}_p$ is simply the vertical component tangent to M. Vectors in $\mathcal{H}_p$ are thus also horizontal for the action on $M \times G$. All the other horizontal vectors in $T_p M \times T_e G$ depend on the choice of λ and have a component of the form $\left(\lambda |X|^2 X|_p, -|X|_p|_g^2 X\right)$. The image of such a horizontal vector under $Q : M \times G \to M$ is given by

$$
\begin{aligned}
DQ\left(\lambda |X|^2 X|_p, -|X|_p|_g^2 X\right) &= \lambda |X|^2 DQ\,(X|_p, 0) - |X|_p|_g^2 DQ\,(0, X) \\
&= \lambda |X|^2 DQ\left(\frac{d}{dt}\,(\exp\,(tX) \cdot p)\,|_{t=0}, 0\right) \\
&\quad - |X|_p|_g^2 DQ\left(0, \frac{d}{dt}\,(\exp\,(tX) \cdot e)\,|_{t=0}\right)
\end{aligned}
$$

$$
\begin{aligned}
&= \ \lambda \left| X \right|^2 \frac{d}{dt} \left(Q \left(\exp \left(tX \right) \cdot p, e \right) \right) \big|_{t=0} \\
&\quad - \left| X \right|_p \big|_g^2 \frac{d}{dt} \left(Q \left(p, \exp \left(tX \right) e \right) \right) \big|_{t=0} \\
&= \ \lambda \left| X \right|^2 \frac{d}{dt} \left(\exp \left(tX \right) \cdot p \right) \big|_{t=0} \\
&\quad - \left| X \right|_p \big|_g^2 \frac{d}{dt} \left(\exp \left(-tX \right) \cdot p \right) \big|_{t=0} \\
&= \ \lambda \left| X \right|^2 X \big|_p + \left| X \right|_p \big|_g^2 X \big|_p \\
&= \ \left(\lambda \left| X \right|^2 + \left| X \right|_p \big|_g^2 \right) X \big|_p
\end{aligned}
$$

The horizontal lift of $X|_p \in \mathcal{V}_p$ is therefore given by

$$
\overline{X|_p} = \frac{\lambda \left| X \right|^2}{\lambda \left| X \right|^2 + \left| X \right|_p \big|_g^2} X \big|_p - \frac{\left| X \right|_p \big|_g^2}{\lambda \left| X \right|^2 + \left| X \right|_p \big|_g^2} X,
$$

and its length in g_λ is given by

$$
\begin{aligned}
\left| X \right|_p \big|_{g_\lambda}^2 \ &= \ \left(\frac{\lambda \left| X \right|^2}{\lambda \left| X \right|^2 + \left| X \right|_p \big|_g^2} \right)^2 \left| X \right|_p \big|_g^2 \\
&\quad + \left(\frac{\left| X \right|_p \big|_g^2}{\lambda \left| X \right|^2 + \left| X \right|_p \big|_g^2} \right)^2 \lambda \left| X \right|^2 \\
&= \ \frac{\lambda \left| X \right|^2}{\lambda \left| X \right|^2 + \left| X \right|_p \big|_g^2} \left| X \right|_p \big|_g^2 .
\end{aligned}
$$

In particular,

$$
0 \leq \left| X \right|_p \big|_{g_\lambda}^2 \leq \left| X \right|_p \big|_g^2,
$$

with limit 0 as $\lambda \to 0$ and limit $\left| X \right|_p \big|_g^2$ as $\lambda \to \infty$. This means that the metric g_λ is gotten from g by squeezing the orbits of the action of G. The squeezing depends on the point, however, according to this formula. The only case where the squeezing is uniform is when the Killing fields generated by the action have constant length on M. The Berger spheres are a special case of this. We have therefore found quite a general context for the Berger spheres.

Using that we know how to compute horizontal lifts and that the metric on $M \times G$ is a product metric it is possible to compute the curvature of g_λ in terms of the curvature of g, λ, the curvature of $(,)$, and the integrability tensor. We will consider two important special cases.

Case a) $X, Y \in \mathcal{H}_p$. In this case the vectors are already horizontal for the action on $M \times G$. Thus we have that $\sec_{g_\lambda} (X, Y) \geq \sec_g (X, Y)$. There is a correction coming from the integrability tensor associated with the action on $M \times G$ that possibly increases these curvatures.

Case b) We assume that X is horizontal and of unit length. We think of Y as given by $Y|_p$ for $Y \in T_e G$ with the additional property that $|Y|_p|_{g_\lambda} = 1$. We can in addition assume that X is extended in the Y direction so that their horizontal lifts commute. This means that the integrability tensor term in the O'Neill formula vanishes. The sectional curvature then satisfies

$$\sec_{g_\lambda}(X,Y) = R_{g+\lambda(,)}\left(X,\overline{Y|_p},\overline{Y|_p},X\right)$$

$$= \frac{\lambda|Y|^2}{\lambda|Y|^2+1}R_g(X,Y|_p,Y|_p,X).$$

An important special case occurs when $G = \mathbb{R}$ and the action is generated by a single Killing field X. In this case Berger originally gave a different way of constructing g_λ. This was also the construction he used for the Berger spheres. We assume that X corresponds to the coordinate vector field ∂_t on $\mathbb{R}$. The idea is to consider actions of $\mathbb{R}$ on $M \times \mathbb{R}$ generated by the Killing fields $X + \mu\partial_t$. If F^t is the flow generated by X, then the flow of $X + \mu\partial_t$ is generated by $(F^t(p), \mu t)$. The submersion map must then be given by

$$Q(p,t) = F^{-\mu^{-1}t}(p).$$

The horizontal field $\mu X - |X|^2\,\partial_t$ is mapped as follows

$$DQ\left(\mu X - |X|^2\,\partial_t\right) = \mu\frac{d}{dt}Q\left(F^t(p),0\right)|_{t=0} - |X|^2\frac{d}{dt}Q(p,t)|_{t=0}$$

$$= \mu\frac{d}{dt}F^t(p)|_{t=0} - |X|^2\frac{d}{dt}F^{-\mu^{-1}t}(p)|_{t=0}$$

$$= \mu X + |X|^2\,\mu^{-1}X$$

$$= \left(\mu + |X|^2\,\mu^{-1}\right)X.$$

The horizontal lift of X is therefore given by

$$\bar{X} = \frac{\mu}{\mu + |X|^2\,\mu^{-1}}X - \frac{|X|^2}{\mu + |X|^2\,\mu^{-1}}\partial_t.$$

If we denote the new metric on M by g_μ we see that

$$g_\mu(X,X) = |\bar{X}|_{g_\mu}^2$$

$$= \left(\frac{\mu}{\mu + |X|_g^2\,\mu^{-1}}\right)^2|X|_g^2 + \left(\frac{|X|_g^2}{\mu + |X|_g^2\,\mu^{-1}}\right)^2$$

$$= \frac{\mu^2 + |X|_g^2}{\left(\mu + |X|_g^2\,\mu^{-1}\right)^2}|X|_g^2$$

$$= \frac{\mu|X|_g^2}{\mu + |X|_g^2\,\mu^{-1}}$$

$$= \frac{\mu^2|X|_g^2}{\mu^2 + |X|_g^2}.$$

This metric g_μ is therefore the same as the metric g_{λ^2} constructed above using Cheeger's method.

6. Further Study

The book by O'Neill [**73**] gives an excellent account of Minkowski geometry and also studies in detail the Schwarzschild metric in the setting of general relativity. It appears to have been the first exact nontrivial solution to the vacuum Einstein field equations. There is also a good introduction to locally symmetric spaces and their properties. This book is probably the most comprehensive elementary text and is good for a first encounter with most of the concepts in differential geometry. The third edition of [**41**] also contains a good number of examples. Specifically they have a lot of material on hyperbolic space. They also have a brief account of the Schwarzschild metric in the setting of general relativity.

Another book, which contains more (actually almost all) advanced examples, is [**11**]. This is also a tremendously good reference on Riemannian geometry in general.

7. Exercises

(1) Show that the Schwarzschild metric doesn't have parallel curvature tensor.
(2) Show that the Berger spheres ($\varepsilon \neq 1$) do not have parallel curvature tensor.
(3) Show that $\mathbb{C}P^2$ has parallel curvature tensor.
(4) The *Heisenberg group* with its Lie algebra is

$$G = \left\{ \begin{bmatrix} 1 & a & c \\ 0 & 1 & b \\ 0 & 0 & 1 \end{bmatrix} : a, b, c \in \mathbb{R} \right\},$$

$$\mathfrak{g} = \left\{ \begin{bmatrix} 0 & x & z \\ 0 & 0 & y \\ 0 & 0 & 0 \end{bmatrix} : a, b, c \in \mathbb{R} \right\}.$$

A basis for the Lie algebra is:

$$X = \begin{bmatrix} 0 & 1 & 0 \\ 0 & 0 & 0 \\ 0 & 0 & 0 \end{bmatrix}, Y = \begin{bmatrix} 0 & 0 & 0 \\ 0 & 0 & 1 \\ 0 & 0 & 0 \end{bmatrix}, Z = \begin{bmatrix} 0 & 0 & 1 \\ 0 & 0 & 0 \\ 0 & 0 & 0 \end{bmatrix}.$$

   (a) Show that the only nonzero brackets are

$$[X, Y] = -[Y, X] = Z.$$

   Now introduce a left-invariant metric on G such that X, Y, Z form an orthonormal frame.
   (b) Show that the Ricci tensor has both negative and positive eigenvalues.
   (c) Show that the scalar curvature is constant.
   (d) Show that the Ricci tensor is not parallel.
(5) Let $\tilde{g} = e^{2\psi} g$ be a metric conformally equivalent to g. Show that
   (a)

$$\tilde{\nabla}_X Y = \nabla_X Y + ((D_X \psi) Y + (D_Y \psi) X - g(X, Y) \nabla \psi)$$

   (b) If X, Y are orthonormal with respect to g, then

$$e^{2\psi} \widetilde{\sec}(X, Y) = \sec(X, Y) - \text{Hess}\psi(X, X) - \text{Hess}\psi(Y, Y)$$
$$- |\nabla \psi|^2 + (D_X \psi)^2 + (D_Y \psi)^2$$

(6) (a) Show that there is a family of Ricci flat metrics on TS^2 of the form

$$dr^2 + \varphi^2(r)\left(\psi^2(r)(\sigma^1)^2 + (\sigma^2)^2 + (\sigma^3)^2\right),$$

$$\dot{\varphi} = \psi,$$
$$\dot{\varphi}^2 = 1 - k\varphi^{-4},$$
$$\varphi(0) = k^{\frac{1}{4}}, \dot{\varphi}(0) = 0,$$
$$\psi(0) = 0, \dot{\psi}(0) = 2.$$

(b) Show that $\varphi(r) \sim r$, $\dot{\varphi}(r) \sim 1$, $\ddot{\varphi}(r) \sim 2kr^{-5}$ as $r \to \infty$. Conclude that all curvatures are of order r^{-6} as $r \to \infty$ and that the metric looks like $(0, \infty) \times \mathbb{R}P^3 = (0, \infty) \times SO(3)$ at infinity. Moreover, show that scaling one of these metrics corresponds to changing k. Thus, we really have only one Ricci flat metric; it is called the *Eguchi-Hanson metric*.

(7) For the general metric

$$dr^2 + \varphi^2(r)\left(\psi^2(r)(\sigma^1)^2 + (\sigma^2)^2 + (\sigma^3)^2\right)$$

show that the $(1,1)$-tensor, which in the orthonormal frame looks like

$$\begin{bmatrix} 0 & -1 & 0 & 0 \\ 1 & 0 & 0 & 0 \\ 0 & 0 & 0 & -1 \\ 0 & 0 & 1 & 0 \end{bmatrix},$$

yields a Hermitian structure.

(a) Show that this structure is Kähler, i.e., parallel, iff $\dot{\varphi} = \psi$.

(b) Find the scalar curvature for such metrics.

(c) Show that there are scalar flat metrics on all the 2-dimensional vector bundles over S^2. The one on TS^2 is the Eguchi-Hanson metric, and the one on $S^2 \times \mathbb{R}^2$ is the Schwarzschild metric.

(8) Show that $\tau(\mathbb{R}P^{n-1})$ admits rotationally symmetric metrics

$$dr^2 + \varphi^2(r)\, ds_{n-1}^2$$

such that $\varphi(r) = r$ for $r > 1$ and the Ricci curvatures are nonpositive. Thus, the Euclidean metric can be topologically perturbed to have nonpositive Ricci curvature. It is not possible to perturb the Euclidean metric in this way to have nonnegative scalar curvature or nonpositive sectional curvature. Try to convince yourself of that by looking at rotationally symmetric metrics on $\mathbb{R}^n$ and $\tau(\mathbb{R}P^{n-1})$.

(9) A Riemannian manifold (M, g) is said to be *locally conformally flat* if every $p \in M$ lies in a coordinate neighborhood U such that

$$g = \varphi^2\left((dx^1)^2 + \cdots + (dx^n)^2\right).$$

(a) Show that the space forms S_k^n are locally conformally flat.

(b) With some help from the literature, show that any 2-dimensional Riemannian manifold is locally conformally flat (isothermal coordinates). In fact, any metric on a closed surface is conformal to a metric of constant curvature. This is called the *uniformization theorem*.

(c) Show that if an Einstein metric is locally conformally flat, then it has constant curvature.

(10) We say that (M,g) admits *orthogonal coordinates* around $p \in M$ if we have coordinates on some neighborhood of p, where

$$g_{ij} = 0 \text{ for } i \neq j,$$

i.e., the coordinate vector fields are perpendicular. Show that such coordinates always exist in dimension 2, while they may not exist in dimension > 3. To find a counterexample, you may want to show that in such coordinates the curvatures $R^l_{ijk} = 0$ if all indices are distinct. What about 3 dimensions?

(11) There is a strange curvature quantity we have not yet mentioned. Its definition is somewhat cumbersome and nonintuitive. First, for two symmetric $(0,2)$-tensors h, k define the *Kulkarni-Nomizu product* as the $(0,4)$-tensor

$$h \circ k \, (v_1, v_2, v_3, v_4) = \quad h\,(v_1, v_3) \cdot k\,(v_2, v_4) + h\,(v_2, v_4) \cdot k\,(v_1, v_3)$$
$$-h\,(v_1, v_4) \cdot k\,(v_2, v_3) - h\,(v_2, v_3) \cdot k\,(v_1, v_4).$$

Note that (M,g) has constant curvature c iff the $(0,4)$-curvature tensor satisfies $R = c \cdot (g \circ g)$. If we use the $(0,2)$ form of the Ricci tensor, then we can decompose the $(0,4)$-curvature tensor as follows in dimensions $n \geq 4$

$$R = \frac{\text{scal}}{2n\,(n-1)} g \circ g + \left(\text{Ric} - \frac{\text{scal}}{n} \cdot g\right) \circ g + W.$$

When $n = 3$ we have instead

$$R = \frac{\text{scal}}{12} g \circ g + \left(\text{Ric} - \frac{\text{scal}}{3} \cdot g\right) \circ g.$$

The $(0,4)$-tensor W defined for $n > 3$ is called the *Weyl tensor*.

(a) Show that these decompositions are orthogonal, in particular:

$$|R|^2 = \left|\frac{\text{scal}}{2n\,(n-1)} g \circ g\right|^2 + \left|\left(\text{Ric} - \frac{\text{scal}}{n} \cdot g\right) \circ g\right|^2 + |W|^2.$$

(b) Show that if we conformally change the metric $\tilde{g} = f \cdot g$, then $\tilde{W} = f \cdot W$.

(c) If (M, g) has constant curvature, then $W = 0$.

(d) If (M, g) is locally conformally equivalent to the Euclidean metric, i.e., locally we can always find coordinates where:

$$g = f \cdot \left((dx^1)^2 + \cdots + (dx^n)^2\right),$$

then $W = 0$. The converse is also true but much harder to prove.

(e) Show that the Weyl tensors for the Schwarzschild metric and the Euguchi-Hanson metrics are not zero.

(f) Show that (M, g) has constant curvature iff $W = 0$ and $\text{Ric} = \frac{\text{scal}}{n}$.

(12) In this problem we shall see that even in dimension 4 the curvature tensor has some very special properties. Throughout we let (M, g) be a 4-dimensional oriented Riemannian manifold. The bi-vectors $\Lambda^2 TM$ come with a natural endomorphism called the Hodge $*$ operator. It is defined as follows: for any oriented orthonormal basis e_1, e_2, e_3, e_4 we define $*\,(e_1 \wedge e_2) = e_3 \wedge e_4$.

(a) Show that his gives a well-defined linear endomorphism which satisfies: $** = I$. (Extend the definition to a linear map: $* : \Lambda^p TM \to \Lambda^q TM$, where $p + q = n$. When $n = 2$, we have: $* : TM \to TM = \Lambda^1 TM$ satisfies: $** = -I$, thus yielding an almost complex structure on any surface.)

(b) Now decompose $\Lambda^2 TM$ into $+1$ and -1 eigenspaces $\Lambda^+ TM$ and $\Lambda^- TM$ for $*$. Show that if e_1, e_2, e_3, e_4 is an oriented orthonormal basis, then

$$e_1 \wedge e_2 \pm e_3 \wedge e_4 \in \Lambda^\pm TM,$$
$$e_1 \wedge e_3 \pm e_4 \wedge e_2 \in \Lambda^\pm TM,$$
$$e_1 \wedge e_4 \pm e_2 \wedge e_3 \in \Lambda^\pm TM.$$

(c) Thus, any linear map $L : \Lambda^2 TM \to \Lambda^2 TM$ has a block decomposition

$$L = \begin{bmatrix} A & D \\ B & C \end{bmatrix},$$

$$
\begin{array}{lll}
A & : & \Lambda^+ TM \to \Lambda^+ TM, \\
D & : & \Lambda^+ TM \to \Lambda^- TM, \\
B & : & \Lambda^- TM \to \Lambda^+ TM, \\
C & : & \Lambda^- TM \to \Lambda^- TM.
\end{array}
$$

In particular, we can decompose the curvature operator $\mathfrak{R} : \Lambda^2 TM \to \Lambda^2 TM$:

$$\mathfrak{R} = \begin{bmatrix} A & D \\ B & C \end{bmatrix}.$$

Since $\mathfrak{R}$ is symmetric, we get that A, C are symmetric and that $D = B^*$ is the adjoint of B. One can furthermore show that

$$A = W^+ + \frac{\mathrm{scal}}{12} I,$$
$$C = W^- + \frac{\mathrm{scal}}{12} I,$$

where the Weyl tensor can be written

$$W = \begin{bmatrix} W^+ & 0 \\ 0 & W^- \end{bmatrix}.$$

Find these decompositions for both of the doubly warped metrics:

$$I \times S^1 \times S^2, dr^2 + \varphi^2(r) d\theta^2 + \psi^2(r) ds_2^2,$$
$$I \times S^3, dr^2 + \varphi^2(r) \left(\psi^2(r)(\sigma^1)^2 + (\sigma^2)^2 + (\sigma^3)^2 \right).$$

Use as basis for TM the natural frames in which we computed the curvature tensors. Now find the curvature operators for the Schwarzschild metric, the Euguchi-Hanson metric, $S^2 \times S^2$, S^4, and $\mathbb{C}P^2$.

(d) Show that (M, g) is Einstein iff $B = 0$ iff for every plane π and its orthogonal complement $\pi^\perp$ we have: $\sec(\pi) = \sec(\pi^\perp)$.

CHAPTER 4

Hypersurfaces

In this chapter we shall explain some of the classical results for hypersurfaces in Euclidean space. First we introduce the Gauss map and show that infinitesimally convex immersions are embeddings of spheres. We then establish a relationship between convexity and positivity of the intrinsic curvatures. This will enable us to see that $\mathbb{C}P^2$ and the Berger spheres are not even locally hypersurfaces in Euclidean space. We give a brief description of some classical existence results for isometric embeddings. Finally, an account of the Gauss-Bonnet theorem and its generalizations is given. One thing one might hope to get out of this chapter is the feeling that positively curved objects somehow behave like convex hypersurfaces, and might therefore have very restricted topological type.

In this chapter we develop the theory of hypersurfaces in general as opposed to just presenting surfaces in 3-space. The reason is that there are some differences depending on the ambient dimension. Essentially, there are three different categories of hypersurfaces that behave very differently from a geometric point of view: curves, surfaces, and hypersurfaces of dimension > 2. We shall see that as the dimension increases, the geometry becomes more and more rigid.

The study of hypersurfaces started as the study of surfaces in Euclidean 3 space. Even before Gauss, both Euler and Meusner made contributions to this area. It was with Gauss, however, that things really picked up speed. One of his most amazing discoveries was that one can detect curvature by measuring angles in polygons.

1. The Gauss Map

We shall suppose that we have a Riemannian manifold (M, g) with $\dim M = n$, and in addition a Riemannian immersion $F : (M, g) \looparrowright (\mathbb{R}^{n+1}, \mathrm{can})$. Locally we have a Riemannian embedding, whence we can find a smooth distance function on some open subset of $\mathbb{R}^{n+1}$ that has the image of M as a level set. Using this we can define the shape operator $S : TM \to TM$ as a locally defined $(1,1)$-tensor, which is well-defined up to sign (we just restrict the Hessian of the distance function to TM). If there is a globally defined normal field for M in $\mathbb{R}^{n+1}$, then we also get a globally defined shape operator. However, it still depends on our choice of normal and is therefore still only well-defined up to sign. Observe that such a global normal field exists exactly when M is orientable. By possibly passing to the orientation cover of M we can always assume that such a normal field exists globally (we can even assume that M is simply connected, although we won't do this). Let $N : M \to T\mathbb{R}^{n+1}$ be such a choice for a unit normal field. Using the trivialization $T\mathbb{R}^{n+1} = \mathbb{R}^{n+1} \times \mathbb{R}^{n+1}$ we then obtain the *Gauss map* $G : M \to S^n(1) \subset \mathbb{R}^{n+1}$, $G(x) = N(x)$ that to each point $x \in M$ assigns our choice of a normal to M at x in $\mathbb{R}^{n+1}$. A picture of the Gauss map for curves and surfaces is presented in Figure 4.1.

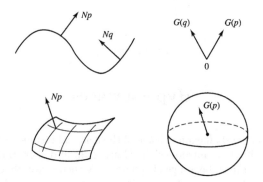

Figure 4.1

Our first important observation is that if we think of TM as a subset of $T\mathbb{R}^{n+1}$ then

$$DG(v) = S(v).$$

This is because $S(v) = \nabla_v N$, and since ∇ is the Euclidean connection we know that this corresponds to our usual notion of the differential of the map $G = N$.

With our first definition of the shape operator as the Hessian of a distance function it is clear that the hypersurface is locally convex (i.e., it lies locally on one side of its tangent space) provided that the shape operator is positive. Below we shall see how positivity of S is actually something that can be measured intrinsically by saying that some curvatures are positive. Before doing this let us use the above interpretation of the shape operator to show

THEOREM 6. (Hadamard, 1897) *Let* $F : (M, g) \looparrowright (\mathbb{R}^{n+1}, \text{can})$ *be an isometric immersion, where* $n > 1$ *and* M *is a closed manifold. If the shape operator is always positive, then* M *is diffeomorphic to a sphere via the Gauss map. Moreover* F *is an embedding.*

PROOF. If the shape operator is positive, then it is in particular nonsingular. Therefore, the Gauss map $G : M \to S^n(1)$ is a local diffeomorphism. When M is closed, it must therefore be a covering map. In case $n = 1$ the degree of this map is the winding number of the curve, while if $n > 1$, then $S^n(1)$ is simply connected, and hence G must be a diffeomorphism.

To show that F is an embedding we use that the Gauss map is a bijection. The proof also works when $n = 1$ as long as we assume that the winding number is ± 1. Fix $x_0 \in M$ and consider the function $f : M \to \mathbb{R}$ that measures the signed distance from $F(x)$ to the tangent plane for $F(M)$ through $F(x_0)$. Thus $f(x)$ is the projection of $F(x) - F(x_0)$ on to the unit normal $N|_{x_0}$. If we think of $N|_{x_0}$ as a vector in Euclidean space and use (v, w) as the inner product, then

$$f(x) = (F(x) - F(x_0), N|_{x_0}).$$

The differential is therefore given by

$$df(v) = (DF(v), N|_{x_0}).$$

As DF is nonsingular this shows that x is a critical point iff $N|_x = \pm N|_{x_0}$. Since the Gauss map is a bijection this shows that f has precisely two critical points x_0 and

x_1. Moreover $f(x_1) \neq f(x_0)$ as f would otherwise vanish everywhere. It follows that f is either nonpositive or nonnegative. If we assume that $F(x) = F(x_0)$ then $f(x) = 0 = f(x_0)$ which is impossible unless $x = x_0$. □

2. Existence of Hypersurfaces

Let us recall the Tangential and Normal curvature equations. The curvature of $\mathbb{R}^{n+1}$ is simply zero everywhere, so if the curvature tensor of M is denoted R, then we have that R is related to S as follows:

$$
\begin{aligned}
0 &= R(X,Y)Z - g(S(Y),Z)S(X) + g(S(X),Z)S(Y), \\
0 &= g(-(\nabla_X S)(Y) + (\nabla_Y S)(X), Z),
\end{aligned}
$$

where X, Y, Z are vector fields on M. We can rewrite these equations as

$$
\begin{aligned}
R(X,Y)Z &= (S(X) \wedge S(Y))(Z) \\
&= g(S(Y),Z)S(X) - g(S(X),Z)S(Y), \\
(\nabla_X S)(Y) &= (\nabla_Y S)(X).
\end{aligned}
$$

The former is the *Gauss equation,* and the latter is the *Codazzi-Mainardi equation.* Thus, R can be computed if we know S. In the Codazzi-Mainardi equation there is of course a question of which connection we use. However, we know that the Euclidean connection when projected down to M gives the Riemannian connection for (M, g), so it actually doesn't matter which connection is used.

We are now ready to show that positive curvature is equivalent to positive shape operator.

PROPOSITION 14. *Suppose we have a Riemannian immersion* $F : (M, g) \hookrightarrow (\mathbb{R}^{n+1}, \mathrm{can})$, *and we fix* $x \in M$. *If* $e_1, \ldots, e_n$ *is an orthonormal eigenbasis for* $S : T_x M \to T_x M$ *with eigenvalues* λ_i, $i = 1, \ldots, n$, *then* $e_i \wedge e_j$, $i < j$ *is an eigenbasis for the curvature operator* $\mathfrak{R} : \Lambda^2(T_x M) \to \Lambda^2(T_x M)$ *with eigenvalues* $\lambda_i \lambda_j$. *In particular, if all sectional curvatures are* $\geq \varepsilon^2 \geq 0$, *then the curvature operator is also* $\geq \varepsilon^2$.

PROOF. Suppose we have an orthonormal eigenbasis $\{e_i\}$ for $T_x M$ with respect to S. Then $S(e_i) = \lambda_i e_i$. Using the Gauss equations we obtain

$$
\begin{aligned}
g(\mathfrak{R}(e_i \wedge e_j), e_k \wedge e_l) &= g(R(e_i, e_j)e_l, e_k) \\
&= g(S(e_j), e_l)g(S(e_i), e_k) - g(S(e_i), e_l)g(S(e_j), e_k) \\
&= \lambda_i \lambda_j (g(e_j, e_l)g(e_i, e_k) - g(e_i, e_l)g(e_j, e_k)) \\
&= \lambda_i \lambda_j g(e_i \wedge e_j, e_k \wedge e_l).
\end{aligned}
$$

Thus we have diagonalized the curvature operator and shown that the eigenvalues are $\lambda_i \lambda_j$, $1 \leq i < j \leq n$. For the last statement we need only observe that the eigenvalues for the curvature operator satisfy

$$
\lambda_i \lambda_j = g(\mathfrak{R}(e_i \wedge e_j), e_i \wedge e_j) = \sec(e_i, e_j) \geq \varepsilon^2.
$$

□

This proposition shows that hypersurfaces have positive curvature operator iff they have positive sectional curvatures. In particular, the standard metric on $\mathbb{C}P^2$ cannot even locally be realized as a hypersurface metric as we saw in the last chapter that it has $\sec \geq 1$ but the curvature operator has 0 as an eigenvalue.

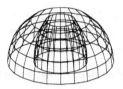

Two surfaces of constant curvature

Figure 4.2

There is a more holistic way of stating the above proposition and with it the Gauss equation:

$$\mathfrak{R} = S \wedge S.$$

The shape operator is therefore a "square root" of the curvature operator. From this interpretation it is tempting to believe that the shape operator can somehow be computed from curvatures. This is always false for surfaces $(n = 2)$, as we shall see below, but to some extent true when $n \geq 3$.

EXAMPLE 26. *Consider a surface $dt^2 + (a \sin(t))^2 d\theta^2$. We know that this can be represented as a surface of revolution in $\mathbb{R}^3$ when $|a| \leq 1$. Such a surface certainly has constant curvature 1. Now it only remains to see how one can represent it as a surface of revolution. We know from chapter 1 that such surfaces look like $(\dot{x}^2 + \dot{y}^2) dt^2 + y^2 d\theta^2$. In our case we therefore have to solve*

$$\dot{x}^2 + \dot{y}^2 = 1,$$
$$y = a \sin t,$$

which implies:

$$x = \int \sqrt{1 - (a \cos t)^2} dt,$$
$$y = a \sin t.$$

The embedding is written as:

$$F(t, \theta) = (x(t), y(t) \cos \theta, y(t) \sin \theta),$$

where

$$DF(\partial_t) = (\dot{x}(t), \dot{y}(t) \cos \theta, \dot{y}(t) \sin \theta),$$
$$DF\left(\frac{1}{y} \partial_\theta\right) = (0, -\sin \theta, \cos \theta)$$

are unit vectors tangent to the surface. Then the normal can be computed as

$$N = DF(\partial_t) \times DF\left(\frac{1}{y} \partial_\theta\right) = (\dot{y}(t), -\dot{x}(t) \cos \theta, -\dot{x}(t) \sin \theta).$$

Since the curvature is $1 = \det S$, either $S = I$ or S has two eigenvalues $\lambda > 1$ and $\lambda^{-1} < 1$. However, if we choose $y = a \sin t$ with $0 < a < 1$, then $S(\partial_t) \neq \partial_t$. Thus, we must be in the second case. The shape operator is therefore really an extrinsic invariant for surfaces. It is not hard to picture these surfaces together with the sphere, although one can't of course see that they actually have the same curvature. In Figure 4.2 we have a picture of the unit sphere together with one of these surfaces.

It turns out that this phenomenon occurs only for surfaces. Having codimension 1 for a surface leaves enough room to bend the surface without changing the metric intrinsically. In higher dimensions, however, we have.

PROPOSITION 15. *Suppose we have a Riemannian immersion* $F : (M, g) \hookrightarrow (\mathbb{R}^{n+1}, \text{can})$, *where* $n \geq 3$. *Fix* $x \in M$ *and suppose the curvature operator* $\mathfrak{R} : \Lambda^2(T_xM) \rightarrow \Lambda^2(T_xM)$ *is positive. Then* $S : T_xM \rightarrow T_xM$ *is intrinsic, i.e., we can compute* S *from information about* (M, g) *alone without knowledge of* F.

PROOF. We shall assume for simplicity that $n = 3$. If e_1, e_2, e_3 is an orthonormal basis for T_xM, then it suffices to compute the matrix $(s_{ij}) = (g(S(e_i), e_j))$. We already know that S is invertible from the above proposition and that all the eigenvalues have the same sign which we can assume to be positive. Thus, it suffices to determine the cofactor matrix (c_{ij}) defined by:

$$c_{ij} = (-1)^{i+j}(s_{i+1,j+1}s_{i+2,j+2} - s_{i+2,j+1}s_{i+1,j+2}).$$

The Gauss equations tell us that

$$
\begin{aligned}
g(\mathfrak{R}(e_i \wedge e_j), e_k \wedge e_l) &= g(R(e_i, e_j)e_l, e_k) \\
&= g(S(e_j), e_l)g(S(e_i), e_k) - g(S(e_i), e_l)g(S(e_j), e_k).
\end{aligned}
$$

Index manipulation will therefore enable us to find c_{ij} from the curvature operator. We also need to find the determinant of S in order to compute S^{-1} from the cofactor matrix. But this can be done using

$$\det(c_{ij}) = (\det S)^{n-1}.$$

$\square$

In case the curvature operator is only nonnegative we can still extract square roots, but they won't be unique. One can find more general conditions under which the shape operator is uniquely defined. As the cofactor matrices can always be found, the only important condition is that $\det S \neq 0$. This will be studied below and used for some very interesting purposes.

This information can be used to rule out even more candidates for hypersurfaces than did the previous result. Namely, when a space has positive curvature operator, then one can find the potential shape operator. However, this shape operator must also satisfy the Codazzi-Mainardi equations. It turns out that in dimensions > 3, these equations are a consequence of the Gauss equations provided the shape operator has nonzero determinant. This was proved by T. Y. Thomas in [91] (see also the exercises to chapter 7). For dimension 3, however, the following example shows that the Codazzi equations can not follow from the Gauss equations.

EXAMPLE 27. Let (M, g) be the *Berger sphere* $\left(S^3, \varepsilon^2\sigma_1^2 + \sigma_2^2 + \sigma_3^2\right)$ with $Y_1 = \varepsilon^{-1}X_1$, $Y_2 = X_2$, $Y_3 = X_3$ as an orthonormal left-invariant frame on $SU(2)$. We computed in chapter 3 that the 2-frame $Y_1 \wedge Y_2$, $Y_2 \wedge Y_3$, $Y_3 \wedge Y_1$ diagonalizes the curvature operator with eigenvalues ε^2, $\left(4 - 3\varepsilon^2\right)$, ε^2. It now follows from our calculations above that if this metric can be locally embedded in $\mathbb{R}^4$, then the shape operator can be computed using this information. If $(s_{ij}) = g(S(Y_i), Y_j)$, then it is easily seen that S must be diagonal, with

$$S(Y_1) = \frac{\varepsilon^2}{\sqrt{4 - 3\varepsilon^2}}Y_1,$$

$$S(Y_2) = \sqrt{4 - 3\varepsilon^2} Y_2,$$
$$S(Y_3) = \sqrt{4 - 3\varepsilon^2} Y_3.$$

We can now get a contradiction by showing that some of the Codazzi-Mainardi equations are not satisfied. For instance, we must have that $(\nabla_{Y_2} S)(Y_3) = (\nabla_{Y_3} S)(Y_2)$. However, these two quantities are not equal

$$
\begin{aligned}
(\nabla_{Y_2} S)(Y_3) &= \sqrt{4 - 3\varepsilon^2} \nabla_{Y_2} Y_3 - S(\nabla_{Y_2} Y_3) \\
&= \sqrt{4 - 3\varepsilon^2} \varepsilon Y_1 - \varepsilon \frac{\varepsilon^2}{\sqrt{4 - 3\varepsilon^2}} Y_1 \\
&= \left(\sqrt{4 - 3\varepsilon^2} - \frac{\varepsilon^2}{\sqrt{4 - 3\varepsilon^2}} \right) \varepsilon Y_1,
\end{aligned}
$$

$$
\begin{aligned}
(\nabla_{Y_3} S)(Y_2) &= \sqrt{4 - 3\varepsilon^2} \nabla_{Y_3} Y_2 - S(\nabla_{Y_3} Y_2) \\
&= -\sqrt{4 - 3\varepsilon^2} \varepsilon Y_1 + \frac{\varepsilon^2}{\sqrt{4 - 3\varepsilon^2}} \varepsilon Y_1 \\
&= \left(-\sqrt{4 - 3\varepsilon^2} + \frac{\varepsilon^2}{\sqrt{4 - 3\varepsilon^2}} \right) \varepsilon Y_1.
\end{aligned}
$$

Now for some positive results.

THEOREM 7. (Fundamental Theorem of Hypersurface Theory) *Suppose we have a Riemannian manifold (M, g) and a symmetric $(1,1)$-tensor S on M that satisfies both the Gauss and the Codazzi-Mainardi Equations on M. Then for every $x \in M$, we can find an isometric embedding $F : (U, g) \hookrightarrow (\mathbb{R}^{n+1}, \mathrm{can})$ on some neighborhood $U \ni x$ with the property that S becomes the shape operator for this embedding.*

PROOF. We shall give an outline of the proof. Our first claim is that we can find a flat metric on $(-\varepsilon, \varepsilon) \times U$, where $U \subset M$ is relatively compact and ε is smaller than $\left| \lambda_i^{-1} \right|$ for any eigenvalue λ_i of S on U. It then follows from material in chapter 5 that any flat metric is locally isometric to a subset of $(\mathbb{R}^{n+1}, \mathrm{can})$. This will then finish the proof.

To construct the metric $\bar{g}$ on $(-\varepsilon, \varepsilon) \times U$ let us assume that it is of the type $\bar{g} = dr^2 + g_r$. If we select coordinates $(x^1, \ldots, x^n)$ on U, then the coordinate fields ∂_i are Jacobi fields. If we evaluate the metric and Hessian of r on these fields the fundamental equations say

$$
\begin{aligned}
\partial_r g_r (\partial_i, \partial_j) &= 2\mathrm{Hess}\, r(\partial_i, \partial_j), \\
\bar{g}_0 &= g.
\end{aligned}
$$

$$
\begin{aligned}
\partial_r (\mathrm{Hess}\, r(\partial_i, \partial_j)) - \mathrm{Hess}^2 r(\partial_i, \partial_j) &= -R(\partial_i, \partial_r, \partial_r, \partial_j) = 0, \\
\mathrm{Hess}\, r(\partial_i, \partial_j)|_{r=0} &= g(S(\partial_i), \partial_j).
\end{aligned}
$$

The latter equation completely determines the Hessian and the former then determines the metric.

We now need to prove that this metric is flat. By construction the radial curvatures vanish. So we need to show that the tangential and mixed curvature equations when evaluated on the chosen coordinates reduce to

$$R^r \left(\partial_i, \partial_j, \partial_k, \partial_l \right) = \operatorname{Hess}r \left(\partial_j, \partial_k \right) \operatorname{Hess}r \left(\partial_i, \partial_l \right) - \operatorname{Hess}r \left(\partial_i, \partial_k \right) \operatorname{Hess}r \left(\partial_j, \partial_l \right),$$

$$\left(\nabla_{\partial_i} \operatorname{Hess}r \right) \left(\partial_j, \partial_k \right) = \left(\nabla_{\partial_j} \operatorname{Hess}r \right) \left(\partial_i, \partial_k \right)$$

where R^r is the intrinsic curvature of g_r on $\{r\} \times U$. At $r = 0$ this is certainly true, since we assumed that S was a solution to these equations. Both the metric and S are given to us as solutions to the fundamental equations. A direct but fairly involved calculation will then show that equality holds for all r. □

We have already seen that positively curved manifolds of dimension $n > 2$ cannot necessarily be represented as hypersurfaces. When $n = 2$, the situation is drastically different.

THEOREM 8. *If (M, g) is 2-dimensional Riemannian manifold with positive curvature, then one can locally isometrically embed (M, g) into $\mathbb{R}^3$. Moreover if M is closed and simply connected then a global embedding exists.*

The proof is beyond what we can cover here, but the previous theorem gives us an idea. Namely, one could simply try to find an appropriate shape operator. This would at least establish the local result. The global result is known as Weyl's problem and was established by Pogorelov and subsequently by Nirenberg.

3. The Gauss-Bonnet Theorem

To finish this chapter we give a description of the global Gauss-Bonnet Theorem and its generalizations. It was shown above that when a hypersurface has positive curvature, then the shape operator is determined by intrinsic data . It turns out that the determinant of the shape operator is always intrinsic. This determinant is also called the *Gauss curvature*.

LEMMA 5. *Let $(M, g) \hookrightarrow (\mathbb{R}^{n+1}, \operatorname{can})$ be an isometric immersion. If n is even, then $\det S$ is intrinsic, and if n is odd, then $|\det S|$ is intrinsic.*

PROOF. Use an eigenbasis for S, $S(e_i) = \lambda_i e_i$; then of course $\det S = \lambda_1 \cdots \lambda_n$. In case $n = 2$ we therefore have $\det S = \sec$. Thus, $\det S$ is intrinsic. In higher dimensions the curvature operator is diagonalized by $e_i \wedge e_j$ with eigenvalues $\lambda_i \lambda_j$. Thus,

$$\det \mathfrak{R} = \prod_{i<j} \lambda_i \lambda_j$$
$$= \left(\lambda_1 \cdots \lambda_n \right)^{n-1}$$
$$= \left(\det S \right)^{n-1}.$$

This clearly proves the lemma. □

The importance of this lemma lies in the fact that $\det S$ is the Jacobian determinant of the Gauss map $G : M \to S^n(1) \subset \mathbb{R}^{n+1}$. When M is a closed manifold we therefore have

$$\deg G = \frac{1}{\operatorname{vol} S^n} \int_M \det S \cdot d\mathrm{vol}$$
$$= \frac{1}{\operatorname{vol} S^n} \int_M \sqrt[n-1]{\det \mathfrak{R}} \cdot d\mathrm{vol}.$$

The degree of the Gauss map is therefore also intrinsic when n is even. This is perhaps less surprising, as W. Dyck, and in greater generality H. Hopf, have shown that closed even-dimensional hypersurfaces have the property that $\deg G$ is related to the Euler characteristic by the formula

$$\deg G = \frac{1}{2}\chi\left(M\right).$$

For an even-dimensional hypersurface we have therefore arrived at the important formula

$$\chi\left(M\right) = \frac{2}{\operatorname{vol}S^n}\int_M \sqrt[n]{\det\mathfrak{R}}\cdot d\mathrm{vol}.$$

As both sides of the formula are intrinsic quantities one might expect this formula to hold for all orientable even-dimensional closed Riemannian manifolds. When $n = 2$, this is the Gauss-Bonnet formula:

$$\chi\left(M\right) = \frac{1}{2\pi}\int_M \sec\cdot d\mathrm{vol}.$$

For higher dimensions, however, we run into trouble. First, observe that the above formula does not give the right answer for manifolds that are not hypersurfaces. A counterexample is $\mathbb{C}P^2$, which has two zero eigenvalues for the curvature operator, and Euler characteristic 3. Thus, a more complicated integrand is necessary. The correct expression is actually a generalized determinant of the curvature operator called the Pfaffian determinant. It is easiest to write it down in an oriented orthonormal frame $E_1,\ldots,E_n$ using the curvature forms defined by

$$R\left(\cdot,\cdot\right)E_i = \Omega_i^j E_j.$$

If we assume that the dimension is $n = 2m$, then the formula looks like this:

$$\begin{aligned}\chi\left(M\right) &= \frac{2}{\operatorname{vol}S^n}\int_M K \\ &= \frac{2\left(2m-1\right)!}{2^{2m}\pi^m\left(m-1\right)!}\int_M K,\end{aligned}$$

where K is defined as

$$K = \frac{1}{n!}\sum \varepsilon^{i_1\cdots i_n}\cdot\Omega_{i_2}^{i_1}\wedge\Omega_{i_4}^{i_3}\wedge\cdots\wedge\Omega_{i_n}^{i_{n-1}},$$

$$\varepsilon^{i_1\cdots i_n} = \text{sign of the permutation }\left(i_1\cdots i_n\right).$$

Allendoerfer and Fenchel independently of each other established this generalized theorem for manifolds that are isometrically embedded in some Euclidean space, but not necessarily of codimension one. Allendoerfer and Weil then established the general case, using some interesting tricks about local isometric embeddability (see [1]). Finally, Chern found a completely intrinsic proof, which makes no mention of isometric embeddings. The theorem is now called the Chern-Gauss-Bonnet Theorem despite the fact that Allendoerfer and Weil were the first to prove it in complete generality in higher dimensions.

We shall give a brief account of how the Gauss-Bonnet theorem can be established for abstract surfaces $\left(M^2,g\right)$. The first proof uses vector fields and is in spirit close to Chern's proof. The second is based on a triangulation of the surface. Both are global in the sense that they do not rely on the classical Gauss-Bonnet formula for the integral of the curvature over a simply connected surface with piecewise smooth boundary.

VECTOR FIELD PROOF. First suppose that M is the torus, and pick your favorite nonzero vector field X. Using the metric, normalize it to have length 1, and then select another field such that we get an orientable orthonormal frame E_1, E_2. Let ω^1, ω^2 be the dual coframe and compute the connection form and curvature form as described in the exercises to chapter 2:

$$
\begin{aligned}
d\omega^1 &= \omega^2 \wedge \omega^1_2, \\
d\omega^1_2 &= \Omega^1_2 = \sec \cdot d\text{vol}.
\end{aligned}
$$

Then we have

$$
\begin{aligned}
\int_M \sec \cdot d\text{vol} &= \int_M \Omega^1_2 \\
&= \int_M d\omega^1_2 \\
&= \int_{\partial M} \omega^1_2 = 0.
\end{aligned}
$$

On other surfaces we can choose a vector field X with isolated zeros at $p_1, \ldots, p_k \in M$. Then we choose the frame E_1, E_2 as above on $M - \{p_1, \ldots, p_k\}$. On a neighborhood U_i around each p_i introduce normal coordinates such that

$$
g = g_{\alpha\beta} = \delta_{\alpha\beta} + O\left(r^2\right).
$$

Here, r is the Euclidean distance from p_i. We can then consider the manifold with boundary $M_\varepsilon = M - \bigcup_{i=1}^k B(p_i, \varepsilon)$, where $B(p_i, \varepsilon)$ is the Euclidean ball of radius ε around p_i. As before we still have

$$
\begin{aligned}
\int_{M_\varepsilon} \sec \cdot d\text{vol} &= \int_{\partial M_\varepsilon} \omega^1_2 \\
&= \sum_{i=1}^k \int_{\partial B(p_i, \varepsilon)} \omega^1_2.
\end{aligned}
$$

Let us now analyze each of the integrals $\int_{\partial B(p_i, \varepsilon)} \omega^1_2$ on U_i. On U_i we could instead find an orientable orthonormal frame $F_1 = \frac{X}{|X|}, F_2$, but this time with respect to the Euclidean metric on U_i. If $\tilde{\omega}^1_2$ is the connection form for this frame, we can construct the integral

$$
\int_{\partial B(p_i, \varepsilon)} \tilde{\omega}^1_2.
$$

Using that the metric is Euclidean up to first order, we obtain that

$$
\omega^1_2 - \tilde{\omega}^1_2 = O(r).
$$

In particular, we must have

$$
\lim_{\varepsilon \to 0} \int_{\partial B(p_i, \varepsilon)} \tilde{\omega}^1_2 = \lim_{\varepsilon \to 0} \int_{\partial B(p_i, \varepsilon)} \omega^1_2.
$$

This proves that the integral $\int_M \sec \cdot d\text{vol}$ does not depend on the metric.

Let us now relate the term $\lim_{\varepsilon \to 0} \int_{\partial B(p_i, \varepsilon)} \tilde{\omega}^1_2$ to the vector field X. We can suppose that we are on a neighborhood $U \subset \mathbb{R}^2$ around the origin and that we have

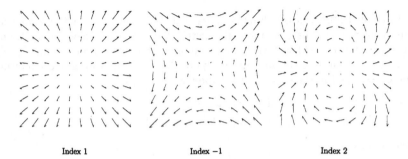

Index 1   Index −1   Index 2

Figure 4.3

a vector field X that vanishes only at the origin. If we normalize X to have unit length $E = X/|X|$, then for each $\varepsilon > 0$ we get a map

$$\partial B(0, \varepsilon) \quad \rightarrow \quad \partial B(0, \varepsilon),$$

$$x \quad \rightarrow \quad \varepsilon \cdot E(x).$$

The degree (see the Appendix) of this map is easily seen to be independent of ε. This degree is known as the *index* of the vector field at the origin and is denoted by $\mathrm{ind}_0 X$. The degree of this map can now be computed as

$$\frac{1}{\ell(\partial B(0, \varepsilon))} \int_{\partial B(0, \varepsilon)} D(\varepsilon \cdot E(x)) = \frac{1}{2\pi} \int_{\partial B(0, \varepsilon)} D(E(x)).$$

One can now easily check that

$$\int_{\partial B(0, \varepsilon)} D(E(x)) = \lim_{\varepsilon \to 0} \int_{\partial B(0, \varepsilon)} \tilde{\omega}_2^1.$$

All in all, we have therefore shown that

$$\frac{1}{2\pi} \int_M \sec \cdot d\mathrm{vol} = \sum_{i=1}^{k} \mathrm{ind}_{p_i}(X).$$

The left-hand side is therefore independent of the metric, while the right-hand side must now be independent of the chosen vector field. Knowing that the right hand side is independent of the vector field, one can easily compute it as the Euler characteristic by choosing a particular vector field on each surface (see also the next proof). Figure 4.3 shows a few pictures of vector fields in the plane. $\square$

TRIANGULATION PROOF. We can also give a proof that ties the curvature integral in with the more standard formula for the Euler characteristic.

A *polygon* in M is a region P which is diffeomorphic to a convex polygon in $\mathbb{R}^2$ via a coordinate chart. The boundary of a polygon is therefore a piecewise smooth curve. Each smooth part of the boundary is called an *edge* or *side*. Using the orientation of M we can assume that ∂P is parametrized counterclockwise as a curve $\gamma : S^1 \to M$, where S^1 is partitioned into intervals $[t_i, t_{i+1}]$ on which γ is smooth. The *exterior angle* at $\gamma(t_i)$ is defined as the angle

$$\sphericalangle \left(\dot{\gamma}(t_i^+), \dot{\gamma}(t_i^-) \right) \in [-\pi, \pi]$$

where $\dot{\gamma}(t_i^+)$ is the right hand derivative and $\dot{\gamma}(t_i^-)$ the left hand derivative. The angle is ≥ 0 when $\dot{\gamma}(t_i^+)$ lies to the left of the line through $\dot{\gamma}(t_i^-)$ and otherwise

≤ 0. Note that this definition also uses the orientation of M. The *interior angle* is defined as

$$\pi - \sphericalangle\left(\dot{\gamma}\left(t_i^+\right), \dot{\gamma}\left(t_i^-\right)\right) \in [0, 2\pi].$$

Now divide M into polygons (or triangles) such that each side of a polygon is the side of precisely one other polygon. In this way each edge in this subdivision is the side of precisely two polygons. Note that all of this can be done without referring to the metric structure of the manifold. Parametrizing the boundary curves of the polygons to run counter clockwise shows that when two polygons have a side in common these sides are parametrized in opposite directions. Finally let F denote the number of polygons in the subdivision (or triangulation), E the number of edges (each edge is the common side of two polygons but only counted once), and V the number of vertices (each vertex is met by any number of polygons, but only counted once). The classical Euler characteristic is then given by

$$\chi = F - E + V.$$

If P_i are the various polygons in the subdivision, then the Gauss-Bonnet integral can be decomposed according to the polygonal subdivision of M

$$\int_M \sec \cdot d\text{vol} = \sum \int_{P_i} \sec \cdot d\text{vol}.$$

In each polygon P_i we now select a positively oriented orthonormal frame E_1, E_2 and coframe ω^1, ω^2. If we use that

$$\begin{aligned} \sec \cdot d\text{vol} &= K\omega^1 \wedge \omega^2 \\ &= d\omega_2^1, \\ d\omega^1 &= \omega^2 \wedge \omega_2^1 \end{aligned}$$

then Green's theorem implies that

$$\sum \int_{P_i} \sec \cdot d\text{vol} = \sum \int_{\partial P_i} \omega_2^1.$$

Each boundary term ∂P_i is further decomposed into oriented edges where each edge is met by one other polygon. If for a given edge γ we denote the frame from the other polygon by $\bar{E}_1, \bar{E}_2$ and $\bar{\omega}^1, \bar{\omega}^2$ then the the integral can be rewritten as

$$\sum \int_{\partial P_i} \omega_2^1 = \sum_\gamma \int_\gamma \left(\omega_2^1 - \bar{\omega}_2^1\right)$$

where there are E terms in the sum on the right hand side. The difference $\omega_2^1 - \bar{\omega}_2^1$ can be understood completely in terms of the angles between $\dot{\gamma}$ and $E_1, \bar{E}_1$.

First define

$$\vartheta = \sphericalangle\left(E_1, \bar{E}_1\right).$$

While this angle is only defined modulo 2π its differential is well-defined. Moreover, we claim that

$$\omega_2^1 - \bar{\omega}_2^1 = -d\vartheta.$$

To see this observe that as E_1, E_2 and $\bar{E}_1, \bar{E}_2$ define the same orientation we must have

$$\begin{aligned} E_1 &= \cos\left(\vartheta\right)\bar{E}_1 + \sin\left(\vartheta\right)\bar{E}_2, \\ E_2 &= -\sin\left(\vartheta\right)\bar{E}_1 + \cos\left(\vartheta\right)\bar{E}_2. \end{aligned}$$

This shows that

$$
\begin{aligned}
\omega^1 &= \cos(\vartheta)\,\bar{\omega}^1 + \sin(\vartheta)\,\bar{\omega}^2, \\
\omega^2 &= -\sin(\vartheta)\,\bar{\omega}^1 + \cos(\vartheta)\,\bar{\omega}^2, \\
\bar{\omega}^1 &= \cos(\vartheta)\,\omega^1 - \sin(\vartheta)\,\omega^2, \\
\bar{\omega}^2 &= \sin(\vartheta)\,\omega^1 + \cos(\vartheta)\,\omega^2
\end{aligned}
$$

Since

$$
\begin{aligned}
\cos^2(\vartheta) + \sin^2(\vartheta) &= 1, \\
\cos(\vartheta)\,d(\cos(\vartheta)) + \sin(\vartheta)\,d(\sin(\vartheta)) &= 0
\end{aligned}
$$

we get

$$
\begin{aligned}
\omega^2 \wedge \omega_2^1 &= d\omega^1 \\
&= d\left(\cos(\vartheta)\,\bar{\omega}^1 + \sin(\vartheta)\,\bar{\omega}^2\right) \\
&= d(\cos(\vartheta)) \wedge \bar{\omega}^1 + \cos(\vartheta)\,d\bar{\omega}^1 \\
&\quad + d(\sin(\vartheta)) \wedge \bar{\omega}^2 + \sin(\vartheta)\,d\bar{\omega}^2 \\
&= d(\cos(\vartheta)) \wedge \left(\cos(\vartheta)\,\omega^1 - \sin(\vartheta)\,\omega^2\right) + \cos(\vartheta)\,\bar{\omega}^2 \wedge \bar{\omega}_2^1 \\
&\quad + d(\sin(\vartheta)) \wedge \left(\sin(\vartheta)\,\omega^1 + \cos(\vartheta)\,\omega^2\right) - \sin(\vartheta)\,\bar{\omega}^1 \wedge \bar{\omega}_2^1 \\
&= \left(\cos(\vartheta)\,d(\sin(\vartheta)) - \sin(\vartheta)\,d(\cos(\vartheta))\right) \wedge \omega^2 \\
&\quad + \cos(\vartheta)\left(\sin(\vartheta)\,\omega^1 + \cos(\vartheta)\,\omega^2\right) \wedge \bar{\omega}_2^1 \\
&\quad - \sin(\vartheta)\left(\cos(\vartheta)\,\omega^1 - \sin(\vartheta)\,\omega^2\right) \wedge \bar{\omega}_2^1 \\
&= \omega^2 \wedge \left(-\cos(\vartheta)\,d(\sin(\vartheta)) + \sin(\vartheta)\,d(\cos(\vartheta)) + \bar{\omega}_2^1\right).
\end{aligned}
$$

Thus

$$
\begin{aligned}
\omega_2^1 - \bar{\omega}_2^1 &= -\cos(\vartheta)\,d(\sin(\vartheta)) + \sin(\vartheta)\,d(\cos(\vartheta)) \\
&= -d\vartheta.
\end{aligned}
$$

Now parametrize $\gamma : [0,1] \to M$ so that

$$
\int_\gamma \left(\omega_2^1 - \bar{\omega}_2^1\right) = \int_0^1 \left(\omega_2^1(\dot{\gamma}(t)) - \bar{\omega}_2^1(\dot{\gamma}(t))\right)\,dt,
$$

and define

$$
\begin{aligned}
\theta &= \sphericalangle\left(\dot{\gamma}(t), E_1|_{\gamma(t)}\right), \\
\bar{\theta} &= \sphericalangle\left(\dot{\gamma}(t), \bar{E}_1|_{\gamma(t)}\right).
\end{aligned}
$$

These angles are only defined modulo 2π, but their derivatives are well-defined. We can check the relationship between these angles and ϑ by doing the calculation

$$
\begin{aligned}
\dot{\gamma} &= \cos(\theta)\,E_1 + \sin(\theta)\,E_2 \\
&= \cos(\theta)\left(\cos(\vartheta)\,\bar{E}_1 + \sin(\vartheta)\,\bar{E}_2\right) \\
&\quad + \sin(\theta)\left(-\sin(\vartheta)\,\bar{E}_1 + \cos(\vartheta)\,\bar{E}_2\right) \\
&= \cos(\theta + \vartheta)\,\bar{E}_1 + \sin(\theta + \vartheta)\,\bar{E}_2 \\
&= \cos(\bar{\theta})\,\bar{E}_1 + \sin(\bar{\theta})\,\bar{E}_2.
\end{aligned}
$$

Thus

$$
\bar{\theta} = \theta + \vartheta.
$$

This shows that

$$
\begin{aligned}
\int_{\gamma} \left(\omega_2^1 - \bar{\omega}_2^1 \right) &= \int_0^1 \left(\omega_2^1 \left(\dot{\gamma}(t) \right) - \bar{\omega}_2^1 \left(\dot{\gamma}(t) \right) \right) dt \\
&= -\int_0^1 \left(d\vartheta \left(\dot{\gamma}(t) \right) \right) dt \\
&= \int_0^1 \frac{d\theta}{dt} dt - \int_0^1 \frac{d\bar{\theta}}{dt} dt.
\end{aligned}
$$

For a given polygon we let $\theta_{ij} = \sphericalangle \left(\dot{\gamma}(t^+), \dot{\gamma}(t^-) \right)$ denote the exterior angles, thus the total angle change for the boundary curve is 2π or in other words

$$
\begin{aligned}
2\pi &= \int_{\partial P_i} \frac{d\theta}{dt} + \sum_j \theta_{ij} \\
&= \int_{\partial P_i} \frac{d\theta}{dt} + \sum_j \left(\pi - \theta_{ij}' \right)
\end{aligned}
$$

where θ_{ij}' is the interior angle.

We can then conclude that

$$
\begin{aligned}
\int_M \sec \cdot d\mathrm{vol} &= \sum_i \int_{\partial P_i} \frac{d\theta}{dt} \\
&= \sum_i \left(2\pi - \sum_j \left(\pi - \theta_{ij}' \right) \right) \\
&= F - \sum_i \sum_j \left(\pi - \theta_{ij}' \right) \\
&= F - \sum_i \sum_j \pi + \sum_i \sum_j \theta_{ij}'.
\end{aligned}
$$

Here the last sum is just the sum of all possible interior angles. As all the interior angles that meet a fixed vertex add up to 2π this sum is simply $2\pi V$. The middle sum is a sum over all sides to each polygon. As each side is the side for two polygons we see that each edge gets counted twice. All in all we have derived the classical Gauss-Bonnet formula

$$
\int_M \sec \cdot d\mathrm{vol} = 2\pi \left(F - E + V \right).
$$

Here the left hand side is independent of the choice of polygonal subdivision and the right hand side is independent of the metric. Combined with the above formula relating the integral to the index sum of a vector field we have also shown that this index sum has the desired relationship to the Euler characteristic. □

4. Further Study

All of the results mentioned in this chapter and much more can be found in Spivak's [**87**, volume 5]. In fact we recommend all of his volumes as a good and thorough introduction to differential geometry. Spivak is also quite careful and complete with references to all the work mentioned here. The only fault Spivak's book has in reference to the generalized Gauss-Bonnet theorem, is that he claims that Allendoerfer and Weil only established this formula for analytic metrics. For a very nice discussion of the Gauss-Bonnet theorem for surfaces see also [**18**].

We can also recommend Stoker's book [**88**]. This book goes from curves to surfaces and ends up with a discussion of general relativity. For the reader who likes old-fashioned well-written books this is a must.

One defect in our treatment here is that we haven't developed submanifold theory in general. This is done in [**87**].

5. Exercises

(1) Consider the hypersurface given by the graph $x^{n+1} = f(x^n)$, where $f : \mathbb{R} \to \mathbb{R}$ is smooth. Show that the shape operator doesn't necessarily vanish but that the hypersurface is isometric to $\mathbb{R}^n$.

(2) If X is a Killing field on an abstract surface (M^2, g) show that the index of any isolated zero is 1.

(3) Assume that we have a Riemannian immersion of an n-manifold into $\mathbb{R}^{n+1}$. If $n \geq 3$, then show that it can't have negative curvature. If $n = 2$ give an example where it does have negative curvature.

(4) Let (M, g) be a closed Riemannian n-manifold, and suppose that there is a Riemannian embedding into $\mathbb{R}^{n+1}$. Show that there must be a point $p \in M$ where the curvature operator $\mathfrak{R} : \Lambda^2 T_p M \to \Lambda^2 T_p M$ is positive. (Hint: Consider $f(x) = |x|^2$ and restrict it to M, then check what happens at a maximum.)

(5) Suppose (M, g) is immersed as a hypersurface in $\mathbb{R}^{n+1}$, with shape operator S.

  (a) Using the Codazzi-Mainardi equations, show that
  $$\mathrm{div} S = d(\mathrm{tr} S).$$

  (b) Show that if $S = f(x) \cdot I$ for some function f, then f must be a constant and the hypersurface must have constant curvature.

  (c) Show that $S = \lambda \cdot \mathrm{Ric}$ iff the metric has constant curvature.

(6) Let g be a metric on S^2 with curvature ≤ 1. Use the Gauss-Bonnet formula to show that $\mathrm{vol}\,(S^2, g) \geq \mathrm{vol} S^2(1) = 4\pi$.

  Show that such a result cannot hold on S^3 by considering the Berger metrics.

(7) Assume that we have an orientable Riemannian manifold with nonzero Euler characteristic and $|\mathfrak{R}| \leq 1$. Find a lower bound for $\mathrm{vol}\,(M, g)$. The one sided curvature bound that we used on surfaces does not suffice in higher dimensions, as one-sided curvature bounds do not necessarily imply one sided bounds on the Chern-Gauss-Bonnet integrand.

(8) Show that in even dimensions, orientable manifolds with positive (or nonnegative) curvature operator have positive (nonnegative) Euler characteristic. Conclude that if in addition, such manifolds have bounded curvature operator, then they have volume bounded from below. What happens when the curvature operator is nonpositive or negative?

(9) In dimension 4 show, using the exercises from chapter 3, that

$$\frac{1}{8\pi^2} \int_M \left(|R|^2 - \left| \mathrm{Ric} - \frac{\mathrm{scal}}{4} g \right|^2 \right) = \frac{1}{8\pi^2} \int_M \mathrm{tr}\,(A^2 - 2BB^* + C^2).$$

It was shown by Allendoerfer and Weil that in dimension 4

$$\chi(M) = \frac{1}{8\pi^2} \int_M \left(|R|^2 - \left| \text{Ric} - \frac{\text{scal}}{4} g \right|^2 \right).$$

You can try to prove this using the above definition of K. If the metric is Einstein, show that

$$\begin{aligned}
\chi(M) &= \frac{1}{8\pi^2} \int_M \text{tr} \left(A^2 - 2BB^* + C^2 \right) \\
&= \frac{1}{8\pi^2} \int_M \left(|W^+|^2 + |W^-|^2 + \frac{\text{scal}^2}{24} \right).
\end{aligned}$$

CHAPTER 5

Geodesics and Distance

We are now ready to introduce the important concept of a geodesic. This will help us define and understand Riemannian manifolds as metric spaces. One is led quickly to two types of "completeness". The first is of standard metric completeness, and the other is what we call geodesic completeness, namely, when all geodesics exist for all time. We shall prove the Hopf-Rinow Theorem, which asserts that these types of completeness for a Riemannian manifold are equivalent. Using the metric structure we can define metric distance functions. We shall study when these distance functions are smooth and show the existence of the smooth distance functions we worked with earlier. In the last section we give some metric characterizations of Riemannian isometries and submersions. We also classify complete simply connected manifolds of constant curvature; showing that they are the ones we have already constructed in chapters 1 and 3.

The idea of thinking of a Riemannian manifold as a metric space must be old, but it wasn't until the early 1920s that first Cartan and then later Hopf and Rinow began to understand the relationship between extendability of geodesics and completeness of the metric. Nonetheless, both Gauss and Riemann had a pretty firm grasp on local geometry, as is evidenced by their contributions: Gauss worked with geodesic polar coordinates and also isothermal coordinates, Riemann was able to give a local characterization of Euclidean space as the only manifold whose curvature tensor vanishes. Surprisingly, it wasn't until Klingenberg's work in the 1950s that one got a thorough understanding of the maximal domain on which one has geodesic polar coordinates in side complete manifolds. This work led to the introduction of the two terms *injectivity radius* and *conjugate radius*. Many of our later results will require a detailed analysis of these concepts. The metric characterization of Riemannian isometries wasn't realized until the late 1930s with the work of Myers and Steenrod. Even more surprising is Berestovskii's much more recent metric characterization of Riemannian submersions.

Another important topic that involves geodesics is the variation of arclength and energy. In this chapter we only develop the first variation formula. This is used to show that curves that minimize length must be geodesics if they are parametrized correctly.

We are also finally getting to results where there is going to be a significant difference between the Riemannian setting and the semi-Riemannian setting. Mixed partials and geodesics easily generalize. However, as there is no norm of vectors in the semi-Riemannian setting we do not have arclength or distances. Nevertheless, the energy functional does make sense so we can still obtain a variational characterization of geodesic as critical points for the energy functional.

1. Mixed Partials

So far we have only worked out the calculus for functions on a Riemannian manifold and have seen that defining the gradient and Hessian requires that we use the metric structure. We are now going to study maps into Riemannian manifolds and how to define meaningful derivatives for such maps. The simplest example is to consider a curve $\gamma : I \to M$ on some interval $I \subset \mathbb{R}$. We know how to define the derivative $\dot{\gamma}$, but not how to define the acceleration in such a way that it also gives us a tangent vector to M. A similar but slightly more general problem is that of defining mixed partial derivatives

$$\frac{\partial^2 \gamma}{\partial t^i \partial t^j}$$

for maps γ with several real variables. As we shall see, covariant differentiation plays a crucial role in the definition of these concepts. In this section we only develop a method that covers second partials. In the next chapter we shall explain how to calculate higher order partials as well. This involves a slightly different approach that is not needed for the developments in this chapter.

Let $\gamma : \Omega \to M$, where $\Omega \subset \mathbb{R}^m$. As we usually reserve x^i for coordinates on M we shall use t^i or s, t, u as coordinates on Ω. The first partials

$$\frac{\partial \gamma}{\partial t^i}$$

are simply defined as the velocity field of $t^i \to \gamma \left(t^1, ..., t^i, ..., t^m\right)$ where the remaining coordinates are fixed. We wish to define the second partials so that they also lie TM as opposed to TTM. In addition we shall also require the following two natural properties:

(1) $\frac{\partial^2 \gamma}{\partial t^i \partial t^j} = \frac{\partial^2 \gamma}{\partial t^j \partial t^i}$,

(2) $\frac{\partial}{\partial t^k} g\left(\frac{\partial \gamma}{\partial t^i}, \frac{\partial \gamma}{\partial t^j}\right) = g\left(\frac{\partial^2 \gamma}{\partial t^k \partial t^i}, \frac{\partial \gamma}{\partial t^j}\right) + g\left(\frac{\partial \gamma}{\partial t^i}, \frac{\partial^2 \gamma}{\partial t^k \partial t^j}\right).$

The first is simply the equality of mixed partials and is similar to assuming that the connection is torsion free. The second is a Leibniz or product rule that is similar to assuming that the connection is metric. Like the Fundamental Theorem of Riemannian Geometry, were we saw that the key properties of the connection in fact also characterized the connection, we can show that these two rules also characterize how we define second partials. More precisely, if we have a way of defining second partials such that these two properties hold, then we claim that there is a Koszul type formula:

$$2g\left(\frac{\partial^2 \gamma}{\partial t^i \partial t^j}, \frac{\partial \gamma}{\partial t^k}\right) = \frac{\partial}{\partial t^i} g\left(\frac{\partial \gamma}{\partial t^j}, \frac{\partial \gamma}{\partial t^k}\right) + \frac{\partial}{\partial t^j} g\left(\frac{\partial \gamma}{\partial t^k}, \frac{\partial \gamma}{\partial t^i}\right) - \frac{\partial}{\partial t^k} g\left(\frac{\partial \gamma}{\partial t^i}, \frac{\partial \gamma}{\partial t^j}\right).$$

This formula is established in the proof of the next lemma.

LEMMA 6. (Uniqueness of mixed partials) *There is at most one way of defining mixed partials so that (1) and (2) hold.*

PROOF. First we show that the Koszul type formula holds if we have a way of defining mixed partials such that (1) and (2) hold:

$$
\frac{\partial}{\partial t^i} g\left(\frac{\partial \gamma}{\partial t^j}, \frac{\partial \gamma}{\partial t^k}\right) + \frac{\partial}{\partial t^j} g\left(\frac{\partial \gamma}{\partial t^k}, \frac{\partial \gamma}{\partial t^i}\right) - \frac{\partial}{\partial t^k} g\left(\frac{\partial \gamma}{\partial t^i}, \frac{\partial \gamma}{\partial t^j}\right)
$$

$$
= g\left(\frac{\partial^2 \gamma}{\partial t^i \partial t^j}, \frac{\partial \gamma}{\partial t^k}\right) + g\left(\frac{\partial \gamma}{\partial t^j}, \frac{\partial^2 \gamma}{\partial t^i \partial t^k}\right)
$$

$$
+ g\left(\frac{\partial^2 \gamma}{\partial t^j \partial t^k}, \frac{\partial \gamma}{\partial t^i}\right) + g\left(\frac{\partial \gamma}{\partial t^k}, \frac{\partial^2 \gamma}{\partial t^j \partial t^i}\right)
$$

$$
- g\left(\frac{\partial^2 \gamma}{\partial t^k \partial t^i}, \frac{\partial \gamma}{\partial t^j}\right) - g\left(\frac{\partial \gamma}{\partial t^i}, \frac{\partial^2 \gamma}{\partial t^k \partial t^j}\right)
$$

$$
= g\left(\frac{\partial^2 \gamma}{\partial t^i \partial t^j}, \frac{\partial \gamma}{\partial t^k}\right) + g\left(\frac{\partial \gamma}{\partial t^k}, \frac{\partial^2 \gamma}{\partial t^j \partial t^i}\right)
$$

$$
+ g\left(\frac{\partial \gamma}{\partial t^j}, \frac{\partial^2 \gamma}{\partial t^i \partial t^k}\right) - g\left(\frac{\partial^2 \gamma}{\partial t^k \partial t^i}, \frac{\partial \gamma}{\partial t^j}\right)
$$

$$
+ g\left(\frac{\partial^2 \gamma}{\partial t^j \partial t^k}, \frac{\partial \gamma}{\partial t^i}\right) - g\left(\frac{\partial \gamma}{\partial t^i}, \frac{\partial^2 \gamma}{\partial t^k \partial t^j}\right)
$$

$$
= 2g\left(\frac{\partial^2 \gamma}{\partial t^i \partial t^j}, \frac{\partial \gamma}{\partial t^k}\right).
$$

Next we observe that if we have a map $\gamma : \Omega \to M$, then we can always add an extra parameter t^{n+1} to get a map $\bar{\gamma} : \Omega \times (-\varepsilon, \varepsilon) \to M$ with the property that

$$
\frac{\partial \bar{\gamma}}{\partial t^{n+1}}\Big|_p = v \in T_p M,
$$

where $v \in T_p M$ is any vector and p is any point in the image of γ. Using $k = n+1$ in the Koszul type formula at p, then shows that $\frac{\partial^2 \gamma}{\partial t^i \partial t^j}$ is uniquely defined as our extension is independent of how mixed partials are defined. $\square$

We can now give a local coordinate definition of mixed partials. As long as the definition gives us properties (1) and (2), the above lemma shows that we have a coordinate independent definition.

Note also that if two different maps $\gamma_1, \gamma_2 : \Omega \to M$ agree on a neighborhood of a point in the domain, then the right hand side of the Koszul type formula will give the same answer for these two maps. Thus there is no loss of generality in assuming that the image of γ lies in a coordinate system.

THEOREM 9. (Existence of mixed partials) *It is possible to define mixed partials in a coordinate system so that (1) and (2) hold.*

PROOF. We assume that we have $\gamma : \Omega \to U \subset M$ where U is a coordinate neighborhood. Furthermore, assume that the parameters in use are called s and t. This avoids introducing more indices than necessary. Finally write $\gamma = (\gamma^1, ..., \gamma^n)$ using the coordinates. The velocity in the s direction is given by

$$
\frac{\partial \gamma}{\partial s} = \frac{\partial \gamma^i}{\partial s} \partial_i
$$

so we can make the suggestive calculation

$$\frac{\partial}{\partial t}\frac{\partial \gamma}{\partial s} = \frac{\partial}{\partial t}\left(\frac{\partial \gamma^i}{\partial s}\partial_i\right)$$

$$= \frac{\partial}{\partial t}\frac{\partial \gamma^i}{\partial s}\partial_i + \frac{\partial \gamma^i}{\partial s}\frac{\partial}{\partial t}\left(\partial_i\right).$$

To make sense of $\frac{\partial}{\partial t}\left(\partial_i\right)$ we define

$$\frac{\partial X}{\partial t}|_p = \nabla_{\dot\gamma(t)}X,$$

where $\gamma\left(t\right) = p$ and X is a vector field defined in a neighborhood of p. With that in mind we have

$$\frac{\partial}{\partial t}\frac{\partial \gamma}{\partial s} = \frac{\partial^2 \gamma^k}{\partial t\partial s}\partial_k + \frac{\partial \gamma^i}{\partial s}\nabla_{\frac{\partial \gamma}{\partial t}}\partial_i$$

$$= \frac{\partial^2 \gamma^k}{\partial t\partial s}\partial_k + \frac{\partial \gamma^i}{\partial s}\frac{\partial \gamma^j}{\partial t}\nabla_{\partial_j}\partial_i$$

$$= \frac{\partial^2 \gamma^k}{\partial t\partial s}\partial_k + \frac{\partial \gamma^i}{\partial s}\frac{\partial \gamma^j}{\partial t}\Gamma^k_{ji}\partial_k$$

Thus we define

$$\frac{\partial^2 \gamma}{\partial t\partial s} = \frac{\partial^2 \gamma^k}{\partial t\partial s}\partial_k + \frac{\partial \gamma^i}{\partial s}\frac{\partial \gamma^j}{\partial t}\Gamma^k_{ji}\partial_k$$

$$= \left(\frac{\partial^2 \gamma^k}{\partial t\partial s} + \frac{\partial \gamma^i}{\partial s}\frac{\partial \gamma^j}{\partial t}\Gamma^k_{ji}\right)\partial_k$$

Since $\frac{\partial^2 \gamma^l}{\partial t\partial s}$ is symmetric in s and t by the usual theorem on equality of mixed partials and the Christoffel symbol Γ^k_{ji} is symmetric in i and j we see that (1) holds.

To check the metric property (2) we use that the Christoffel symbols satisfy the metric property

$$\partial_k g_{ij} = \Gamma_{ki,j} + \Gamma_{kj,i}.$$

With that in mind we calculate

$$\frac{\partial}{\partial t}g\left(\frac{\partial \gamma}{\partial s},\frac{\partial \gamma}{\partial u}\right)$$

$$= \frac{\partial}{\partial t}\left(g_{ij}\frac{\partial \gamma^i}{\partial s}\frac{\partial \gamma^j}{\partial u}\right)$$

$$= \frac{\partial g_{ij}}{\partial t}\frac{\partial \gamma^i}{\partial s}\frac{\partial \gamma^j}{\partial u} + g_{ij}\frac{\partial^2 \gamma^i}{\partial t\partial s}\frac{\partial \gamma^j}{\partial u} + g_{ij}\frac{\partial \gamma^i}{\partial s}\frac{\partial^2 \gamma^j}{\partial t\partial u}$$

$$= g_{ij}\left(\frac{\partial^2 \gamma^i}{\partial t\partial s} + \frac{\partial \gamma^k}{\partial s}\frac{\partial \gamma^l}{\partial t}\Gamma^i_{kl}\right)\frac{\partial \gamma^j}{\partial u} + g_{ij}\frac{\partial \gamma^i}{\partial s}\left(\frac{\partial^2 \gamma^j}{\partial t\partial u} + \frac{\partial \gamma^k}{\partial u}\frac{\partial \gamma^l}{\partial t}\Gamma^j_{kl}\right)$$

$$+\frac{\partial g_{ij}}{\partial t}\frac{\partial \gamma^i}{\partial s}\frac{\partial \gamma^j}{\partial u} - g_{ij}\frac{\partial \gamma^k}{\partial s}\frac{\partial \gamma^l}{\partial t}\frac{\partial \gamma^j}{\partial u}\Gamma^i_{kl} - g_{ij}\frac{\partial \gamma^i}{\partial s}\frac{\partial \gamma^k}{\partial u}\frac{\partial \gamma^l}{\partial t}\Gamma^j_{kl}$$

$$= g\left(\frac{\partial^2\gamma}{\partial t\partial s}, \frac{\partial \gamma}{\partial u}\right) + g\left(\frac{\partial \gamma}{\partial s}, \frac{\partial^2\gamma}{\partial t\partial u}\right)$$

$$+\frac{\partial g_{ij}}{\partial t}\frac{\partial \gamma^i}{\partial s}\frac{\partial \gamma^j}{\partial u} - \frac{\partial \gamma^k}{\partial s}\frac{\partial \gamma^l}{\partial t}\frac{\partial \gamma^j}{\partial u}\Gamma_{kl,j} - \frac{\partial \gamma^i}{\partial s}\frac{\partial \gamma^k}{\partial u}\frac{\partial \gamma^l}{\partial t}\Gamma^j_{kl,i}$$

$$= g\left(\frac{\partial^2\gamma}{\partial t\partial s}, \frac{\partial \gamma}{\partial u}\right) + g\left(\frac{\partial \gamma}{\partial s}, \frac{\partial^2\gamma}{\partial t\partial u}\right)$$

$$+\partial_k g_{ij}\frac{\partial \gamma^k}{\partial t}\frac{\partial \gamma^i}{\partial s}\frac{\partial \gamma^j}{\partial u} - \frac{\partial \gamma^i}{\partial s}\frac{\partial \gamma^k}{\partial t}\frac{\partial \gamma^j}{\partial u}\Gamma_{ki,j} - \frac{\partial \gamma^i}{\partial s}\frac{\partial \gamma^j}{\partial u}\frac{\partial \gamma^k}{\partial t}\Gamma_{kj,i}$$

$$= g\left(\frac{\partial^2\gamma}{\partial t\partial s}, \frac{\partial \gamma}{\partial u}\right) + g\left(\frac{\partial \gamma}{\partial s}, \frac{\partial^2\gamma}{\partial t\partial u}\right).$$

$\square$

In case $M \subset N$ it is often convenient to calculate the mixed partials in N first and then project them onto M. For each $p \in M$ we use the orthogonal projection $\mathrm{proj}_M : T_pN \to T_pM$. The next proposition shows that this is a valid way of calculating mixed partials.

PROPOSITION 16. (Mixed partials in submanifolds) *If* $\gamma : \Omega \to M \subset N$ *and* $\frac{\partial^2\gamma}{\partial t^i\partial t^j} \in T_pN$ *is the mixed partial in* N, *then*

$$\mathrm{proj}_M\left(\frac{\partial^2\gamma}{\partial t^i\partial t^j}\right) \in T_pM$$

is the mixed partial in M.

PROOF. Let $\bar{g}$ be the Riemannian metric in N and g its restriction to the submanifold M. We know that $\frac{\partial^2\gamma}{\partial t^j\partial t^i} \in TN$ satisfies

$$2\bar{g}\left(\frac{\partial^2\gamma}{\partial t^i\partial t^j}, \frac{\partial \gamma}{\partial t^k}\right) = \frac{\partial}{\partial t^i}\bar{g}\left(\frac{\partial \gamma}{\partial t^j}, \frac{\partial \gamma}{\partial t^k}\right) + \frac{\partial}{\partial t^j}\bar{g}\left(\frac{\partial \gamma}{\partial t^k}, \frac{\partial \gamma}{\partial t^i}\right)$$
$$-\frac{\partial}{\partial t^k}\bar{g}\left(\frac{\partial \gamma}{\partial t^i}, \frac{\partial \gamma}{\partial t^j}\right).$$

As $\frac{\partial \gamma}{\partial t^i}, \frac{\partial \gamma}{\partial t^j}, \frac{\partial \gamma}{\partial t^k} \in TM$ this shows that

$$2\bar{g}\left(\frac{\partial^2\gamma}{\partial t^i\partial t^j}, \frac{\partial \gamma}{\partial t^k}\right) = \frac{\partial}{\partial t^i}g\left(\frac{\partial \gamma}{\partial t^j}, \frac{\partial \gamma}{\partial t^k}\right) + \frac{\partial}{\partial t^j}g\left(\frac{\partial \gamma}{\partial t^k}, \frac{\partial \gamma}{\partial t^i}\right)$$
$$-\frac{\partial}{\partial t^k}g\left(\frac{\partial \gamma}{\partial t^i}, \frac{\partial \gamma}{\partial t^j}\right).$$

Next use that $\frac{\partial \gamma}{\partial t^k} \in TM$ to alter the left hand side to

$$2\bar{g}\left(\frac{\partial^2\gamma}{\partial t^i\partial t^j}, \frac{\partial \gamma}{\partial t^k}\right) = 2g\left(\mathrm{proj}_M\left(\frac{\partial^2\gamma}{\partial t^i\partial t^j}\right), \frac{\partial \gamma}{\partial t^k}\right).$$

This shows that $\mathrm{proj}_M\left(\frac{\partial^2\gamma}{\partial t^i\partial t^j}\right)$ is the correct mixed partial in M. $\square$

We shall use this way of calculating mixed partials in several situations below.

2. Geodesics

We can now define the acceleration of a curve $\gamma : I \to M$ by the formula

$$\ddot\gamma = \frac{d^2\gamma}{dt^2}.$$

In local coordinates this becomes

$$\ddot\gamma = \frac{d^2\gamma^k}{dt^2}\partial_k + \frac{d\gamma^i}{dt}\frac{d\gamma^j}{dt}\Gamma_{ij}^k\partial_k.$$

A C^∞ curve $\gamma : I \to M$ is called a *geodesic* if $\ddot\gamma = 0$. If γ is a geodesic, then the speed $|\dot\gamma| = \sqrt{g(\dot\gamma,\dot\gamma)}$ is constant, as

$$\frac{d}{dt}g(\dot\gamma,\dot\gamma) = 2g(\ddot\gamma,\dot\gamma) = 0.$$

So a geodesic is a constant-speed curve, or phrased differently, it is parametrized proportionally to arc length. If $|\dot\gamma| \equiv 1$, one says that γ is parametrized by arc length.

If $r : U \to \mathbb{R}$ is a distance function, then we know that for $\partial_r = \nabla r$ we have $\nabla_{\partial_r}\partial_r = 0$. The integral curves for $\nabla r = \partial_r$ are therefore geodesics. Below we shall develop a theory for geodesics independently of distance functions and then use this to show the existence of distance functions.

Geodesics are fundamental in the study of the geometry of Riemannian manifolds in the same way that straight lines are fundamental in Euclidean geometry. At first sight, however, it is not even clear that there are going to be any nonconstant geodesics to study on a general Riemannian manifold. In this section we are going to establish that every Riemannian manifold has many non-constant geodesics. Informally speaking, we can find a unique one at each point with a given tangent vector at that point. However, the question of how far it will extend from that point is subtle. To deal with the existence and uniqueness questions, we need to use some information from differential equations.

In local coordinates on $U \subset M$ the equation for a curve to be a geodesic is:

$$\begin{aligned}
0 &= \ddot\gamma \\
&= \frac{d^2\gamma^k}{dt^2}\partial_k + \frac{d\gamma^i}{dt}\frac{d\gamma^j}{dt}\Gamma_{ij}^k\partial_k
\end{aligned}$$

Thus, the curve $\gamma : I \to U$ is a geodesic if and only if the coordinate components γ^k satisfy:

$$\ddot\gamma^k(t) = -\dot\gamma^i(t)\dot\gamma^j(t)\Gamma_{ji}^k|_{\gamma(t)}$$

for $k = 1,\ldots,n$. Because this is a second-order system of differential equations, we expect an existence and a uniqueness result for the initial value problem of specifying value and first derivative, i.e.,

$$\begin{aligned}
\gamma(0) &= q, \\
\dot\gamma(0) &= \dot\gamma^i(0)\,\partial_i|_q.
\end{aligned}$$

But because the system is nonlinear, we are not entitled to expect that solutions will exist for all t.

The precise statements obtained from the theory of ordinary differential equations are a bit of a mouthful, but we might as well go for the whole thing right off the bat, since we shall need it all eventually. Still working in our coordinate situation, we get the following facts:

THEOREM 10. (Local Uniqueness) *Let I_1 and I_2 be intervals with $t_0 \in I_1 \cap I_2$, if $\gamma_1 : I_1 \to U$ and $\gamma_2 : I_2 \to U$ are geodesics with $\gamma_1(t_0) = \gamma_2(t_0)$ and $\dot{\gamma}_1(t_0) = \dot{\gamma}_2(t_0)$, then $\gamma_1|_{I_1 \cap I_2} = \gamma_2|_{I_1 \cap I_2}$.*

THEOREM 11. (Existence) *For each $p \in U$ and $v \in \mathbb{R}^n$, there is a neighborhood V_1 of p, a neighborhood V_2 of v, and an $\varepsilon > 0$ such that for each $q \in V_1$ and $w \in V_2$, there is a geodesic $\gamma_{q,w} : (-\varepsilon, \varepsilon) \to U$ with*

$$\gamma(0) = q,$$
$$\dot{\gamma}(0) = w^i \partial_i|_q.$$

Moreover, the mapping

$$(q, w, t) \to \gamma_{q,w}(t)$$

is C^∞ on $V_1 \times V_2 \times (-\varepsilon, \varepsilon)$.

It is worthwhile to consider what these assertions become in informal terms. The existence statement includes not only "small-time" existence of a geodesic with given initial point and initial tangent, it also asserts a kind of local uniformity for the interval of existence. If you vary the initial conditions but don't vary them too much, then there is a fixed interval $(-\varepsilon, \varepsilon)$ on which all the geodesics with the various initial conditions are defined. Some or all may be defined on larger intervals, but all are defined at least on $(-\varepsilon, \varepsilon)$.

The uniqueness assertion amounts to saying that geodesics cannot be tangent at one point without coinciding. Just as two straight lines that intersect and have the same tangent (at the point of intersection) must coincide, so two geodesics with a common point and equal tangent at that point must coincide.

Both of the differential equations statements are for geodesics with image in a fixed coordinate chart. By relatively simple covering arguments these statements can be extended to geodesics not necessarily contained in a coordinate chart. Let us begin with the uniqueness question:

LEMMA 7. (Global Uniqueness) *Let I_1 and I_2 be open intervals with $t_0 \in I_1 \cap I_2$, if $\gamma_1 : I_1 \to M$ and $\gamma_2 : I_2 \to M$ are geodesics with $\gamma_1(t_0) = \gamma_2(t_0)$ and $\dot{\gamma}_1(t_0) = \dot{\gamma}_2(t_0)$, then $\gamma_1|_{(I_1 \cap I_2)} = \gamma_2|_{(I_1 \cap I_2)}$.*

PROOF. Define

$$A = \{t \in I_1 \cap I_2 : \gamma_1(t) = \gamma_2(t), \ \dot{\gamma}_1(t) = \dot{\gamma}_2(t)\}.$$

Then $t_0 \in A$. Also, A is closed in $I_1 \cap I_2$ by continuity of γ_1, γ_2, $\dot{\gamma}_1$, and $\dot{\gamma}_2$. Finally, A is open, by virtue of the local uniqueness statement for geodesics in coordinate charts: if $t_1 \in A$, then choose a coordinate chart U around $\gamma_1(t_1) = \gamma_2(t_1)$. Then $(t_1 - \varepsilon, t_1 + \varepsilon) \subset I_1 \cap I_2$ and $\gamma_i|_{(t_1-\varepsilon,t_1+\varepsilon)}$ both have images contained in U. The coordinate uniqueness result then shows that $\gamma_1|_{(t_1-\varepsilon,t_1+\varepsilon)} = \gamma_2|_{(t_1-\varepsilon,t_1+\varepsilon)}$, so that $(t_1 - \varepsilon, t_1 + \varepsilon) \subset A$. $\square$

The coordinate-free global existence picture is a little more subtle. The first, and easy, step is to notice that if we start with a geodesic, then we can enlarge its interval of definition to be maximal. This follows from the uniqueness assertions: If we look at all geodesics $\gamma : I \to M$, $0 \in I$, $\gamma(0) = p$, $\dot{\gamma}(0) = v$, p and v fixed, then the union of all their domains of definition is a connected open subset of $\mathbb{R}$ on which such a geodesic is defined. And clearly its domain of definition is maximal.

The next observation, also straightforward, is that if $\widehat{K}$ is a compact subset of TM, then there is an $\varepsilon > 0$ such that for each $(q, v) \in \widehat{K}$, there is a geodesic $\gamma : (-\varepsilon, \varepsilon) \to M$ with $\gamma(0) = q$ and $\dot{\gamma}(0) = v$. This is an immediate application of the local uniformity part of the differential equations existence statement together with the usual covering-of-compact-set argument.

The next point to ponder is what happens when the maximal domain of definition is not all of $\mathbb{R}$. For this, let I be a connected open subset of $\mathbb{R}$ that is bounded above, i.e., I has the form $(-\infty, b)$, $b \in \mathbb{R}$ or (a, b), $a, b \in \mathbb{R}$. Suppose $\gamma : I \to M$ is a maximal geodesic. Then $\gamma(t)$ must have a specific kind of behavior as t approaches b : If K is any compact subset of M, then there is a number $t_K < b$ such that if $t_K < t < b$, then $\gamma(t) \in M - K$. We say that γ *leaves every compact set* as $t \to b$.

To see why γ must leave every compact set, suppose K is a compact set it doesn't leave, i.e., suppose there is a sequence $t_1, t_2, \ldots \in I$ with $\lim t_j = b$ and $\gamma(t_j) \in K$ for each j. Now $|\dot{\gamma}(t_j)|$ is independent of j, since geodesics have constant speed. So $\{\dot{\gamma}(t_j) : j = 1, \ldots\}$ lies in a compact subset of TM, namely,

$$\widehat{K} = \{v_q : q \in K, v \in T_q M, |v| \leq |\dot{\gamma}|\}.$$

Thus there is an $\varepsilon > 0$ such that for each $v_q \in \widehat{K}$, there is a geodesic $\gamma : (-\varepsilon, \varepsilon) \to M$ with $\gamma(0) = q$, $\dot{\gamma}(0) = v$. Now choose t_j such that $b - t_j < \varepsilon/2$. Then $\gamma_{q,v}$ patches together with γ to extend γ: beginning at t_j we can continue γ by ε, which takes us beyond b, since t_j is within $\varepsilon/2$ of b. This contradicts the maximality of I.

One important consequence of these observations is what happens when M itself is compact:

LEMMA 8. *If M is a compact Riemannian manifold, then for each $p \in M$ and $v \in T_p M$, there is a geodesic $\gamma : \mathbb{R} \to M$ with $\gamma(0) = p$, $\dot{\gamma}(0) = v$. In other words, geodesics exist for all time.*

A Riemannian manifold where all geodesics exist for all time is called *geodesically complete*.

A slightly trickier point is the following: Suppose $\gamma : I \to M$ is a geodesic and $0 \in I$, where I is a bounded connected open subset of $\mathbb{R}$. Then we would like to say that for $q \in M$ near enough to $\gamma(0)$ and $v \in T_q M$ near enough to $\dot{\gamma}(0)$ there is a geodesic $\gamma_{q,v}$ with q, v as initial position and tangent, respectively, *and* with $\gamma_{q,v}$ defined on an interval almost as big as I. This is true, and it is worth putting in formal language:

LEMMA 9. *Suppose $\gamma : [a, b] \to M$ is a geodesic on a compact interval. Then there is a neighborhood U in TM of $\dot{\gamma}(0)$ such that if $v \in U$, then there is a geodesic*

$$\gamma_v : [a, b] \to M$$

with $\dot{\gamma}_v(0) = v$.

PROOF. Subdivide the interval $a = b_0 < b_1 < \cdots < b_k = b$ in such a way that we have neighborhoods V_i of $\dot{\gamma}(b_i)$ where any geodesic $\gamma : [b_i, b_i + \varepsilon) \to M$ with $\dot{\gamma}(b_i) \in V_i$ is defined on $[b_i, b_{i+1}]$. Using that the map $(t, v) \to \gamma_v(t)$ is continuous, where γ is the geodesic with $\dot{\gamma}(0) = v$ we can select a new neighborhood $U_0 \subset V_0$ of $\dot{\gamma}(b_0)$ such that $\dot{\gamma}_v(b_1) \in V_1$ for $v \in U_0$. Next select $U_1 \subset U_0$ so that $\dot{\gamma}_v(b_2) \in V_2$ for $v \in U_1$ etc. In this way we get the desired neighborhood $U = U_{k-1}$ in at most k steps. $\square$

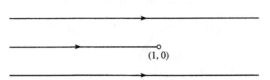

Figure 5.1

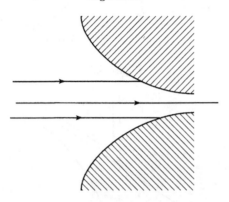

Figure 5.2

All this seems a bit formal, pedantic and perhaps abstract as well, in the absence of explicitly computed examples. First, one can easily check that geodesics in Euclidean space are straight lines. Using this observation it is simple to give examples of the above ideas by taking M to be open subsets of $\mathbb{R}^2$ with its usual metric.

EXAMPLE 28. *In the plane* $\mathbb{R}^2$ *minus one point, say* $\mathbb{R}^2 - \{(1,0)\}$ *the unit speed geodesic from* $(0,0)$ *with tangent* $(1,0)$ *is defined on* $(-\infty,1)$ *only. But nearby geodesics from* $(0,0)$ *with tangents* $(1+\varepsilon_1, \varepsilon_2)$, $\varepsilon_1, \varepsilon_2$ *small,* $\varepsilon_2 \neq 0$, *are defined on* $(-\infty, \infty)$. *Thus maximal intervals of definition can jump up in size, but, as already noted, not down. See also Figure 5.1.*

EXAMPLE 29. *On the other hand, for the region*

$$\{(x,y) : |xy| < 1\},$$

the curve $t \to (t,0)$ *is a geodesic defined on all of* $\mathbb{R}$ *that is a limit of unit speed geodesics* $t \to (t, \varepsilon)$, $\varepsilon \to 0$, *each of which is defined only on a finite interval* $\left(-\frac{1}{\varepsilon}, \frac{1}{\varepsilon}\right)$. *Note that as required, the endpoints of these intervals go to infinity (in both directions). See also Figure 5.2.*

The reader should think through these examples and those in the exercises very carefully, since geodesic behavior is a fundamental topic in all that follows.

EXAMPLE 30. *We think of the spheres* $(S^n(r), \mathrm{can}) = S^n_{r^{-2}}$ *as being in* $\mathbb{R}^{n+1}$. *The acceleration of a curve* $\gamma : I \to S^n(r)$ *can be computed as the Euclidean acceleration projected onto* $S^n(r)$. *Thus* γ *is a geodesic iff* $\ddot{\gamma}$ *is normal to* $S^n(r)$. *This means that* $\ddot{\gamma}$ *and* γ *should be proportional as vectors in* $\mathbb{R}^{n+1}$. *Great circles* $\gamma(t) = a\cos(\alpha t) + b\sin(\alpha t)$, *where* $a, b \in \mathbb{R}^{n+1}$, $|a| = |b| = r$, *and* $a \perp b$, *clearly*

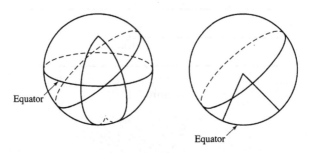

Figure 5.3

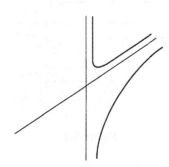

Figure 5.4

have this property. Furthermore, since $\gamma(0) = a \in S^n(r)$ and $\dot{\gamma}(0) = \alpha b \in T_a S^n(r)$, we see that we have a geodesic for each initial value problem.

We can easily picture great circles on spheres as depicted in Figure 5.3. Still, it is convenient to have a different way of understanding this. For this we project the sphere orthogonally onto the plane containing the equator. Thus the north and south poles are mapped to the origin. As all geodesics are great circles, they must project down to ellipses that have the origin as center and whose greater axis has length r. Of course, this simply describes exactly the way in which we draw three-dimensional pictures on paper.

EXAMPLE 31. *We think of $S^n_{-r^{-2}}$ as a hypersurface in Minkowski space $\mathbb{R}^{1,n}$. In this case the acceleration is still the projection of the acceleration in Minkowski space. In Minkowski space the acceleration in the usual coordinates is the same as the Euclidean acceleration. Thus we just have to find the Minkowski projection onto the hypersurface. By analogy with the sphere, one might guess that the hyperbolae $\gamma(t) = a\cosh(\alpha t) + b\sinh(\alpha t)$, $a, b \in \mathbb{R}^{1,n}$, $|a|^2 = -r^2$, $|b|^2 = r^2$, and $a \perp b$ all in the Minkowski sense, are our geodesics. And indeed this is true.*

This time the geodesics are hyperbolae. Drawing several of them on the space itself as seen in Minkowski space is not so easy. However, as with the sphere we can resort to the trick of projecting hyperbolic space onto the plane containing the last n coordinates. The geodesics there can then be seen to be hyperbolae whose asymptotes are straight lines through the origin. See also Figure 5.4.

EXAMPLE 32. *On a Lie group G with a left-invariant metric one might suspect that the geodesics are the integral curves for the left-invariant vector fields. This in turn is equivalent to the assertion that $\nabla_X X \equiv 0$ for all left-invariant vector fields. But our Lie group model for the upper half plane does not satisfy this. However, we did show in chapter 3 that $\nabla_X X = \frac{1}{2}[X, X] = 0$ when the metric is bi-invariant and X is left-invariant. Moreover, all compact Lie groups admit bi-invariant metrics (see exercises to chapter 1).*

3. The Metric Structure of a Riemannian Manifold

The positive definite inner product structures on the tangent space of a Riemannian manifold automatically give rise to a concept of lengths of tangent vectors. From this one can obtain an idea of the length of a curve as the integral of the length of its velocity vector field. This is a direct extension of the usual calculus concept of the length of curves in Euclidean space. Indeed, the definition of Riemannian manifolds is motivated from the beginning by lengths of curves. The situation is turned around a bit from that of $\mathbb{R}^n$, though: On Euclidean spaces, we have in advance a concept of distance between points. Thus, the definition of lengths of curves is justified by the fact that the length of a curve should be approximated by sums of distances for a fine subdivision (e.g., a fine polygonal approximation). For Riemannian manifolds, there is no immediate idea of distance between points. Instead, we have a natural idea of (tangent) vector length, hence curve length, and we shall use the length-of-curve idea to define distance between points. The goal of this section is to carry out these constructions in detail.

First, recall that a mapping $\gamma : [a, b] \to M$ is a piecewise C^∞ curve if γ is continuous and if there is a partition $a = a_1 < a_2 < \ldots < a_k = b$ of $[a, b]$ such that $\gamma|_{[a_i, a_{i+1}]}$ is C^∞ for $i = 1, \ldots, k-1$. Occasionally it will be convenient to work with curves that are merely absolutely continuous. A curve $\gamma : [a, b] \to \mathbb{R}^n$ is absolutely continuous if the derivative exists almost everywhere and $\gamma(t) = \gamma(a) + \int_a^t \dot\gamma(s)\, ds$. If $F : \mathbb{R}^n \to \mathbb{R}^n$ is a diffeomorphism, then we see that also $F \circ \gamma$ is absolutely continuous. Thus it makes sense to work with absolutely continuous curves in smooth manifolds.

Let $\gamma : [a, b] \to M$ be a piecewise C^∞ (or merely absolutely continuous) curve in a Riemannian manifold. Then the *length* $\ell(\gamma)$ is defined as follows:

$$\ell(\gamma) = \int_a^b |\dot\gamma(t)|\, dt = \int_a^b \sqrt{g\left(\dot\gamma(t), \dot\gamma(t)\right)}\, dt.$$

It is clear from the definition that the function $t \to |\dot\gamma(t)|$ is integrable in the Riemann (or Lebesgue) integral sense, so $\ell(\gamma)$ is a well-defined finite, nonnegative number. The chain and substitution rules show that $\ell(\gamma)$ is invariant under reparametrization. A curve $\gamma : [a, b] \to M$ is said to be parametrized by arc length if $\ell(\gamma|_{[a,t]}) = t - a$ for all $t \in [a, b]$, or equivalently, if $|\dot\gamma(t)| = 1$ at all smooth points $t \in [a, b]$. A curve $\gamma : [a, b] \to M$ such $|\dot\gamma(t)| > 0$ wherever it is smooth can be reparametrized by arc length without changing the length of the curve. To see this consider

$$s = \varphi(t) = \int_a^t |\dot\gamma(\tau)|\, d\tau.$$

Thus φ is strictly increasing on $[a, b]$, and the curve $\gamma \circ \varphi^{-1} : [0, \ell(\gamma)] \to M$ has tangent vectors of unit length at all points where it is smooth. A slightly stickier, and often ignored, point is what happens to curves that have stationary points. We can still construct the integral:

$$s = \varphi(t) = \int_a^t |\dot{\gamma}(\tau)| \, d\tau,$$

but we can't find a smooth inverse to φ if $\dot{\gamma}$ is zero somewhere. We can, however, find a curve $\sigma : [0, \ell(\gamma)] \to M$ such that

$$\gamma(t) = (\sigma \circ \varphi)(t) = \sigma(s).$$

To ensure that σ is well-defined we just have to check that $\gamma(t_1) = \gamma(t_2)$ if $\varphi(t_1) = \varphi(t_2)$. The latter equality, however, implies that $|\dot{\gamma}| = 0$ (almost everywhere) on $[t_1, t_2]$ so it does follow that $\gamma(t_1) = \gamma(t_2)$. We now need to check that σ has unit speed. This is straightforward at points where $\dot{\gamma} \neq 0$, but at the stationary points for γ it is not even clear that σ is differentiable. In fact it need not be if γ has a cusp-like singularity. The set of trouble points is the set of critical values for φ so it is at least a set of measure zero (this is simply Sard's theorem for functions $\mathbb{R} \to \mathbb{R}$). This shows that we can still define the length of σ as

$$\ell(\sigma) = \int_0^{\ell(\gamma)} |\dot{\sigma}| \, ds$$

and that σ is parametrized by arclength. In this way we have constructed a generalized reparametrization of γ, that is parametrized by arclength. Note that even if we start with a smooth curve γ the reparametrized curve σ might just be absolutely continuous. It is therefore quite natural to work with the larger class of absolutely continuous curves. Nevertheless, we have chosen to mostly stay with the more mundane piecewise smooth curves as they suffice for developing the theory Riemannian manifolds.

We are now ready to introduce the idea of distance between points. First, for each pair of points $p, q \in M$ we define the path space

$$\Omega_{p,q} = \{\gamma : [0,1] \to M : \gamma \text{ is piecewise } C^\infty \text{ and } \gamma(0) = p, \gamma(1) = q\}.$$

We can then define the distance $d(p, q)$ between points $p, q \in M$ as

$$d(p,q) = \inf\{\ell(\gamma) : \gamma \in \Omega_{p,q}\}.$$

It follows immediately from this condition that $d(p,q) = d(q,p)$ and $d(p,q) \leq d(p,r) + d(r,q)$. The fact that $d(p,q) = 0$ only when $p = q$ will be established below. Thus, $d(\,,\,)$ satisfies all the properties of a metric.

As for metric spaces, we have various metric balls defined via the metric

$$
\begin{aligned}
B(p,r) &= \{x \in M : d(p,x) < r\}, \\
\bar{B}(p,r) &= D(p,r) = \{x \in M : d(p,x) \leq r\}.
\end{aligned}
$$

More generally, we can define the distance between subsets $A, B \subset M$ as

$$d(A,B) = \inf\{d(p,q) : p \in A, q \in B\}.$$

With this we then have

$$
\begin{aligned}
B(A,r) &= \{x \in M : d(A,x) < r\}, \\
\bar{B}(A,r) &= D(A,r) = \{x \in M : d(A,x) \leq r\}.
\end{aligned}
$$

The infimum of curve lengths in the definition of $d(p,q)$ can fail to be realized. This is illustrated, for instance, by the "punctured plane" $\mathbb{R}^2 - \{(0,0)\}$

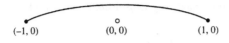

$(-1,0)$ $(0,0)$ $(1,0)$

Figure 5.5

with the usual Riemannian metric of $\mathbb{R}^2$ restricted to $\mathbb{R}^2 - \{(0,0)\}$. The distance $d((-1,0),(1,0)) = 2$, but this distance is not realized by any curve, since every curve of length 2 in $\mathbb{R}^2$ from $(-1,0)$ to $(1,0)$ passes through $(0,0)$ (see Figure 5.5). In a sense that we shall explore later, $\mathbb{R}^2 - \{(0,0)\}$ is incomplete. For the moment, we introduce some terminology for the cases where the infimum $d(p,q)$ is realized.

A curve $\sigma \in \Omega(p,q)$ is a *segment* if $\ell(\sigma) = d(p,q)$ and σ is parametrized proportionally to arc length, i.e., $|\dot{\sigma}|$ is constant on the set where σ is smooth.

EXAMPLE 33. *In Euclidean space $\mathbb{R}^n$, segments according to this definition are straight line segments parametrized with constant speed, i.e. curves of the form $t \rightarrow p + t \cdot v$. In $\mathbb{R}^n$, each pair of points p,q is joined by a segment $t \rightarrow p + t(q - p)$ that is unique up to reparametrization.*

EXAMPLE 34. *In $S^2(1)$ segments are portions of great circles with length $\leq \pi$. (We assume for the moment some basic observations of spherical geometry: these will arise later as special cases of more general results.) Every two points are joined by a segment, but there may be more than one segment joining a given pair if the pair are far enough apart, i.e., each pair of antipodal points is joined by infinitely many distinct segments.*

EXAMPLE 35. *In $\mathbb{R}^2 - \{(0,0)\}$, as already noted, not every pair of points is joined by a segment.*

Later we shall show that segments are always geodesics. Conversely, geodesics are segments if they are short enough; precisely, if γ is a geodesic defined on an open interval containing 0, then $\gamma|_{[0,\varepsilon]}$ is a segment for all sufficiently small $\varepsilon > 0$. Furthermore, we shall show that each pair of points in a Riemannian manifold can be joined by at least one segment provided that the Riemannian manifold is complete as a metric space in the metric just defined. This result explains what is "wrong" with the punctured plane. It also explains why spheres have to have segments between each pair of points: compact spaces are always complete in any metric compatible with the (compact) topology.

Some work needs to be done before we can prove these general statements. To start with, let us dispose of the question of compatibility of topologies.

THEOREM 12. *The metric topology obtained from the distance $d(\ ,\ )$ on a Riemannian manifold is the same as the manifold topology.*

PROOF. Fix $p \in M$ and a coordinate neighborhood U of p such that $x^i(p) = 0$. We assume in addition that $g_{ij}|_p = \delta_{ij}$. On U we have the given Riemannian metric g and also the Euclidean metric g_0 defined by

$$g_0(\partial_i, \partial_j) = \delta_{ij}.$$

Thus g_0 is constant and equal to g at p. Finally we can after possibly shrinking U also assume that

$$
\begin{aligned}
U &= B^{g_0}(p, \varepsilon) \\
&= \{x \in U : d_{g_0}(p, x) < \varepsilon\} \\
&= \left\{x \in U : \sqrt{(x^1)^2 + \cdots + (x^n)^2} < \varepsilon\right\}.
\end{aligned}
$$

Thus the Euclidean distance is

$$
d_{g_0}(p, x) = \sqrt{(x^1)^2 + \cdots + (x^n)^2}.
$$

For $x \in U$ we can compare these two metrics as follows: There are continuous functions: $\lambda, \mu : U \to (0, \infty)$ such that if $v \in T_x M$, then

$$
\lambda(x) |v|_{g_0} \leq |v|_g \leq \mu(x) |v|_{g_0}.
$$

Moreover, $\lambda(x), \mu(x) \to 1$ as $x \to p$.

Now let $c : [0, 1] \to M$ be a curve from p to $x \in U$.

1: If c is a straight line in the Euclidean metric, then it lies in U and

$$
\begin{aligned}
d_{g_0}(p, x) &= \ell_{g_0}(c) \\
&= \int_0^1 |\dot{c}|_{g_0} \, dt \\
&\geq \frac{1}{\max \mu(c(t))} \int_0^1 |\dot{c}|_g \, dt \\
&= \frac{1}{\max \mu(c(t))} \ell_g(c) \\
&\geq \frac{1}{\max \mu(c(t))} d_g(p, x).
\end{aligned}
$$

2: If c lies entirely in U then

$$
\begin{aligned}
\ell_g(c) &= \int_0^1 |\dot{c}|_g \, dt \\
&\geq (\min \lambda(c(t))) \int_0^1 |\dot{c}|_{g_0} \, dt \\
&\geq (\min \lambda(c(t))) \, d_{g_0}(p, x).
\end{aligned}
$$

3: If c leaves U, then there will be a smallest t_0 such that $c(t_0) \notin U$, then

$$
\begin{aligned}
\ell_g(c) &\geq \int_0^{t_0} |\dot{c}|_g \, dt \\
&\geq (\min \lambda(c(t))) \int_0^{t_0} |\dot{c}|_{g_0} \, dt \\
&\geq (\min \lambda(c(t))) \, \varepsilon \\
&\geq (\min \lambda(c(t))) \, d_{g_0}(p, x).
\end{aligned}
$$

By possibly shrinking U again we can now guarantee that $\min \lambda \geq \lambda_0 > 0$ and $\max \mu \leq \mu_0 < \infty$. We have then proven that

$$
d_g(p, x) \leq \mu_0 d_{g_0}(p, x)
$$

and

$$\lambda_0 d_{g_0}(p,x) \leq \inf \ell_g(c)$$
$$= d_g(p,x).$$

Thus the Euclidean and Riemannian distances are comparable on a neighborhood of p. This shows that the metric topology and the manifold topology (coming from the Euclidean distance) are equivalent. It also shows that $p = q$ if $d(p,q) = 0$. Finally note that

$$\lim_{x \to p} \frac{d_g(p,x)}{d_{g_0}(p,x)} = 1$$

since $\lambda(x), \mu(x) \to 1$ as $x \to p$. □

Just as compact Riemannian manifolds are automatically geodesically complete, this theorem also shows that such spaces are metrically complete.

COROLLARY 2. *If M is a compact manifold and g is a Riemannian metric on M, then (M, d_g) is a complete metric space, where d_g is the Riemannian distance function determined by g.*

Let us relate these new concepts to our distance functions from chapter 2.

LEMMA 10. *Suppose $r : U \to \mathbb{R}$ is a smooth distance function and $U \subset (M, g)$ is open, then the integral curves for ∇r are segments in (U, g).*

PROOF. Fix $p, q \in U$ and let $\gamma(t) : [0, b] \to U$ be a curve from p to q. Then

$$\ell(\gamma) = \int_0^b |\dot{\gamma}| dt$$
$$= \int_0^b |\nabla r| \cdot |\dot{\gamma}| dt$$
$$\geq \int_0^b |g(\nabla r, \dot{\gamma})| \, dt$$
$$\geq \left| \int_0^b d(r \circ \gamma) \, dt \right|$$
$$= |r(q) - r(p)|.$$

Here the first inequality is the Cauchy-Schwarz inequality. This shows that

$$d(p,q) \geq |r(q) - r(p)|.$$

If we choose γ as an integral curve for ∇r, i.e., $\dot{\gamma} = \nabla r \circ \gamma$, then equality holds in the Cauchy-Schwarz inequality and $d(r \circ \gamma) > 0$. Thus

$$\ell(\gamma) = |r(q) - r(p)|.$$

This shows that integral curves must be segments. Notice that we only considered curves in U, and therefore only established the result for (U, g) and not (M, g). □

EXAMPLE 36. Let $M = S^1 \times \mathbb{R}$ and $U = (S^1 - \{e^{i0}\}) \times \mathbb{R}$. On U we have the distance function $u(\theta, x) = \theta$, $\theta \in (0, 2\pi)$. The previous lemma shows that any curve $\gamma(t) = (e^{it}, r_0)$, $t \in I$, where I does not contain 0 is a segment in U. If, however, the length of I is $> \pi$, then such curves can clearly not be segments in M.

The *functional distance* d_F between points in a manifold is defined as

$$d_F(p, q) = \sup\{|f(p) - f(q)| : f : M \to \mathbb{R} \text{ has } |\nabla f| \leq 1 \text{ on } M\}.$$

This distance is always smaller than the arclength distance. One can, however, show as before that it generates the standard manifold topology. In fact, after we have established the existence of smooth distance functions, it will become clear that the two distances are equal provided p and q are sufficiently close to each other.

4. First Variation of Energy

In this section we shall study the arclength functional

$$\ell(\gamma) = \int_0^1 |\dot{\gamma}| \, dt,$$
$$\gamma \in \Omega_{p,q}$$

in further detail. The minima, if they exist, are pre-segments. That is, they have minimal length but we are not guaranteed that they have the correct parametrization. We also saw that in some cases suitable geodesics minimize this functional. One problem with this functional is that it is invariant under change of parametrization. Minima, if they exist, therefore do not come with a fixed parameter. This problem can be overcome, at the expense of geometric intuition, by considering the energy functional

$$E(\gamma) = \frac{1}{2} \int_0^1 |\dot{\gamma}|^2 \, dt,$$
$$\gamma \in \Omega_{p,q}.$$

This functional measures the total kinetic energy of a particle traveling along γ with the speed dictated by γ. We start by showing that these two functionals have the same minima.

PROPOSITION 17. *If $\sigma \in \Omega_{p,q}$ is a constant speed curve that minimizes $\ell :$ $\Omega_{p,q} \to [0, \infty)$, then σ also minimizes $E : \Omega_{p,q} \to [0, \infty)$. Conversely if $\sigma \in \Omega_{p,q}$ minimizes $E : \Omega_{p,q} \to [0, \infty)$, then σ also minimizes $\ell : \Omega_{p,q} \to [0, \infty)$.*

PROOF. The Cauchy-Schwarz inequality for functions tells us that

$$\ell(\gamma) = \int_0^1 |\dot{\gamma}| \cdot 1 \, dt$$
$$\leq \sqrt{\int_0^1 |\dot{\gamma}|^2 \, dt} \sqrt{\int_0^1 1^2 \, dt}$$
$$= \sqrt{\int_0^1 |\dot{\gamma}|^2 \, dt}$$
$$= \sqrt{2E(\gamma)},$$

with equality holding iff $|\dot{\gamma}| = c \cdot 1$ for some constant c, i.e., γ has constant speed. In case γ is only absolutely continuous this inequality still holds. Moreover, when equality holds the speed is constant wherever it is defined. Let $\sigma \in \Omega_{p,q}$ be a constant speed curve that minimizes ℓ and $\gamma \in \Omega_{p,q}$. Then

$$
\begin{aligned}
E\left(\sigma\right) \;&=\; \frac{1}{2}\left(\ell\left(\sigma\right)\right)^2 \\
&\leq\; \frac{1}{2}\left(\ell\left(\gamma\right)\right)^2 \\
&\leq\; E\left(\gamma\right),
\end{aligned}
$$

so σ also minimizes E.

Conversely let $\sigma \in \Omega_{p,q}$ minimize E and $\gamma \in \Omega_{p,q}$. If γ does not have constant speed we can without changing its length reparametrize it to an absolutely continuous curve $\bar{\gamma}$ that has constant speed almost everywhere. Then

$$
\begin{aligned}
\ell\left(\sigma\right) \;&\leq\; \sqrt{2E\left(\sigma\right)} \\
&\leq\; \sqrt{2E\left(\bar{\gamma}\right)} \\
&=\; \ell\left(\bar{\gamma}\right) \\
&=\; \ell\left(\gamma\right).
\end{aligned}
$$

$\square$

Our next goal is to show that minima of E must be geodesics. To do this we have to develop the *first variation formula for energy*. A variation of a curve $\gamma : I \to M$ is a family of curves $\bar{\gamma} : (-\varepsilon, \varepsilon) \times [a, b] \to M$, such that $\bar{\gamma}\left(0, t\right) = \gamma\left(t\right)$ for all $t \in [a, b]$. We say that such a variation is piecewise smooth if it is continuous and we can partition $[a, b]$ in to intervals $[a_i, a_{i+1}]$, $i = 0, ..., m - 1$, in such a way that $\bar{\gamma} : (-\varepsilon, \varepsilon) \times [a_i, a_{i+1}] \to M$ is smooth. Thus the curves $t \to \gamma_s\left(t\right) = \bar{\gamma}\left(s, t\right)$ are all piecewise smooth, while the curves $s \to \bar{\gamma}\left(s, t\right)$ are smooth. The velocity field for this variation is the field $\frac{\partial \bar{\gamma}}{\partial t}$ which is well-defined on each interval $[a_i, a_{i+1}]$. At the break points a_i, there are two possible values for this field; a right derivative and a left derivative:

$$
\begin{aligned}
\frac{\partial \bar{\gamma}}{\partial t^+}\big|_{(s, a_i)} \;&=\; \frac{\partial \bar{\gamma}|_{[a_i, a_{i+1}]}}{\partial t}\big|_{(s, a_i)}, \\
\frac{\partial \bar{\gamma}}{\partial t^-}\big|_{(s, a_i)} \;&=\; \frac{\partial \bar{\gamma}|_{[a_{i-1}, a_i]}}{\partial t}\big|_{(s, a_i)}.
\end{aligned}
$$

The *variational field* is defined as $\frac{\partial \bar{\gamma}}{\partial s}$. This field is well-defined everywhere. It is smooth on each $(-\varepsilon, \varepsilon) \times [a_i, a_{i+1}]$ and continuous on $(-\varepsilon, \varepsilon) \times I$. The special case where $a = 0$, $b = 1$, $\bar{\gamma}\left(s, 0\right) = p$ and $\bar{\gamma}\left(s, 1\right) = q$ for all s is of special importance as all of the curves $\gamma_s \in \Omega_{p,q}$. Such variations are called *proper variations* of γ.

LEMMA 11. (The First Variation Formula) *Let* $\bar{\gamma} : (-\varepsilon, \varepsilon) \times [a, b] \to M$ *be a piecewise smooth variation, then*

$$
\begin{aligned}
\frac{dE\left(\gamma_s\right)}{ds} \;=\; & -\int_a^b g\left(\frac{\partial^2 \bar{\gamma}}{\partial t^2}, \frac{\partial \bar{\gamma}}{\partial s}\right) dt + g\left(\frac{\partial \bar{\gamma}}{\partial t^-}, \frac{\partial \bar{\gamma}}{\partial s}\right)\bigg|_{(s,b)} - g\left(\frac{\partial \bar{\gamma}}{\partial t^+}, \frac{\partial \bar{\gamma}}{\partial s}\right)\bigg|_{(s,a)} \\
& + \sum_{i=1}^{m-1} g\left(\frac{\partial \bar{\gamma}}{\partial t^-} - \frac{\partial \bar{\gamma}}{\partial t^+}, \frac{\partial \bar{\gamma}}{\partial s}\right)\bigg|_{(s,a_i)}.
\end{aligned}
$$

PROOF. It suffices to prove the formula for smooth variations as we can otherwise split up the integral into parts that are smooth:

$$E\left(\gamma_s\right) = \int_a^b \left|\frac{\partial\bar{\gamma}}{\partial t}\right|^2 dt$$

$$= \sum_{i=0}^{m-1} \int_{a_i}^{a_{i+1}} \left|\frac{\partial\bar{\gamma}}{\partial t}\right|^2 dt$$

and apply the formula to each part of the variation.

For a smooth variation $\bar{\gamma} : (-\varepsilon, \varepsilon) \times [a, b] \to M$ we have

$$\frac{dE\left(\gamma_s\right)}{ds} = \frac{d}{ds}\frac{1}{2} \int_a^b g\left(\frac{\partial\bar{\gamma}}{\partial t}, \frac{\partial\bar{\gamma}}{\partial t}\right) dt$$

$$= \frac{1}{2} \int_a^b \frac{\partial}{\partial s} g\left(\frac{\partial\bar{\gamma}}{\partial t}, \frac{\partial\bar{\gamma}}{\partial t}\right) dt$$

$$= \int_a^b g\left(\frac{\partial^2\bar{\gamma}}{\partial s\partial t}, \frac{\partial\bar{\gamma}}{\partial t}\right) dt$$

$$= \int_a^b g\left(\frac{\partial^2\bar{\gamma}}{\partial t\partial s}, \frac{\partial\bar{\gamma}}{\partial t}\right) dt$$

$$= \int_a^b \frac{\partial}{\partial t} g\left(\frac{\partial\bar{\gamma}}{\partial s}, \frac{\partial\bar{\gamma}}{\partial t}\right) dt - \int_a^b g\left(\frac{\partial\bar{\gamma}}{\partial s}, \frac{\partial^2\bar{\gamma}}{\partial t^2}\right) dt$$

$$= g\left(\frac{\partial\bar{\gamma}}{\partial s}, \frac{\partial\bar{\gamma}}{\partial t}\right)\Big|_a^b - \int_a^b g\left(\frac{\partial\bar{\gamma}}{\partial s}, \frac{\partial^2\bar{\gamma}}{\partial t^2}\right) dt$$

$$= -\int_a^b g\left(\frac{\partial\bar{\gamma}}{\partial s}, \frac{\partial^2\bar{\gamma}}{\partial t^2}\right) dt + g\left(\frac{\partial\bar{\gamma}}{\partial s}, \frac{\partial\bar{\gamma}}{\partial t}\right)\Big|_{(s,b)} - g\left(\frac{\partial\bar{\gamma}}{\partial s}, \frac{\partial\bar{\gamma}}{\partial t}\right)\Big|_{(s,a)}.$$

$\square$

We can now completely characterize the local minima for the energy functional.

THEOREM 13. (Characterization of local minima) If $\gamma \in \Omega_{p,q}$ is a local minimum for $E : \Omega_{p,q} \to [0, \infty)$, then γ is a smooth geodesic.

PROOF. The assumption guarantees that

$$\frac{dE\left(\gamma_s\right)}{ds} = 0$$

for any proper variation of γ. The trick now is to find appropriate variations. In fact, if $V(t)$ is any vector field along $\gamma(t)$, i.e., $V(t) \in T_{\gamma(t)}M$, then there is a variation so that $V(t) = \frac{\partial\gamma}{\partial s}|_{(0,t)}$. One such variation is gotten by declaring that the variational curves $s \to \gamma(s, t)$ are geodesics with $\frac{\partial\gamma}{\partial s}|_{(0,t)} = V(t)$. As geodesics vary nicely with respect to the initial data this variation will be as smooth as V is. Finally, if $V(a) = 0$ and $V(b) = 0$, then the variation is proper.

Using such a variational field the first variation formula at $s = 0$ only depends on γ itself and the variational field V

$$
\begin{aligned}
\frac{dE(\gamma_s)}{ds}\Big|_{s=0} &= -\int_a^b g(\ddot{\gamma}, V)\, dt + g\left(\frac{d\gamma}{dt^-}(b), V(b)\right) - g\left(\frac{d\gamma}{dt^+}(a), V(a)\right) \\
&\quad + \sum_{i=1}^{m-1} g\left(\frac{d\gamma}{dt^-}(a_i) - \frac{d\gamma}{dt^+}(a_i), V(a_i)\right) \\
&= -\int_a^b g(\ddot{\gamma}, V)\, dt + \sum_{i=1}^{m-1} g\left(\frac{d\gamma}{dt^-}(a_i) - \frac{d\gamma}{dt^+}(a_i), V(a_i)\right).
\end{aligned}
$$

We now specify V further. First select $V(t) = \lambda(t)\ddot{\gamma}(t)$, where $\lambda(a_i) = 0$ at the break points a_i where γ might not be smooth, and also $\lambda(a) = \lambda(b) = 0$. Finally assume that $\lambda(t) > 0$ elsewhere. Then

$$
\begin{aligned}
0 &= \frac{dE(\gamma_s)}{ds}\Big|_{s=0} \\
&= -\int_a^b g(\ddot{\gamma}, \lambda(t)\ddot{\gamma})\, dt \\
&= -\int_a^b \lambda(t)|\ddot{\gamma}|^2\, dt.
\end{aligned}
$$

Since $\lambda(t) > 0$ where $\ddot{\gamma}$ is defined it must follow that $\ddot{\gamma} = 0$ at those points. Thus γ is a broken geodesic. Next select a general variational field V such that

$$
\begin{aligned}
V(a_i) &= \frac{d\gamma}{dt^-}(a_i) - \frac{d\gamma}{dt^+}(a_i), \\
V(a) &= V(b) = 0
\end{aligned}
$$

and otherwise arbitrary, then we obtain

$$
\begin{aligned}
0 &= \frac{dE(\gamma_s)}{ds}\Big|_{s=0} \\
&= \sum_{i=1}^{m-1} g\left(\frac{d\gamma}{dt^-}(a_i) - \frac{d\gamma}{dt^+}(a_i), V(a_i)\right) \\
&= \sum_{i=1}^{m-1} \left|\frac{d\gamma}{dt^-}(a_i) - \frac{d\gamma}{dt^+}(a_i)\right|^2.
\end{aligned}
$$

This forces

$$
\frac{d\gamma}{dt^-}(a_i) = \frac{d\gamma}{dt^+}(a_i)
$$

and hence the broken geodesic has the same velocity from the left and right at the places where it is potentially broken. Uniqueness of geodesics then shows that γ is a smooth geodesic. $\qquad\square$

This also shows:

COROLLARY 3. (Characterization of segments) *Any piecewise smooth segment is a geodesic.*

While this result shows precisely what the local minima of the energy functional must be it does not guarantee that geodesics are local minima. In Euclidean space all geodesics are minimal as they are the integral curves for globally defined distance

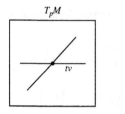
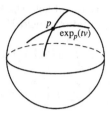

Figure 5.6

functions: $u(x) = v \cdot x$, where v is a unit vector. On the unit sphere, however, no geodesic of length $> \pi$ can be locally minimizing. Since such geodesics always form part of a great circle, we see that the complement of the geodesic in the great circle has length $< \pi$. This shows that the geodesic can't be an absolute minimum. However, we can also easily construct a variation where the nearby curves are all shorter. We shall spend much more time on these issues in the subsequent sections as well as the next chapter. Certainly much more work has to be done before we can say more about when geodesics are minimal. The above proof does, however, tell us that a geodesic $\gamma \in \Omega_{p,q}$ is always a *stationary point* for $E : \Omega_{p,q} \to [0, \infty)$, in the sense that

$$\frac{dE(\gamma_s)}{ds}\Big|_{s=0} = 0$$

for all proper variations of γ.

5. The Exponential Map

For a tangent vector $v \in T_pM$, let γ_v be the unique geodesic with $\gamma(0) = p$ and $\dot{\gamma}(0) = v$, and $[0, \ell_v)$ the nonnegative part of the maximal interval on which γ is defined. Notice that $\gamma_{\alpha v}(t) = \gamma_v(\alpha t)$ for all $\alpha > 0$ and $t < \ell_{\alpha v}$. In particular, $\ell_{\alpha v} = \alpha^{-1}\ell_v$. Let $O_p \subset T_pM$ be the set of vectors v such that $1 < \ell_v$, so that $\gamma_v(t)$ is defined on $[0, 1]$. Then define the *exponential map* at p by

$$\exp_p \;\; : \;\; O_p \to M$$
$$\exp_p(v) \;\; = \;\; \gamma_v(1).$$

In the exercises to this chapter we have a problem that elucidates the relationship between the just defined exponential map and the Lie group exponential map introduced earlier. In Figure 5.6 we have shown how radial lines in the tangent space are mapped to radial geodesics in M via the exponential map. The "homogeneity property" $\gamma_v(t) = \gamma_{tv}(1)$ shows that $\exp_p(tv) = \gamma_v(t)$. Given that, it is natural to think of $\exp_p(v)$ in a polar coordinate representation: From p ones goes "distance" $|v|$ in the direction $v/|v|$. This gives the point $\exp_p(v)$, since $\gamma_{v/|v|}(|v|) = \gamma_v(1)$.

The individual $\exp_p$ maps can be combined to form a map $\exp : \bigcup O_p \to M$ by setting $\exp|_{O_p} = \exp_p$. This map $\exp$ is also called the *exponential map*.

The standard theory of ordinary differential equations that we have already discussed tells us that the set $O = \bigcup O_p$ is open in TM and that $\exp : O \to M$ is smooth. In addition $O_p \subset T_pM$ is open, and $\exp_p : O_p \to M$ is also smooth. It is an important property that $\exp_p$ is in fact a local diffeomorphism around $0 \in T_pM$. The details of this are given in the following:

PROPOSITION 18. *If $p \in M$, then*
(1)
$$D \exp_p : T_0(T_pM) \to T_pM$$
is nonsingular at the origin of T_pM. Consequently $\exp_p$ is a local diffeomorphism.
(2) Define $E : O \to M \times M$ by $E(v) = (\pi(v), \exp v)$, where $\pi(v)$ is the base point of v, i.e., $v \in T_{\pi(v)}M$. Then for each $p \in M$ and with it the zero vector, $0_p \in T_pM$,
$$DE : T_{(p,0_p)}(TM) \to T_{(p,p)}(M \times M)$$
is nonsingular. Consequently, E is a diffeomorphism from a neighborhood of the zero section of TM onto an open neighborhood of the diagonal in $M \times M$.

PROOF. The proofs of both statements are an immediate application of the inverse function theorem, once a crucial observation has been made. This observation is as follows: Let $I_0 : T_pM \to T_0T_pM$ be the canonical isomorphism, i.e., $I_0(v) = \frac{d}{dt}(tv)|_{t=0}$. Now we recall that if $v \in O_p$, then $\gamma_v(t) = \gamma_{tv}(1)$ for all $t \in [0,1]$. Thus,

$$
\begin{aligned}
D \exp_p(I_0(v)) &= \frac{d}{dt} \exp_p(tv)|_{t=0} \\
&= \frac{d}{dt} \gamma_{tv}(1)|_{t=0} \\
&= \frac{d}{dt} \gamma_v(t)|_{t=0} \\
&= \dot{\gamma}_v(0) \\
&= v.
\end{aligned}
$$

In other words $D \exp_p \circ I_0$ is the identity map on T_pM. This shows that $D \exp_p$ is nonsingular. The second statement of (1) follows from the inverse function theorem.

The proof of (2) is again an exercise in unraveling tangent spaces and identifications. The tangent space $T_{(p,p)}(M \times M)$ is naturally identified with $T_pM \times T_pM$. The tangent space $T_{(p,0_p)}(TM)$ is also naturally identified to $T_pM \times T_{0_p}(T_pM) \simeq T_pM \times T_pM$.

We know that E takes (p, v) to $(p, \exp_p(v))$. Note that varying p is just the identity in the first coordinate, but something unpredictable in the second. While if we fix p and vary v in T_pM, then the first coordinate is fixed and we simply have $\exp_p(v)$ in the second coordinate. This explains what the differential $DE|_{(p,0_p)}$ is. If we consider it as a linear map $T_pM \times T_pM \to T_pM \times T_pM$, then it is the identity on the first factor to the first factor, identically 0 from the second factor to the first, and the identity from the second fact to the second factor as it is $D \exp_p \circ I_{0_p}$. Thus it looks like

$$
\begin{bmatrix}
I & 0 \\
* & I
\end{bmatrix}
$$

which is clearly nonsingular.

Now, the inverse function theorem gives (local) diffeomorphisms via E of neighborhoods of $(p, 0_p) \in TM$ onto neighborhoods of $(p, p) \in M \times M$. Since E maps the zero section of TM diffeomorphically to the diagonal in $M \times M$ and the zero section is a properly embedded submanifold of TM, it is easy to see that these local diffeomorphisms fit together to give a diffeomorphism of a neighborhood of the zero section in TM onto a neighborhood of the diagonal in $M \times M$. □

132 5. GEODESICS AND DISTANCE

All this formalism with the exponential maps yields some results with geometric meaning. First, we get a coordinate system around p by identifying T_pM with $\mathbb{R}^n$ via an isomorphism, and using that the exponential map $\exp_p : T_pM \to M$ is a diffeomorphism on a neighborhood of the origin. Such coordinates are called *normal (exponential) coordinates* at p. They are unique up to how we choose to identify T_pM with $\mathbb{R}^n$. Requiring this identification to be a linear isometry gives uniqueness up to an orthogonal transformation of $\mathbb{R}^n$. Later in the chapter we show that they are indeed normal in the sense that the Christoffel symbols vanish at p.

The second item of geometric interest is the following idea: Thinking about S^2 and great circles (which we know are geodesics), it is clear that we cannot say that two points that are close together are joined by a unique geodesic. On S^2 there will be a short geodesic connection, but there will be other, long ones, too. What might be hoped is that points that are close together would have a unique "short" geodesic connecting them. This is exactly what (2) in the proposition says! As long as we keep q_1 and q_2 near p, there is only one way to go from q_1 to q_2 via a geodesic that isn't very long, i.e., has the form $\exp_{q_1} tv$, $v \in T_{q_1}M$, with $|v|$ small. This will be made more useful and clear in the next section, where we show that such short geodesics in fact are segments.

Suppose N is an embedded submanifold of M. The normal bundle of N in M is the vector bundle over N consisting of the orthogonal complements of the tangent spaces $T_pN \subset T_pM$.

$$TN^\perp = \{v \in T_pM : p \in N, v \in (T_pN)^\perp \subset T_pM\}.$$

So for each $p \in N$, $T_pM = T_pN \oplus (T_pN)^\perp$ is an orthogonal direct sum. Define the *normal exponential map* $\exp^\perp$ by restricting $\exp$ to $O \cap TN^\perp$ so $\exp^\perp : O \cap TN^\perp \to M$. As in part (2) of the previous proposition, one can show that $D\exp^\perp$ is nonsingular at 0_p, $p \in N$. Then it follows that there is an open neighborhood U of the zero section in $TN^\perp$ on which $\exp^\perp$ is a diffeomorphism onto its image in M. Such an image $\exp^\perp(U)$ is called a *tubular neighborhood* of N in M, because if N is a curve in $\mathbb{R}^3$ it looks like a solid tube around the curve.

6. Why Short Geodesics Are Segments

In the previous section, we saw that points that are close together on a Riemannian manifold are connected by a short geodesic, and by exactly one short geodesic in fact. But so far, we don't have any real evidence that such short geodesics are segments. In this section we shall prove that short geodesics are segments. Incidentally, several different ways of saying that a curve is a segment are in common use: "minimal geodesic," "minimizing curve," "minimizing geodesic," and even "minimizing geodesic segment."

The precise result we want to prove in this section is this:

THEOREM 14. *Suppose M is a Riemannian manifold, $p \in M$, and $\varepsilon > 0$ is such that*

$$\exp_p : B(0,\varepsilon) \to U \subset M$$

is a diffeomorphism onto its image in M. Then $U = B(p,\varepsilon)$ and for each $v \in B(0,\varepsilon)$, the geodesic $\gamma_v : [0,1] \to M$ defined by

$$\gamma_v(t) = \exp_p(tv)$$

is the unique segment in M from p to $\exp_p v$.

On $U = \exp_p(B(0, \varepsilon))$ we have the function $r(x) = |\exp_p^{-1}(x)|$. That is, r is simply the Euclidean distance function from the origin on $B(0, \varepsilon) \subset T_pM$ in exponential coordinates. We know that $\nabla r = \partial_r = \frac{1}{r}(x^i \partial_i)$ in Cartesian coordinates on T_pM. The goal here is to establish:

LEMMA 12. (The Gauss Lemma) *On* (U, g) *the function* r *satisfies* $\nabla r = \partial_r$, *where* $\partial_r = D\exp_p(\partial_r)$.

Let us see how this implies the Theorem.

PROOF OF THEOREM. First observe that in $B(0, \varepsilon)$ the integral curves for ∂_r are the line segments $\gamma(s) = s \cdot \frac{v}{|v|}$ of unit speed. The integral curves for ∂_r on U are therefore the unit speed geodesics $\gamma(s) = \exp\left(s \cdot \frac{v}{|v|}\right)$. Thus the Lemma implies that r is a distance function on U. This shows that among curves from p to $q = \exp(x)$ in $U - \{p\}$, the geodesic from p to q is the shortest curve, furthermore, it has length $< \varepsilon$. In particular, $U \subset B(p, \varepsilon)$. To see that this geodesic is a segment in M, we must show that any curve that leaves U has length $> \varepsilon$. Suppose we have a curve $\gamma : [0, b] \to M$ from p to q that leaves U. Let $a \in [0, b]$ be the largest value so that $\gamma(a) = p$. Then $\gamma|_{[a,b]}$ is a shorter curve from p to q. Next let $t_0 \in (a, b)$ be the first value for which $\gamma(t_0) \notin U$. Then $\gamma|_{(a,t_0)}$ lies entirely in $U - \{p\}$ and is shorter than the original curve. We now see

$$
\begin{aligned}
\ell\left(\gamma|_{(a,t_0)}\right) &= \int_a^{t_0} |\dot\gamma| \, dt \\
&= \int_a^{t_0} |\nabla r| \cdot |\dot\gamma| \, dt \\
&\geq \int_a^{t_0} dr(\dot\gamma) \, dt \\
&= r(\gamma(t_0)) - r(\gamma(a)) \\
&= \varepsilon,
\end{aligned}
$$

since $r(p) = 0$ and the values of r converge to ε as we approach ∂U. Thus γ is not a segment from p to q.

Finally we have to show that $B(p, \varepsilon) = U$. We already have $U \subset B(p, \varepsilon)$. Conversely if $q \in B(p, \varepsilon)$ then it is joined to p by a curve of length $< \varepsilon$. The above argument then shows that this curve lies in U. Whence $B(p, \varepsilon) \subset U$. $\qquad\square$

PROOF OF GAUSS LEMMA. We select an orthonormal basis for T_pM and introduce Cartesian coordinates. These coordinates are then also used on $B(p, \varepsilon)$ via the exponential map. Denote these coordinates by $(x^1, \ldots, x^n)$ and the coordinate vector fields by $\partial_1, \ldots, \partial_n$. Then

$$
\begin{aligned}
r^2 &= (x^1)^2 + \cdots + (x^n)^2, \\
\partial_r &= \frac{1}{r} x^i \partial_i.
\end{aligned}
$$

To show that this is the gradient for $r(x)$ on (M, g), we must prove that $dr(v) = g(\partial_r, v)$. We already know that

$$
dr = \frac{1}{r}(x^1 dx^1 + \cdots + x^n dx^n),
$$

but we have no knowledge of g, since it is just some abstract metric.

We prove that $dr(v) = g(\partial_r, v)$ by using suitable vector fields in place of v. In fact we are going to use Jacobi fields for r. Let us start with $v = \partial_r$. The right hand side is 1 as the integral curves for ∂_r are unit speed geodesics. The left hand side is also quickly computed to be 1. Next we take a rotation vector field $J = -x^i \partial_j + x^j \partial_i$, $i, j = 2, \ldots, n$, $i < j$. In dimension 2 this is simply the angular field ∂_θ. We immediately see that the left hand side vanishes: $dr(J) = 0$. For the right hand side we first note that J really is a Jacobi field as $L_{\partial_r} J = [\partial_r, J] = 0$. Using that $\nabla_{\partial_r} \partial_r = 0$ we then get

$$
\begin{aligned}
\partial_r g(\partial_r, J) &= g(\nabla_{\partial_r} \partial_r, J) + g(\partial_r, \nabla_{\partial_r} J) \\
&= 0 + g(\partial_r, \nabla_{\partial_r} J) \\
&= -g(\partial_r, \nabla_J \partial_r) \\
&= -\frac{1}{2} D_J g(\partial_r, \partial_r) \\
&= 0.
\end{aligned}
$$

Thus $g(\partial_r, J)$ is constant along geodesics emanating from p. Next observe that

$$
\begin{aligned}
|g(\partial_r, J)| &\leq |\partial_r||J| \\
&= |J| \\
&\leq |x^i| |\partial_j| + |x^j| |\partial_i| \\
&\leq r(x)(|\partial_i| + |\partial_j|)
\end{aligned}
$$

Continuity of $D\exp_p$ shows that ∂_i, ∂_j are bounded on $B(p, \varepsilon)$. Thus $|g(\partial_r, J)| \to 0$ as $r \to 0$. This shows that $g(\partial_r, J) = 0$. Finally we observe that any vector v is a linear combination of ∂_r and rotation vector fields. This proves the claim. □

There is an equivalent statement of the Gauss Lemma asserting that

$$
\exp_p : B(0, \varepsilon) \to B(p, \varepsilon)
$$

is a *radial isometry*:

$$
g\left(D\exp_p(\partial_r), D\exp_p(v)\right) = g_p(\partial_r, v)
$$

on $T_p M$. A careful translation process of the previous proof shows that this is exactly what we have proved.

The next corollary is also an immediate consequence of the above theorem and its proof.

COROLLARY 4. *If $x \in M$ and $\varepsilon > 0$ is such that $\exp_x : B(0, \varepsilon) \to B(p, \varepsilon)$ is defined and a diffeomorphism, then for each $\delta < \varepsilon$,*

$$
\exp_x(B(0, \delta)) = B(x, \delta),
$$

and

$$
\exp_x(\bar{B}(0, \delta)) = \bar{B}(x, \delta).
$$

7. Local Geometry in Constant Curvature

Let us restate what we have done in this chapter so far. Given $p \in (M, g)$ we found coordinates near p using the exponential map such that the distance function $r(x) = d(p, x)$ to p has the formula

$$
r(x) = \sqrt{(x^1)^2 + \cdots + (x^n)^2}.
$$

Furthermore, we showed that $\nabla r = \partial_r$. By calculating each side in the formula in coordinates we get

$$\sum_{i,j} \frac{1}{r} g^{ij} x^j \partial_i = \nabla r = \partial_r = \frac{1}{r} x^i \partial_i.$$

After equating the coefficients of these vector fields we obtain the following curious relationship between the coordinates and the metric coefficients

$$\sum_j g^{ij} x^j = x^i$$

which is equivalent to

$$\sum_j g_{ij} x^j = x^i.$$

This relationship, as we shall see, fixes the behavior of g_{ij} around p up to first order and shows that the coordinates are normal in the sense used in chapter 2.

LEMMA 13.

$$g_{ij} = \delta_{ij} + O\left(r^2\right).$$

PROOF. The fact that $g_{ij}|_p = \delta_{ij}$ follows from taking one partial derivative on both sides of the above relation

$$
\begin{aligned}
\delta_k^i &= \partial_k x^i \\
&= \partial_k \sum_j g_{ij} x^j \\
&= \left(\partial_k g_{ij}\right) x^j + g_{ij} \partial_k x^j \\
&= \left(\partial_k g_{ij}\right) x^j + g_{ik}.
\end{aligned}
$$

As $x^j(p) = 0$, the claim follows. The fact that $\partial_k g_{ij}|_p = 0$ comes about by taking two partial derivatives on both sides

$$
\begin{aligned}
0 &= \partial_l \partial_k x^i \\
&= \partial_l \left(\left(\partial_k g_{ij}\right) x^j\right) + \partial_l g_{ik} \\
&= \left(\partial_l \partial_k g_{ij}\right) x^j + \partial_k g_{ij} \partial_l x^j + \partial_l g_{ik} \\
&= \left(\partial_l \partial_k g_{ij}\right) x^j + \partial_k g_{il} + \partial_l g_{ik}.
\end{aligned}
$$

Evaluating at p then gives us

$$\partial_k g_{il}|_p + \partial_l g_{ik}|_p = 0.$$

Now combine this with the relations for the Christoffel symbols of the first kind:

$$
\begin{aligned}
\partial_i g_{kl} &= \Gamma_{ik,l} + \Gamma_{il,k}, \\
\Gamma_{ij,k} &= \frac{1}{2}\left(\partial_j g_{ik} + \partial_i g_{jk} - \partial_k g_{ji}\right)
\end{aligned}
$$

to get

$$\Gamma_{kl,i}|_p = \frac{1}{2}\left(\partial_k g_{il}|_p + \partial_l g_{ik}|_p - \partial_i g_{kl}|_p\right)$$

$$= -\frac{1}{2}\partial_i g_{kl}|_p$$

$$= -\frac{1}{2}\left(\Gamma_{ik,l}|_p + \Gamma_{il,k}|_p\right)$$

$$= \frac{1}{4}\left(\partial_l g_{ik}|_p + \partial_k g_{il}|_p\right)$$

$$= 0$$

which is what we wanted to prove.   □

In polar coordinates around p any Riemannian metric therefore has the form

$$g = dr^2 + g_r$$

where g_r is a metric on S^{n-1}. The Euclidean metric looks like

$$\delta_{ij} = dr^2 + r^2 ds_{n-1}^2,$$

where ds_{n-1}^2 is the canonical metric on S^{n-1}. Since these two metrics agree up to first order we obtain

$$\lim_{r\to 0} g_r = \lim_{r\to 0}\left(r^2 ds_{n-1}^2\right) = 0,$$

$$\lim_{r\to 0}\left(\partial_r g_r - \frac{2}{r}g_r\right) = \lim_{r\to 0}\left(\partial_r\left(r^2 ds_{n-1}^2\right) - \frac{2}{r}\left(r^2 ds_{n-1}^2\right)\right) = 0.$$

As

$$\partial_r g_r = 2\mathrm{Hess}r$$

this translates into

$$\lim_{r\to 0}\left(\mathrm{Hess}r - \frac{1}{r}g_r\right) = 0.$$

This can also be seen by computing the Hessian of $\frac{1}{2}r^2$ at p. Just note that this function has a critical point at p. Thus the coordinate formula for the Hessian is independent of the metric and must therefore be the identity map at p.

THEOREM 15. (Riemann, 1854) *If a Riemannian n-manifold (M,g) has constant sectional curvature k, then every point in M has a neighborhood that is isometric to an open subset of the space form S_k^n.*

PROOF. We use polar coordinates around $p \in M$ and the asymptotic behavior of g_r and $\mathrm{Hess}r$ near p that was just established. We shall also use the fundamental equations that were introduced in chapter 2. On the same neighborhood we can also introduce a metric of constant curvature k

$$\tilde{g} = dr^2 + \mathrm{sn}_k^2(r)\, ds_{n-1}^2,$$

$$\mathrm{Hess}_{\tilde{g}}r = \frac{\mathrm{sn}_k'(r)}{\mathrm{sn}_k(r)}\tilde{g}_r.$$

Since the curvature is k for both of these metrics we see that $\mathrm{Hess}_{\tilde{g}}r$ and $\mathrm{Hess}_g r$ solve the same equation when evaluated on unit parallel fields perpendicular to ∂_r.

Note, however, that g and $\tilde{g}$ most likely have different parallel fields X and $\tilde{X}$:

$$\partial_r \left(\text{Hess}_g r \left(X, X \right) \right) - \text{Hess}_g^2 r \left(X, X \right) = -\sec \left(X, \partial_r \right) = -k,$$

$$\lim_{r \to 0} \left(\text{Hess}_g r \left(X, X \right) - \frac{1}{r} \right) = 0,$$

$$\partial_r \left(\text{Hess}_{\tilde{g}} r \left(\tilde{X}, \tilde{X} \right) \right) - \text{Hess}_{\tilde{g}}^2 r \left(\tilde{X}, \tilde{X} \right) = -\sec \left(\tilde{X}, \partial_r \right) = -k,$$

$$\lim_{r \to 0} \left(\text{Hess} r_{\tilde{g}} \left(\tilde{X}, \tilde{X} \right) - \frac{1}{r} \right) = 0,$$

Hence

$$\text{Hess}_g r \left(X, X \right) = \text{Hess}_{\tilde{g}} r \left(\tilde{X}, \tilde{X} \right) = \frac{\text{sn}_k' \left(r \right)}{\text{sn}_k \left(r \right)}.$$

This shows that

$$\text{Hess}_g r = \frac{\text{sn}_k' \left(r \right)}{\text{sn}_k \left(r \right)} g_r$$

When evaluating on Jacobi fields instead we see that both g_r and $\text{sn}_k^2 \left(r \right) ds_{n-1}^2$ solve the equation

$$\partial_r g_r = 2 \frac{\text{sn}_k' \left(r \right)}{\text{sn}_k \left(r \right)} g_r,$$

$$\lim_{r \to 0} g_r = 0.$$

This shows that

$$g = dr^2 + \text{sn}_k^2 \left(r \right) ds_{n-1}^2.$$

In other words we have found a coordinate system on a neighborhood around $p \in M$ where the metric is the same as the constant curvature metric. $\qquad \square$

8. Completeness

One of the foundational centerpieces of Riemannian geometry is the Hopf-Rinow theorem. This theorem states that all concepts of completeness are equivalent. This should not be an unexpected result for those who have played around with open subsets of Euclidean space. For it seems that in these examples, geodesic and metric completeness break down in exactly the same places. As with most foundational theorems, the proof is slightly intricate.

THEOREM 16. (H. Hopf-Rinow, 1931) *The following statements are equivalent:*
(1) M is geodesically complete, i.e., all geodesics are defined for all time.
(2) M is geodesically complete at p, i.e., all geodesics through p are defined for all time.
(3) M satisfies the Heine-Borel property, i.e., every closed bounded set is compact.
(4) M is metrically complete.

PROOF. $(1) \Rightarrow (2)$, $(3) \Rightarrow (4)$ are trivial.
$(4) \Rightarrow (1)$ Recall that every geodesic $\gamma : [0, b) \to M$ defined on a maximal interval must leave every compact set if $b < \infty$. This violates metric completeness as $\gamma(t_i)$, $t_i \to b$ is a Cauchy sequence.
$(2) \Rightarrow (3)$ Consider $\exp_p : T_p M \to M$. It suffices to show that

$$\exp_p \left(\overline{B}(0, r) \right) = \overline{B}(p, r)$$

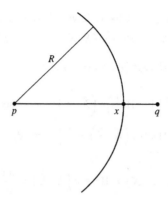

Figure 5.7

for all r (note that $\subset$ always holds). Consider

$$I = \{r : \exp(\overline{B}(0,r) = \overline{B}(p,r)\}.$$

(i)   We have already seen that I contains all r close to zero.

(ii) I is closed: Let $r_i \in I$ converge to r and select $q \in \overline{B}(p,r)$ and $q_i \in \overline{B}(p,r_i)$ converging to q. We can find $v_i \in \overline{B}(0,r_i)$ with $q_i = \exp_p(v_i)$. Then (v_i) will subconverge to some $v \in \overline{B}(0,r)$. Continuity of $\exp_p$ then implies that $\exp_p(v) = q$. (You should think about why it is possible to choose the q_i's.)

(iii) I is open: We show that if $R \in I$, then $R+\varepsilon \in I$ for all small ε. First, choose a compact set K that contains $\overline{B}(p,R)$ in its interior. Then fix $\varepsilon > 0$ such that all points in K of distance $\leq \varepsilon$ can be joined by a unique geodesic segment. Given

$$q \in \overline{B}(p, R + \varepsilon) - \overline{B}(p, R)$$

select for each $\delta > 0$ a curve $\gamma_\delta : [0,1] \to M$ with

$$\begin{aligned}
\gamma_\delta(0) &= p, \\
\gamma_\delta(1) &= q, \\
L(\gamma_\delta) &\leq d(p,q) + \delta.
\end{aligned}$$

Suppose t_δ is the first value such that $\gamma_\delta(t_\delta) \in \partial\overline{B}(p,R)$. If x is an accumulation point for $\gamma_\delta(t_\delta)$, then we must have that

$$R + d(x,q) = d(p,x) + d(x,q) = d(p,q).$$

Now choose a segment from q to x and a segment from p to x of the form $\exp_p(tv)$, see also Figure 5.7. These two geodesics together form a curve from p to q of length $d(p,q)$. Hence, it is a segment. Consequently, it is smooth and by uniqueness of geodesics is the continuation of $\exp_p(tv)$, $0 \leq t \leq 1 + \frac{\varepsilon}{|\dot\gamma|}$. This shows that $q \in \exp_p\left(\overline{B}(0, R + \varepsilon)\right)$.

Statements (i), (ii), and (iii) together imply that $I = [0, \infty)$, which is what we wanted to prove.   □

From part (ii) of (2) $\Rightarrow$ (3) we get the additional result:

COROLLARY 5. *If M is complete in any of the above ways, then any two points in M can be joined by a segment.*

COROLLARY 6. *Suppose M admits a proper (preimages of compact sets are compact) Lipschitz function $f : M \to \mathbb{R}$. Then M is complete.*

PROOF. We establish the Heine-Borel property. Let $C \subset M$ be bounded and closed. Since f is Lipschitz the image $f(C)$ is also bounded. Thus $f(C) \subset [a, b]$ and $C \subset f^{-1}([a, b])$. As f is proper the preimage $f^{-1}([a, b])$ is compact. Since C is closed and a subset of a compact set it must itself be compact. $\qquad\square$

This corollary makes it easy to check completeness for all of our examples. In these examples, the distance function can be extended to a proper continuous function on the entire space.

From now on, virtually all Riemannian manifolds will automatically be assumed to be connected and complete.

9. Characterization of Segments

In this section we will try to determine when a geodesic is a segment and then use this to find a maximal domain in $T_p M$ on which the exponential map is an embedding. These issues can be understood through a systematic investigation of when distance functions to points are smooth. All Riemannian manifolds are assumed to be complete in this section.

9.1. The Segment Domain. Fix $p \in (M, g)$ and let $r(x) = d(x, p)$. We know that r is smooth near p and that the integral curves for r are geodesics emanating from p. Since M is complete, these integral curves can be continued indefinitely beyond the places where r is smooth. These geodesics could easily intersect after some time, thus they don't generate a flow on M, but just having them at points where r might not be smooth helps us understand why r is not smooth at these places. We know from chapter 2 that another obstruction to r being smooth is the possibility of conjugate points (we use the notation *conjugate points* instead of focal point for distance functions to a point).

To clarify matters we introduce some terminology: The *segment domain* is

$$\mathrm{seg}(p) = \left\{ v \in T_p M : \exp_p(tv) : [0, 1] \to M \text{ is a segment} \right\}.$$

The Hopf-Rinow Theorem implies that $M = \exp_p(\mathrm{seg}(p))$. We see that $\mathrm{seg}(p)$ is a closed star-shaped subset of $T_p M$. The star "interior" of $\mathrm{seg}(p)$ is

$$\mathrm{seg}^0(p) = \{ sv : s \in [0, 1), \ v \in \mathrm{seg}(p) \}.$$

We shall show below that this set is in fact the interior of $\mathrm{seg}(p)$, but this requires that we know the set is open. We start by proving

PROPOSITION 19. *If $x \in \exp_p(\mathrm{seg}^0(p))$, then it joined to p by a unique segment. In particular $\exp_p$ is injective on $\mathrm{seg}^0(p)$.*

PROOF. To see this note that there is a segment $\sigma : [0, 1) \to M$ with $\sigma(0) = p$, $\sigma(t_0) = x$, $t_0 < 1$. Therefore, if $\hat{\sigma} : [0, t_0] \to M$ is another segment from p to x, we could construct a nonsmooth segment

$$\gamma(s) = \begin{cases} \hat{\sigma}(s), & s \in [0, t_0], \\ \sigma(s), & s \in [t_0, 1], \end{cases}$$

and we know that this is impossible. $\qquad\square$

On the image $U_p = \exp_p(\text{seg}^0(p))$ we can define $\partial_r = D\exp_p(\partial_r)$, which is, we hope, the gradient for

$$r(x) = d(x, p) = |\exp_p^{-1}(x)|.$$

From our earlier observations we know that r would be smooth on U_p with gradient ∂_r if we could show that $\exp_p : \text{seg}^0(p) \to U_p$ is a diffeomorphism. This requires in addition to injectivity that the map is nonsingular and $\text{seg}^0(p)$ is open. Nonsingularity is taken care of in the next lemma.

LEMMA 14. $\exp_p : \text{seg}^0(p) \to U_p$ is nonsingular everywhere, or, in other words, $D\exp_p$ is nonsingular at every point in $\text{seg}^0(p)$.

PROOF. If $\exp_p$ is singular somewhere, then we can find v such that $\exp_p$ is singular at v and nonsingular at all points tv, $t \in [0, 1)$. We claim that $v \notin \text{seg}^0(p)$. As $\gamma(t) = \exp_p(tv)$ is an embedding on $[0, 1]$ we can find neighborhoods U around $[0, 1)v \subset T_pM$ and V around $\gamma([0, 1)) \subset M$ such that $\exp_p : U \to V$ is a diffeomorphism. Note that $v \notin U$ and $\gamma(1) \notin V$. If we take a tangent vector $w \in T_vT_pM$, then we can extend it to a Jacobi field J on T_pM, i.e., $[\partial_r, J] = 0$. Next J can be pushed forward via $\exp_p$ to a vector field, also called J, that also commutes with ∂_r on V. If $D\exp_p|_v w = 0$, then

$$\lim_{t \to 1} J|_{\exp(tv)} = \lim_{t \to 1} D\exp_p(J)|_{\exp(tv)} = 0.$$

In particular, we see that $D\exp_p$ is singular at v iff $\exp_p(v)$ is a conjugate point for r. This characterization of course assumes that r is smooth on a region that has $\exp_p(v)$ as a accumulation point.

The fact that

$$\lim_{t \to 1} g(J, J)|_{\exp(tv)} \searrow 0 \text{ as } t \to 1$$

implies that there must be a sequence of numbers $t_n \to 1$ such that

$$\frac{\partial_r g(J, J)}{g(J, J)}\Big|_{\exp(t_n v)} \to -\infty \text{ as } n \to \infty.$$

Now use the first fundamental equation evaluated on the Jacobi field J

$$\partial_r g(J, J) = 2\text{Hess}r(J, J)$$

to conclude that $\text{Hess}r$ satisfies

$$\frac{\text{Hess}r(J, J)}{g(J, J)}\Big|_{\exp(t_n v)} \to -\infty \text{ as } n \to \infty.$$

If we assume that $v \in \text{seg}^0(p)$, then $\gamma(t) = \exp_p(tv)$ is a segment on some interval $[0, 1 + \varepsilon]$, $\varepsilon > 0$. Choose ε so small that $\tilde{r}(x) = d(x, \gamma(1 + \varepsilon))$ is smooth on a ball $B(\gamma(1 + \varepsilon), 2\varepsilon)$ (which contains $\gamma(1)$). Then consider the function

$$e(x) = r(x) + \tilde{r}(x).$$

From the triangle inequality, we know that

$$e(x) \geq 1 + \varepsilon = d(p, \gamma(1 + \varepsilon))$$

Furthermore, $e(x) = 1 + \varepsilon$ whenever $x = \gamma(t)$, $t \in [0, 1 + \varepsilon]$. Thus, e has an absolute minimum along $\gamma(t)$ and must therefore have nonnegative Hessian at all the points $\gamma(t)$. On the other hand,

$$\frac{\text{Hesse}(J, J)}{g(J, J)}\Big|_{\exp(t_n v)} = \frac{\text{Hess}r(J, J)}{g(J, J)}\Big|_{\exp(t_n v)} + \frac{\text{Hess}\tilde{r}(J, J)}{g(J, J)}\Big|_{\exp(t_n v)} \xrightarrow[n \to \infty]{} -\infty$$

since Hess$\tilde{r}$ is bounded in a neighborhood of $\gamma(1)$ and the term involving Hess$r$ goes to $-\infty$ as $n \to \infty$. □

We have now shown that $\exp_p$ is injective and has nonsingular differential on $\mathrm{seg}^0(p)$. Before showing that $\mathrm{seg}^0(p)$ is open we characterize elements in the star "boundary" of $\mathrm{seg}^0(p)$ as points that fail to have one of these properties.

LEMMA 15. *If $v \in \mathrm{seg}(p) - \mathrm{seg}^0(p)$, then either*
(1) $\exists w\,(\neq v) \in \mathrm{seg}(p) : \exp_p(v) = \exp_p(w)$, *or*
(2) $D\exp_p$ *is singular at v.*

PROOF. Let $\gamma(t) = \exp_p(tv)$. For $t > 1$ choose segments

$$\sigma_t(s) \quad : \quad [0,1] \to M,$$
$$\sigma_t(0) \quad = \quad p,$$
$$\sigma_t(1) \quad = \quad \gamma(t).$$

Since we have assumed that $\gamma : [0,t]$ is not a segment for $t > 1$ we see that $\dot{\sigma}_t(0)$ is never proportional to $\dot{\gamma}(0)$. Now choose $t_n \to 1$ such that $\dot{\sigma}_{t_n}(0) \to w \in T_pM$. We have that

$$\ell(\sigma_{t_n}) = |\dot{\sigma}_{t_n}(0)| \to \ell(\gamma|_{[0,1]}) = |\dot{\gamma}(0)|,$$

so $|w| = |\dot{\gamma}(0)|$. Now either $w = \dot{\gamma}(0)$ or $w \neq \dot{\gamma}(0)$. In the latter case, we note that w is not a positive multiple of $\dot{\gamma}(0)$ since $|w| = |\dot{\gamma}(0)|$. Therefore, we have found the promised w from (1). If the former happens, we must show that $D\exp_p$ is singular at v. If, in fact, $D\exp_p$ is nonsingular at v, then $\exp_p$ is an embedding near v. Thus,

$$\dot{\sigma}_{t_n}(0) \quad \to \quad v = \dot{\gamma}(0),$$
$$\exp_p(\dot{\sigma}_{t_n}(0)) \quad = \quad \exp_p(t_n\dot{\gamma}(0)),$$

implies $\dot{\sigma}_{t_n}(0) = t_n \cdot v$, showing that γ is a segment on some interval $[0, t_n]$, $t_n > 1$. This, however, contradicts our choice of γ. □

Notice that in the first case the gradient ∂_r on M becomes undefined at $x = \exp_p(v)$, since it would be either $D\exp_p(v)$ or $D\exp_p(w)$, while in the second case the Hessian of r becomes undefined, since it is forced to go to $-\infty$ along certain fields. Finally we show

PROPOSITION 20. $\mathrm{seg}^0(p)$ *is open.*

PROOF. If we fix $v \in \mathrm{seg}^0(p)$, then there is going to be a neighborhood V around v on which $\exp_p$ is a diffeomorphism onto its image. If $v_i \in V$ converge to v, then we know that $D\exp_p$ is also nonsingular at v_i. Now assume that $w_i \in \mathrm{seg}(p)$ satisfy

$$\exp_p(v_i) = \exp_p(w_i).$$

In case w_i has an accumulation point $w \neq v$, we get $v \notin \mathrm{seg}^0(p)$. Hence $w_i \to v$, showing that $w_i \in V$ for large i. As $\exp_p$ is a diffeomorphism on V this implies that $w_i = v_i$. Thus we have shown that $v^i \in \mathrm{seg}^0(p)$. □

All of this implies that $r(x) = d(x,p)$ is smooth on the open and dense subset $U_p - \{p\} \subset M$ and in addition that it is not smooth on $M - U_p$.

The set $\mathrm{seg}(p) - \mathrm{seg}^0(p)$ is called the *cut locus* of p in T_pM. Thus, being inside the cut locus means that we are on the region where r is smooth. Going back to our characterization of segments, we have

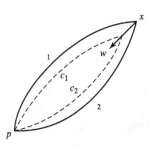

Figure 5.8

COROLLARY 7. *Let $\gamma : [0, \infty) \to M$ be a geodesic with $\gamma(0) = p$. If*

$$\text{cut}(\dot{\gamma}(0)) = \max\{t : \gamma|_{[0,t]} \text{ is a segment}\},$$

then r is smooth at $\gamma(t)$, $t < \text{cut}(\dot{\gamma}(0))$, but not smooth at $x = \gamma(\text{cut}(\dot{\gamma}(0)))$. Furthermore, the failure of r to be smooth at x is because $\exp_p : \text{seg}(p) \to M$ either fails to be one-to-one at x or has x as a critical value.

9.2. The Injectivity Radius. The largest radius ε for which

$$\exp_p : B(0, \varepsilon) \to B(p, \varepsilon)$$

is a diffeomorphism is called the *injectivity radius* inj(p) at p. If $v \in \text{seg}(p) - \text{seg}^0(p)$ is the closest point to 0 in this set, then we have that inj$(p) = |v|$. It turns out that such v can be characterized as:

LEMMA 16. (Klingenberg): *Suppose $v \in \text{seg}(p) - \text{seg}^0(v)$ and that $|v| = \text{inj}(p)$. Then either*

(1) there is precisely one other vector w with

$$\exp_p(w) = \exp_p(v),$$

and it is characterized by

$$\frac{d}{dt}\Big|_{t=1} \exp_p(tw) = -\frac{d}{dt}\Big|_{t=1} \exp_p(tv),$$

or

(2) $x = \exp_p(v)$ is a critical value for $\exp_p : \text{seg}(p) \to M$.

In the first case there are exactly two segments from p to $x = \exp_p(v)$, and they fit smoothly together at x to form a geodesic loop.

PROOF. Suppose x is a regular value for $\exp_p : \text{seg}(p) \to M$ and that $\gamma_1, \gamma_2 : [0, 1] \to M$ are segments from p to $x = \exp_p(v)$. If $\dot{\gamma}_1(1) \neq -\dot{\gamma}_2(1)$, then we can find $w \in T_x M$ such that $g(w, \dot{\gamma}_1(1))$, $g(w, \dot{\gamma}_2(1)) < 0$, i.e., w forms an angle $> \frac{\pi}{2}$ with both $\dot{\gamma}_1(1)$ and $\dot{\gamma}_2(1)$. Next select $c(s)$ with $\dot{c}(0) = w$. As $D\exp_p$ is nonsingular at $\dot{\gamma}_i(0)$ there are unique curves $v_i(s) \in T_p M$ with $v_i(0) = \dot{\gamma}_i(0)$ and $D\exp_p(v_i(s)) = c(s)$ (see also Figure 5.8). But then the curves $t \to \exp_p(tv_i(s))$ have length

$$\begin{aligned} |v_i| &= d(p, c(s)) \\ &< d(p, x) \\ &= |v|. \end{aligned}$$

This implies that $\exp_p$ is not one-to-one on $\text{seg}^0(p)$, a contradiction. □

10. Riemannian Isometries

We are now ready to explain the key properties of Riemannian isometries. Much of theory is local, so we shall not necessarily assume that the Riemannian manifolds being investigated are complete. After this thorough discussion of Riemannian isometries we classify all complete simply connected Riemannian manifolds of constant sectional curvature.

10.1. Local Isometries. We say that a map $F : (M, g) \to (N, \bar{g})$ is a *local Riemannian isometry* if for each $p \in M$ the differential $DF_p : T_pM \to T_{F(p)}N$ is a linear isometry. A special and trivial example of such a map is a local coordinate system $\varphi : U \to \Omega \subset \mathbb{R}^n$ where we use the induced metric g on U and its coordinate representation $g_{ij}dx^i dx^j$ on Ω.

PROPOSITION 21. *Let $F : (M, g) \to (N, \bar{g})$ be a local Riemannian isometry.*

(1) F maps geodesics to geodesics.

(2) $F \circ \exp_p(v) = \exp_{F(p)} \circ DF_p(v)$ if $\exp_p(v)$ is defined. In other words

$$
\begin{array}{ccc}
O_p \subset T_pM & \xrightarrow{DF} & O_{F(p)} \subset T_{F(p)}N \\
\exp_p \downarrow & & \exp_{F(p)} \downarrow \\
M & \xrightarrow{F} & N
\end{array}
$$

(3) F is distance decreasing.

(4) If F is also a bijection, then it is distance preserving.

PROOF. (1) The first property is completely obvious. We know that geodesics depend only on the metric and not on any given coordinate system. However, a local Riemannian isometry is locally nothing but a change of coordinates.

(2) If $\exp_p(v)$ is defined, then $t \to \exp_p(tv)$ is a geodesic. Thus also $t \to F(\exp_p(tv))$ is a geodesic. Since

$$
\begin{aligned}
\frac{d}{dt} F(\exp_p(tv))|_{t=0} &= DF\left(\frac{d}{dt} \exp_p(tv)|_{t=0}\right) \\
&= DF(v),
\end{aligned}
$$

we have that $F(\exp_p(tv)) = \exp_{F(p)}(tDF(v))$. Setting $t = 1$ then proves the claim.

(3) This is also obvious as F must preserve the length of curves.

(4) Both F and F^{-1} are distance decreasing so they must both be distance preserving. $\qquad \square$

This proposition quickly yields two important results for local Riemannian isometries.

PROPOSITION 22. (Uniqueness of Riemannian Isometries) *Let $F, G : (M, g) \to (N, \bar{g})$ be local Riemannian isometries. If M is connected and $F(p) = G(p)$, $DF_p = DG_p$, then $F = G$ on M.*

PROOF. Let

$$
A = \{x \in M : F(x) = G(x), \ DF_x = DG_x\}.
$$

We know that $p \in A$ and that A is closed. Property (2) from the above proposition tells us that

$$
\begin{aligned}
F \circ \exp_x (v) &= \exp_{F(x)} \circ DF_x (v) \\
&= \exp_{G(x)} \circ DG_x (v) \\
&= G \circ \exp_x (v),
\end{aligned}
$$

if $x \in A$. Since $\exp_x$ maps onto a neighborhood of x it follows that some neighborhood of x also lies in A. This shows that A is open and hence all of M as M is connected. $\qquad\square$

PROPOSITION 23. *Let $F : (M, \bar{g}) \to (N, g)$ be a Riemannian covering map. $(M, \bar{g})$ is complete if and only if (N, g) is complete.*

PROOF. Let $\gamma : (-\varepsilon, \varepsilon) \to N$ be a geodesic with $\gamma (0) = p$ and $\dot{\gamma} (0) = v$. For any $\bar{p} \in F^{-1} (p)$ there is a unique lift $\bar{\gamma} : (-\varepsilon, \varepsilon) \to M$, i.e., $F \circ \bar{\gamma} = \gamma$, with $\bar{\gamma} (0) = \bar{p}$. Since F is a local isometry, the inverse is locally defined and also an isometry. Thus $\bar{\gamma}$ is also a geodesic.

If we assume N is complete, then γ and also $\bar{\gamma}$ will exist for all time. As all geodesics in M must be of the form $\bar{\gamma}$ this shows that all geodesics in M exist for all time.

If, conversely, we suppose that M is complete, then $\bar{\gamma}$ can be extended to be defined for all time. Then $F \circ \bar{\gamma}$ is a geodesic defined for all time that extends γ. Thus N is geodesically complete. $\qquad\square$

LEMMA 17. *Let $F : (M, g) \to (N, \bar{g})$ be a local Riemannian isometry. If M is complete, then F is a Riemannian covering map.*

PROOF. Fix $q \in N$ and assume that $\exp_q : B (0, \varepsilon) \to B (q, \varepsilon)$ is a diffeomorphism. We claim that $F^{-1} (B (q, \varepsilon))$ is evenly covered by the sets $B (p, \varepsilon)$ where $F (p) = q$. Completeness of M guarantees that $\exp_p : B (0, \varepsilon) \to B (p, \varepsilon)$ is defined and property (2) that

$$
F \circ \exp_p (v) = \exp_q \circ DF_p (v)
$$

for all $v \in B (0, \varepsilon) \subset T_p M$. As $\exp_q : B (0, \varepsilon) \to B (q, \varepsilon)$ and $DF_p : B (0, \varepsilon) \to B (0, \varepsilon)$ are diffeomorphisms it follows that $F \circ \exp_p : B (0, \varepsilon) \to B (q, \varepsilon)$ is a diffeomorphism. Thus each of the maps $\exp_p : B (0, \varepsilon) \to B (p, \varepsilon)$ and $F : B (p, \varepsilon) \to B (q, \varepsilon)$ are diffeomorphisms as well. Finally we need to make sure that

$$
F^{-1} (B (q, \varepsilon)) = \bigcup_{F(p)=q} B (p, \varepsilon).
$$

If $x \in F^{-1} (B (q, \varepsilon))$, then we can join q and $F (x)$ by a unique geodesic $\gamma (t) = \exp_q (tv)$, $v \in B (0, \varepsilon)$. Completeness of M again guarantees a geodesic $\sigma : [0, 1] \to M$ with $\sigma (1) = x$ and $DF_x (\dot{\sigma} (1)) = \dot{\gamma} (1)$. Since $F \circ \sigma$ is a geodesic with the same initial values at $t = 1$ as γ we must have $F (\sigma (t)) = \gamma (t)$ for all t. As $q = \gamma (0)$ we have therefore proven that $F (\sigma (0)) = q$ and hence that $x \in B (\sigma (0), \varepsilon)$. $\qquad\square$

If $S \subset \mathrm{Iso} (M, g)$ is a set of isometries, then the *fixed point set* of S is defined as those points in M that are fixed by all isometries in S

$$
\mathrm{Fix} (S) = \{x \in M : F (x) = x \text{ for all } F \in S\}.
$$

While the fixed point set for a general set of diffeomorphisms can be quite complicated, the situation for isometries is much more manageable. A submanifold

$N \subset (M, g)$ is said to be *totally geodesic* if for each $p \in N$ a neighborhood of $0 \in T_pN$ is mapped into N via the exponential map $\exp_p$. This means that geodesics in N are also geodesics in M and conversely that any geodesic in M which is tangent to N at some point must lie in N for a short time.

PROPOSITION 24. *Let $S \subset \mathrm{Iso}\,(M, g)$ be a set of isometries, then each connected component of the fixed point set is a totally geodesic submanifold.*

PROOF. Let $p \in \mathrm{Fix}\,(S)$ and consider the subspace $V \subset T_pM$ that is fixed by the linear isometries $DF_p : T_pM \to T_pM$, where $F \in S$. Note that each such F fixes p so we know that $DF_p : T_pM \to T_pM$. If $v \in V$, then $t \to \exp_p(tv)$ must be fixed by each of the isometries in S as the initial position and velocity is fixed by these isometries. Thus $\exp_p(tv) \in \mathrm{Fix}\,(S)$ as long as it is defined. This shows that $\exp_p : V \to \mathrm{Fix}\,(S)$.

Next let $\varepsilon > 0$ be chosen so that $\exp_p : B\,(0, \varepsilon) \to B\,(p, \varepsilon)$ is a diffeomorphism. If $q \in \mathrm{Fix}\,(S) \cap B\,(p, \varepsilon)$, then the unique geodesic $\gamma : [0, 1] \to B\,(p, \varepsilon)$ from p to q has the property that its endpoints are fixed by each $F \in S$. Now $F \circ \gamma$ is also a geodesic from p to q which in addition lies in $B\,(p, \varepsilon)$ as the length is unchanged. Thus $F \circ \gamma = \gamma$ and hence γ lies in $\mathrm{Fix}\,(S) \cap B\,(p, \varepsilon)$.

Thus we have shown that $\exp_p : V \cap B\,(0, \varepsilon) \to \mathrm{Fix}\,(S) \cap B\,(p, \varepsilon)$ is a bijection. This establishes the lemma.   □

10.2. Constant Curvature Revisited. We just saw that isometries are uniquely determined by their differential. What about the existence question? Given any linear isometry $L : T_pM \to T_qN$, is there an isometry $F : M \to N$ such that $DF_p = L$? If we let $M = N$, this would in particular mean that if π is a 2-plane in T_pM and $\tilde{\pi}$ a 2-plane in T_qM, then there should be an isometry $F : M \to M$ such that $F(\pi) = \tilde{\pi}$. But this would imply that M has constant sectional curvature. The above problem can therefore not be solved in general. If we go back and inspect our knowledge of $\mathrm{Iso}(S_k^n)$, we see that these spaces have enough isometries so that any linear isometry $L : T_pS_k^n \to T_qS_k^n$ can be extended to a global isometry $F : S_k^n \to S_k^n$ with $DF_p = L$. In some sense these are the only spaces with this property, as we shall see.

THEOREM 17. *Suppose (M, g) is a Riemannian manifold of dimension n and constant curvature k. If M is simply connected and $L : T_pM \to T_qS_k^n$ is a linear isometry, then there is a unique local Riemannian isometry called the* monodromy *map $F : M \to S_k^n$ with $DF_p = L$. Furthermore, this map is a diffeomorphism if (M, g) is complete.*

Before giving the proof, let us look at some examples.

EXAMPLE 37. *Suppose we have an immersion $M^n \to S_k^n$. Then F will be one of the maps described in the theorem if we use the pullback metric on M. Such maps can fold in wild ways when $n \geq 2$ and need not resemble covering maps in any way whatsoever.*

EXAMPLE 38. *If $U \subset S_k^n$ is a contractible bounded open set with ∂U a smooth hypersurface, then one can easily construct a diffeomorphism $F : M = S_k^n - \{pt\} \to S_k^n - U$. Near the missing point in M the metric will necessarily look pretty awful, although it has constant curvature.*

EXAMPLE 39. *If $M = \mathbb{R}P^n$ or $(\mathbb{R}^n - \{0\})/antipodal$ map, then M is not simply connected and does not admit an immersion into S_k^n.*

EXAMPLE 40. *If M is the universal covering of the constant curvature sphere with a pair of antipodal point removed $S^2 - \{\pm p\}$, then the monodromy map is not one-to-one. In fact it must be the covering map $M \to S^2 - \{\pm p\}$.*

COROLLARY 8. *If M is a closed simply connected manifold with constant-curvature k, then $k > 0$ and $M = S^n$. Thus, $S^p \times S^q, \mathbb{C}P^n$ do not admit any constant curvature metrics.*

COROLLARY 9. *If M is geodesically complete and noncompact with constant curvature k, then $k \leq 0$ and the universal covering is diffeomorphic to $\mathbb{R}^n$. In particular, $S^2 \times \mathbb{R}^2$ and $S^n \times \mathbb{R}$ do not admit any geodesically complete metrics of constant curvature.*

Now for the proof of the theorem. A different proof of the case where M is complete is developed in the exercises to this chapter.

PROOF. We know that M can be covered by sets U_α such that each U_α admits a Riemannian embedding $F_\alpha : U_\alpha \to S_k^n$. Furthermore, if $q \in U_\alpha$, $\bar{q} \in S_k^n$ and $L : T_q U_\alpha \to T_{\bar{q}} S_k^n$ is a linear isometry, then there is a unique F_α such that $F_\alpha(q) = \bar{q}$ and $DF_\alpha|_p = L$.

The construction of F now proceeds in the same way one does analytic continuation on simply connected domains. We fix base points $p \in M$, $\bar{p} \in S_k^n$ and a linear isometry $L : T_p M \to T_{\bar{p}} S_k^n$. Next let $x \in M$ be an arbitrary point. If $\gamma : [0,1] \to M$ is a curve from p to x, then we can cover it by a string of sets $U_{\alpha_0}, ..., U_{\alpha_k}$, where $p \in U_{\alpha_0}$, $x \in U_{\alpha_k}$, and $\gamma(t_i) \in U_{\alpha_i} \cap U_{\alpha_{i+1}}$. Define F on U_{α_0} so that $F(p) = \bar{p}$ and $DF_p = L$. Then define $F|_{U_{\alpha_{i+1}}}$ successively such that it agrees with $F|_{U_{\alpha_i}}$ and $DF|_{U_{\alpha_i}}$ at $\gamma(t_i)$. This defines F uniquely on all of the sets U_{α_i} and hence also at x. If we covered γ by a different string of sets, then uniqueness of isometries tell us that we have to get the same answer along γ as we assume that $F(p) = \bar{p}$ and $DF_p = L$. If we used a different path $\bar{\gamma}$ which was also covered by the same string of sets U_{α_i} we would clearly also end up with the same answer at x. Finally we use that M is simply connected to connect any two paths γ_0, γ_1 from p to x by a family of paths $H(s,t)$ such that each $\gamma_s(t) = H(s,t)$ is a path from p to x. If F_{γ_s} is the map we obtain near x by using the path γ_s, then we have just seen that $F_{\gamma_s}(x)$ is fixed as long as s is so small that all the curves are covered by the same string of sets. This shows that $s \to F_{\gamma_s}(x)$ is locally constant and hence that $F(x)$ is well-defined by our construction.

If M is complete we know that F has to be a covering map. As S_k^n is simply connected it must be a diffeomorphism. $\qquad \square$

We can now give the classification of complete simply connected Riemannian manifolds with constant curvature. This result was actually proven before the issues of completeness were completely understood. Killing first proved the result assuming in effect that the manifold has an $\varepsilon > 0$ such that for all p the map $\exp_p : B(0,\varepsilon) \to B(p,\varepsilon)$ is a diffeomorphism. Hopf then realized that it was sufficient to assume that the manifold was geodesically complete. Since metric completeness immediately implies geodesic completeness this is clearly the best result one could have expected at the time.

COROLLARY 10. (Classification of Constant Curvature Spaces, Killing, 1893 and H. Hopf, 1926) *If (M, g) is a connected, geodesically complete Riemannian manifold with constant curvature k , then the universal covering is isometric to S_k^n.*

This result shows how important the completeness of the metric is. A large number of open manifolds admit immersions into Euclidean space of the same dimension (e.g., $S^n \times \mathbb{R}^k$) and therefore carry incomplete metrics with zero curvature. Carrying a complete Riemannian metric of a certain type, therefore, often implies various topological properties of the underlying manifold. Riemannian geometry at its best tries to understand this interplay between metric and topological properties.

10.3. Metric Characterization of Maps. As promised we shall in this section give some metric characterizations of Riemannian isometries and Riemannian submersions. For a Riemannian manifold (M, g) we let the corresponding metric space be denoted by (M, d_g) or simply (M, d) if only one metric is in play. It is natural to ask whether one can somehow recapture the Riemannian metric g from the distance d_g. If for instance $v, w \in T_p M$, then we would like to be able to compute $g(v, w)$ from knowledge of d_g. One way of doing this is by taking two curves α, β such that $\dot{\alpha}(0) = v$ and $\dot{\beta}(0) = w$ and observe that

$$|v| = \lim_{t \to 0} \frac{d(\alpha(t), \alpha(0))}{t},$$

$$|w| = \lim_{t \to 0} \frac{d(\beta(t), \beta(0))}{t},$$

$$\cos \angle (v, w) = \frac{g(v, w)}{|v| |w|} = \lim_{t \to 0} \frac{d(\alpha(t), \beta(t))}{t}.$$

Thus, g can really be found from d given that we use the differentiable structure of M. It is perhaps then not so surprising that many of the Riemannian maps we consider have synthetic characterizations, that is, characterizations that involve only knowledge of the metric space (M, d).

Before proceeding with our investigations, let us introduce a new type of coordinates. Using geodesics we have already introduced one set of geometric coordinates via the exponential map. We shall now use the distance functions to construct *distance coordinates*. For a point $p \in M$ fix a neighborhood $U \ni p$ such that for each $x \in U$ we have that $B(q, \text{inj}(q)) \supset U$. Thus, for each $q \in U$ the distance function $r_q(x) = d(x, q)$ is smooth on $U - \{q\}$. Now choose $q_1, \ldots, q_n \in U - \{p\}$, where $n = \dim M$. If the vectors $\nabla r_{q_1}(p), \ldots, \nabla r_{q_n}(p) \in T_p M$ are linearly independent, the inverse function theorem tells us that $\varphi = (r_{q_1}, \ldots, r_{q_n})$ can be used as coordinates on some neighborhood V of p. The size of the neighborhood will depend on how these gradients vary. Thus, an explicit estimate for the size of V can be gotten from bounds on the Hessians of the distance functions. Clearly, one can arrange for the gradients to be linearly independent or even orthogonal at any given point.

We just saw that bijective Riemannian isometries are distance preserving. The next result shows that the converse is also true.

THEOREM 18. (Myers-Steenrod, 1939) *If (M, g) and $(N, \bar{g})$ are Riemannian manifolds and $F : M \to N$ a bijection, then F is a Riemannian isometry if F is distance-preserving, i.e., $d_{\bar{g}}(F(p), F(q)) = d_g(p, q)$ for all $p, q \in M$.*

PROOF. Let F be distance-preserving. First let us show that F is differentiable. Fix $p \in M$ and let $q = F(p)$. Near q introduce distance coordinates $(r_{q_1}, \ldots, r_{q_n})$

and find p_i such that $F(p_i) = q_i$. Now observe that

$$
\begin{aligned}
r_{q_i} \circ F(x) &= d(F(x), q_i) \\
&= d(F(x), F(p_i)) \\
&= d(x, p_i).
\end{aligned}
$$

Since $d(p, p_i) = d(q, q_i)$, we can assume that the q_is and p_is are chosen such that $r_{p_i}(x) = d(x, p_i)$ are smooth at p. Thus, $(r_{q_1}, \ldots, r_{q_n}) \circ F$ is smooth at p, showing that F must be smooth at p.

To show that F is a Riemannian isometry it suffices to check that $|DF(v)| = |v|$ for all tangent vectors $v \in TM$. For a fixed $v \in T_p M$ let $\gamma(t) = \exp_p(tv)$. For small t we know that γ is a constant speed segment. Thus, for small t, s we can conclude

$$
|t - s| \cdot |v| = d_g(\gamma(t), \gamma(s)) = d_{\bar{g}}(F \circ \gamma(t), F \circ \gamma(s)),
$$

implying

$$
\begin{aligned}
|DF(v)| &= \left| \frac{d(F \circ \gamma)}{dt} \right|_{t=0} \\
&= \lim_{t \to 0} \frac{d_{\bar{g}}(F \circ \gamma(t), F \circ \gamma(0))}{|t|} \\
&= \lim_{t \to 0} \frac{d_g(\gamma(t), \gamma(0))}{|t|} \\
&= |\dot{\gamma}(0)| \\
&= |v|.
\end{aligned}
$$

$\square$

Our next goal is to find a characterization of Riemannian submersions. Unfortunately, the description only gives us functions that are C^1, but there doesn't seem to be a better formulation. Let $F : (M, \bar{g}) \to (N, g)$ be a function. We call F a *submetry* if for every $p \in M$ we can find $r > 0$ such that for each $\varepsilon \le r$ we have $F(B(p, \varepsilon)) = B(F(p), \varepsilon)$. Submetries are locally distance-nonincreasing and therefore also continuous. In addition, we have that the composition of submetries (or Riemannian submersions) are again submetries (or Riemannian submersions). We can now prove

THEOREM 19. (Berestovski, 1995) *If $F : (M, \bar{g}) \to (N, g)$ is a surjective submetry, then F is a C^1 Riemannian submersion.*

PROOF. Fix points $q \in N$ and $p \in M$ with $F(p) = q$. Then select distance coordinates $(r_1, \ldots, r_k)$ around q. Now observe that all of the r_is are Riemannian submersions and therefore also submetries. Then the compositions $r_i \circ F$ are also submetries. Thus, F is C^1 iff all the maps $r_i \circ F$ are C^1. Therefore, it suffices to prove the result in the case of functions $r : (U \subset M, g) \to ((a, b), \text{can})$.

Let $x \in M$. By restricting r to a small convex neighborhood of x, we can assume that the fibers of r are closed and that any two points in the domain are joined by a unique geodesic. We now wish to show that r has a continuous unit gradient field ∇r. We know that the integral curves for ∇r should be exactly the unit speed geodesics that are mapped to unit speed geodesics by r. Since r is distance-nonincreasing, it is clear that any piecewise smooth unit speed curve that is mapped to a unit speed geodesic must be a smooth unit speed geodesic. Thus, these integral curves are unique and vary continuously to the extent that they exist. To establish the existence of these curves we use the submetry property. First fix $p \in M$ and let $\gamma(t) : [0, r] \to (a, b)$ be the unit speed segment with $\gamma(0) = r(p)$. Let U_t denote the fiber of r above $\gamma(t)$. Now select a unit speed segment $\bar{\gamma} : [0, r] \to M$ with $\bar{\gamma}(0) = p$ and $\gamma(r) \in U_r$. This is possible since $r(B(p, \varepsilon)) = B(\gamma(0), \varepsilon)$. It is now easy to check, again using the submetry property, that $\gamma(t) = r \circ \bar{\gamma}(t)$, as desired. $\qquad \square$

11. Further Study

There are many textbooks on Riemannian geometry that treat all of the basic material included in this chapter. Some of the better texts are [**19**], [**20**], [**41**], [**56**] and [**73**]. All of these books, as is usual, emphasize the variational approach as being *the* basic technique used to prove every theorem. To see how the variational approach works the text [**68**] is also highly recommended.

12. Exercises

(1) Assume that (M, g) has the property that all geodesics exist for a fixed time $\varepsilon > 0$. Show that (M, g) is geodesically complete.

(2) A Riemannian manifold is said to be homogeneous if the isometry group acts transitively. Show that homogeneous manifolds are geodesically complete.

(3) Assume that we have coordinates in a Riemannian manifold so that $g_{1i} = \delta_{1i}$. Show that x^1 is a distance function.

(4) Let γ be a geodesic in a Riemannian manifold (M, g). Let g' be another Riemannian metric on M with the properties: $g'(\dot{\gamma}, \dot{\gamma}) = g(\dot{\gamma}, \dot{\gamma})$ and $g'(X, \dot{\gamma}) = 0$ iff $g(X, \dot{\gamma}) = 0$. Show that γ is also a geodesic with respect to g'.

(5) Show that if we have a vector field X on a Riemannian manifold (M, g) that vanishes at $p \in M$, then for any tensor T we have $L_X T = \nabla_X T$ at p. Conclude that the Hessian of a function is independent of the metric at a critical point. Can you find an interpretation of $L_X T$ at p?

(6) Show that any Riemannian manifold (M, g) admits a conformal change $(M, \lambda^2 g)$, where $\lambda : M \to (0, \infty)$, such that $(M, \lambda^2 g)$ is complete.

(7) On an open subset $U \subset \mathbb{R}^n$ we have the induced distance from the Riemannian metric, and also the induced distance from $\mathbb{R}^n$. Show that the two can agree even if U isn't convex.

(8) Let $N \subset (M, g)$ be a submanifold. Let ∇^N denote the connection on N that comes from the metric induced by g. Define the second fundamental form of N in M by

$$\mathrm{II}(X, Y) = \nabla_X^N Y - \nabla_X Y$$

   (a) Show that $\mathrm{II}(X, Y)$ is symmetric and hence tensorial in X and Y.
   (b) Show that $\mathrm{II}(X, Y)$ is always normal to N.
   (c) Show that $\mathrm{II} = 0$ on N iff N is totally geodesic.
   (d) If R^N is the curvature tensor for N, then

$$\begin{aligned} g(R(X, Y)Z, W) &= g(R^N(X, Y)Z, W) \\ &\quad - g(\mathrm{II}(Y, Z), \mathrm{II}(X, W)) + g(\mathrm{II}(X, Z), \mathrm{II}(Y, W)). \end{aligned}$$

(9) Let $f : (M, g) \to \mathbb{R}$ be a smooth function on a Riemannian manifold.

    (a) If $\gamma : (a, b) \to M$ is a geodesic, compute the first and second derivatives of $f \circ \gamma$.

    (b) Use this to show that at a local maximum (or minimum) for f the gradient is zero and the Hessian nonpositive (or nonnegative).

    (c) Show that f has everywhere nonnegative Hessian iff $f \circ \gamma$ is convex for all geodesics γ in (M, g).

(10) Let $N \subset M$ be a submanifold of a Riemannian manifold (M, g).

    (a) The distance from N to $x \in M$ is defined as

$$d(x, N) = \inf \{d(x, p) : p \in N\}.$$

A unit speed curve $\sigma : [a, b] \to M$ with $\sigma(a) \in N, \sigma(b) = x$, and $\ell(\sigma) = d(x, N)$ is called a segment from x to N. Show that σ is also a segment from N to any $\sigma(t)$, $t < b$. Show that $\dot{\sigma}(a)$ is perpendicular to N.

    (b) Show that if N is a closed subspace of M and (M, g) is complete, then any point in M can be joined to N by a segment.

    (c) Show that in general there is an open neighborhood of N in M where all points are joined to N by segments.

    (d) Show that $d(\cdot, N)$ is smooth on a neighborhood of N and that the integral curves for its gradient are the geodesics that are perpendicular to N.

(11) Compute the cut locus on a square torus $\mathbb{R}^2/\mathbb{Z}^2$.

(12) Compute the cut locus on a sphere and real projective space with the constant curvature metrics.

(13) In a metric space (X, d) one can measure the length of continuous curves $\gamma : [a, b] \to X$ by

$$\ell(\gamma) = \sup \left\{ \sum d(\gamma(t_i), \gamma(t_{i+1})) : a = t_1 \le t_2 \le \cdots \le t_{k-1} \le t_k = b \right\}.$$

    (a) Show that a curve has finite length iff it is absolutely continuous. Hint: Use the characterization that $\gamma : [a, b] \to X$ is absolutely continuous if and only if for each $\varepsilon > 0$ there is a $\delta > 0$ so that $\sum d(\gamma(s_i), \gamma(s_{i+1})) \le \varepsilon$ provided $\sum |s_i - s_{i+1}| \le \delta$.

    (b) Show that this definition gives back our previous definition for smooth curves on Riemannian manifolds.

    (c) Let $\gamma : [a, b] \to M$ be an absolutely continuous curve whose length is $d(\gamma(a), \gamma(b))$. Show that $\gamma = \sigma \circ \varphi$ for some segment σ and reparametrization φ.

(14) Show that in a Riemannian manifold,

$$d(\exp_p(tv), \exp_p(tw)) = |t| \cdot |v - w| + O(t^2).$$

(15) Assume that we have coordinates x^i around a point $p \in (M, g)$ such that $x^i(p) = 0$ and $g_{ij}x^j = x^i$. Show that these must be exponential coordinates. Hint: Define

$$r = \sqrt{(x^1)^2 + \cdots + (x^n)^2}$$

and show that it is a smooth distance function away from p, and that the integral curves for the gradient are geodesics emanating from p.

(16) If $N_1, N_2 \subset M$ are totally geodesic submanifolds, show that each component of $N_1 \cap N_2$ is a submanifold which is totally geodesic. Hint: The potential tangent space at $p \in N_1 \cap N_2$ should be $T_p N_1 \cap T_p N_2$.

(17) Show that for a complete manifold the functional distance is the same as the distance. What about incomplete manifolds?

(18) Let $\gamma : [0, 1] \to M$ be a geodesic such that $\exp_{\gamma(0)}$ is regular at all $t\dot{\gamma}(0)$, for $t \leq 1$. Show that γ is a local minimum for the energy functional. Hint: Show that the lift of γ via $\exp_{\gamma(0)}$ is a minimizing geodesic in a suitable metric.

(19) Show, using the exercises on Lie groups from chapters 1 and 2, that on a Lie group G with a bi-invariant metric the geodesics through the identity are exactly the homomorphisms $\mathbb{R} \to G$. Conclude that the Lie group exponential map coincides with the exponential map generated by the bi-invariant Riemannian metric. Hint: First show that homomorphisms $\mathbb{R} \to G$ are precisely the integral curves for left invariant vector fields through $e \in G$.

(20) Repeat the previous exercise assuming that the metric is a bi-invariant semi-Riemannian metric. Show that the matrix group $Gl_n(\mathbb{R})$ of invertible $n \times n$ matrices admits a bi-invariant semi-Riemannian metric. Hint: for $X, Y \in T_I Gl_n(\mathbb{R})$ define

$$g(X, Y) = -\mathrm{tr}(XY).$$

(21) Construct a Riemannian metric on the tangent bundle to a Riemannian manifold (M, g) such that $\pi : TM \to M$ is a Riemannian submersion and the metric restricted to the tangent spaces is the given Euclidean metric.

(22) For a Riemannian manifold (M, g) let FM be the frame bundle of M. This is a fiber bundle $\pi : FM \to M$ whose fiber over $p \in M$ consists of orthonormal bases for $T_p M$. Find a Riemannian metric on FM that makes π into a Riemannian submersion and such that the fibers are isometric to $O(n)$.

(23) Show that a Riemannian submersion is a submetry.

(24) (Hermann) Let $f : (M, \bar{g}) \to (N, g)$ be a Riemannian submersion.
    (a) Show that (N, g) is complete if $(M, \bar{g})$ is complete.
    (b) Show that f is a fibration if $(M, \bar{g})$ is complete i.e., for every $p \in N$ there is a neighborhood $p \in U$ such that $f^{-1}(U)$ is diffeomorphic to $U \times f^{-1}(p)$. Give a counterexample when $(M, \bar{g})$ is not complete.

CHAPTER 6

Sectional Curvature Comparison I

In the last chapter we classified spaces with constant curvature. The goal of this chapter is to compare manifolds to spaces with constant curvature. We shall for instance prove the Hadamard-Cartan theorem, which says that a simply connected manifold with sec ≤ 0 is diffeomorphic to $\mathbb{R}^n$. There are also some interesting restrictions on the topology in positive curvature that we shall investigate, notably, Synge's theorem, which says that an orientable even-dimensional manifold with positive curvature is simply connected. The results in this chapter basically comprise everything that was know about the relationship between topology and curvature prior to 1945. In Chapters 7, 9, 11 we shall deal with some more advanced and modern topics in the theory of manifolds with lower curvature bounds.

We start by introducing the concept of differentiation of vector fields along curves. This generalizes and ties in nicely with our mixed second partials from the last chapter and also allows us to define higher order partials. This is then used to develop the second variation formula of Synge. The second variation formula is used to prove most of the results in this chapter. However, in non-positive curvature we also show how the fundamental equations can be used to give alternate and simpler proofs of several results.

At the end of the chapter we establish some basic comparison estimates that are needed later in the text. These results are used to show how geodesics and curvature can help in estimating the injectivity, conjugate and convexity radius. This is used to give Berger's proof of the classical quarter pinched sphere theorem.

1. The Connection Along Curves

Recall that in chapter 2 we introduced Jacobi and parallel fields for a smooth distance function. Here we are going to generalize these concepts so that we can talk about Jacobi and parallel fields along just one geodesic, rather than the whole family of geodesics associated to a distance function. This will be quite useful when we study variations.

1.1. Vector Fields Along Curves. Let $\gamma : I \to M$ be a curve in M. A vector field V along γ is by definition a map $V : I \to TM$ with $V(t) \in T_{\gamma(t)}M$ for all $t \in I$. We want to define the covariant derivative

$$\dot{V}(t) = \frac{d}{dt}V(t) = \nabla_{\dot{\gamma}}V$$

of V along γ, assuming γ and V have appropriate smoothness. We know that V can be thought of as the variational field for a variation $\bar{\gamma} : (-\varepsilon, \varepsilon) \times I \to M$. It is therefore natural to define

$$\frac{d}{dt}V(t) = \frac{\partial^2 \bar{\gamma}}{\partial t \partial s}(0, t).$$

Doing the calculation in local coordinates gives

$$\begin{aligned} V\left(t\right) & = V^{k}\left(t\right)\partial_{k} \\ & = \frac{\partial\bar{\gamma}^{k}}{\partial s}\left(0,t\right)\partial_{k} \end{aligned}$$

and

$$\begin{aligned} \frac{\partial^{2}\bar{\gamma}}{\partial t\partial s}\left(0,t\right) & = \frac{\partial^{2}\bar{\gamma}^{k}}{\partial t\partial s}\left(0,t\right)\partial_{k} + \frac{\partial\bar{\gamma}^{i}}{\partial s}\left(0,t\right)\frac{\partial\bar{\gamma}^{j}}{\partial t}\left(0,t\right)\Gamma_{ij}^{k}\partial_{k} \\ & = \frac{dV^{k}}{dt}\left(t\right)\partial_{k} + V^{i}\left(t\right)\frac{d\gamma^{j}}{dt}\left(t\right)\Gamma_{ij}^{k}\partial_{k}. \end{aligned}$$

Thus $\dot{V}$ does not depend on how the variation was chosen. As the variation can be selected independently of the coordinate system we see that the local coordinate formula is independent of the coordinate system. This formula also shows that if $V\left(t\right) = X_{\gamma(t)}$ for some vector field X defined in a neighborhood of $\gamma\left(t_{0}\right)$, then

$$\dot{V}\left(t_{0}\right) = \nabla_{\dot{\gamma}(t_{0})}X.$$

Some caution is necessary when thinking of $\dot{V}$ in this way as it is not in general true that $\dot{V}\left(t_{0}\right) = 0$ when $\dot{\gamma}\left(t_{0}\right) = 0$. It could, e.g., happen that γ is the constant curve. In this case $V\left(t\right)$ is simply a curve in $T_{\gamma(t_{0})}M$ and as such has a well-defined velocity that doesn't have to be zero.

Using this definition we also see that

$$\frac{d}{dt}g\left(V,W\right) = g\left(\dot{V},W\right) + g\left(V,\dot{W}\right)$$

for vector fields V,W along γ. This follows from the product rule for mixed partials by selecting a two-parameter variation $\bar{\gamma}\left(s,u,t\right)$ such that

$$\begin{aligned} \frac{\partial\bar{\gamma}}{\partial s}\left(0,0,t\right) & = V\left(t\right), \\ \frac{\partial\bar{\gamma}}{\partial u}\left(0,0,t\right) & = W\left(t\right). \end{aligned}$$

In addition the local coordinate formula shows that

$$\begin{aligned} \frac{d}{dt}\left(V\left(t\right) + W\left(t\right)\right) & = \frac{d}{dt}V\left(t\right) + \frac{d}{dt}W\left(t\right), \\ \frac{d}{dt}\left(\lambda\left(t\right)V\left(t\right)\right) & = \frac{d\lambda}{dt}\left(t\right)V\left(t\right) + \lambda\left(t\right)\frac{dV}{dt}\left(t\right) \end{aligned}$$

where $\lambda : I \to \mathbb{R}$ is a function.

As with second partials we see that differentiation along curves can be done in a larger space and then projected on to M. Specifically, if $M \subset N$ and $\gamma : I \to M$ is a curve and $V : I \to TM$ a vector field along γ, then we can compute $\dot{V} \in TN$ and then project $\mathrm{proj}_{M}\left(\dot{V}\right) \in TM$ to get the derivative of V along γ in M. In the next subsection we shall give a cautionary example.

1.2. Third Partials. One of the uses of taking derivatives of vector fields along curves is that we can now define third and higher order partial derivatives. If we wish to compute

$$\frac{\partial^{3}\gamma}{\partial s\partial t\partial u}\left(s_{0},t_{0},u_{0}\right)$$

then we consider the vector field $s \to \frac{\partial^2 \gamma}{\partial t \partial u}(s, t_0, u_0) = V(s)$ and define

$$\frac{\partial^3 \gamma}{\partial s \partial t \partial u}(s_0, t_0, u_0) = \frac{dV}{ds}(s_0).$$

Something rather interesting happens when we consider third partials. We expected and proved that second partials commute. This, however, does not carry over to third partials. It is true that

$$\frac{\partial^3 \gamma}{\partial s \partial t \partial u} = \frac{\partial^3 \gamma}{\partial s \partial u \partial t}$$

but if we switch the first two variables the derivatives might be different. The reason we are not entitled to have these derivatives commute lies in the fact that they were defined in a slightly different manner.

EXAMPLE 41. *Let*

$$\gamma(t, \theta) = \begin{bmatrix} \cos(t) \\ \sin(t)\cos(\theta) \\ \sin(t)\sin(\theta) \end{bmatrix}$$

be the standard parametrization of $S^2(1) \subset \mathbb{R}^3$ as a surface of revolution around the x-axis. We can compute all derivatives in $\mathbb{R}^3$ and then project them on to $S^2(1)$ in order to find the intrinsic partial derivatives. The curves $t \to \gamma(t, \theta)$ are geodesics. We can see this by direct calculation as

$$\frac{\partial \gamma}{\partial t} = \begin{bmatrix} -\sin(t) \\ \cos(t)\cos(\theta) \\ \cos(t)\sin(\theta) \end{bmatrix} \in TS^2(1),$$

$$\frac{\partial^2 \gamma}{\partial t^2} = \begin{bmatrix} -\cos(t) \\ -\sin(t)\cos(\theta) \\ -\sin(t)\sin(\theta) \end{bmatrix} \in T\mathbb{R}^3.$$

Thus the Euclidean acceleration is proportional to the base point γ and therefore has zero projection on to $S^1(1)$. Next we compute

$$\frac{\partial^2 \gamma}{\partial \theta \partial t} = \begin{bmatrix} 0 \\ -\cos(t)\sin(\theta) \\ \cos(t)\cos(\theta) \end{bmatrix} \in T\mathbb{R}^3.$$

This vector is perpendicular to γ and therefore represents the actual intrinsic mixed partial. Finally we calculate

$$\frac{\partial^3 \gamma}{\partial t \partial \theta \partial t} = \begin{bmatrix} 0 \\ \sin(t)\sin(\theta) \\ -\sin(t)\cos(\theta) \end{bmatrix} \in T\mathbb{R}^3,$$

$$\frac{\partial^3 \gamma}{\partial \theta \partial t^2} = \begin{bmatrix} 0 \\ \sin(t)\sin(\theta) \\ -\sin(t)\cos(\theta) \end{bmatrix} \in T\mathbb{R}^3.$$

These are equal as we would expect in $\mathbb{R}^3$. They are also both tangent to $S^2(1)$. The first term is thus $\frac{\partial^3 \gamma}{\partial t \partial \theta \partial t}$ as computed in $S^2(1)$. The second has no meaning in $S^2(1)$ as we are supposed to first project $\frac{\partial^2 \gamma}{\partial t^2}$ on to $S^2(1)$ before computing $\frac{\partial}{\partial \theta}\frac{\partial^2 \gamma}{\partial t^2}$ in $\mathbb{R}^3$ and then again project to $S^2(1)$. In $S^2(1)$ we have therefore seen that $\frac{\partial^3 \gamma}{\partial \theta \partial t^2} = 0$ while $\frac{\partial^3 \gamma}{\partial t \partial \theta \partial t} \neq 0$.

In this example it is also interesting to note that the equator $t = 0$ given by $\theta \to \gamma(0, \theta)$ is a geodesic and that $\frac{\partial^2 \gamma}{\partial \theta \partial t} = 0$ along this equator.

We are now ready to prove what happens when the first two partials in a third order partial are interchanged.

LEMMA 18.
$$\frac{\partial^3 \gamma}{\partial u \partial s \partial t} - \frac{\partial^3 \gamma}{\partial s \partial u \partial t} = R\left(\frac{\partial \gamma}{\partial u}, \frac{\partial \gamma}{\partial s}\right) \frac{\partial \gamma}{\partial t}.$$

PROOF. This result is hardly surprising if we recall the definition of curvature and think of these partial derivatives as covariant derivatives. It is, however, not so clear what happens when these derivatives are not covariant derivatives. We are therefore forced to do the calculation in local coordinates. To simplify matters let us assume that we are at a point $p = \gamma(u, s, t)$ where $g_{ij}|_p = \delta_{ij}$ and $\Gamma_{ij}^k|_p = 0$. This implies that

$$\frac{\partial}{\partial u}(\partial_i)|_p = 0.$$

Thus

$$
\begin{aligned}
\frac{\partial^3 \gamma}{\partial u \partial s \partial t}\Big|_p &= \frac{\partial}{\partial u}\left(\frac{\partial^2 \gamma^l}{\partial s \partial t}\partial_l + \frac{\partial \gamma^i}{\partial t}\frac{\partial \gamma^j}{\partial s}\Gamma_{ij}^l \partial_l\right) \\
&= \frac{\partial^3 \gamma^l}{\partial u \partial s \partial t}\partial_l + \frac{\partial \gamma^i}{\partial t}\frac{\partial \gamma^j}{\partial s}\frac{\partial}{\partial u}\left(\Gamma_{ij}^l\right)\partial_l \\
&= \frac{\partial^3 \gamma^l}{\partial u \partial s \partial t}\partial_l + \frac{\partial \gamma^i}{\partial t}\frac{\partial \gamma^j}{\partial s}\frac{\partial \gamma^k}{\partial u}\left(\partial_k \Gamma_{ij}^l\right)\partial_l
\end{aligned}
$$

$$\frac{\partial^3 \gamma}{\partial s \partial u \partial t}\Big|_p = \frac{\partial^3 \gamma^l}{\partial s \partial u \partial t}\partial_l + \frac{\partial \gamma^i}{\partial t}\frac{\partial \gamma^j}{\partial u}\frac{\partial \gamma^k}{\partial s}\left(\partial_k \Gamma_{ij}^l\right)\partial_l$$

and using our formula for R_{ijk}^l in terms of the Christoffel symbols from chapter 2 gives

$$
\begin{aligned}
\frac{\partial^3 \gamma}{\partial u \partial s \partial t}\Big|_p - \frac{\partial^3 \gamma}{\partial s \partial u \partial t}\Big|_p &= \frac{\partial \gamma^i}{\partial t}\frac{\partial \gamma^j}{\partial s}\frac{\partial \gamma^k}{\partial u}\left(\partial_k \Gamma_{ij}^l\right)\partial_l - \frac{\partial \gamma^i}{\partial t}\frac{\partial \gamma^j}{\partial u}\frac{\partial \gamma^k}{\partial s}\left(\partial_k \Gamma_{ij}^l\right)\partial_l \\
&= \frac{\partial \gamma^i}{\partial t}\frac{\partial \gamma^j}{\partial s}\frac{\partial \gamma^k}{\partial u}\left(\partial_k \Gamma_{ij}^l\right)\partial_l - \frac{\partial \gamma^i}{\partial t}\frac{\partial \gamma^k}{\partial u}\frac{\partial \gamma^j}{\partial s}\left(\partial_j \Gamma_{ik}^l\right)\partial_l \\
&= \frac{\partial \gamma^i}{\partial t}\frac{\partial \gamma^j}{\partial s}\frac{\partial \gamma^k}{\partial u}\left(\partial_k \Gamma_{ij}^l - \partial_j \Gamma_{ik}^l\right)\partial_l \\
&= \frac{\partial \gamma^i}{\partial t}\frac{\partial \gamma^j}{\partial s}\frac{\partial \gamma^k}{\partial u}\left(\partial_k \Gamma_{ji}^l - \partial_j \Gamma_{ki}^l\right)\partial_l \\
&= \frac{\partial \gamma^i}{\partial t}\frac{\partial \gamma^j}{\partial s}\frac{\partial \gamma^k}{\partial u}R_{kji}^l \partial_l \\
&= R\left(\frac{\partial \gamma}{\partial u}, \frac{\partial \gamma}{\partial s}\right)\frac{\partial \gamma}{\partial t}.
\end{aligned}
$$

$\square$

1.3. Parallel Fields. A vector field V along γ is said to be *parallel along* γ provided that $\dot{V} \equiv 0$. We know that the tangent field $\dot{\gamma}$ along a geodesic is parallel. We also just saw above that the unit field perpendicular to a great circle in $S^2(1)$ is a parallel field.

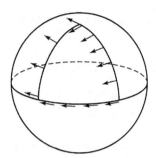

Figure 6.1

If V, W are two parallel fields along γ, then we clearly have that $g(V, W)$ is constant along γ. Parallel fields along a curve therefore neither change their lengths nor their angles relative to each other, just as parallel fields in Euclidean space are of constant length and make constant angles. Based on the above example we can pictorially describe parallel translation along certain triangles in $S^2(1)$ see Figure 6.1 The exercises to this chapter will cover some features of parallel translation on surfaces to aid the reader's geometric understanding.

THEOREM 20. (Existence and Uniqueness of Parallel fields) *If $t_0 \in I$ and $v \in T_{\gamma(t_0)}M$, then there is a unique parallel field $V(t)$ defined on all of I with $V(t_0) = v$.*

PROOF. Choose vector fields $E_1(t), \ldots, E_n(t)$ along γ forming a basis for $T_{\gamma(t)}M$ for all $t \in I$. This is always possible. Any vector field $V(t)$ along γ can then be written $V(t) = \lambda^i(t)E_i(t)$ for $\lambda^i : I \to \mathbb{R}$. Thus,

$$
\begin{aligned}
\dot{V} &= \nabla_{\dot{\gamma}}V = \sum \dot{\lambda}^i(t)E_i(t) + \lambda^i(t)\nabla_{\dot{\gamma}}E_i \\
&= \sum \dot{\lambda}^j(t)E_j(t) + \sum_{i,j} \lambda^i(t) \cdot \alpha_i^j(t)E_j(t), \text{ where } \nabla_{\dot{\gamma}}E_i = \sum \alpha_i^j(t)E_j \\
&= \sum_j (\dot{\lambda}^j(t) + \lambda^i(t)\alpha_i^j(t))E_j(t).
\end{aligned}
$$

Hence, V is parallel iff $\lambda^1(t), \ldots, \lambda^n(t)$ satisfy the system of first-order linear differential equations

$$
\dot{\lambda}^j(t) = -\sum_{i=1}^n \alpha_i^j(t)\lambda^i(t) \qquad j = 1, \ldots, n.
$$

Such differential equations have the property that given initial values $\lambda^1(t_0), \ldots, \lambda^n(t_0)$, there are unique solutions defined on all of I with these initial values. $\square$

The existence and uniqueness assertion that concluded this proof is a standard theorem in differential equations that we take for granted. The reader should recall that linearity of the equations is a crucial ingredient in showing that the solution exists on all of I. Nonlinear equations can fail to have solutions over a whole given interval as we saw with geodesics in chapter 5.

Parallel fields can be used as a substitute for Cartesian coordinates. Namely, if we choose a parallel orthonormal framing $E_1(t), \ldots, E_n(t)$ along the curve $\gamma(t)$:

$I \to (M, g)$, then we've seen that any vector field $V(t)$ along γ has the property that

$$
\begin{aligned}
\frac{dV}{dt} &= \frac{d}{dt} \left(\alpha^i(t) E_i(t) \right) \\
&= \dot{\alpha}^i(t) E_i(t) + \alpha^i(t) \cdot \dot{E}_i(t) \\
&= \dot{\alpha}^i(t) E_i(t).
\end{aligned}
$$

So $\frac{d}{dt} V$, when represented in coordinates of the frame, is exactly what we would expect. We could more generally choose a tensor T along $\gamma(t)$ of type $(0, p)$ or $(1, p)$ and compute $\frac{d}{dt} T$. For the sake of simplicity, choose a $(1, 1)$ tensor S. Then write $S(E_i(t)) = \alpha_i^j(t) E_j(t)$. Thus S is represented by the matrix $(\alpha_i^j(t))$ along the curve. As before, we see that $\frac{d}{dt} S$ is represented by $(\dot{\alpha}_i^j(t))$.

This makes it possible to understand equations involving only one differentiation of the type ∇_X. Let F^t be the local flow near some point $p \in M$ and H a hypersurface in M through p that is perpendicular to X. Next choose vector fields $E_1, \ldots, E_n$ on H which form an orthonormal frame for the tangent space to M. Finally, construct an orthonormal framing in a neighborhood of p by parallel translating $E_1, \ldots, E_n$ along the integral curves for X. Thus, $\nabla_X E_i = 0$, $i = 1, \ldots, n$. Therefore, if we have a vector field Y near p, we can write $Y = \alpha^i \cdot E_i$ and $\nabla_X Y = D_X(\alpha^i) \cdot E_i$. Similarly, if S is a $(1, 1)$-tensor, we have $S(E_i) = \alpha_i^j E_j$, and $\nabla_X S$ is represented by $(D_X(\alpha_i^j))$.

In this way parallel frames make covariant derivatives look like standard derivatives in the same fashion that coordinate vector fields make Lie derivatives look like standard derivatives.

2. Second Variation of Energy

Recall that all geodesics are stationary points for the energy functional. To better understand what happens near a geodesic we do exactly what we would do in calculus, namely, compute the second derivative of any variation of a geodesic.

THEOREM 21. (Synge's second variation formula, 1926) *Let* $\bar{\gamma} : (-\varepsilon, \varepsilon) \times [a, b]$ *be a smooth variation of a geodesic* $\gamma(t) = \bar{\gamma}(0, t)$. *Then*

$$
\frac{d^2 E(\gamma_s)}{ds^2}\Big|_{s=0} = \int_a^b \left| \frac{\partial^2 \bar{\gamma}}{\partial t \partial s} \right|^2 dt - \int_a^b g \left(R \left(\frac{\partial \bar{\gamma}}{\partial s}, \frac{\partial \bar{\gamma}}{\partial t} \right) \frac{\partial \bar{\gamma}}{\partial t}, \frac{\partial \bar{\gamma}}{\partial s} \right) dt + g \left(\frac{\partial^2 \bar{\gamma}}{\partial s^2}, \frac{\partial \bar{\gamma}}{\partial t} \right) \Big|_a^b
$$

PROOF. The first variation formula tells us that

$$
\frac{dE(\gamma_s)}{ds} = -\int_a^b g \left(\frac{\partial \bar{\gamma}}{\partial s}, \frac{\partial^2 \bar{\gamma}}{\partial t^2} \right) dt + g \left(\frac{\partial \bar{\gamma}}{\partial s}, \frac{\partial \bar{\gamma}}{\partial t} \right) \Big|_{(s,a)}^{(s,b)}.
$$

For the purposes of the proof it is easier to work with a two parameter variation $\bar{\gamma}(u, s, t)$. We then obtain the desired result by setting $u = s$ in the variation $\bar{\gamma}(u + s, t)$ coming from our original one parameter variation. With this in mind we can calculate

$$\frac{\partial^2 E(\gamma_s)}{\partial u \partial s} = -\frac{\partial}{\partial u}\int_a^b g\left(\frac{\partial \bar{\gamma}}{\partial s}, \frac{\partial^2 \bar{\gamma}}{\partial t^2}\right) dt + \frac{\partial}{\partial u} g\left(\frac{\partial \bar{\gamma}}{\partial s}, \frac{\partial \bar{\gamma}}{\partial t}\right)\Big|_{(u,s,a)}^{(u,s,b)}$$

$$= -\int_a^b g\left(\frac{\partial^2 \bar{\gamma}}{\partial u \partial s}, \frac{\partial^2 \bar{\gamma}}{\partial t^2}\right) dt - \int_a^b g\left(\frac{\partial \bar{\gamma}}{\partial s}, \frac{\partial^3 \bar{\gamma}}{\partial u \partial t^2}\right) dt$$

$$+ g\left(\frac{\partial^2 \bar{\gamma}}{\partial u \partial s}, \frac{\partial \bar{\gamma}}{\partial t}\right)\Big|_{(u,s,a)}^{(u,s,b)} + g\left(\frac{\partial \bar{\gamma}}{\partial s}, \frac{\partial^2 \bar{\gamma}}{\partial u \partial t}\right)\Big|_{(u,s,a)}^{(u,s,b)}$$

Setting $s = u = 0$ and using that $\gamma(0,0,t)$ is a geodesic we get

$$\frac{\partial^2 E(\gamma_s)}{\partial u \partial s}\Big|_{s=0, u=0}$$

$$= -\int_a^b g\left(\frac{\partial \bar{\gamma}}{\partial s}, \frac{\partial^3 \bar{\gamma}}{\partial u \partial t^2}\right) dt + g\left(\frac{\partial^2 \bar{\gamma}}{\partial u \partial s}, \frac{\partial \bar{\gamma}}{\partial t}\right)\Big|_a^b + g\left(\frac{\partial \bar{\gamma}}{\partial s}, \frac{\partial^2 \bar{\gamma}}{\partial u \partial t}\right)\Big|_a^b$$

$$= -\int_a^b g\left(\frac{\partial \bar{\gamma}}{\partial s}, R\left(\frac{\partial \bar{\gamma}}{\partial u}, \frac{\partial \bar{\gamma}}{\partial t}\right)\frac{\partial \bar{\gamma}}{\partial t}\right) dt - \int_a^b g\left(\frac{\partial \bar{\gamma}}{\partial s}, \frac{\partial^3 \bar{\gamma}}{\partial t \partial u \partial t}\right) dt$$

$$+ g\left(\frac{\partial^2 \bar{\gamma}}{\partial u \partial s}, \frac{\partial \bar{\gamma}}{\partial t}\right)\Big|_a^b + g\left(\frac{\partial \bar{\gamma}}{\partial s}, \frac{\partial^2 \bar{\gamma}}{\partial u \partial t}\right)\Big|_a^b$$

$$= -\int_a^b g\left(\frac{\partial \bar{\gamma}}{\partial s}, R\left(\frac{\partial \bar{\gamma}}{\partial u}, \frac{\partial \bar{\gamma}}{\partial t}\right)\frac{\partial \bar{\gamma}}{\partial t}\right) dt + \int_a^b g\left(\frac{\partial^2 \bar{\gamma}}{\partial t \partial s}, \frac{\partial^2 \bar{\gamma}}{\partial u \partial t}\right) dt$$

$$- \int_a^b \frac{\partial}{\partial t} g\left(\frac{\partial \bar{\gamma}}{\partial s}, \frac{\partial^2 \bar{\gamma}}{\partial u \partial t}\right) dt + g\left(\frac{\partial^2 \bar{\gamma}}{\partial u \partial s}, \frac{\partial \bar{\gamma}}{\partial t}\right)\Big|_a^b + g\left(\frac{\partial \bar{\gamma}}{\partial s}, \frac{\partial^2 \bar{\gamma}}{\partial u \partial t}\right)\Big|_a^b$$

$$= -\int_a^b g\left(\frac{\partial \bar{\gamma}}{\partial s}, R\left(\frac{\partial \bar{\gamma}}{\partial u}, \frac{\partial \bar{\gamma}}{\partial t}\right)\frac{\partial \bar{\gamma}}{\partial t}\right) dt + \int_a^b g\left(\frac{\partial^2 \bar{\gamma}}{\partial t \partial s}, \frac{\partial^2 \bar{\gamma}}{\partial u \partial t}\right) dt$$

$$- g\left(\frac{\partial \bar{\gamma}}{\partial s}, \frac{\partial^2 \bar{\gamma}}{\partial u \partial t}\right)\Big|_a^b + g\left(\frac{\partial^2 \bar{\gamma}}{\partial u \partial s}, \frac{\partial \bar{\gamma}}{\partial t}\right)\Big|_a^b + g\left(\frac{\partial \bar{\gamma}}{\partial s}, \frac{\partial^2 \bar{\gamma}}{\partial u \partial t}\right)\Big|_a^b$$

$$= \int_a^b g\left(\frac{\partial^2 \bar{\gamma}}{\partial t \partial s}, \frac{\partial^2 \bar{\gamma}}{\partial t \partial u}\right) dt - \int_a^b g\left(R\left(\frac{\partial \bar{\gamma}}{\partial u}, \frac{\partial \bar{\gamma}}{\partial t}\right)\frac{\partial \bar{\gamma}}{\partial t}, \frac{\partial \bar{\gamma}}{\partial s}\right) dt + g\left(\frac{\partial^2 \bar{\gamma}}{\partial u \partial s}, \frac{\partial \bar{\gamma}}{\partial t}\right)\Big|_a^b$$

$$\square$$

The formula is going to be used in different ways below. First we observe that for proper variations the last terms drops out and the formula only depends on the variational field $V(t) = \frac{\partial \bar{\gamma}}{\partial s}(0,t)$ and the velocity field $\dot{\gamma}$ of the original geodesic:

$$\frac{d^2 E(\gamma_s)}{ds^2}\Big|_{s=0} = \int_a^b \left|\dot{V}\right|^2 dt - \int_a^b g\left(R(V,\dot{\gamma})\dot{\gamma}, V\right) dt$$

Another special case occurs when the variational field is parallel $\dot{V} = 0$. In this case the first term drops out:

$$\frac{d^2 E(\gamma_s)}{ds^2}\Big|_{s=0} = -\int_a^b g\left(R(V,\dot{\gamma})\dot{\gamma}, V\right) dt + g\left(\frac{\partial^2 \bar{\gamma}}{\partial s^2}, \dot{\gamma}\right)\Big|_a^b$$

but the formula still depends on the variation and not just on V. If, however, we select the variation where $s \to \bar{\gamma}(s,t)$ are geodesics, then the last term also drops out.

Another variational field that is often quite useful is the variational field that comes from a *geodesic variation*, i.e., $t \to \bar{\gamma}(s,t)$ is a geodesic for all s and not just for $s = 0$. We encountered these fields in chapter 2 as vector fields satisfying $L_{\partial_r} J = 0$. Here they are only defined along a single geodesic so the Lie derivative equation no longer makes sense. The second order Jacobi equation, however, does make sense in this context. We can now check that the variational field indeed does satisfy such a second order equation.

$$
\begin{aligned}
0 &= \frac{\partial^3 \bar{\gamma}}{\partial s \partial^2 t} \\
&= R\left(\frac{\partial \bar{\gamma}}{\partial s}, \frac{\partial \bar{\gamma}}{\partial t}\right) \frac{\partial \bar{\gamma}}{\partial t} + \frac{\partial^3 \bar{\gamma}}{\partial t \partial s \partial t} \\
&= R\left(\frac{\partial \bar{\gamma}}{\partial s}, \frac{\partial \bar{\gamma}}{\partial t}\right) \frac{\partial \bar{\gamma}}{\partial t} + \frac{\partial^3 \bar{\gamma}}{\partial^2 t \partial s} \\
&= R\left(\frac{\partial \bar{\gamma}}{\partial s}, \frac{\partial \bar{\gamma}}{\partial t}\right) \frac{\partial \bar{\gamma}}{\partial t} + \frac{\partial^2}{\partial t^2} \frac{\partial \bar{\gamma}}{\partial s}.
\end{aligned}
$$

So if the variational field along γ is $J(t) = \frac{\partial \bar{\gamma}}{\partial s}(0,t)$, then this field solves the linear second order *Jacobi Equation*

$$
\ddot{J} + R(J, \dot{\gamma})\dot{\gamma} = 0.
$$

Given $J(0)$ and $\dot{J}(0)$ such a field is therefore uniquely defined. These variational fields are called *Jacobi fields* along γ. In case $J(0) = 0$, these fields can be constructed via the geodesic variation

$$
\bar{\gamma}(s,t) = \exp_p\left(t\left(\dot{\gamma}(0) + s\dot{J}(0)\right)\right).
$$

Since $\bar{\gamma}(s,0) = p$ for all s we must have $J(0) = \frac{\partial \bar{\gamma}}{\partial s}(0,0) = 0$. The derivative is computed as follows

$$
\begin{aligned}
\frac{\partial^2 \bar{\gamma}}{\partial t \partial s}(0,0) &= \frac{\partial^2 \bar{\gamma}}{\partial s \partial t}(0,0) \\
&= \frac{\partial}{\partial s}\left(\dot{\gamma}(0) + s\dot{J}(0)\right)|_{s=0} \\
&= \dot{J}(0).
\end{aligned}
$$

What is particularly interesting about these Jacobi fields is that they control two things we are interesting in studying.

First we see that they tie in with the differential of the exponential map as

$$
\begin{aligned}
J(t) &= \frac{\partial \bar{\gamma}}{\partial s}(0,t) \\
&= \frac{\partial}{\partial s}\exp_p\left(t\left(\dot{\gamma}(0) + s\dot{J}(0)\right)\right)|_{(0,t)} \\
&= D\exp_p\left(\frac{\partial}{\partial s}\left(t\left(\dot{\gamma}(0) + s\dot{J}(0)\right)\right)|_{(0,t)}\right) \\
&= D\exp_p\left(t\dot{J}(0)\right),
\end{aligned}
$$

where we think of $t\dot{J}(0) \in T_{t\dot{\gamma}(0)}T_pM$. Thus we have that

$$
D\exp_p(w) = J(1)
$$

where $w \in T_v T_p M$ and J is the Jacobi field along $\gamma(t) = \exp_p(tv)$ such that $J(0) = 0$ and $\dot{J}(0) = w$. This shows, in particular, that if $D \exp_p$ is nonsingular at v, then there is a Jacobi field $J(t)$ such that $J(0) = 0$ and $J(1)$ is any specified vector in $T_q M$, $q = \exp_p(v)$.

Second we can see that Jacobi fields tie in with the differential and Hessian of the function $f(x) = \frac{1}{2}(d(x,p))^2$. This is a consequence of the second variation formula. We assume that f is smooth near q and construct a geodesic variation

$$\bar{\gamma}(s,t) = \exp_p\left(t\left(\dot{\gamma}(0) + s\dot{J}(0)\right)\right)$$

where $\exp_p(\dot{\gamma}(0)) = q$. We then observe that the the curves $t \to \bar{\gamma}(s,t)$ are minimizing geodesics from p to $\bar{\gamma}(s,1)$ and hence measure the distance to those points. In particular,

$$
\begin{aligned}
f(\bar{\gamma}(s,1)) &= \frac{1}{2}\left(\int_0^1 \left|\frac{\partial}{\partial t}\bar{\gamma}(s,t)\right| dt\right)^2 \\
&= \frac{1}{2}\left|\frac{\partial}{\partial t}\bar{\gamma}(s,t)\right|^2 \\
&= \frac{1}{2}\int_0^1 \left|\frac{\partial}{\partial t}\bar{\gamma}(s,t)\right|^2 dt \\
&= E(\gamma_s).
\end{aligned}
$$

Let $J(s,t) = \frac{\partial}{\partial s}\bar{\gamma}(s,t)$ be the variational Jacobi field along any of the geodesics $t \to \gamma_s(t) = \bar{\gamma}(s,t)$. Then the first variation formula tells us

$$
\begin{aligned}
df(J(s,1)) &= \frac{\partial}{\partial s}f(\bar{\gamma}(s,1)) \\
&= \frac{\partial E(\gamma_s)}{\partial s} \\
&= g(J(s,1),\dot{\gamma}_s(1)).
\end{aligned}
$$

Showing that

$$\nabla f|_{\bar{\gamma}(s)} = \dot{\gamma}_s(1).$$

The Hessian of f is given by

$$
\begin{aligned}
\mathrm{Hess}f(J(1),J(1)) &= g\left(\nabla_{J(1)}\nabla f, J(1)\right) \\
&= g\left(\nabla_{J(1)}\left(\frac{\partial\bar{\gamma}}{\partial t}\right), J(1)\right) \\
&= g\left(\frac{\partial}{\partial s}\left(\frac{\partial\bar{\gamma}}{\partial t}\right), J\right)|_{(0,1)} \\
&= g\left(\frac{\partial^2\bar{\gamma}}{\partial s\partial t}, J\right)|_{(0,1)} \\
&= g\left(\frac{\partial^2\bar{\gamma}}{\partial t\partial s}, J\right)|_{(0,1)} \\
&= g\left(\dot{J}(1), J(1)\right).
\end{aligned}
$$

3. Nonpositive Sectional Curvature

In this section we shall use everything we have learned so far, and then some, to show that the exponential map $\exp_p : T_pM \to M$ is a covering map, provided (M, g) has nonpositive sectional curvature everywhere. This implies, in particular, that no compact simply connected manifold admits such a metric. We shall also prove some interesting results about the fundamental groups of such manifolds.

The first observation about manifolds with nonpositive curvature is that any geodesic from p to q must be a local minimum for $E : \Omega(p, q) \to [0, \infty)$ by our second variation formula. This is in sharp contrast to what we shall prove in positive curvature, where sufficiently long geodesics can never be local minima.

In this section we show how both the variational techniques and the fundamental equations can be used to prove the necessary qualitative and quantitative estimates. Recall from our discussion of the fundamental equations in chapter 2 that Jacobi fields seem particularly well-suited for the task of studying nonpositive curvature. This will be born out by what we do below.

3.1. Manifolds Without Conjugate Points. First some generalities:

LEMMA 19. *Suppose* $\exp_p : T_pM \to M$ *is nonsingular everywhere (i.e., has no critical points); then it is a covering map.*

PROOF. By definition $\exp_p$ is an immersion, so on T_pM choose the pullback metric to make it into a local Riemannian isometry. We then know from chapter 5 that $\exp_p$ is a covering map provided this new metric on T_pM is complete. To see this, simply observe that the metric is geodesically complete at the origin, since straight lines through the origin are still geodesics.   □

We can now prove our first big result. It was originally established by Mangoldt for surfaces. Hadamard in a survey article then also gave a different proof. It appears that Cartan only knew of Hadamard's paper and gave credit only to him for this result on surfaces. Cartan proved a generalization to higher dimensions under the assumption that the manifold is metrically complete.

THEOREM 22. (v. Mangoldt, 1881, Hadamard, 1889, and Cartan, 1925) *If (M, g) is complete, connected, and has* sec ≤ 0, *then the universal covering is diffeomorphic to* $\mathbb{R}^n$.

We are going to give two proofs of this result. One is more classical and uses Jacobi fields along geodesics. The other uses the fundamental equations from chapter 2.

JACOBI FIELD PROOF. The goal is to show that

$$\left| D\exp_p(w) \right| \geq 0$$

for all $w \in T_vT_pM$, with equality holding only for $w = 0$. This shows that $\exp_p$ is nonsingular everywhere and hence a covering map. To check this we select a Jacobi field J along $\gamma(t) = \exp_p(tv)$ such that $J(0) = 0$ and $\dot{J}(0) = w$. Then

$|D\exp_p(w)| = |J(1)|$. We now consider $t \to \frac{1}{2}|J(t)|^2$. The first and second derivative of this function is

$$\frac{d}{dt}\left(\frac{1}{2}|J(t)|^2\right) = g\left(\dot{J}, J\right),$$

$$\frac{d^2}{dt^2}\left(\frac{1}{2}|J(t)|^2\right) = \frac{d}{dt}g\left(\dot{J}, J\right)$$

$$= g\left(\ddot{J}, J\right) + g\left(\dot{J}, \dot{J}\right)$$

$$= -g\left(R\left(J, \dot{\gamma}\right)\dot{\gamma}, J\right) + \left|\dot{J}\right|^2$$

$$\geq \left|\dot{J}\right|^2$$

as we assumed that $g(R(x,y)y,x) \leq 0$ for all tangent vectors x, y. Integrating this inequality gives

$$g\left(\dot{J}, J\right) \geq \int_0^t \left|\dot{J}\right|^2 dt + g\left(\dot{J}(0), J(0)\right)$$

$$= \int_0^t \left|\dot{J}\right|^2 dt$$

$$> 0$$

unless $\dot{J}(t) = 0$ for all t, in which case $\dot{J}(0) = w = 0$. So we can assume $w \neq 0$. Integrating the last inequality yields

$$\frac{1}{2}|J(t)|^2 > 0,$$

which is what we wanted to prove. $\qquad\square$

FUNDAMENTAL EQUATION PROOF. We consider a maximal ball $B(0, R) \subset T_pM$ on which $\exp_p$ is nonsingular. The goal is to show that $R = \infty$. Inspecting the proof of the characterization of the segment domain from chapter 5 we see that if $R < \infty$ and $x \in \partial B(0, R)$ is a singular point for $\exp_p$ then we can find a Jacobi field J such that $\mathrm{Hess}r(J, J)$ becomes negative definite as we approach x. The fundamental equations show that

$$\partial_r\left(\mathrm{Hess}r(J, J)\right) = -R(J, \partial_r, \partial_r, J) + \mathrm{Hess}^2r(J, J) \geq 0.$$

Moreover as J isn't tangent to ∂_r we also have that $\mathrm{Hess}r(J, J)$ is positive near the origin. Thus $\mathrm{Hess}r(J, J)$ stays positive. $\qquad\square$

No similar theorem is true for Riemannian manifolds with $\mathrm{Ric} \leq 0$ or $\mathrm{scal} \leq 0$, since we have Ricci flat metrics on $\mathbb{R}^2 \times S^2$ and scalar flat metrics on $\mathbb{R} \times S^p$, $p \geq 1$.

3.2. The Fundamental Group in Nonpositive Curvature. We are going to prove two results on the structure of the fundamental group for manifolds with nonpositive curvature. The interested reader is referred to the book by Eberlein [34] for further results on manifolds with nonpositive curvature.

First we need a little preparation. Let (M, g) be a complete simply connected Riemannian manifold of nonpositive curvature. The two key properties we shall use are that any two points in M lie on a unique geodesic, and that distance functions are everywhere smooth and convex.

We just saw that $\exp_p : T_p M \to M$ is a diffeomorphism for all $p \in M$. This shows as in Euclidean space that there is only one geodesic through p and $q \, (\neq p)$.

This also shows that the distance function $d(x, p)$ is smooth on $M - \{p\}$. The modified distance function

$$x \to f_{0,p}(x) = \frac{1}{2} (d(x, p))^2$$

is therefore smooth everywhere. If $J(t)$ is a Jacobi field along a geodesic emanating from p with $J(0) = 0$, then we know that

$$\begin{aligned}
\mathrm{Hess} f\,(J(1), J(1)) &= g\left(\dot{J}(1), J(1)\right) \\
&\geq \int_0^1 \left|\dot{J}\right|^2 dt \\
&> 0.
\end{aligned}$$

Since $J(1)$ can be arbitrary we have shown that the Hessian is positive definite. If σ is a geodesic, this implies that $f \circ \sigma$ is convex as

$$\begin{aligned}
\frac{d}{dt} f \circ \sigma &= g(\nabla f, \dot{\sigma}), \\
\frac{d^2}{dt^2} f \circ \sigma &= \frac{d}{dt} g(\nabla f, \dot{\sigma}) \\
&= g(\nabla_{\dot{\sigma}} \nabla f, \dot{\sigma}) + g(\nabla f, \ddot{\sigma}) \\
&= \mathrm{Hess} f\,(\dot{\sigma}, \dot{\sigma}) \\
&> 0.
\end{aligned}$$

With this in mind we can generalize the idea of convexity slightly (see also chapter 9) to mean that the function is convex or strictly convex when restricted to any geodesic. One sees that the maximum of any number of convex functions is again convex (you only need to prove this in dimension 1, as we can restrict to geodesics). Given a finite collection of points $p_1, \ldots, p_k \in M$, we can then consider the strictly convex function

$$x \to \max \{f_{0,p_1}(x), \ldots f_{0,p_k}(x)\}.$$

In general, any proper nonnegative strictly convex proper function has a unique minimum. To see this, first observe that there must be a minimum. If there were two minima, then the function would be strictly convex when restricted to a geodesic joining these two minima. But then the function would have smaller values on the interior of this segment than at the endpoints. The uniquely defined minimum for

$$x \to \max \{f_{0,p_1}(x), \ldots f_{0,p_k}(x)\}$$

is denoted by $\mathrm{cm}_\infty \{p_1, \ldots, p_k\}$ and called the L^∞ *center of mass* of $\{p_1, \ldots, p_k\}$. If, instead of taking the maximum, we had taken the average we would have arrived at the usual center of mass also known as the L^2 center of mass.

The first theorem is concerned with fixed points of isometries.

THEOREM 23. (E. Cartan, 1925) *If (M, g) is a complete simply connected Riemannian manifold of nonpositive curvature, then any isometry $F : M \to M$ of finite order has a fixed point.*

PROOF. The idea, which is borrowed from Euclidean space, is that the center of mass of any orbit must be a fixed point. First, define the period k of F as the smallest integer such that $F^k = id$. Second, for any $p \in M$ consider the orbit $\{p, F(p), \ldots, F^{k-1}(p)\}$ of p. Then construct the center of mass

$$q = \mathrm{cm}_\infty \{p, F(p), \ldots, F^{k-1}(p)\}.$$

We claim that $F(q) = q$. This is because the function

$$x \to f(x) = \max\{f_{0,p}(x), \ldots f_{0,F^{k-1}(p)}(x)\}$$

has not only q as a minimum, but also $F(q)$. To see this just observe that since F is an isometry, we have

$$
\begin{aligned}
f(F(q)) &= \max\{f_{0,p}(F(q)), \ldots f_{0,F^{k-1}(p)}(F(q))\} \\
&= \frac{1}{2}\left(\max\{d(F(q),p), \ldots, d(F(q), F^{k-1}(p))\}\right)^2 \\
&= \frac{1}{2}\left(\max\{d(F(q), F^k(p)), \ldots, d(F(q), F^{k-1}(p))\}\right)^2 \\
&= \frac{1}{2}\left(\max\{d(q, F^{k-1}(p)), \ldots, d(q, F^{k-2}(p))\}\right)^2 \\
&= f(q).
\end{aligned}
$$

Therefore, the uniqueness of minima for strictly convex functions implies that $F(q) = q$. $\qquad\square$

COROLLARY 11. *If (M, g) is a complete Riemannian manifold of nonpositive curvature, then the fundamental group is torsion free, i.e., all nontrivial elements have infinite order.*

The second theorem requires more preparation and more careful analysis of distance functions. Suppose again that (M, g) is complete, simply connected and of non-positive curvature. Let us fix a modified distance function: $x \to f_{0,p}(x)$ and a unit speed geodesic $\gamma : \mathbb{R} \to M$. The Hessian estimate from above only implies that

$$\frac{d^2}{dt^2}(f_{0,p} \circ \gamma) > 0.$$

However, we know that this second derivative is 1 in Euclidean space. It is therefore not surprising that we have a much better quantitative estimate.

LEMMA 20.

$$\frac{d^2}{dt^2}(f_{0,p} \circ \gamma) \geq 1.$$

Again we give two proofs of this.

JACOBI FIELD PROOF. This result would follow if we could prove that

$$
\begin{aligned}
\mathrm{Hess} f_{0,p}(J(1), J(1)) &= g\left(\dot{J}(1), J(1)\right) \\
&\geq g(J(1), J(1)).
\end{aligned}
$$

The reason behind the proof of this is slightly tricky and is known as *Jacobi field comparison*. We consider the ratio

$$\lambda(t) = \frac{g\left(\dot{J}(t), J(t)\right)}{|J(t)|^2}.$$

Using that the sectional curvature is nonpositive and the Cauchy-Schwarz inequality we see that the derivative satisfies

$$\dot{\lambda}(t) = \frac{-g\left(R\left(J,\dot{\gamma}\right)\dot{\gamma},J\right)|J|^2 + \left|\dot{J}\right|^2|J|^2 - 2\left(g\left(J,\dot{J}\right)\right)^2}{|J|^4}$$

$$\geq \frac{\left|\dot{J}\right|^2|J|^2 - 2\left(g\left(J,\dot{J}\right)\right)^2}{|J|^4}$$

$$\geq \frac{\left(g\left(J,\dot{J}\right)\right)^2 - 2\left(g\left(J,\dot{J}\right)\right)^2}{|J|^4}$$

$$= -\lambda^2.$$

Hence

$$\frac{\dot{\lambda}}{\lambda^2} + 1 \geq 0.$$

We know that $\lambda(t) \to \infty$ as $t \to 0$, so when integrating this from 0 to 1 we get

$$0 \leq \int_0^1 \left(\frac{\dot{\lambda}}{\lambda^2} + 1\right) dt$$

$$= -\lambda^{-1}\big|_0^1 + 1$$

$$= -\lambda^{-1}(1) + 1.$$

This implies the desired inequality

$$1 \leq \lambda(1) = \frac{g\left(\dot{J}(1), J(1)\right)}{|J(1)|^2}.$$

□

FUNDAMENTAL EQUATION PROOF: The fundamental equations restricted to S^{n-1} in our polar coordinate expression of $g = dr^2 + g_r$ on $(0, \infty) \times S^{n-1}$ tell us that

$$\partial_r g_r = 2\text{Hess}r,$$

$$\partial_r \text{Hess}r \geq \text{Hess}^2 r,$$

$$\lim_{r \to 0}\left(\text{Hess}r - \frac{1}{r}g_r\right) = 0.$$

Next observe that

$$\partial_r\left(\text{Hess}r - \frac{1}{r}g_r\right) = \partial_r\text{Hess}r + \frac{1}{r^2}g_r - \frac{2}{r}\text{Hess}r$$

$$\geq \text{Hess}^2 r - \frac{2}{r}\text{Hess}r + \frac{1}{r^2}g_r$$

$$= \left(\text{Hess}r - \frac{1}{r}g_r\right)^2$$

$$\geq 0.$$

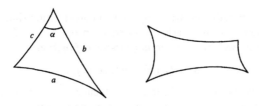

Figure 6.2

This, together with $\lim_{r \to 0} \left(\text{Hess} r - \frac{1}{r} g_r \right) = 0$ implies that $\text{Hess} r \geq \frac{1}{r} g_r$. It is then a simple calculation to see that

$$\text{Hess} \frac{1}{2} r^2 \geq g,$$

which implies the above lemma. $\qquad \qquad \square$

Integrating the inequality from the lemma twice yields

$$\begin{aligned}
\left(d \left(p, \gamma \left(t \right) \right) \right)^2 &\geq \left(d \left(p, \gamma \left(0 \right) \right) \right)^2 + 2g \left(\nabla f_{0,p}, \dot{\gamma} \left(0 \right) \right) \cdot t + t^2 \\
&= \left(d \left(p, \gamma \left(0 \right) \right) \right)^2 + \left(d \left(\gamma \left(0 \right), \gamma \left(t \right) \right) \right)^2 \\
&\quad - 2d \left(p, \gamma \left(0 \right) \right) d \left(\gamma \left(0 \right), \gamma \left(t \right) \right) \cos \angle \left(\nabla f_{0,p}, \dot{\gamma} \left(0 \right) \right).
\end{aligned}$$

Thus, if we have a triangle in M with sides lengths a, b, c and where the angle opposite a is α, then

$$a^2 \geq b^2 + c^2 - 2bc \cos \alpha.$$

From this, one can conclude that the angle sum in any triangle is $\leq \pi$, and more generally that the angle sum in any quadrilateral is $\leq 2\pi$. See Figure 6.2.

Now suppose that (M, g) has negative curvature. Then it must follow that all of the above inequalities are strict, unless p lies on the geodesic γ. In particular, the angle sum in any nondegenerate quadrilateral is $< 2\pi$. With this we can now show

THEOREM 24. (Preissmann, 1943) *If (M, g) is a compact manifold of negative curvature, then any Abelian subgroup of the fundamental group is cyclic. In particular, no compact product manifold $M \times N$ admits a metric with negative curvature.*

PROOF. We think of the fundamental group $\pi_1 (M)$ as acting by isometries on the universal covering $\tilde{M}$, and fix $\alpha \in \pi_1 (M)$. An *axis* for α is a geodesic $\gamma : \mathbb{R} \to \tilde{M}$ such that $\alpha (\gamma)$ is a reparametrization of γ. Since isometries map geodesics to geodesics, it must follow that

$$\alpha \circ \gamma \left(t \right) = \gamma \left(t + a \right).$$

Namely, α translates the geodesic either forward or backward. It is not possible for α to reverse the orientation of γ so that

$$\alpha \circ \gamma \left(t \right) = \gamma \left(-t + a \right),$$

as this would yield a fixed point

$$\alpha \left(\gamma \left(\frac{a}{2} \right) \right) = \gamma \left(\frac{a}{2} \right).$$

The uniquely defined number a is called the *period* of α along γ.

We now claim two things: first, that axes exist for the given α, and second, that they are unique when the curvature is negative.

To prove the first claim we are going to do a construction that will also be used later in this chapter. Given a deck transformation $\alpha : \tilde{M} \to \tilde{M}$ for a Riemannian covering $\pi : \tilde{M} \to M$ we consider the *displacement function*

$$x \to \delta_\alpha (x) = d (x, \alpha (x)).$$

Note that $\delta_\alpha (x)$ is the length of the shortest curve from x to $\alpha (x)$. Each such curve is a loop in M that lies in the homotopy class defined by α and based at $\pi (x) \in M$. In particular, we see that $\delta_\alpha (x) = \delta_\alpha (y)$ if $\pi (x) = \pi (y)$. This shows that $x \to \delta_\alpha (x)$ is the lift of a function on M that is never zero. In fact

$$\delta_\alpha (x) \geq 2 \mathrm{inj}_{\pi(x)} \geq 2 \mathrm{inj}_M > 0$$

as M is compact. Compactness of M then shows that there must be a point $q = \pi (p) \in M$ where this function attains its minimum. Let $\sigma = \pi \circ \gamma : [0, l] \to M$ be the unit speed loop at q that corresponds to the minimal geodesic from p to $\alpha (p)$ in $\tilde{M}$. We claim that σ is the shortest noncontractible loop in M corresponding to α. This is simply because any loop $c : [0, b] \to M$ that represents α lifts to a curve $\bar{c} : [0, b] \to \tilde{M}$ such that $\alpha (\bar{c} (0)) = \bar{c} (b)$. Thus

$$\ell (c) = \ell (\bar{c}) \geq d (\bar{c} (0), \bar{c} (b)) = \delta_\alpha (\bar{c} (0)) \geq \delta_\alpha (p).$$

The loop σ also corresponds to α if we think of it as based at any other point $q' = \pi (p') = \sigma (a)$ on itself. This means that we have a possibly piecewise smooth curve from p' to $\alpha (p')$ of length $\delta_\alpha (p)$ given by

$$\gamma' (t) = \begin{cases} \gamma (t + a) & 0 \leq t \leq l - a \\ \alpha \circ \gamma (t - l + a) & l - a \leq t \leq l \end{cases}$$

In particular,

$$\delta_\alpha (q') = \delta_\alpha (p') \leq \delta_\alpha (p) = \delta_\alpha (q).$$

But q was a global minimum for δ_α so $\delta_\alpha (q') = \delta_\alpha (q)$. This shows that γ' is also a smooth geodesic. Since it agrees with γ on $[a, 1]$ it is simply an extension of γ. Thus

$$\alpha (p') = \alpha \circ \gamma (a) = \gamma (l + a).$$

As p' was arbitrary this shows that γ is a geodesic such that $\alpha \circ \gamma (t) = \gamma (t + l)$. This gives the construction of the axis.

To see that axes are unique in negative curvature, assume that we have two different axes γ_1 and γ_2 for α. If these intersect in one point, they must, by virtue of being invariant under α, intersect in at least two points. But then they must be equal. We can therefore assume that they do not intersect. Then pick $p_1 \in \gamma_1$ and $p_2 \in \gamma_2$, and join these points by a segment σ. Then $\alpha \circ \sigma$ is a segment from $\alpha (p_1)$ to $\alpha (p_2)$. Since α is an isometry that preserves γ_1 and γ_2, we see that the adjacent angles along the two axes formed by the quadrilateral $p_1, p_2, \alpha (p_1), \alpha (p_2)$ must add up to π (see also Figure 6.3). But then the angle sum is 2π, which is not possible unless the quadrilateral is degenerate. That is, all points lie on one geodesic.

Finally pick an element $\beta \in \pi_1 (M)$ that commutes with α. First, note that β preserves the unique axis γ for α, since

$$\begin{aligned} \beta (\gamma) &= \beta (\alpha (\gamma)) \\ &= \alpha (\beta (\gamma)) \end{aligned}$$

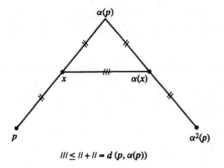

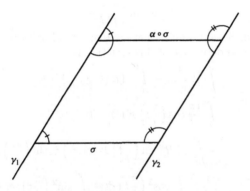

Figure 6.3

implies that $\beta \circ \gamma$ is an axis for α, and must therefore be γ itself. Next consider the group H generated by α, β. Any element in this group has γ as an axis. Thus we get a map

$$H \to \mathbb{R}$$

that sends an isometry to its uniquely defined period. Clearly, this map is a homomorphism with no kernel. Now, it is easy to check that any subgroup of $\mathbb{R}$ must either be cyclic or dense (like $\mathbb{Q}$). In the present case $H \subset \mathbb{R}$ must be discrete as the displacement and hence period is always larger than $\mathrm{inj}_M > 0$.   □

4. Positive Curvature

In this section we shall prove some of the classical results for manifolds with positive curvature. In contrast to the previous section, it is not possible to carry Euclidean geometry over to this setting. So while we try to imitate the results, new techniques are necessary.

In our discussion of the fundamental equations from chapter 2 we saw that using parallel fields most easily gave useful information about Hessians of distance functions. While we shall use parallel fields as opposed to Jacobi fields in our study of positive curvature we shall not be using the fundamental equations yet. We stick to the more classical approach of using variational techniques here. In the section

"More on Positive Curvature" below we shall show how some more sophisticated techniques can be used in conjunction with the developments here to establish some more specific results.

4.1. The Diameter Estimate. Our first restriction on positively curved manifolds is an estimate on how long minimal geodesics can be. It was first proven by Bonnet for surfaces and later by Synge for general Riemannian manifolds as an application of his second variation formula.

LEMMA 21. (Bonnet, 1855 and Synge, 1926) *Suppose* (M, g) *satisfies* sec $\geq k > 0$. *Then geodesics of length* $> \frac{\pi}{\sqrt{k}}$ *cannot be (locally) minimizing.*

PROOF. Let $\gamma : [0, l] \to M$ be a unit speed geodesic of length $l > \frac{\pi}{\sqrt{k}}$. Along γ consider the variational field

$$V(t) = \sin\left(\frac{\pi}{l}t\right) E(t),$$

where E is parallel. Since V vanishes at $t = 0$ and $t = l$, it corresponds to a proper variation. The second derivative of this variation is

$$
\begin{aligned}
\frac{d^2 E}{ds^2}\Big|_{s=0} &= \int_0^l \left|\dot{V}\right|^2 dt - \int_0^l g\left(R(V, \dot{\gamma})\dot{\gamma}, V\right) dt \\
&= \int_0^l \left|\frac{\pi}{l}\cos\left(\frac{\pi}{l}t\right) E(t)\right|^2 dt \\
&\quad - \int_0^l g\left(R\left(\sin\left(\frac{\pi}{l}t\right) E(t), \dot{\gamma}\right)\dot{\gamma}, \sin\left(\frac{\pi}{l}t\right) E(t)\right) dt \\
&= \left(\frac{\pi}{l}\right)^2 \int_0^l \cos^2\left(\frac{\pi}{l}t\right) dt - \int_0^l \sin^2\left(\frac{\pi}{l}t\right) \sec(E, \dot{\gamma}) dt \\
&\leq \left(\frac{\pi}{l}\right)^2 \int_0^l \cos^2\left(\frac{\pi}{l}t\right) dt - k \int_0^l \sin^2\left(\frac{\pi}{l}t\right) dt \\
&< k \int_0^l \cos^2\left(\frac{\pi}{l}t\right) dt - k \int_0^l \sin^2\left(\frac{\pi}{l}t\right) dt \\
&= 0.
\end{aligned}
$$

Thus all nearby curves in the variation are shorter than γ. □

The next result is a very interesting and completely elementary consequence of the above result. It seems to have first been pointed out by Hopf-Rinow for surfaces in their famous paper on completeness and then by Myers for general Riemannian manifolds.

COROLLARY 12. (Hopf-Rinow, 1931 and Myers, 1932) *Suppose* (M, g) *is complete and satisfies* sec $\geq k > 0$. *Then* M *is compact and satisfies* diam$(M, g) \leq \frac{\pi}{\sqrt{k}} = $ diamS_k^n. *In particular*, M *has finite fundamental group.*

PROOF. As no geodesic of length $> \frac{\pi}{\sqrt{k}}$ can realize the distance between endpoints and M is complete, the diameter cannot exceed $\frac{\pi}{\sqrt{k}}$. Finally use that the universal cover has the same curvature condition to conclude that it must also be compact. Thus, the fundamental group is finite. □

These results were later extended to manifolds with positive Ricci curvature by Myers.

THEOREM 25. (Myers, 1941) *Suppose (M, g) is a complete Riemannian manifold with* $\mathrm{Ric} \geq (n-1)k > 0$. *Then* $\mathrm{diam}(M, g) \leq \pi/\sqrt{k}$. *Furthermore,* (M, g) *has finite fundamental group.*

PROOF. It suffices to show as before that no geodesic of length $> \frac{\pi}{\sqrt{k}}$ can be minimal. If $\gamma : [0, l] \to M$ is the geodesic we now select $n - 1$ variational fields

$$V_i(t) = \sin\left(\frac{\pi}{l}t\right) E_i(t), i = 2, ..., n$$

as before. This time we also assume that $\dot\gamma, E_2, ...E_n$ form an orthonormal basis for $T_{\gamma(t)}M$. By adding up the contributions to the second variation formula for each variational field we get

$$\sum_{i=2}^{n} \frac{d^2 E}{ds^2}|_{s=0} = \sum_{i=2}^{n} \int_0^l \left|\dot{V}_i\right|^2 dt - \int_0^l g\left(R\left(V_i, \dot\gamma\right)\dot\gamma, V_i\right) dt$$

$$= (n-1)\left(\frac{\pi}{l}\right)^2 \int_0^l \cos^2\left(\frac{\pi}{l}t\right)$$

$$- \sum_{i=2}^{n} \int_0^l \sin^2\left(\frac{\pi}{l}t\right) \sec\left(E_i, \dot\gamma\right) dt$$

$$= (n-1)\left(\frac{\pi}{l}\right)^2 \int_0^l \cos^2\left(\frac{\pi}{l}t\right) dt - \int_0^l \sin^2\left(\frac{\pi}{l}t\right) \mathrm{Ric}\left(\dot\gamma, \dot\gamma\right) dt$$

$$< (n-1)k \int_0^l \cos^2\left(\frac{\pi}{l}t\right) dt - (n-1)k \int_0^l \sin^2\left(\frac{\pi}{l}t\right) dt$$

$$< 0.$$

$\square$

EXAMPLE 42. *The incomplete Riemannian manifold* $S^2 - \{\pm p\}$ *clearly has constant curvature 1 and infinite fundamental group. To make things worse; the universal covering also has diameter* π.

EXAMPLE 43. $S^1 \times \mathbb{R}^3$ *admits a complete doubly warped product metric*

$$dr^2 + \varphi^2(r)d\theta^2 + \psi^2(r)ds_2^2,$$

which has $\mathrm{Ric} > 0$ *everywhere. For* $t \geq 1$ *just let* $\varphi(t) = t^{-1/4}$ *and* $\psi(t) = t^{3/4}$ *and then adjust* φ *and* ψ *near* $t = 0$ *to make things work out.*

4.2. The Fundamental Group in Even Dimensions. For the next result we need to study what happens when we have a closed geodesic in a Riemannian manifold of positive curvature.

Let $\gamma : [0, l] \to M$ be a closed unit speed geodesic, i.e., $\dot\gamma(0) = \dot\gamma(l)$. Let $p = \gamma(0) = \gamma(l)$ and consider parallel translation along γ. This defines a linear isometry $P : T_p M \to T_p M$. Since γ is a closed geodesic, we have that $P(\dot\gamma(0)) = \dot\gamma(l) = \dot\gamma(0)$. Thus, P preserves the orthogonal complement to $\dot\gamma(0)$ in $T_p M$. Now recall that linear isometries $L : \mathbb{R}^k \to \mathbb{R}^k$ with $\det L = (-1)^{k+1}$ have 1 as an eigenvalue $\left(L(v) = v \text{ for some } v \in \mathbb{R}^k\right)$. We can use this to construct a closed parallel field around γ. Namely,

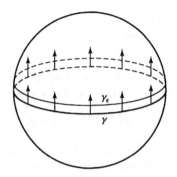

Figure 6.4

(1) If M is orientable and even-dimensional, then parallel translation around a closed geodesic preserves orientation and therefore has det $= 1$. Since the complement to $\dot{\gamma}(t)$ in T_pM is odd dimensional we can therefore find a closed parallel field around γ.

(2) If M is not orientable, has odd dimension, and furthermore, γ is a nonorientable loop (this means that the orientation changes as we go around this loop), then parallel translation around γ is orientation reversing and therefore has det $= -1$. Now, the complement to $\dot{\gamma}(t)$ in T_pM is even-dimensional, and since $P(\dot{\gamma}(0)) = \dot{\gamma}(0)$, we have that the restriction of P to this even-dimensional subspace still has det $= -1$. Thus, we get a closed parallel field in this case as well.

In Figure 6.4 we have sketched what happens when the closed geodesic is the equator on the standard sphere. In this case there is only one choice for the parallel field, and the shorter curves are the latitudes close to the equator.

We shall now prove an interesting and surprising topological result for positively curved manifolds.

THEOREM 26. (Synge, 1936) *Let M be a compact manifold with* sec > 0.
(1) If M is even-dimensional and orientable, then M is simply connected.
(2) If M is odd-dimensional, then M is orientable.

PROOF. The proof goes by contradiction. So in either case we have a nontrivial universal covering $\pi : \tilde{M} \to M$. We are now going to use the concepts of displacement and axis that we developed in the proof of Preissmann's Theorem in the last section. Among the finite selection of nontrivial deck transformations, which in the odd dimensional case are also assumed to be orientation reversing, we select α such that it has the smallest possible displacement. We now get a geodesic $\tilde{\gamma} : \mathbb{R} \to \tilde{M}$ that is mapped to itself by α. Moreover, $\gamma = \pi \circ \tilde{\gamma}$ is the shortest noncontractible loop in M corresponding to α. Finally, our choice of α ensures that it is the shortest possible noncontractible curve in M, which in the odd dimensional case also reverses orientation.

In both cases our assumptions are such that these loops are closed geodesics with perpendicular parallel fields by our discussion above. We can now use the second variation formula with this parallel field as variational field. Note that the variation isn't proper, but since the geodesic is closed the end point terms cancel each other

$$
\begin{aligned}
\frac{d^2 E\left(\gamma_s\right)}{ds^2}\Big|_{s=0} &= -\int_a^b g\left(R\left(E, \dot{\gamma}\right) \dot{\gamma}, E\right) dt + g\left(\frac{\partial^2 \bar{\gamma}}{\partial s^2}, \dot{\gamma}\right)\Big|_a^b \\
&= -\int_a^b g\left(R\left(E, \dot{\gamma}\right) \dot{\gamma}, E\right) dt \\
&= -\int_a^b \sec\left(E, \dot{\gamma}\right) dt \\
&< 0.
\end{aligned}
$$

Thus all nearby curves in this variation are closed curves whose lengths are shorter than γ. This contradicts our choice of γ as the shortest noncontractible curve. $\square$

The first important conclusion we get from this result is that while $\mathbb{R}P^2 \times \mathbb{R}P^2$ has positive Ricci curvature (its universal cover $S^2 \times S^2$ has positive Ricci curvature), it cannot support a metric of positive sectional curvature. It is, however, completely unknown whether $S^2 \times S^2$ admits a metric of positive sectional curvature. This is known as the Hopf problem (there is also the other Hopf problem from chapter 4 about the Euler characteristic). Recall that above we showed, using fundamental group considerations, that no product manifold admits negative curvature. In this case, fundamental group considerations cannot take us as far, since positively curved manifolds are often simply connected, something that never happens for compact negatively curved manifolds.

5. Basic Comparison Estimates

In this section we shall prove most of the comparison estimates that will be needed later in the text. These results are an enhancement of the case by case estimates we have derived by using the second variation formula or Jacobi field comparison. Jacobi field comparison can in fact be generalized to cover the present results. The case of upper curvature bounds generalizes immediately, while lower curvature bounds require another idea as we can't use the Cauchy-Schwarz inequality. Instead we revert to the fundamental equations to get more explicit and general information about the metric. The advantage is that all proofs are the same regardless of what curvature inequalities we have.

PROPOSITION 25. (Riccati Comparison Estimate) *Suppose we have real numbers $k \leq K$ and an absolutely continuous function $\lambda : (0, b) \to \mathbb{R}$ which satisfies*

$$
-K \leq \dot{\lambda} + \lambda^2 \leq -k.
$$

If the initial condition for λ is $\lambda(r) = \frac{1}{r} + O(r)$, as $r \to 0$, then

$$
\frac{\operatorname{sn}_K'(r)}{\operatorname{sn}_K(r)} \leq \lambda(r) \leq \frac{\operatorname{sn}_k'(r)}{\operatorname{sn}_k(r)}
$$

for as long as $\operatorname{sn}_K(r) > 0$.

PROOF. The two inequalities are proved in a similar manner. Recall that we covered the case where $K = 0$ when doing the Jacobi field comparison in nonpositive curvature. The reason for assuming $\operatorname{sn}_K(r) > 0$ is that we have to make sure that the comparison function is defined on $[0, r]$. In case $K \leq 0$, this is always true, while if $K > 0$, one must assume that $r < \frac{\pi}{\sqrt{K}}$. The next thing to observe is that the functions we wish to compare with are the solutions for the initial value problems

$$\dot{\lambda}_k + (\lambda_k)^2 = -k,$$

$$\lambda_k(r) = \frac{1}{r} + O(r),$$

$$\dot{\lambda}_K + (\lambda_K)^2 = -K,$$

$$\lambda_K(r) = \frac{1}{r} + O(r),$$

Thus, we are simply comparing a function satisfying a differential inequality to the solution for the corresponding differential equation. The result should therefore not come as any surprise (see also below).

For convenience, we shall concentrate on showing only

$$\lambda(r) \leq \lambda_k(r) = \frac{\mathrm{sn}_k'(r)}{\mathrm{sn}_k(r)}.$$

The idea is simply that the inequality

$$\dot{\lambda} + \lambda^2 \leq -k$$

can be separated to yield

$$\frac{\dot{\lambda}}{\lambda^2 + k} \leq -1 = \frac{\dot{\lambda}_k}{(\lambda_k)^2 + k}.$$

Thus

$$\int_0^r \frac{\dot{\lambda}}{\lambda^2 + k} dt \leq \int_0^r \frac{\dot{\lambda}_k}{(\lambda_k)^2 + k} dt,$$

and

$$F(\lambda(r)) \leq F(\lambda_k(r)),$$

where F is the antiderivative of $\frac{1}{\lambda^2+k}$ that satisfies $\lim_{\lambda \to \infty} F(\lambda) = 0$. Since F has positive derivative we can conclude that $\lambda(r) \leq \lambda_k(r)$. $\qquad \square$

It is worthwhile pointing out that there is a much more general comparison principle for first order differential equations. If $\lambda : [0, b)$ satisfies

$$\dot{\lambda} \leq f(\lambda)$$

then

$$\lambda(r) \leq \mu(r),$$

where μ is the solution to

$$\dot{\mu} = f(\mu),$$

$$\mu(0) = \lambda(0).$$

This is quite simple to establish and often very useful. The problem in the above case is that we have a singular initial value. It is a special feature of Riccati equations that we can also work with such initial conditions.

Let us now apply these results to one of the most commonly occurring geometric situations. Suppose that on a Riemannian manifold (M, g) we have introduced exponential coordinates around a point $p \in M$ so that $g = dr^2 + g_r$ on a star shaped open set in $T_pM - \{0\} = (0, \infty) \times S^{n-1}$. Along any given geodesic from p

the metric g_r is thought of as being on S^{n-1}. It is not important for the next result that M be complete as it is essentially local in nature.

THEOREM 27. *Assume that (M, g) satisfies $k \leq \sec \leq K$. If g_r represents the metric in the polar coordinates, then we have*

$$\operatorname{sn}_K^2 (r) \, ds_{n-1}^2 \;\; \leq \;\; g_r \leq \operatorname{sn}_k^2 (r) \, ds_{n-1}^2,$$

$$\frac{\operatorname{sn}_K' (r)}{\operatorname{sn}_K (r)} g_r \;\; \leq \;\; \operatorname{Hess} r \leq \frac{\operatorname{sn}_k' (r)}{\operatorname{sn}_k (r)} g_r.$$

PROOF. We first need to observe that $\operatorname{Hess} r$ has the initial values

$$\operatorname{Hess} r = \frac{1}{r} g_r + O(r).$$

The first thought is to use the fundamental equations on a parallel field X

$$\partial_r (\operatorname{Hess} r (X, X)) + \operatorname{Hess}^2 r (X, X) = -\sec (X, \partial_r).$$

This runs into a bit of trouble, though, as we don't necessarily have that

$$\begin{aligned}
\operatorname{Hess}^2 r (X, X) & = g (\nabla_X \partial_r, \nabla_X \partial_r) \\
& = (g (\nabla_X \partial_r, X))^2 \\
& = (\operatorname{Hess} r (X, X))^2
\end{aligned}$$

unless X is an eigenvector for $\nabla . \partial_r$. As we can't ensure that this happens we must resort to a slight trick. For fixed $\theta \in S^{n-1}$ define

$$\begin{aligned}
\lambda_{\min} (r, \theta) & = \min_{v \perp \partial_r} \frac{\operatorname{Hess} r (v, v)}{g_r (v, v)}, \\
\lambda_{\max} (r, \theta) & = \max_{v \perp \partial_r} \frac{\operatorname{Hess} r (v, v)}{g_r (v, v)}
\end{aligned}$$

These functions must be Lipschitz in r and hence absolutely continuous, as they are given through a minimum or maximum procedure. We now claim that

$$\begin{aligned}
\dot{\lambda}_{\max} (r) + \lambda_{\max}^2 (r) & \leq \; -k, \\
\lambda_{\max} (r) & = \; \frac{1}{r} + O(r), \\
\dot{\lambda}_{\min} (r) + \lambda_{\min}^2 (r) & \geq \; -K, \\
\lambda_{\min} (r) & = \; \frac{1}{r} + O(r).
\end{aligned}$$

The initial conditions are obvious from $\operatorname{Hess} r$. To establish the first inequality at a point r_0 where $\lambda_{\max} (r)$ is differentiable, select a unit v such that

$$\begin{aligned}
\operatorname{Hess} r (v, v) & = \; \lambda_{\max} (r_0, \theta) \, g_r (v, v), \\
\operatorname{Hess}^2 r (v, v) & = \; \lambda_{\max}^2 (r_0, \theta) \, g_r (v, v).
\end{aligned}$$

Then extend v to a parallel field V along the geodesic through p and (r_0, θ) and consider the function $\phi (r) = \operatorname{Hess} r (V, V)$. Then $\phi (r) \leq \lambda_{\max} (r)$ and $\phi (r_0) = \lambda_{\max} (r_0)$. Thus λ and ϕ have the same derivative at $r = r_0$. Using that $\nabla_{\partial_r} V = 0$

this yields

$$
\begin{aligned}
\dot{\lambda}_{\max}(r_0) + \lambda_{\max}^2(r_0) &= \phi'(r_0) + \lambda_{\max}^2(r_0) \\
&= \partial_r \mathrm{Hess}\, r\,(v,v) + \mathrm{Hess}^2 r\,(v,v) \\
&= (\nabla_{\partial_r} \mathrm{Hess}\, r)\,(V,V) + \mathrm{Hess}^2 r\,(v,v) \\
&= -g\,(R(v,\partial_r)\,\partial_r, v) \\
&\leq -k.
\end{aligned}
$$

The analysis is similar for the smallest eigenvalue.

Thus, we obtain the desired estimated for the Hessian. For the metric itself we switch to Jacobi fields and use the differential equation

$$
\partial_r g_r = 2\mathrm{Hess}\, r.
$$

The estimates for the Hessian then imply that

$$
2\frac{\mathrm{sn}_K'(r)}{\mathrm{sn}_K(r)} g_r \;\leq\; \partial_r g_r \leq 2\frac{\mathrm{sn}_k'(r)}{\mathrm{sn}_k(r)} g_r,
$$

$$
g_r \;=\; O\left(r^2\right).
$$

If we compare this to what happens in constant curvature k or K where we have

$$
\begin{aligned}
\partial_r\left(\mathrm{sn}_k^2(r)\,ds_{n-1}^2\right) &= 2\frac{\mathrm{sn}_k'(r)}{\mathrm{sn}_k(r)}ds_{n-1}^2, \\
\mathrm{sn}_k^2(r)\,ds_{n-1}^2 &= O\left(r^2\right) \\[4pt]
\partial_r\left(\mathrm{sn}_K^2(r)\,ds_{n-1}^2\right) &= 2\frac{\mathrm{sn}_K'(r)}{\mathrm{sn}_K(r)}ds_{n-1}^2, \\
\mathrm{sn}_K^2(r)\,ds_{n-1}^2 &= O\left(r^2\right)
\end{aligned}
$$

we see that the desired inequality for the metric g_r also holds. $\square$

6. More on Positive Curvature

In this section we shall show some further restrictions on the topology of manifolds with positive curvature. The highlight will be the classical quarter pinched sphere theorem of Rauch, Berger and Klingenberg. To prove this theorem requires considerable preparations. We shall have much more to say about this theorem and its generalizations in chapter 11.

6.1. The Conjugate Radius. As in the case where $\sec \leq 0$ we are going to find domains in the tangent space on which the exponential map is nonsingular.

EXAMPLE 44. *Consider S_K^n, $K > 0$. If we fix $p \in S_K^n$ and use polar coordinates, then the metric looks like $dr^2 + \mathrm{sn}_K^2 ds_{n-1}^2$. At distance $\frac{\pi}{\sqrt{K}}$ from p we therefore hit a conjugate point no matter what direction we go in.*

As a generalization of our result on no conjugate points when $\sec \leq 0$ we can show

THEOREM 28. *If (M,g) has $\sec \leq K$, $K > 0$, then*

$$
\exp_p : B\left(0, \frac{\pi}{\sqrt{K}}\right) \to M
$$

has no critical points.

PROOF. As before, pick a ball

$$B\left(0, R\right) \subset B\left(0, \frac{\pi}{\sqrt{K}}\right)$$

that contains no critical points for $\exp_p$. The comparison estimate for the pull-back metric with $\sec \leq K$ then yields

$$g_r \geq \operatorname{sn}_K^2\left(r\right) ds_{n-1}^2 \text{ for } r \in (0, R).$$

If $R < \pi/\sqrt{K}$, we further see that the metric does vanish on the boundary of $B\left(0, R\right)$. This shows that the pull-back metric cannot degenerate in $B\left(0, \pi/\sqrt{K}\right)$. Consequently $D \exp_p$ is nonsingular at any such point. $\qquad\square$

Next we turn our attention to convexity radius.

THEOREM 29. *Suppose R satisfies*
(1) $R \leq \frac{1}{2} \cdot \operatorname{inj}(x)$, $x \in B(p, R)$,
(2) $R \leq \frac{1}{2} \cdot \frac{\pi}{\sqrt{K}}$, $K = \sup\{\sec(\pi) : \pi \subset T_x M, \ x \in B(p, R)\}$.
Then $r(x) = d(x, p)$ is smooth and convex on $B(p, R)$, and any two points in $B(p, R)$ are joined by a unique segment that lies in $B(p, R)$.

PROOF. The first condition tells us that any two points in $B(p, R)$ are joined by a unique segment in M, and that $r(x)$ is smooth on $B(p, 2 \cdot R)$. The second condition ensures us that $\operatorname{Hess} r \geq 0$ on $B(p, R)$. It then remains to be shown that if $x, y \in B(p, R)$, and $\gamma : [0, 1] \to M$ is the unique segment joining them, then $\gamma \subset B(p, R)$. For fixed $x \in B(p, R)$, define C_x to be the set of ys for which this holds. Certainly $x \in C_x$ and C_x is open. If $y \in B(p, R) \cap \partial C_x$, then the segment $\gamma : [0, 1] \to M$ joining x to y must lie in $\overline{B(p, R)}$ by continuity. Now consider $\varphi(t) = r(\gamma(t))$. By assumption

$$\varphi(0), \varphi(1) \ < \ R,$$
$$\ddot{\varphi}(t) \ = \ \operatorname{Hess} r\left(\dot{\gamma}(t), \ \dot{\gamma}(t)\right) \geq 0.$$

Thus, φ is convex, and consequently

$$\max \varphi(t) \leq \max \left\{\varphi(0), \varphi(1)\right\} < R,$$

showing that $\gamma \subset B(p, R)$. $\qquad\square$

The largest R such that $r(x)$ is convex on $B(p, R)$ and any two points in $B(p, R)$ are joined by unique segments in $B(p, R)$ is called the *convexity radius* at p. Globally,

$$\operatorname{conv.rad}(M, g) = \inf_{p \in M} \operatorname{conv.rad}(p).$$

The previous result tell us

$$\operatorname{conv.rad}(M, g) \geq \min\left\{\frac{\operatorname{inj}(M, g)}{2}, \frac{\pi}{2\sqrt{K}}\right\}, \quad K = \sup \sec(M, g).$$

In non-positive curvature this simplifies to

$$\operatorname{conv.rad}(M, g) = \frac{\operatorname{inj}(M, g)}{2}.$$

Now that we can control conjugate points, we also get estimates for the injectivity radius. For Riemannian manifolds with $\sec \leq 0$ the injectivity radius satisfies

$$\operatorname{inj}(p) = \frac{1}{2} \cdot (\text{length of shortest geodesic loop based at } p).$$

This is because there are no conjugate points whatsoever. On a closed Riemannian manifold with sec ≤ 0 we get that

$$\text{inj}(M) = \inf_{p \in M} \text{inj}(p) = \frac{1}{2} \cdot (\text{length of shortest closed geodesic}).$$

Since M is closed, the infimum must be a minimum (this is not obvious, since we haven't shown that $p \to \text{inj}(p)$ is continuous, but you can prove this for yourself using that $\exp : TM \to M \times M$ is continuous). If $p \in M$ realizes this infimum, and $\gamma : [0,1] \to M$ is the geodesic loop realizing $\text{inj}(p)$, then we can split γ into two equal segments joining p and $\gamma\left(\frac{1}{2}\right)$. Thus, $\text{inj}\left(\gamma\left(\frac{1}{2}\right)\right) \leq \text{inj}(p)$, but this means that γ must also be a geodesic loop as seen from $\gamma\left(\frac{1}{2}\right)$. In particular, it is smooth at p and forms a closed geodesic.

More generally, we have that if (M,g) has sec $\leq K$, where $K > 0$, then

$$\text{inj}\,(p) \;\geq\; \min\left\{ \frac{\pi}{\sqrt{K}}, \frac{1}{2} \cdot (\text{length of shortest geodesic loop based at } p) \right\},$$

$$\text{inj}(M) \;=\; \inf_{p \in M} \text{inj}(p) = \min\left\{ \frac{\pi}{\sqrt{K}}, \frac{1}{2} \cdot (\text{length of shortest closed geodesic}) \right\}.$$

These estimates will be used in the next section.

6.2. The Injectivity Radius in Even Dimensions.
We get another interesting restriction on the geometry of positively curved manifolds.

THEOREM 30. (Klingenberg, 1959) *Suppose (M,g) is an orientable even-dimensional manifold with $0 < \text{sec} \leq 1$. Then $\text{inj}\,(M,g) \geq \pi$. If M is not orientable, then $\text{inj}\,(M,g) \geq \frac{\pi}{2}$.*

PROOF. The nonorientable case follows from the orientable case, as the orientation cover has $\text{inj}\,(M,g) \geq \pi$.

From our discussion above, we know that the upper curvature bound implies that if $\text{inj}M < \pi$, then it must be realized by a closed geodesic. So let us assume that we have a closed geodesic $\gamma : [0, 2\text{inj}M] \to M$ parametrized by arclength, where $2\text{inj}M < 2\pi$. Since M is orientable and even dimensional, we know that for all small $\varepsilon > 0$ there are curves $\gamma_\varepsilon : [0, 2\text{inj}M] \to M$ that converge to γ as $\varepsilon \to 0$ and with

$$\ell\left(\gamma_\varepsilon\right) < \ell\left(\gamma\right) = 2\text{inj}M.$$

Since

$$\gamma_\varepsilon \subset B\left(\gamma_\varepsilon\left(0\right), \text{inj}M\right).$$

there is a unique segment from $\gamma_\varepsilon\left(0\right)$ to $\gamma_\varepsilon\left(t\right)$. Thus, if $\gamma_\varepsilon\left(t_\varepsilon\right)$ is the point at maximal distance from $\gamma_\varepsilon\left(0\right)$ on γ_ε, we get a segment σ_ε joining these points that in addition is perpendicular to γ_ε at $\gamma_\varepsilon\left(t_\varepsilon\right)$. As $\varepsilon \to 0$ we have that $t_\varepsilon \to \text{inj}M$, and thus the segments σ_ε must subconverge to a segment from $\gamma\left(0\right)$ to $\gamma\left(\text{inj}M\right)$ that is perpendicular to γ at $\gamma\left(\text{inj}M\right)$. However, as the conjugate radius is $\geq \pi > \text{inj}M$, and γ is a geodesic loop realizing the injectivity radius at $\gamma\left(0\right)$, we know from chapter 5 that there can only be two segments from $\gamma\left(0\right)$ to $\gamma\left(\text{inj}M\right)$. Thus, we have a contradiction with our assumption $\pi > \text{inj}M$. $\qquad\square$

In Figure 6.5 we have pictured a fake situation, which still gives the correct idea of the proof. The closed geodesic is the equator on the standard sphere, and σ_ε converges to a segment going through the north pole.

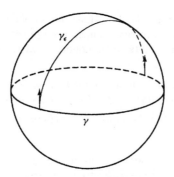

Figure 6.5

A similar result can clearly not hold for odd-dimensional manifolds. In dimension 3 we have the quotients of spheres $S^3/\mathbb{Z}_k$ for all positive integers k. Here the image of the Hopf fiber via the covering map $S^3 \to S^3/\mathbb{Z}_k$ is a closed geodesic of length $\frac{2\pi}{k}$ which goes to 0 as $k \to \infty$. Also, the Berger spheres $\left(S^3, g_\varepsilon\right)$ give counterexamples, as the Hopf fiber is a closed geodesic of length $2\pi\varepsilon$. In this case the curvatures lie in $\left[\varepsilon^2, 4 - 3\varepsilon^2\right]$. So if we rescale the upper curvature bound to be 1, the length of the Hopf fiber becomes $2\pi\varepsilon\sqrt{4 - 3\varepsilon^2}$ and the curvatures will lie in the interval $\left[\frac{\varepsilon^2}{4-3\varepsilon^2}, 1\right]$. When $\varepsilon < \frac{1}{\sqrt{3}}$, the Hopf fibers have length $< 2\pi$. In this case the lower curvature bound becomes smaller than $\frac{1}{9}$.

A much deeper result by Klingenberg asserts that if a simply connected manifold has all its sectional curvatures in the interval $(\frac{1}{4}, 1]$, then the injectivity radius is still $\geq \pi$. This result has been improved first by Klingenberg-Sakai and Cheeger-Gromoll to allow for the curvatures to be in $\left[\frac{1}{4}, 1\right]$. More recently, Abresch-Meyer showed that the injectivity radius estimate still holds if the curvatures are in $\left[\frac{1}{4} - 10^{-6}, 1\right]$. The Berger spheres show that such an estimate will not hold if the curvatures are allowed to be in $\left[\frac{1}{9} - \varepsilon, 1\right]$. Notice that the hypothesis on the fundamental group being trivial is necessary in order to eliminate all the constant-curvature spaces with small injectivity radius.

These injectivity radius estimates can be used to prove some fascinating sphere theorems. This will we discussed next.

6.3. Applications of Index Estimation. Some notions and results from topology are needed to explain the material here.

We say that $A \subset X$ is l-connected if the relative homotopy groups $\pi_k(X, A)$ vanish for $k \leq l$. A theorem of Hurewicz then shows that the relative homology groups $H_k(X, A)$ also vanish for $k \leq l$. Using the long exact sequence for the pair (X, A)

$$H_{k+1}(X, A) \to H_k(A) \to H_k(X) \to H_k(X, A)$$

then shows that $H_k(A) \to H_k(X)$ is an isomorphism for $k < l$ and surjective for $k = l$.

We say that a critical point $p \in M$ for a smooth function $f : M \to \mathbb{R}$ has index $\geq m$ if the Hessian of f is negative definite on a m-dimensional subspace in T_pM. Note that if $m \geq 1$, then p can't be a local minimum for f as the function must decrease in the directions where the Hessian is negative definite. The index of a critical point gives us information about how the topology of M changes as we pass through this point. In Morse theory a much more precise statement is proven, but it also requires the critical points to be nondegenerate, an assumption we do not make here (see [**68**]).

THEOREM 31. *Let $f : M \to \mathbb{R}$ be a smooth proper function. If b is not a critical value for f and all critical points in $f^{-1}([a, b])$ have index $\geq m$, then*

$$f^{-1}((-\infty, a]) \subset f^{-1}((-\infty, b])$$

is $(m-1)$-connected.

OUTLINE OF PROOF. If there are no critical points in $f^{-1}([a, b])$, then the gradient flow will deform $f^{-1}((-\infty, b])$ to $f^{-1}((-\infty, a])$. This is proved in detail in chapter 11. If there are critical points, then we can cover the set of critical points by finitely many sets U_i, where $\bar{U}_i \subset V_i$ and V_i is a coordinate chart where the first m coordinates correspond to directions where f decreases. These V_is exist because of our assumption about the index of the critical points.

Now consider a map $\phi : S^{k-1} \to f^{-1}([a, b])$. The negative gradient flow of f will deform ϕ to a map ϕ' that meets the regions V_i. Using transversality we can, after possibly making another small perturbation to ϕ', assume that not all of the first m coordinate functions on V_i vanish on ϕ'. Having done that ϕ' can be flowed in those directions until the image of ϕ' in U_i lies below the critical value. In this way it is possible to continuously deform ϕ until it its image lies in $f^{-1}((-\infty, a])$. $\square$

In analogy with $\Omega_{p,q}(M)$ define

$$\Omega_{A,B}(M) = \{\gamma : [0, 1] \to M : \gamma(0) \in A, \gamma(1) \in B\}.$$

If $A, B \subset M$ are compact, then the energy functional

$$E : \Omega_{A,B}(M) \to [0, \infty)$$

is reasonably nice in the sense that it behaves like a proper smooth function on a manifold. If in addition A and B are submanifolds then the variational fields for variations in $\Omega_{A,B}(M)$ consist of fields along the curve that are tangent to A and B at the endpoints. Therefore, critical points are naturally identified with geodesic that are perpendicular to A and B at the endpoints. We say that the index of such a geodesic $\geq k$ if there is a k-dimensional space of fields along the geodesic such that the second variation of the these fields is negative.

THEOREM 32. *Let M be a complete Riemannian manifold and $A \subset M$ a compact submanifold. If every geodesic in $\Omega_{A,A}(M)$ has index $\geq k$, then $A \subset M$ is k-connected.*

PROOF. See also [**58**, Theorem 2.5.16] for a proof. Identify $A = E^{-1}(0)$ and use the above as a guide for what should happen. This shows that $A \subset \Omega_{A,A}(M)$ is $(k-1)$-connected. Next we note that

$$\pi_l(\Omega_{A,A}(M), A) = \pi_{l+1}(M, A).$$

This gives the result. $\square$

This theorem can be used to prove a sphere theorem by Berger.

THEOREM 33. (Berger, 1958) *Let M be a closed n-manifold with $\sec \geq 1$ and $\operatorname{inj}_p > \pi/2$ for some $p \in M$, then M is $(n-1)$-connected and hence a homotopy sphere.*

PROOF. First note that every geodesic loop at p is either the constant curve or has length $> \pi$ since $\text{inj}_p > \pi/2$. We showed earlier in the chapter that geodesics of length $> \pi$ have proper variations whose second derivative is negative. In fact each parallel along the geodesic could be modified to create such a variation. As there is an $(n-1)$-dimensional space of such parallel fields we conclude that the index of such geodesics is $\geq (n-1)$. This shows that $p \in M$ is $(n-1)$-connected and consequently that M is $(n-1)$-connected.

Finally, to see that M is a homotopy sphere we take a map $F : M \to S^n$ of degree 1. A theorem of H. Hopf guarantees that this is always possible. Since M is $(n-1)$-connected this map must be an isomorphism on π_k for $k < n$ as S^n is also $(n-1)$-connected. This implies that

$$\pi_n(M) \simeq H_n(M) \to H_n(S^n) \simeq \pi_n(S^n)$$

is an isomorphism. Hurewicz' result shows that the homotopy and homology groups are isomorphic, while the fact that F has degree 1 implies that $H_n(S^n) \to H_n(M)$ is an isomorphism. A theorem of Whitehead then implies that F is a homotopy equivalence. $\square$

This theorem is even more interesting in view of the injectivity radius estimate in positive curvature that we discussed in the above section. Thus we get

COROLLARY 13. (Rauch-Berger-Klingenberg, 1951-61) *Let M be a closed simply connected n-manifold with $4 > \sec \geq 1$, then M is $(n-1)$-connected and hence a homotopy sphere.*

The conclusion can be strengthened to say that M is homeomorphic to a sphere. This follows from the solution to the generalized Poincaré conjecture in dimensions $n \neq 3$ given what we have already proven. In chapter 11 we shall describe an explicit homeomorphism.

Using a similar analysis to the above theorem one also gets the less interesting result.

COROLLARY 14. *Let M be a closed n-manifold with $\text{Ric} \geq (n-1)$ and $\text{inj}_p > \pi/2$ for some $p \in M$, then M is simply connected.*

Finally we mention a significant result that allows us to make strong conclusions about connectedness in positive curvature. The result will be used and enhanced in chapter 7.

LEMMA 22. (The Connectedness Principle, Wilking, 2003) *Let M^n be a compact n-manifold with positive sectional curvature.*

(a) If $N^{n-k} \subset M^n$ is a codimension k totally geodesic submanifold, then $N \subset M$ is $(n-2k+1)$-connected.

(b) If $N_1^{n-k_1}$ and $N_2^{n-k_2}$ are totally geodesic submanifolds of M with $k_1 \leq k_2$ and $k_1 + k_2 \leq n$, then $N_1 \cap N_2$ is a nonempty totally geodesic submanifold and $N_1 \cap N_2 \to N_2$ is $(n-k_1-k_2)$-connected.

PROOF. (a) Let $\gamma \in \Omega_{N,N}(M)$ be a geodesic and E a parallel field along γ such that E is tangent to N at the endpoints. Then we can construct a variation $\bar{\gamma}(s,t)$ such that $\bar{\gamma}(0,t) = \gamma(t)$ and $s \to \bar{\gamma}(s,t)$ is a geodesic with initial velocity $E|_{\gamma(t)}$. Since N is totally geodesic we see that $\bar{\gamma}(s,0), \bar{\gamma}(s,1) \in N$. Thus the variational curves lie in $\Omega_{N,N}(M)$. The second variation formula for this variation now tells us that

$$\frac{d^2 E(\gamma_s)}{ds^2}\Big|_{s=0} = \int_0^1 \left|\dot{E}\right|^2 dt - \int_a^b g\left(R\left(\dot{E},\dot{\gamma}\right)\dot{\gamma},\dot{E}\right) dt + g\left(\frac{\partial^2 \bar{\gamma}}{\partial s^2},\dot{\gamma}\right)\Big|_0^1$$

$$= -\int_0^1 g\left(R\left(\dot{E},\dot{\gamma}\right)\dot{\gamma},\dot{E}\right) dt$$

$$< 0$$

since $\dot{E} = 0$, $\frac{\partial^2 \bar{\gamma}}{\partial s^2} = 0$, and E is perpendicular to $\dot{\gamma}$. Thus each such parallel field gives us a negative variation. This shows that the index of γ is bigger than the set of variational parallel fields.

Let $V \subset T_{\gamma(1)}M$ be the subspace of vectors $v = E(1)$ where E is a parallel field along γ such that $E(0) \in T_{\gamma(0)}N$. The space of parallel fields used to get negative variations is then identified with $V \cap T_{\gamma(1)}N$. To find the dimension of that space we note that T_pN and hence also V have dimension $n-k$. Moreover, V and $T_{\gamma(1)}N$ lie in the orthogonal complement to $\dot{\gamma}(1)$. Putting this together gives us

$$2n - 2k = \dim\left(T_{\gamma(1)}N\right) + \dim\left(V\right)$$

$$= \dim\left(V \cap T_{\gamma(1)}N\right) + \dim\left(V + T_{\gamma(1)}N\right)$$

$$\leq \dim\left(V \cap T_{\gamma(1)}N\right) + n - 1.$$

(b) It is easy to show that $N_1 \cap N_2$ is also totally geodesic. The key is to guess that for $p \in N_1 \cap N_2$ we have

$$T_p\left(N_1 \cap N_2\right) = T_pN_1 \cap T_pN_2.$$

To see that $N_1 \cap N_2 \neq \varnothing$ select a geodesic from N_1 to N_2. The dimension conditions imply that there is a $(n - k_1 - k_2 + 1)$-dimensional space of parallel field along this geodesic that are tangent to N_1 and N_2 at the end points. Since $k_1 + k_2 \leq n$ we get a variation with negative second derivative, thus nearby variational curves are shorter. This shows that there can't be a nontrivial geodesic of shortest length joining N_1 and N_2.

Using $E : \Omega_{N_1,N_2}(M) \to [0,\infty)$ we can identify $N_1 \cap N_2 = E^{-1}(0)$. So we have in fact shown that $N_1 \cap N_2 \subset \Omega_{N_1,N_2}(M)$ is $(n - k_1 - k_2)$-connected . Using that $N_1 \subset M$ is $(n - 2k_1 + 1)$-connected shows that $\Omega_{N_1,N_2}(M) \subset \Omega_{M,N_2}(M)$ is also $(n - 2k_1 + 1)$-connected. Since $k_1 \leq k_2$ this shows that $N_1 \cap N_2 \subset \Omega_{M,N_2}(M)$ is $(n - k_1 - k_2)$-connected. Finally observe that $\Omega_{M,N_2}(M)$ can be retracted to N_2 and therefore is homotopy equivalent to N_2. This proves the claim.   □

What is commonly known as Frankel's theorem is included in part (b). The statement is simply that under the conditions in (b) the intersection is nonempty.

7. Further Study

Several textbooks treat the material mentioned in this chapter, and they all use variational calculus. We especially recommend [19], [41], and [56]. The latter also discusses in more detail closed geodesics and, more generally, minimal maps and surfaces in Riemannian manifolds.

As we won't discuss manifolds of nonpositive curvature in detail later in the text some references for this subject should be mentioned here. With the knowledge we have right now, it shouldn't be too hard to read the books [9] and [7]. For a more

advanced account we recommend the survey by Eberlein-Hammenstad-Schroeder in
[**45**]. At the moment the best, most complete, and up to date book on the subject
is probably [**34**].

For more information about the injectivity radius in positive curvature the
reader should consult the article by Abresch and Meyer in [**50**].

All of the necessary topological background material used in this chapter can
be found in [**68**] and [**86**].

8. Exercises

(1) Show that in even dimensions the sphere and real projective space are the
only closed manifolds with constant positive curvature.

(2) Suppose we have a rotationally symmetric metric $dr^2 + \varphi^2(r)\,d\theta^2$. We
wish to understand parallel translation along a latitude, i.e., a curve with
$r = a$. To do this we construct a cone $dr^2 + (\varphi(a) + \dot\varphi(a)(r - a))^2\,d\theta^2$
that is tangent to this surface at the latitude $r = a$. In case the surface
really is a surface of revolution, this cone is a real cone that is tangent to
the surface along the latitude $r = a$.

   (a) Show that in the standard coordinates (r, θ) on these surfaces, the
covariant derivative ∇_{∂_θ} is the same along the curve $r = a$. Conclude
that parallel translation is the same along this curve on these two
surfaces.

   (b) Now take a piece of paper and try to figure out what parallel translations along a latitude on a cone looks like. If you unwrap the paper it
is flat; thus parallel translation is what it is in the plane. Now rewrap
the paper and observe that parallel translation along a latitude does
not necessarily generate a closed parallel field.

   (c) Show that in the above example the parallel field along $r = a$ closes
up if $\dot\varphi(a) = 0$.

(3) *(Fermi-Walker transport)* Related to parallel transport there is a more
obscure type of transport that is sometimes used in physics. Let γ :
$[a, b] \to M$ be a curve into a Riemannian manifold whose speed never
vanishes and

$$T = \frac{\dot\gamma}{|\dot\gamma|}$$

the unit tangent of γ. We say that V is a Fermi-Walker field along γ if

$$\dot V = g(V, T)\dot T - g\left(V, \dot T\right)T$$
$$= \left(\dot T \wedge T\right)(V).$$

   (a) Show that given $V(t_0)$ there is a unique Fermi-Walker field V along
γ whose value at t_0 is $V(t_0)$.

   (b) Show that T is a Fermi-Walker field along γ.

   (c) Show that if V, W are Fermi-Walker fields along γ, then $g(V, W)$ is
constant along γ.

   (d) If γ is a geodesic, then Fermi-Walker fields are parallel.

(4) Let (M, g) be a complete n-manifold of constant curvature k. Select a
linear isometry $L : T_pM \to T_{\bar p}S_k^n$. When $k \leq 0$ show that

$$\exp_p \circ L^{-1} \circ \exp_{\bar p}^{-1} : S_k^n \to M$$

is a Riemannian covering map. When $k > 0$ show that

$$\exp_p \circ L^{-1} \circ \exp_{\bar p}^{-1} : S_k^n - \{-\bar p\} \to M$$

extends to a Riemannian covering map $S_k^n \to M$. (Hint: Use that the differential of the exponential maps is controlled by the metric, which in turn can be computed when the curvature is constant. You should also use the conjugate radius ideas presented in connection with the Hadamard-Cartan theorem.)

(5) Let $\gamma : [0,1] \to M$ be a geodesic. Show that $\exp_{\gamma(0)}$ has a critical point at $t\dot\gamma(0)$ iff there is a Jacobi field J along γ such that $J(0) = 0$, $\dot J(0) \neq 0$, and $J(t) = 0$.

(6) Let $\gamma(s,t) : [0,1]^2 \to (M,g)$ be a variation such that $R\left(\frac{\partial\gamma}{\partial s}, \frac{\partial\gamma}{\partial t}\right) = 0$ everywhere. Show that for each $v \in T_{\gamma(0,0)}M$, there is a parallel field $V : [0,1]^2 \to TM$ along γ, i.e., $\frac{\partial V}{\partial s} = \frac{\partial V}{\partial t} = 0$ everywhere.

(7) Using

$$R\left(\frac{\partial\gamma}{\partial s}, \frac{\partial\gamma}{\partial t}\right)\frac{\partial\gamma}{\partial u} = \frac{\partial^3\gamma}{\partial s\partial t\partial u} - \frac{\partial^3\gamma}{\partial t\partial s\partial u}$$

show that the two skew-symmetry properties and Bianchi's first identity hold for the curvature tensor.

(8) Let γ be a geodesic and X a Killing field in a Riemannian manifold. Show that the restriction of X to γ is a Jacobi field.

(9) Let γ be a geodesic in a Riemannian manifold and J_1, J_2 Jacobi fields along γ. Show that

$$g\left(\dot J_1, J_2\right) - g\left(J_1, \dot J_2\right)$$

is constant. A special case is when $J_2 = \dot\gamma$.

(10) A Riemannian manifold is said to be k-*point homogeneous* if for all pairs of points $(p_1, \ldots, p_k)$ and $(q_1, \ldots, q_k)$ with $d(p_i, p_j) = d(q_i, q_j)$ there is an isometry F with $F(p_i) = q_i$. When $k = 1$ we simply say that the space is homogeneous.

(a) Show that a homogenous space has constant scalar curvature.

(b) Show that if $k > 2$ and (M,g) is k-point homogeneous, then M is also $(k-1)$-point homogeneous.

(c) Show that if (M,g) is two-point homogeneous, then (M,g) is an Einstein metric.

(d) Show that if (M,g) is three-point homogeneous, then (M,g) has constant curvature.

(e) Classify all three-point homogeneous spaces. Hint: The only one that isn't simply connected is the real projective space.

(11) Show that if G is an infinite Abelian group that is the subgroup of the fundamental group of a manifold with constant curvature, then either the manifold is flat or G is cyclic.

(12) Let $M \to N$ be a Riemannian k-fold covering map. Show, $\text{vol}M = k \cdot \text{vol}N$.

(13) Starting with a geodesic on a two-dimensional space form, discuss how the equidistant curves change as they move away from the original geodesic.

(14) Introduce polar coordinates $(r, \theta) \in (0, \infty) \times S^{n-1}$ on a neighborhood U around a point $p \in (M,g)$. If (M,g) has $\sec \geq 0$ ($\sec \leq 0$), show that any

curve $\gamma(t) = (r(t), \theta(t))$ is shorter (longer) in the metric g than in the Euclidean metric on U.

(15) Around an orientable hypersurface $H \hookrightarrow (M, g)$ introduce the usual coordinates $(r, x) \in \mathbb{R} \times H$ on some neighborhood U around H. On U we have aside from the given metric g, also the radially flat metric $dt^2 + g_0$, where g_0 is the restriction of g to H. If M has sec ≥ 0 (sec ≤ 0) and $\gamma(t) = (r(t), x(t))$ is a curve, where $r \geq 0$ and the shape operator is ≤ 0 (≥ 0) at $x(t)$ for all t, show that γ is shorter (longer) with respect to g than with respect to the radially flat metric $dt^2 + g_0$.

(16) *(Frankel)* Let M be an n-dimensional Riemannian manifold of positive curvature and A, B two totally geodesic submanifolds. Show that A and B must intersect if $\dim A + \dim B \geq n - 1$. Hint: assume that A and B do not intersect. Then find a segment of shortest length from A to B. Show that this segment is perpendicular to each submanifold. Then use the dimension condition to find a parallel field along this geodesic that is tangent to A and B at the endpoints to the segments. Finally use the second variation formula to get a shorter curve from A to B.

(17) Let M be a complete n-dimensional Riemannian manifold and $A \subset M$ a compact submanifold. Without using Wilking's connectedness principle establish the following statements directly.

    (a) Show that curves in $\Omega_{A,A}(M)$ that are not stationary for the energy functional can be deformed to shorter curves in $\Omega_{A,A}(M)$.

    (b) Show that the stationary curves for the energy functional on $\Omega_{A,A}(M)$ consists of geodesics that are perpendicular to A at the end points.

    (c) If M has positive curvature, $A \subset M$ is totally geodesic, and $2\dim A \geq \dim M$, then the stationary curves can be deformed to shorter curves in $\Omega_{A,A}(M)$.

    (d) *(Wilking)* Conclude from c. that any curve $\gamma : [0,1] \to M$ that starts and ends in A is homotopic through such curves to a curve in A, i.e., $\pi_1(M, A)$ is trivial.

(18) Generalize Preissmann's theorem to show that any solvable subgroup of the fundamental group must be cyclic.

(19) Let (M, g) be an oriented manifold of positive curvature and suppose we have an isometry $F : M \to M$ of finite order without fixed points. Show that if $\dim M$ is even, then F must be orientation reversing, while if $\dim M$ is odd, it must be orientation preserving. Weinstein has proven that this holds even if we don't assume that F has finite order.

(20) Use an analog of Cartan's result on isometries of finite order in nonpositive curvature to show that any closed manifold of constant curvature $= 1$ must either be the standard sphere or have diameter $\leq \frac{\pi}{2}$. Generalize this to show that any closed manifold with sec ≥ 1 is either simply connected or has diameter $\leq \frac{\pi}{2}$. In chapter 11 we shall show the stronger statement that a closed manifold with sec ≥ 1 and diameter $> \frac{\pi}{2}$ must in fact be homeomorphic to a sphere.

(21) *(The Index Form)* Below we shall use the second variation formula to prove several results established in chapter 5. If V, W are vector fields along a geodesic $\gamma : [0,1] \to (M, g)$, then the *index form* is the symmetric

bilinear form

$$I_0^1(V, W) = I(V, W) = \int_0^1 \left(g\left(\dot{V}, \dot{W}\right) - g\left(R(V, \dot{\gamma})\,\dot{\gamma}, W\right) \right) dt.$$

In case the vector fields come from a proper variation of γ this is equal to the second variation of energy. Assume below that $\gamma : [0, 1] \to (M, g)$ locally minimizes the energy functional. This implies that $I(V, V) \geq 0$ for all proper variations.

(a) If $I(V, V) = 0$ for a proper variation, then V is a Jacobi field. Hint: Let W be any other variational field that also vanishes at the end points and use that

$$0 \leq I(V + \varepsilon W, V + \varepsilon W) = I(V, V) + 2\varepsilon I(V, W) + \varepsilon^2 I(W, W)$$

for all small ε to show that $I(V, W) = 0$. Then use that this holds for all W to show that V is a Jacobi field.

(b) Let V and J be variational fields along γ such that $V(0) = J(0)$ and $V(1) = J(1)$. If J is a Jacobi field show that

$$I(V, J) = I(J, J).$$

(c) *(The Index Lemma)* Assume in addition that there are no Jacobi fields along γ that vanish at both end points. If V and J are as in b. show that $I(V, V) \geq I(J, J)$ with equality holding only if $V = J$ on $[0, 1]$. Hint: Prove that if $V \neq J$, then

$$0 < I(V - J, V - J) = I(V, V) - I(J, J).$$

(d) Assume that there is a nontrivial Jacobi field J that vanishes at 0 and 1, show that $\gamma : [0, 1 + \varepsilon] \to M$ is not locally minimizing for $\varepsilon > 0$. Hint: For sufficiently small ε there is a Jacobi field $K : [1 - \varepsilon, 1 + \varepsilon] \to TM$ such that $K(1 + \varepsilon) = 0$ and $K(1 - \varepsilon) = J(1 - \varepsilon)$. Let V be the variational field such that $V|_{[0, 1-\varepsilon]} = J$ and $V|_{[1-\varepsilon, 1+\varepsilon]} = K$. Finally extend J to be zero on $[1, 1 + \varepsilon]$. Now show that

$$\begin{aligned} 0 &= I_0^1(J, J) = I_0^{1+\varepsilon}(J, J) = I_0^{1-\varepsilon}(J, J) + I_{1-\varepsilon}^{1+\varepsilon}(J, J) \\ &> I_0^{1-\varepsilon}(J, J) + I_{1-\varepsilon}^{1+\varepsilon}(K, K) = I(V, V). \end{aligned}$$

CHAPTER 7

The Bochner Technique

Aside from the variational techniques we saw used in the last chapter one of the oldest and most important techniques in modern Riemannian geometry is that of the Bochner technique. In this chapter we shall prove some of the classical theorems Bochner proved about obstructions to the existence of Killing fields and harmonic 1-forms. We also explain briefly how the Bochner technique extends to forms. This will in the next chapter lead us to a classification of compact manifolds with nonnegative curvature operator. To establish the relevant Bochner formula for forms, we have used the language of Clifford multiplication. It is, in our opinion, much easier to work consistently with Clifford multiplication, rather than trying to keep track of wedge products, interior products, Hodge star operators, exterior derivatives, and their dual counter parts. In addition, it has the effect of preparing one for the world of spinors, although we won't go into this here. In the last section we give a totally different application of the Bochner technique. In effect, we try to apply it to the curvature tensor itself. The outcome will be used in the next chapter in our classification of manifolds with nonnegative curvature operator.

It should be noted that we have not used a unified approach to the Bochner technique. There are many equivalent approaches and we have tried to discuss a few of them here. It is important to learn how it is used in its various guises, as one otherwise could not prove some of the results we present. We have for pedagogical reasons used Stokes' theorem throughout rather than the maximum principle. The reason is that one can then cover the material with minimum knowledge of geodesic geometry. The maximum principle in the strongest possible form is established and used in chapter 9. The interested reader is encouraged to learn about it there, and then go back and try it out in connection with the Bochner technique.

The Bochner technique was, as the name indicates, invented by Bochner. However, Bernstein knew about it for harmonic functions on domains in Euclidean space. Specifically, he used

$$\Delta \frac{1}{2} |\nabla u|^2 = |\nabla^2 u|^2 ,$$

where $u : \Omega \subset \mathbb{R}^n \to \mathbb{R}$ and $\Delta u = 0$. It was Bochner who realized that when the same trick is attempted on Riemannian manifolds, a curvature term also appears. Namely, for $u : (M, g) \to \mathbb{R}$ with $\Delta_g u = 0$ one has

$$\Delta \frac{1}{2} |\nabla u|^2 = |\nabla^2 u|^2 + \mathrm{Ric}\,(\nabla u, \nabla u) .$$

With this it is clear that curvature influences the behavior of harmonic functions. The next nontrivial step Bochner took was to realize that one can compute $\Delta \frac{1}{2} |\omega|^2$ for any harmonic form ω and then try to get information about the topology of the manifold. The key ingredient here is of course Hodge's theorem, which states that any homology class can be represented by a harmonic form. Yano further refined the Bochner technique, but it seems to be Lichnerowicz who really put things into gear, when around 1960 he presented his formulae for the Laplacian on forms and spinors. After this work, Berger, D. Meyer, Gallot, Gromov-Lawson, Witten, and many others have made significant contributions to this tremendously important subject.

Prior to Bochner's work Weitzenböck also developed a formula very similar to the Bochner formula. We shall also explain this related formula and how it can be used to establish the Bochner formulas we use. It appears that Weitzenböck never realized that his work could have an impact on geometry and only thought of his work as an application of algebraic invariant theory.

1. Killing Fields

We shall see how Killing fields interact with curvature in various settings. But first we need to establish some general properties (see also the appendix for explanations of how flows and Lie derivatives are related).

1.1. Killing Fields in General. A vector field X on a Riemannian manifold (M, g) is called a *Killing field* if the local flows generated by X act by isometries. This translates into the following simple characterization:

PROPOSITION 26. *A vector field X on a Riemannian manifold (M, g) is a Killing field if and only if $L_X g = 0$.*

PROOF. Let F^t be the local flow for X. Recall that

$$(L_X g)(v, w) = \frac{d}{dt} g\left(DF^t(v), DF^t(w)\right)|_{t=0}.$$

Thus we have

$$
\begin{aligned}
\frac{d}{dt} g\left(DF^t(v), DF^t(w)\right)|_{t=t_0} &= \frac{d}{dt} g\left(DF^{t-t_0} DF^{t_0}(v), DF^{t-t_0} DF^{t_0}(w)\right)|_{t=t_0} \\
&= \frac{d}{ds} g\left(DF^s DF^{t_0}(v), DF^s DF^{t_0}(w)\right)|_{s=0} \\
&= L_X g\left(DF^{t_0}(v), DF^{t_0}(w)\right).
\end{aligned}
$$

This shows that $L_X g = 0$ if and only if $t \to g\left(DF^t(v), DF^t(w)\right)$ is constant. Since F^0 is the identity map this is equivalent to assuming the flow acts by isometries. □

We can now use this characterization to show

PROPOSITION 27. *X is a Killing field iff $v \to \nabla_v X$ is a skew symmetric $(1,1)$-tensor.*

PROOF. Let $\theta_X(v) = g(X, v)$ be the 1-form dual to X and recall that

$$d\theta_X(V, W) + (L_X g)(V, W) = 2g(\nabla_V X, W).$$

Thus $L_X g \equiv 0$ iff $v \to \nabla_v X$ is skew-symmetric. □

PROPOSITION 28. *For a given $p \in M$, a Killing field X is uniquely determined by $X|_p$ and $(\nabla X)|_p$.*

PROOF. The equation $L_X g \equiv 0$ is linear in X, so the space of Killing fields is a vector space. Therefore, it suffices to show that $X \equiv 0$ on M provided $X|_p = 0$ and $(\nabla X)|_p = 0$. Using an open-closed argument, we can reduce our considerations to a neighborhood of p.

Let F^t be the local flow for X near p. The condition $X|_p = 0$ implies that $F^t(p) = p$ for all t. Thus $DF^t : T_p M \to T_p M$. We claim that also $DF^t = I$. By assumption we know that that X commutes with any vector field at p

$$
\begin{aligned}
[X, Y]|_p &= \nabla_{X(p)} Y - \nabla_{Y(p)} X \\
&= \nabla_0 Y - 0 = 0.
\end{aligned}
$$

If we have $Y|_p = v$, then the definition of the Lie derivative implies

$$
0 = L_X Y|_p = \frac{d}{dt} \left(DF^t(v) - v \right) |_{t=0}.
$$

Consequently

$$
\begin{aligned}
0 &= L_X DF^{t_0}(Y)|_p \\
&= \frac{d}{ds} \left(DF^s DF^{t_0}(v) - DF^{t_0}(v) \right) |_{s=0} \\
&= \frac{d}{dt} \left(DF^{t-t_0} DF^{t_0}(v) - DF^{t_0}(v) \right) |_{t=t_0} \\
&= \frac{d}{dt} \left(DF^t(v) - DF^{t_0}(v) \right) |_{t=t_0}.
\end{aligned}
$$

In other words $t \to DF^t(v)$ is constant. As $DF^0(v) = v$ we have that $DF^t = I$.

Since the flow diffeomorphisms act by isometries, we can conclude that they must be the identity map, and hence $X = 0$ in a neighborhood of p.

We could also have used that X, when restricted to a geodesic γ must be a Jacobi field as the flow of X generates a geodesic variation. Thus $X = 0$ along this geodesic if $X|_{\gamma(0)}$ and $\nabla_{\dot\gamma(0)} X$ both vanish. □

These properties lead us to two important general results about Killing fields.

THEOREM 34. *The zero set of a Killing field is a disjoint union of totally geodesic submanifolds each of even codimension.*

PROOF. The flow generated by a Killing field X on (M, g) acts by isometries so we know from chapter 5 that the fixed point set of these isometries is a union of totally geodesic submanifolds. We next observe that the fixed point set of these isometries is precisely the set of points where the Killing field vanishes. Finally assume that $X|_p = 0$ and let $V \oplus W = T_p M$ be the orthogonal decomposition where V is the tangent space to the zero set for X. We know from the above proof that $\nabla_v X = 0$ iff $v \in V$. Thus $w \to \nabla_w X$ is an isomorphism on W. As it is also a skew-symmetric map it follows that W is even-dimensional. □

THEOREM 35. *The set of Killing fields* iso(M, g) *is a Lie algebra of dimension* $\leq \frac{(n+1)n}{2}$. *Furthermore, if M is compact (or complete), then* iso(M, g) *is the Lie algebra of* Iso(M, g).

PROOF. Note that $L_{[X,Y]} = [L_X, L_Y]$. So if $L_X g = L_Y g = 0$, we also have that $L_{[X,Y]} g = 0$. Thus, iso(M, g) does form a Lie algebra. We have just seen that the map $X \to (X|_p, (\nabla X)|_p)$ is linear and has trivial kernel. So

$$\dim(\mathfrak{iso}(M,g)) \leq \dim T_p M + \dim(\text{skew-symmetric transformations of } T_p M)$$
$$= n + \frac{n(n-1)}{2} = \frac{(n+1)n}{2}.$$

The last statement is not easy to prove. Observe, however, that since M is compact, each vector field generates a global flow on M. Each Killing field therefore generates a 1-parameter subgroup of $\text{Iso}(M,g)$. If we take it for granted that $\text{Iso}(M,g)$ is a Lie group, then the identity component is of course generated by the 1-parameter subgroups, and each such group by definition generates a Killing field.

In case M is complete one can show that Killing fields generate global flows as in the compact case. With that in mind a similar proof can be carried through. □

Recall that $\dim(\text{Iso}(S_k^n)) = \frac{(n+1)n}{2}$. Thus, all space forms have maximal dimension for their isometry groups. If we consider other complete spaces with constant curvature, then we know they look like S_k^n/Γ, where $\Gamma \subset \text{Iso}(S_k^n)$ acts freely and discontinuously on S_k^n. The isometries on the quotient S_k^n/Γ can now be identified with those isometries of $\text{Iso}(S_k^n)$ that commute with all elements in Γ. So if $\dim(\text{Iso}(S_k^n/\Gamma))$ is maximal, then the elements in Γ must commute with all the elements in the connected component of $\text{Iso}(S_k^n)$ containing the identity. This shows that Γ has to consist of homotheties. Thus, Γ can essentially only be $\{I, -I\}$ if it is nontrivial. But $-I$ acts freely only on the sphere. Thus, only one other constant-curvature space form has maximal dimension for the isometry group, namely $\mathbb{R}P^n$.

More generally, one can prove that if (M,g) is complete and

$$\dim(\text{Iso}(M,g)) = \frac{n(n+1)}{2},$$

then (M,g) has constant curvature. To see this, we need a new construction. The frame bundle FM of (M,g) is the set of $(p, e_1, \ldots, e_n)$, where $p \in M$ and $e_1, \ldots, e_n$ forms an orthonormal basis for $T_p M$. It is not hard to see that this is a manifold of dimension $\frac{n(n+1)}{2}$. Any isometry $F : M \to M$ induces a map of FM by sending $(p, e_1, \ldots, e_n)$ to $(F(p), DF(e_1), \ldots, DF(e_n))$. The uniqueness theorem for isometries shows that the induced action of $\text{Iso}(M)$ on FM cannot have any fixed points. Thus each orbit

$$\{(F(p), DF(e_1), \ldots, DF(e_n)) : F \in \text{Iso}(M)\} \subset FM$$

is a properly embedded submanifold of FM diffeomorphic to $\text{Iso}(M)$. When

$$\dim(\text{Iso}(M)) = \frac{n(n+1)}{2}$$

it must therefore follow that the orbit is a component of FM. FM itself has either one or two components depending on whether M is nonorientable or orientable. In either case any two 2-dimensional subspaces admit bases that are mapped to each other by an isometry of M. This clearly shows that M has constant curvature.

1.2. Killing Fields in Negative Ricci Curvature.

For a $(1,1)$-tensor T the norm is

$$|T|^2 = \text{tr}(T \circ T^*) = \sum_{i=1}^{n} g(T(E_i), T(E_i))$$

where T^* is the adjoint and E_i is any orthonormal frame.

PROPOSITION 29. *Let X be a Killing field on (M, g) and consider the function*
$f = \frac{1}{2} g(X, X) = \frac{1}{2} |X|^2$, *then*
(1) $\nabla f = -\nabla_X X$.
(2) $\operatorname{Hess} f (V, V) = g(\nabla_V X, \nabla_V X) - R(V, X, X, V)$.
(3) $\Delta f = |\nabla X|^2 - \operatorname{Ric}(X, X)$.

PROOF. To see (1), observe that

$$
\begin{aligned}
g(V, \nabla f) &= D_V f \\
&= g(\nabla_V X, X) \\
&= -g(V, \nabla_X X).
\end{aligned}
$$

For (2), we repeatedly use that $V \to \nabla_V X$ is skew-symmetric:

$$
\begin{aligned}
\operatorname{Hess} f (V, V) &= g(\nabla_V (-\nabla_X X), V) \\
&= -g(R(V, X) X, V) - g(\nabla_X \nabla_V X, V) - g(\nabla_{[V,X]} X, V) \\
&= -R(V, X, X, V) - g(\nabla_X \nabla_V X, V) \\
&\quad + g(\nabla_{\nabla_X V} X, V) - g(\nabla_{\nabla_V X} X, V) \\
&= -R_X(V) + g(\nabla_V X, \nabla_V X) - g(\nabla_X \nabla_V X, V) - g(\nabla_V X, \nabla_X V) \\
&= -R_X(V) + g(\nabla_V X, \nabla_V X) - D_X g(\nabla_V X, V) \\
&= -R_X(V) + g(\nabla_V X, \nabla_V X).
\end{aligned}
$$

For (3) we select an orthonormal frame E_i and see that

$$
\begin{aligned}
\Delta f &= \sum_{i=1}^{n} \operatorname{Hess} f(E_i, E_i) \\
&= \sum_{i=1}^{n} g(\nabla_{E_i} X, \nabla_{E_i} X) - \sum_{i=1}^{n} R(E_i, X, X, E_i) \\
&= \sum_{i=1}^{n} g(\nabla_{E_i} X, \nabla_{E_i} X) - \operatorname{Ric}(X, X) \\
&= |\nabla X|^2 - \operatorname{Ric}(X, X).
\end{aligned}
$$

$\square$

THEOREM 36. (Bochner, 1946) *Suppose (M, g) is compact, oriented, and has* $\operatorname{Ric} \leq 0$. *Then every Killing field is parallel. Furthermore, if* $\operatorname{Ric} < 0$, *then there are no nontrivial Killing fields.*

PROOF. If we define $f = \frac{1}{2} |X|^2$ for a Killing field X, then using Stokes' theorem and the condition $\operatorname{Ric} \leq 0$ gives us

$$
\begin{aligned}
0 &= \int_M \Delta f \cdot dvol \\
&= \int_M \left(-\operatorname{Ric}(X, X) + |\nabla X|^2 \right) \cdot dvol \\
&\geq \int_M |\nabla X|^2 \cdot dvol \\
&\geq 0.
\end{aligned}
$$

Thus $|\nabla X| \equiv 0$ and X must be parallel. In addition

$$\mathrm{Ric}(X, X) \quad \leq \quad 0,$$

$$\int_M -\mathrm{Ric}(X, X) \cdot d\mathrm{vol} \quad = \quad 0$$

so $\mathrm{Ric}\,(X, X) \equiv 0$. If $\mathrm{Ric} < 0$ this implies that $X \equiv 0$. $\qquad\square$

COROLLARY 15. *With* (M, g) *as in the theorem, we have*

$$\dim(\mathfrak{iso}(M, g)) = \dim(\mathrm{Iso}(M, g)) \leq \dim M,$$

and $\mathrm{Iso}(M, g)$ *is finite if* $\mathrm{Ric}(M, g) < 0$.

PROOF. Since any Killing field is parallel, the linear map: $X \to X|_p$ from $\mathfrak{iso}(M, g)$ to $T_p M$ is injective. This gives the result. For the second part use that $\mathrm{Iso}(M, g)$ is compact, since M is compact, and that the identity component is trivial. $\qquad\square$

COROLLARY 16. *With* (M, g) *as before and* $p = \dim(\mathfrak{iso}(M, g))$, *we have that the universal covering splits isometrically as* $\widetilde{M} = \mathbb{R}^p \times N$.

PROOF. On $\widetilde{M}$ there are p linearly independent parallel vector fields, which we can assume to be orthonormal. Since $\widetilde{M}$ is simply connected, each of these vector fields is the gradient field for a distance function. If we consider just one of these distance function we see that the metric splits as $g = dr^2 + g_r = dr^2 + g_0$ since the Hessian of this distance function vanishes. As we get such a splitting for p distance functions with orthonormal gradients we get the desired splitting of $\widetilde{M}$. $\qquad\square$

The result about nonexistence of Killing fields can actually be slightly improved to yield

THEOREM 37. *Suppose* (M, g) *is a compact manifold with quasi-negative Ricci curvature, i.e.,* $\mathrm{Ric} \leq 0$ *and* $\mathrm{Ric}(v, v) < 0$ *for all* $v \in T_p M - \{0\}$ *for some* $p \in M$. *Then* (M, g) *admits no nontrivial Killing fields.*

PROOF. We already know that any Killing field is parallel. Thus a Killing field is always zero or never zero. If the latter holds, then $\mathrm{Ric}(X, X)(p) < 0$, but this contradicts

$$0 = \Delta f(p) = -\mathrm{Ric}(X, X)(p) > 0.$$

$\qquad\square$

Bochner's theorem has been generalized by X. Rong to a more general statement asserting that a closed Riemannian manifold with negative Ricci curvature can't admit a pure F-structure of positive rank (see [83] for the definition of F structure and proof of this). An F-structure on M is essentially a finite covering of open sets U_i on some finite covering space $\hat{M} \to M$, such that we have a Killing field X_i on each U_i. Furthermore, these Killing fields must commute whenever they are defined at the same point, i.e., $[X_i, X_j] = 0$ on $U_i \cap U_j$. The idea of the proof is to consider the function

$$f = \det\left(g\left(X_i, X_j\right)\right)_{i,j}$$

If only one vector field is given on all of M, then this reduces to the function

$$f = g\left(X, X\right)$$

that we considered above. For the above expression one must show that it is a reasonably nice function that has a similar Bochner formula.

1.3. Killing Fields in Positive Curvature. We can actually also say quite a bit about Killing fields in positive sectional curvature. This is a much more recent development in Riemannian geometry. We shall mostly study how Killing fields can be used to make conclusions about the Euler characteristic, but we shall also mention several stronger results without giving their proofs.

Recall that any vector field on an even-dimensional sphere has a zero, since the Euler characteristic is 2 ($\neq 0$). At some point H. Hopf conjectured that in fact any even-dimensional manifold with positive sectional curvature has positive Euler characteristic. If the curvature operator is positive, this is certainly true, as we shall see below. From chapter 6 we know that $|\pi_1| < \infty$, provided just the Ricci curvature is positive. Thus $H_1(M, \mathbb{R}) = 0$. This shows that the conjecture holds in dimension 2. In dim $= 4$, Poincaré duality implies that

$$H_1(M, \mathbb{R}) = H_3(M, \mathbb{R}) = 0.$$

Hence

$$\chi(M) = 1 + \dim H_2(M, \mathbb{R}) + 1 \geq 2.$$

In higher dimensions we have the following partial justification for the Hopf conjecture.

THEOREM 38. (Berger, 1965) *If (M, g) is a compact, even-dimensional manifold of positive sectional curvature, then every Killing field has a zero.*

PROOF. Consider as before $f = \frac{1}{2}|X|^2$. If X has no zeros, f will have a positive minimum at some point $p \in M$. Then of course Hess$f|_p \geq 0$. We also know that

$$\text{Hess}f(V, V) = g(\nabla_V X, \nabla_V X) - R(V, X, X, V)$$

and by assumption, $g(R(V, X)X, V) > 0$ if X and V are linearly independent. Using this, we shall find V such that Hess$f(V, V) < 0$ near p, thus arriving at a contradiction.

Recall that the linear endomorphism $v \to \nabla_v X$ is skew-symmetric. Furthermore, $(\nabla_X X)|_p = 0$, since $\nabla f|_p = -(\nabla_X X)|_p$, and f has a minimum at p. Thus, we have a skew-symmetric map $T_p M \to T_p M$ with at least one zero eigenvalue. But then, even dimensionality of $T_p M$ ensures us that there must be at least one more zero eigenvector $v \in T_p M$ linearly independent from X. Thus,

$$\begin{aligned}
\text{Hess}f(v, v) &= g(\nabla_v X, \nabla_v X) - R(v, X, X, v) \\
&= -R(v, X, X, v) < 0.
\end{aligned}$$

$\square$

In odd dimensions we can get completely different information from the existence of a nontrivial Killing field.

THEOREM 39. (X. Rong, 1995) *If a closed Riemannian n-manifold (M, g) admits a nontrivial Killing field, then the fundamental group has a cyclic subgroup of index $\leq c(n)$.*

The reader should consult [**82**] for a more general statement and the proof. Observe, however, that should such a manifold admit a free isometric action by

S^1, then the quotient M/S^1 is a positively curved manifold as well (by the O'Neill formula from chapter 3.) Thus, we have a fibration,

$$S^1 \quad \rightarrow \quad M$$
$$\downarrow$$
$$M/S^1.$$

In case M is even-dimensional, we know that $\pi_1(M)$ is either trivial or $\mathbb{Z}_2$. So in this case there is nothing to prove. Otherwise, the quotient is even-dimensional. The long exact sequence for the homotopy groups is

$$\pi_2(M/S^1) \rightarrow \mathbb{Z} \rightarrow \pi_1(M) \rightarrow \pi_1(M/S^1) = \left\{ \begin{array}{l} \{1\}, \\ \mathbb{Z}_2. \end{array} \right.$$

Thus, $\pi_1(M)$ must either be a finite cyclic group or contain a finite cyclic subgroup of index 2. This result is enhanced by the surprising observation of R. Shankar that there are 7-manifolds with positive curvature and fundamental groups of the form $\mathbb{Z}_n \times \mathbb{Z}_2$, i.e., Abelian but noncyclic fundamental groups. This is in contrast to Preissmann's theorem for compact manifolds of negative curvature and also gives a counterexample to a conjecture of Chern.

Next there are several results about the structure of positively curved manifolds with effective torus actions. These will also partially support the Hopf conjectures.

Having an isometric torus action implies that $\mathrm{iso}(M,g)$ contains a certain number of linearly independent commuting Killing fields. By Berger's result we know that in even dimensions these Killing fields must vanish somewhere. Moreover, the structure of these zero sets is so that each component is a totally geodesic submanifold of even codimension. A type of induction on dimension can now be used to extract information about these manifolds. To understand how this works some important topological results on the zero set for a Killing field are also needed. To state these results we define the Betti numbers of an n-manifold M as

$$b_p(M) = \dim H_p(M, \mathbb{R}) = \dim H^p(M, \mathbb{R})$$

and the Euler characteristic as the alternating sum

$$\chi(M) = \sum_{p=0}^{n} (-1)^p b_p(M).$$

It is a key result in algebraic topology that $H_p(M, \mathbb{R})$ and $H^p(M, \mathbb{R})$ have the same dimension when we use real coefficients. Note next that Poincaré duality implies that $b_p(M) = b_{n-p}(M)$.

THEOREM 40. (Conner, 1957) *Let X be a Killing field on a compact Riemannian manifold. If $N_i \subset M$ are the components of the zero set for X, then*
1) $\chi(M) = \sum_i \chi(N_i)$,
2) $\sum_p b_{2p}(M) \geq \sum_i \sum_p b_{2p}(N_i)$,
3) $\sum_p b_{2p+1}(M) \geq \sum_i \sum_p b_{2p+1}(N_i)$.

For proofs see [59] and [15]. There is a trick to how one gets the results for Killing fields given the theorems in [15] which work for torus actions. Given $X \in \mathrm{iso}(M,g)$ let $T \subset \mathrm{Iso}(M,g)$ be the closure of the group generated by the flow of X. As $\mathrm{Iso}(M,g)$ is compact we see that T is a compact Abelian Lie group and hence a torus T^k. The set of points that are fixed by the action of T on M is still the zero set for X so results from [15] can now be used.

Using part 1) of this theorem we can easily prove something about 6-manifolds with positive sectional curvature.

COROLLARY 17. *If M is a compact 6-manifold with positive sectional curvature that admits Killing field, then $\chi(M) > 0$.*

PROOF. We know that the zero set for a Killing field is non-empty and that each component has even codimension. Thus each component is a 0, 2, or 4-dimensional manifold with positive sectional curvature. This shows that M has positive Euler characteristic.   □

If we consider 4-manifolds we get a much stronger result (see also [54]).

THEOREM 41. (W. Y. Hsiang-Kleiner, 1989) *If M^4 is a compact orientable positively curved 4-manifold that admits a Killing field, then the Euler characteristic is ≤ 3. In particular, M is topologically equivalent to S^4 or $\mathbf{C}P^2$.*

This has been partially generalized to higher dimensions by Grove-Searle, Rong and most recently Wilking. The best results, however, do not generalize the Hsiang-Kleiner classification as they require more isometries.

To explain these results we introduce some more notation. The *rank* of a compact Lie group is the maximal dimension of an Abelian subalgebra in the corresponding Lie algebra. The *symmetry rank* of a compact Riemannian manifold is the rank of the isometry group. To state and prove the results below we need some notation related to the symmetry rank. Fix an Abelian subalgebra $\mathfrak{h}(M, g) \subset \mathrm{iso}(M, g)$ whose dimension is the symmetry rank and define $\mathcal{Z}(\mathfrak{h}(M, g))$ as the collection of $N \subsetneqq M$ that are components for the zero sets of Killing fields in $\mathfrak{h}(M, g)$. We have the following important general properties.

PROPOSITION 30. *1) If $N \in \mathcal{Z}(\mathfrak{h}(M, g))$, then all Killing fields in $\mathfrak{h}(M, g)$ are tangent to N.*

2) $N \in \mathcal{Z}(\mathfrak{h}(M, g))$ is maximal with respect to inclusion iff the restriction of $\mathfrak{h}(M, g)$ to N has dimension $\dim(\mathfrak{h}(M, g)) - 1$.

3) If $N \in \mathcal{Z}(\mathfrak{h}(M, g))$, then N is contained in finitely many maximal sets $N_1, ..., N_l$ and $N = N_1 \cap \cdots \cap N_l$.

4) If $N, N' \in \mathcal{Z}(\mathfrak{h}(M, g))$, then $N \cap N' \in \mathcal{Z}(\mathfrak{h}(M, g))$.

PROOF. 1) Assume that N is a component for the zero set of $X \in \mathfrak{h}(M, g)$ and that $Y \in \mathfrak{h}(M, g)$. Since $L_Y X = 0$ and Y is a Killing field we get

$$
\begin{aligned}
0 &= (L_Y g)(X, X) \\
&= D_Y |X|^2 - 2g(L_Y X, X) \\
&= D_Y |X|^2.
\end{aligned}
$$

The flow of Y therefore preserves the level sets for $|X|^2$. In particular, Y is tangent to N.

2) Assume that N is a component of the zero set of $X \in \mathfrak{h}(M, g)$ and that $Y \in \mathfrak{h}(M, g)$ vanishes on N. Fix $p \in N$ and consider the skew-symmetric linear transformations $(\nabla X)|_p$ and $(\nabla Y)|_p$. We know that they vanish on $T_p N$ and hence preserve the orthogonal complement $V \subset T_p M$. This gives us two commuting skew-symmetric linear transformations

$$
(\nabla X)|_p, (\nabla Y)|_p : V \to V
$$

on an even dimensional space. Since they commute there is a decomposition of

$$V = E_1 \oplus \cdots \oplus E_l$$

into 2-dimensional subspaces E_i such that both $(\nabla X)|_p$ and $(\nabla Y)|_p$ preserve these subspaces. As the space of skew-symmetric transformations on E_1, say, is one-dimensional there must be some linear combination of $(\nabla X)|_p$ and $(\nabla Y)|_p$ that vanishes on E_1. This shows that some linear combination $\alpha X + \beta Y$ not only vanishes on N but on a larger set as $(\nabla(\alpha X + \beta Y))|_p$ vanishes on a subspace that is larger than T_pN. This Killing field $\alpha X + \beta Y$ must therefore vanish on a component that includes N. Since this violates maximality of N we conclude that Y can't vanish on N unless it is a multiple of X.

Conversely suppose that the restriction of $\mathfrak{h}(M, g)$ to N is $\dim(\mathfrak{h}(M, g)) - 1$ dimensional and let $X \in \mathfrak{h}(M, g)$ be a Killing field that vanishes on N. If

$$N \subset N' \in \mathcal{Z}(\mathfrak{h}(M, g)),$$

then some $Y \in \mathfrak{h}(M, g)$ must vanish on N' and hence also on N. Our assumption then implies that X and Y are linearly dependent. Thus also X vanishes on N' and $N = N'$ as desired.

3) Fix $p \in N$ and let

$$\mathfrak{h}_0 = \{X \in \mathfrak{h}(M, g) : X|_N = 0\}.$$

Then $\mathfrak{h}_0$ acts on T_pM by the skew-symmetric transformations $(\nabla X)|_p$ and this action completely determines $X \in \mathfrak{h}_0$. These transformations vanish on T_pN, and as $\mathfrak{h}_0$ is Abelian we can decompose

$$T_pM = T_pN + E_1 + \cdots + E_m$$

into an orthogonal sum of invariant 2-dimensional subspaces E_j. If $X' \in \mathfrak{h}_0$ vanishes on $N' \supset N$, then $T_pN' = \ker((\nabla X')|_p)$. As all of the spaces in our decomposition of T_pM are invariant under $(\nabla X')|_p$ its kernel will be a sum of T_pN and a finite collection of E_js. Since T_pN' completely determines N', this shows that only finitely many sets in $\mathcal{Z}(\mathfrak{h}(M, g))$ contain N.

Let

$$N_1, ..., N_l \in \mathcal{Z}(\mathfrak{h}(M, g))$$

be the maximal sets that contain N and $X_i \in \mathfrak{h}_0$ corresponding nontrivial Killing fields vanishing on N_i. Also select $X \in \mathfrak{h}_0$ so that N is a component for its zero set. Thus $\ker(\nabla X)|_p = T_pN$ and $(\nabla X)|_p$ is nontrivial on each E_j. We claim that $(\nabla X)|_p$ is a linear combination of $(\nabla X_i)|_p$. This shows that

$$\bigcap \ker(\nabla X_i)|_p = \ker(\nabla X)|_p = T_pN$$

and hence that

$$N = N_1 \cap \cdots \cap N_l.$$

To establish the claim we can first assume that after reindexing the X_is that $(\nabla X_1)|_p, ..., (\nabla X_k)|_p$ form a basis for

$$\text{span}\{(\nabla X_1)|_p, ..., (\nabla X_l)|_p\}$$

Now consider

$$Y = \alpha_0 X + \alpha_1 X_1 + \cdots + \alpha_k X_k.$$

If we can arrange matters so that $\ker(\nabla Y)|_p$ is not contained in any of the kernels $\ker(\nabla X_i)|_p$, then we have found a nontrivial $Y \in \mathfrak{h}_0$ that vanishes on a set $N' \supset N$

which is not contained in any of the maximal sets $N_1, ..., N_l \supset N$. As there are no other maximal sets containing N, this shows that $Y = 0$ and in particular that

$$(\nabla X)\big|_p \in \text{span}\left\{(\nabla X_1)\big|_p, ..., (\nabla X_k)\big|_p\right\}$$

as $(\nabla X_1)\big|_p$, ..., $(\nabla X_k)\big|_p$ are linearly independent. The desired Y is found as follows. After possibly rearranging the E_js we can assume that for some $s \leq k$, none of the operators $(\nabla X_i)\big|_p$, $i = 1, ..., k$ vanish on all of $E_1 + \cdots + E_s$. Insuring that $(\nabla Y)\big|_p$ vanishes on $E_1 + \cdots + E_s$ can be divided into s linear equations by checking that it vanishes on each E_j. As there are $k + 1 > s$ variables $\alpha_0, ..., \alpha_k$ in our definition of Y it is possible to find nontrivial scalars that force $(\nabla Y)\big|_p$ to vanish on $E_1 + \cdots + E_s$.

4) Let N be a component for the zero set of X and N' one for X'. Clearly $\alpha X + \beta X'$ vanishes on $N \cap N'$. If $p \in N \cap N'$, then we also know that $(\nabla X)\big|_p$ and $(\nabla X')\big|_p$ simultaneously vanish on

$$T_p N \cap T_p N' = T_p (N \cap N'),$$

but not on any nonzero vector in the orthogonal complement. We decompose the orthogonal complement into two dimensional subspaces E_i that are invariant under both $(\nabla X)\big|_p$ and $(\nabla X')\big|_p$. Since these operators don't simultaneously vanish on these subspaces we can adjust α and β appropriately so that $(\nabla (\alpha X + \beta X'))\big|_p$ doesn't vanish on the orthogonal complement. This shows that $\alpha X + \beta X'$ has $N \cap N'$ as a component for its zero set. $\qquad\square$

When M is even dimensional and has positive curvature Berger's results tells us that $\mathcal{Z}(\mathfrak{h}(M, g))$ is nonempty as long as $\mathfrak{h}(M, g)$ is nontrivial. In odd dimensions this is not quite true as the unit vector field that generates the Hopf fibration $S^3(1) \to S^2\left(\frac{1}{4}\right)$ is a Killing field. A similar result, however, does hold if we assume that $\dim(\mathfrak{h}(M, g)) \geq 2$.

We start by explaining the Grove-Searle result.

THEOREM 42. (Grove-Searle, 1994) *Let M be a compact n-manifold with positive sectional curvature and symmetry rank k. If $k \geq \frac{n}{2}$, then M is diffeomorphic to either a sphere, complex projective space or a cyclic quotient of a sphere $S^n/\mathbb{Z}_q$, where $\mathbb{Z}_q$ is a cyclic group of order q acting by isometries on the unit sphere.*

PROOF. The above proposition shows that maximal elements of $\mathcal{Z}(\mathfrak{h}(M, g))$ also satisfy the hypotheses. If there is no codimension 2 set in $\mathcal{Z}(\mathfrak{h}(M, g))$, then we can inductively construct a 1- or 2-manifold $N \in \mathcal{Z}(\mathfrak{h}(N, g))$ with $\dim\left(\mathfrak{h}\left(N^2, g\right)\right) \geq 2$ by successively choosing maximal elements inside other maximal elements. A 1-manifold has 1-dimensional isometry group so that does not happen. The 2-dimensional case is eliminated as follows. Select $X, Y \in \mathfrak{h}\left(N^2, g\right)$. Since X is nontrivial we can find a point p such that X vanishes at p but not near p. As Y preserves the component $\{p\}$ of the zero set for X it also vanishes at p. Then

$$(\nabla X)\big|_p, (\nabla Y)\big|_p : T_p M \to T_p M$$

completely determine the Killing fields. As the set of skew-symmetric transformations on $T_p M$ is 1-dimensional they must be linearly dependent. This shows that $\dim\left(\mathfrak{h}\left(N^2, g\right)\right) = 1$.

The topological classification is now established using the next theorem. $\qquad\square$

THEOREM 43. (Grove-Searle, 1994) *Let M be a closed n-manifold with positive sectional curvature. If M admits a Killing field such that the zero set has a component N of codimension 2, then M is diffeomorphic to S^n, $\mathbb{C}P^{\frac{n}{2}}$ or a cyclic quotient of a sphere $S^n/\mathbb{Z}_q$.*

Below we prove a weaker version of this theorem that only addresses what the homology groups are. The latter theorem can also be used to extend the corollary about the Euler characteristic of 6-manifolds under fairly weak symmetry rank conditions.

THEOREM 44. (Püttmann-Searle, 2002) *If M^{2n} is a compact $2n$-manifold with positive sectional curvature and symmetry rank $k \geq \frac{2n-4}{4}$, then $\chi(M) > 0$.*

PROOF. When $2n = 2, 4$ we make no assumptions about the symmetry rank and we know the theorem holds. When $2n = 6$ it is the above corollary. Next consider the case where M is 8-dimensional. The proof is as in the 6-dimensional situation unless the zero set for the Killing field has a 6-dimensional component. In that case the above theorem establishes the claim.

We actually prove something a bit stronger so that we can use a stronger induction hypothesis: If M is as in the theorem, then each $N \in \mathcal{Z}(\mathfrak{h}(M,g))$ has positive Euler characteristic.

Note that this definitely holds when $2n = 2, 4$.

Now for the induction step. Let $N \in \mathcal{Z}(\mathfrak{h}(M,g))$ and assume that N is contained in a maximal component:

$$N \subset N' \in \mathcal{Z}(\mathfrak{h}(M,g)).$$

There are now two cases. If N' has codimension ≥ 4, then we see that N' together with the restriction of $\mathfrak{h}(M,g)$ to N' satisfy the statement of the theorem. Thus our induction hypothesis ensures us that $\chi(N) > 0$. In case N' has codimension 2 we have to use more specific information. The Grove-Searle result tells us that M is diffeomorphic to S^{2n}, $\mathbb{R}P^{2n}$, or $\mathbb{C}P^n$ as it is even dimensional. In all of these cases the odd dimensional homology groups of M vanish. If N_i are the components of the zero set for $X \in \mathfrak{h}(M,g)$ part 3 of Conner's theorem tells us

$$0 = \sum_p b_{2p+1}(N') \geq \sum_i \sum_p b_{2p+1}(N_i)$$

Therefore, the components N_i can't have odd dimensional homology groups either. In particular, $\chi(N) > 0$. This completes the induction step. □

There is unfortunately a positively curved 24-manifold $F_4/Spin(8)$ which has symmetry rank 4 and is therefore not covered by the conditions of the theorem. However, Rong and Su have much better version as we shall see below that does cover this case (see also [84]).

It is tempting to suppose that one could show that the odd homology groups with real coefficients vanish given the assumptions of the previous theorem. In fact, all known even dimensional manifolds with positive sectional curvature have vanishing odd dimensional homology groups.

Next we mention without proof an extension of the Grove-Searle result by Wilking, (see also [95]).

THEOREM 45. (Wilking, 2003) *Let M is a compact simply connected positively curved n-manifold with symmetry rank k. If $k \geq \frac{n}{4} + 1$, then M has the topology of a sphere, complex projective space or quaternionic projective space.*

The proof is considerably more complicated than the above theorems. An important tool is Wilking's connectedness principle which we discussed in chapter 6. This principle also needs to be improved in case we have a group action. We present a simplified version, which suffices for our other applications below.

LEMMA 23. (Connectedness Principle with symmetries, Wilking, 2003) *Let M^n be a compact n-manifold with positive sectional curvature and X a Killing field. If N^{n-k} is a component for the zero set of X, then $N \subset M$ is $(n - 2k + 2)$-connected.*

PROOF. Consider a unit speed geodesic γ that is perpendicular to N at the endpoints. It is hard to extract information directly, using only parallel fields along γ as we did without symmetry assumptions.

Instead consider the fields E along γ such that

$$
\begin{aligned}
E(0) &\in T_{\gamma(0)}N, \\
\dot{E} &= -\frac{g(E, \nabla_{\dot{\gamma}}X)}{|X|^2}X.
\end{aligned}
$$

Note that the second equation isn't singular at $t = 0$ as

$$g(E, \nabla_{\dot{\gamma}}X) = -g(\dot{\gamma}, \nabla_E X) = 0$$

since we assumed that $E(0) \in T_{\gamma(0)}N$. These fields in addition have the properties

$$
\begin{aligned}
g(E, X) &= 0, \\
g\left(E(1), \nabla_{\dot{\gamma}(1)}X\right) &= 0, \\
g(E, \dot{\gamma}) &= 0.
\end{aligned}
$$

The first condition follows from $X|_{\gamma(0)} = 0$ and

$$
\begin{aligned}
\frac{d}{dt}g(E, X) &= g\left(\dot{E}, X\right) + g(E, \nabla_{\dot{\gamma}}X) \\
&= g\left(-\frac{g(E, \nabla_{\dot{\gamma}}X)}{|X|^2}X, X\right) + g(E, \nabla_{\dot{\gamma}}X) \\
&= 0.
\end{aligned}
$$

As $X|_{\gamma(1)} = 0$ this also implies the second property. Finally we show that $E \perp \dot{\gamma}$. This follows from

$$
\begin{aligned}
\frac{d}{dt}g(E, \dot{\gamma}) &= g\left(\dot{E}, \dot{\gamma}\right) \\
&= g\left(-\frac{g(E, \nabla_{\dot{\gamma}}X)}{|X|^2}X, \dot{\gamma}\right) \\
&= -\frac{g(E, \nabla_{\dot{\gamma}}X)}{|X|^2}g(X, \dot{\gamma})
\end{aligned}
$$

but

$$
\begin{aligned}
\frac{d}{dt}g(X, \dot{\gamma}) &= g(\nabla_{\dot{\gamma}}X, \dot{\gamma}) = 0, \\
X|_{\gamma(0)} &= 0.
\end{aligned}
$$

so $g(X, \dot{\gamma}) = 0$.

In particular,

$$E(1) \perp \mathrm{span}\left\{\dot{\gamma}(1), \nabla_{\dot{\gamma}(1)}X\right\}.$$

The space span $\left\{\dot{\gamma}(1), \nabla_{\dot{\gamma}(1)}X\right\}$ is 2-dimensional as $\dot{\gamma}$ is perpendicular to the component N of the zero set for X. This means that the space of such fields E, where in addition $E(1)$ is tangent to N, must have dimension at least $n - 2k + 2$ (see also the proof of part (a) of the connectedness principle in chapter 6).

We now need to check that such fields give us negative second variation. This is not immediately obvious as $\left|\dot{E}\right|$ doesn't vanish. However, we can resort to a trick that forces it down in size without losing control of the curvatures $\sec(E, \dot{\gamma})$. Recall from "Riemannian Submersions" in chapter 3 that we can perturb the metric g to a metric g_λ where X has been squeezed to have size $\to 0$ as $\lambda \to 0$. At the same time, directions orthogonal to X remain unchanged and the curvatures $\sec(E, \dot{\gamma})$ can only become larger as both E and $\dot{\gamma}$ are perpendicular to X. Finally γ remains a geodesic as the metric is only altered in a direction perpendicular to γ (see also exercises to chapter 5).

The second variation formula looks like

$$
\begin{aligned}
\frac{d^2 E}{ds^2}\Big|_{s=0} &= \int_a^b \left|\dot{E}(t)\right|^2_{g_\lambda} dt - \int_a^b g_\lambda\left(R(E, \dot{\gamma})\dot{\gamma}, E\right) dt \\
&\leq \int_a^b \left|\frac{g(E, \nabla_{\dot{\gamma}}X)}{|X|^2_g}X\right|^2_{g_\lambda} dt - \int_a^b \sec_g(E, \dot{\gamma})|E|^2_g dt \\
&= \int_a^b \frac{\left|g(E, \nabla_{\dot{\gamma}}X)\right|^2_{g_\lambda}}{|X|^4_g}|X|^2_{g_\lambda} dt - \int_a^b \sec_g(E, \dot{\gamma})|E|^2_g dt \\
&\to -\int_a^b \sec_g(E, \dot{\gamma})|E|^2_g dt \text{ as } \lambda \to 0.
\end{aligned}
$$

This shows that all of the fields E must have negative second variation in the metric g_λ for sufficiently small λ.

This perturbation is of course independent of $\gamma \in \Omega_{N,N}(M)$ and so we have found a new metric where all such geodesics have index $\geq n - 2k + 2$. This shows that $N \subset M$ is $(n - 2k + 2)$-connected. $\qquad\square$

One can get a bit of a feeling for how the connectedness principle comes in handy by proving a homological version of the Grove-Searle result about having a codimension 2 zero set for a Killing field. The version we present is quite weak, but strong enough to be used for the Püttmann-Searle result we proved above.

LEMMA 24. *Let M^n be a positively curved manifold and X a Killing field on M. If the zero set of X has a component of codimension 2, then*

$$
\begin{aligned}
b_{2p+1}(M) &= 0 \text{ for } 2p+1 < n, \\
b_{2p}(M) &\approx b_{2p+2}(M) \text{ for } 2p+2 \leq n.
\end{aligned}
$$

PROOF. The previous lemma shows that $N \subset M$ is $(n-2)$-connected. Using this together with Poincaré duality shows that

$$b_p(M) = b_p(N) = b_{n-p-2}(N) = b_{n-p-2}(M) = b_{p+2}(M)$$

as long as $p < n - 2$ and $n - p - 2 < n - 2$. Thus we need $p < n - 2$ and $p > 0$. We already know $b_1(M) = 0$ so this proves the claim.   □

This Lemma has been partly generalized by Rong and Su to the case where the codimension is 4 or 6.

LEMMA 25. *Let M^n be a positively curved manifold and $N \subset M$ a component of the zero set for a Killing field X.*

1) If N has codimension 4, then

$$\chi(M) \geq \chi(N) > 0.$$

2) If N has codimension 6, and is $(n - 6)$-connected, then

$$\chi(M) \geq \chi(N) > 0.$$

PROOF. Part of the proof is purely topological and won't be proven here. The statement is as follows. Assume that $N \subset M$ has codimension ≤ 6 and is $(n - 6)$-connected. If $\chi(M) \geq \chi(N)$, then $\chi(N) > 0$.

We then need to see that these conditions are satisfied. As $N \subset M$ does have codimension ≤ 6 Frankel's theorem shows that the other components of the zero set must have dimension 0, 2 or 4. This shows that $\chi(M) \geq \chi(N)$. In part two we are assuming that $N \subset M$ is $(n - 6)$-connected so we need only establish this in the codimension 4 situation. There the connectedness principle gives us that $N \subset M$ is $(n - 2 \cdot 4 + 2)$-connected.   □

THEOREM 46. (Rong-Su, 2005) *If M^{2n} is a compact $2n$-manifold with positive sectional curvature and symmetry rank $k > \frac{2n-4}{8}$, then $\chi(M) > 0$.*

PROOF. The induction step is again that each set in $\mathcal{Z}(\mathfrak{h}(M, g))$ has positive Euler characteristic.

Let $N \in \mathcal{Z}(\mathfrak{h}(M, g))$. If N has dimension ≤ 4 or codimension ≤ 4, then we already know that $\chi(N) > 0$. This takes care of the situations where $2n \leq 10$. In particular, we have $k \geq 2$.

If N is contained in a maximal set with codimension ≥ 8, then induction implies that $\chi(N) > 0$.

Next consider the situation where N isn't maximal and all maximal sets containing N have codimension ≤ 6. We know that if $N_1, ..., N_l$ are the maximal sets containing N, then

$$N = N_1 \cap \cdots \cap N_l.$$

The above lemma and part (b) of the connectedness principle from chapter 6 now establishes our claim in the following manner. If necessary reorder N_j so that N_l has maximal codimension. If we let $\dim(N_j) = 2n - k_j$ the connectedness principle then tells us that $N_{l-1} \cap N_l \subset N_l$ is $(2n - k_{l-1} - k_l)$-connected and in particular $(\dim(N_l) - 6)$-connected. If $l = 2$, this finishes the argument. Otherwise note that all of the intersections $N_l \cap N_j$ have codimension ≤ 6 in N_l. We can then after possibly rearranging the sets use the same argument with M replaced by N_l to see that

$$N_{l-2} \cap N_{l-1} \cap N_l \subset N_{l-1} \cap N_l$$

is $(\dim(N_{l-1} \cap N_l) - 6)$-connected. Continuing in this fashion finally shows that

$$N = N_1 \cap \cdots \cap N_l \subset N_2 \cap \cdots \cap N_l$$

is $(\dim(N_2 \cap \cdots \cap N_l) - 6)$-connected.

Finally we have to address what happens if N is maximal and has codimension 6. If there is some other maximal N' with codimension ≤ 6 then as above $N \cap N' \subset N$ is $(\dim(N) - 6)$-connected. Next note that $N \cap N'$ is part of the zero set for $Y|_N$, where Y is the Killing field that vanishes on N'. Hence $\chi(N) \geq \chi(N \cap N') > 0$.

Thus we have to consider the case were all other maximal sets have codimension ≤ 8. Each of the other maximal sets then satisfy the induction hypothesis. Thus all sets in $\mathcal{Z}(\mathfrak{h}(M, g))$ except possibly N itself must have positive Euler characteristic as they are all intersections of maximal sets and hence contained in some maximal set of codimension ≤ 8. Since $k \geq 2$ there is a Killing field which is nontrivial when restricted to N. All of its zero sets in N have positive Euler characteristic as we just saw, so Conner's theorem shows that N also has positive Euler characteristic. $\square$

The theorem also holds if we only assume that $k \geq \frac{2n-4}{8}$ as well as $k \geq 2$ when $2n = 12$. The proof is basically the same except the induction breaks down when $2n = 20$ and either N is itself maximal and has dimension 12 or if N is 6 dimensional and all of the maximal sets containing N have dimension 12. In these cases the 12 dimensional sets inherit only one Killing field, rather than two as hypothesized in the theorem. To settle this case requires a slightly more delicate analysis. Just like the theorem of Wilking one has to use isometric involutions coming from the isometric action generated by the Killing fields.

2. Hodge Theory

The reader who is not familiar with de Rham cohomology might wish to consult the appendix before proceeding.

Recall that on a manifold M we have the *de Rham complex*

$$0 \ \to \ \Omega^0(M) \ \overset{d^0}{\to} \ \Omega^1(M) \ \overset{d^1}{\to} \ \Omega^2(M) \ \to \ \cdots \ \overset{d^{n-1}}{\to} \ \Omega^n(M) \ \to \ 0,$$

where $\Omega^k(M)$ denotes the space of k-forms on M and

$$d^k : \Omega^k(M) \to \Omega^{k+1}(M)$$

is exterior differentiation. The *de Rham cohomology* groups

$$H^k(M) = \frac{\ker(d^k)}{\mathrm{im}(d^{k-1})}$$

compute the real cohomology of M. We know that $H^0(M) \simeq \mathbb{R}$ if M is connected, and $H^n(M) = \mathbb{R}$ if M is orientable and compact. In this case we have a pairing,

$$\Omega^k(M) \times \Omega^{n-k}(M) \ \to \ \mathbb{R},$$

$$(\omega_1, \omega_2) \ \to \ \int_M \omega_1 \wedge \omega_2,$$

inducing a nondegenerate pairing

$$H^k(M) \times H^{n-k}(M) \to \mathbb{R}$$

on the cohomology groups. The two vector spaces $H^k(M)$ and $H^{n-k}(M)$ are therefore dual to each other and in particular have the same dimension.

Now suppose M is endowed with a Riemannian metric g. Then each of the spaces $\Omega^k(M)$ is also endowed with a pointwise inner product structure:

$$\Omega^k(M) \times \Omega^k(M) \to \Omega^0(M).$$

This structure is obtained by declaring that if $E_1, \ldots, E_n$ is an orthonormal frame, then the dual coframe $\sigma^1, \ldots, \sigma^n$ is also orthonormal, and furthermore that all the k-forms $\sigma^{i_n} \wedge \ldots \wedge \sigma^{i_k}$, $i_1 < \cdots < i_k$, are orthonormal. A different way of introducing this inner product structure is by first observing that k-forms of the type $f_0 \cdot df_1 \wedge \ldots \wedge df_k$, where $f_0, f_1, \ldots f_k \in \Omega^0(M)$, actually span $\Omega^k(M)$. The product on $\Omega^0(M)$ is obviously just multiplication of functions. On $\Omega^1(M)$ we declare that

$$\begin{aligned} g\left(f_0 \cdot df_1, \ h_0 dh_1\right) &= f_0 \cdot h_0 \cdot g\left(df_1, dh_1\right) \\ &= f_0 \cdot h_0 \cdot g\left(\nabla f_1, \nabla h_1\right), \end{aligned}$$

and on $\Omega^k(M)$ we define

$$\begin{aligned} & g\left(f_0 \cdot df_1 \wedge \ldots \wedge df_k, \ h_0 dh_1 \wedge \ldots \wedge dh_k\right) \\ =\ & f_0 \cdot h_0 \ g\left(df_1 \wedge \ldots \wedge df_k, \ dh_1 \wedge \ldots \wedge dh_k\right) \\ =\ & f_0 \cdot h_0 \ \det\left(g\left(df_i, dh_j\right)_{1 \leq i,j \leq k}\right) \\ =\ & f_0 \cdot h_0 \ \det\left(g\left(\nabla f_i, \nabla h_j\right)_{1 \leq i,j \leq k}\right). \end{aligned}$$

By integrating this pointwise inner product we get an inner product on $\Omega^k(M)$:

$$(\omega_1, \omega_2) = \int_M g(\omega_1, \omega_2) d\mathrm{vol}, \quad \omega_1, \omega_2 \in \Omega^k(M).$$

Using this inner product we can implicitly define the *Hodge star operator*

$$* : \Omega^k(M) \to \Omega^{n-k}(M)$$

by the formula

$$(*\omega_1, \omega_2) = \int_M g(*\omega_1, \omega_2) d\mathrm{vol} = \int_M \omega_1 \wedge \omega_2$$

In other words, the Hodge operator gives us an explicit isomorphism

$$* : H^k(M) \to H^{n-k}(M)$$

that depends on the metric g. The fact that it is an isomorphism is a consequence of the next lemma.

LEMMA 26. *The square of the Hodge star* $*^2 : \Omega^k \to \Omega^k$ *is simply multiplication by* $(-1)^{k(n-k)}$.

PROOF. The simplest way of showing this is to prove that if $\sigma^1, \ldots, \sigma^n$ is a positively oriented orthonormal coframe then

$$* \left(\sigma^1 \wedge \cdots \wedge \sigma^k\right) = \sigma^{k+1} \wedge \cdots \wedge \sigma^n.$$

This follows easily from the above definition of $*$, the fact that the forms $\sigma^{i_{k+1}} \wedge \cdots \wedge \sigma^{i_n}$, $i_{k+1} < \cdots < i_n$ are orthonormal, and that $\sigma^1 \wedge \cdots \wedge \sigma^n$ represents the volume form on (M, g). $\square$

The inner product structures on $\Omega^k(M)$ allow us to define the adjoint

$$\delta^k : \Omega^{k+1}(M) \to \Omega^k(M)$$

to d^k via the formula

$$(\delta^k \omega_1, \omega_2) = (\omega_1, d^k \omega_2).$$

From the definition of the Hodge star operator, we see that δ can be computed from d as follows (see also the appendix for a different formula).

LEMMA 27.

$$\delta^k = (-1)^{(n-k)(k+1)} * d^{n-k-1} * .$$

PROOF. If $\omega_1 \in \Omega^{k+1}(M)$, $\omega_2 \in \Omega^k(M)$, then

$$
\begin{aligned}
(\delta^k \omega_1, \omega_2) &= (\omega_1, d^k \omega_2) \\
&= (-1)^{k(n-k)} (* * \omega_1, d^k \omega_2) \\
&= (-1)^{k(n-k)} \int_M * \omega_1 \wedge d^k \omega_2 \\
&= (-1)^{n-k-1+k(n-k)} \int_M d^{n-1}(* \omega_1 \wedge \omega_2) \\
&\quad -(-1)^{n-k-1+k(n-k)} \int_M (d^{n-k-1} * \omega_1) \wedge \omega_2 \\
&= (-1)^{n-k+k(n-k)} \int_M (d^{n-k-1} * \omega_1) \wedge \omega_2 \\
&= (-1)^{(k+1)(n-k)} (* d^{n-k-1} * \omega_1, \omega_2).
\end{aligned}
$$

$\square$

We now have a diagram of complexes,

$$
\begin{array}{ccccccccc}
0 & \to & \Omega^0(M) & \xrightarrow{d} & \Omega^1(M) & \xrightarrow{d} & \cdots & \xrightarrow{d} & \Omega^n(M) & \to & 0 \\
& & \updownarrow * & & \updownarrow * & & & & \updownarrow * & & \\
0 & \to & \Omega^n(M) & \xrightarrow{\delta} & \Omega^{n-1}(M) & \xrightarrow{\delta} & & \xrightarrow{\delta} & \Omega^0(M) & \to & 0,
\end{array}
$$

where each square commutes up to some sign.

The *Laplacian on forms*, also called the *Hodge Laplacian*, is defined as

$$
\begin{aligned}
\triangle : & \quad \Omega^k(M) \to \Omega^k(M), \\
\triangle \omega &= (d\delta + \delta d)\omega.
\end{aligned}
$$

In the next section we shall see that on functions, the Hodge Laplacian is the negative of the previously defined Laplacian, hence the need for a slightly different symbol $\triangle$ as opposed to Δ.

LEMMA 28. $\triangle \omega = 0$ iff $d\omega = 0$ and $\delta \omega = 0$.

PROOF.

$$
\begin{aligned}
(\triangle \omega, \omega) &= (d\delta \omega, \omega) + (\delta d \omega, \omega) \\
&= (\delta \omega, \delta \omega) + (d\omega, d\omega)
\end{aligned}
$$

Thus, $\triangle \omega = 0$ implies $(\delta \omega, \delta \omega) = (d\omega, d\omega) = 0$, which shows that $\delta \omega = 0, d\omega = 0$. The opposite direction is obvious. $\square$

We can now introduce the *Hodge cohomology*:

$$\mathcal{H}^k(M) = \{\omega \in \Omega^k(M) : \triangle\omega = 0\}.$$

Since all harmonic forms are closed, we have a natural map:

$$\mathcal{H}^k(M) \to H^k(M).$$

THEOREM 47. (Hodge, 1935) *This map is an isomorphism.*

PROOF. The proof of this theorem actually requires a lot of work and can't really be proved in detail here. A good reference for a rigorous treatment is [**81**]. We'll try to give the essential idea. The claim, in other words, is that for any closed form ω we can find a unique exact form $d\theta$ such that $\triangle(\omega + d\theta) = 0$. The uniqueness part is obviously equivalent to the statement that the harmonic exact forms are zero everywhere.

To establish this decomposition one can do something a little more general. First observe that since

$$H^k(M) = \frac{\ker d^k}{\operatorname{im}(d^{k-1})},$$

we would be done if we could only prove that

$$\Omega^k(M) = \operatorname{im} d^{k-1} \oplus \ker(\delta^{k-1}).$$

This statement is actually quite reasonable from the point of view of linear algebra in finite dimensions. There we know that the Fredholm alternative guarantees the decomposition

$$W = \operatorname{im}(L) \oplus \ker(L^*),$$

where $L : V \to W$ is a linear map between inner product spaces and $L^* : W \to V$ is the adjoint. This theorem extends to infinite dimensions with some modifications. Notice that such a decomposition is necessarily orthogonal.

Let us see how this implies the theorem. If $\omega \in \Omega^k$, then we can write $\omega = d\theta + \widetilde{\omega}$, where $\delta\widetilde{\omega} = 0$. Therefore, if $d\omega = 0$, then $d\widetilde{\omega} = 0$ as well. But then $\widetilde{\omega}$ must be harmonic, and we have obtained the desired decomposition. To check uniqueness, we must show that $d\theta = 0$ if $\triangle d\theta = 0$. The equation $\triangle d\theta = 0$ reduces to $\delta d\theta = 0$. This shows that $d\theta = 0$, since

$$
\begin{aligned}
0 &= (\delta d\theta, \theta) \\
&= (d\theta, d\theta) \\
&= \int_M g(d\theta, d\theta) \\
&\geq 0.
\end{aligned}
$$

$\square$

3. Harmonic Forms

We shall now see how Hodge theory can be used to get information about the Betti numbers $b_i(M) = \dim\mathcal{H}^i(M)$ given various curvature inequalities.

3.1. 1-Forms. Suppose that θ is a harmonic 1-form on (M, g). We shall consider $f = \frac{1}{2}g(\theta, \theta)$ in analogy with our analysis of Killing fields. One of the technical problems is that we don't have a really good feeling for what $g(\theta, \theta)$ is. If X is the vector field dual to θ, i.e., $\theta(v) = g(X, v)$ for all v, then, of course,

$$f = \frac{1}{2}g(\theta, \theta) = \frac{1}{2}g(X, X) = \frac{1}{2}\theta(X).$$

We now have to figure out what the harmonicity of θ is good for.

PROPOSITION 31. *If X is a vector field on (M, g) and $\theta_X = g(X, \cdot)$ is the dual 1-form, then*

$$\mathrm{div}X = -\delta\left(\theta_X\right)$$

PROOF. (See also the appendix.) We shall prove this in the case where M is compact and oriented. If $f \in \Omega^0(M)$, then δ is defined by the relationship

$$\int_M g(df, \omega) = \int_M f \cdot \delta\omega.$$

So we need to show that

$$\int_M g(df, \theta_X)d\mathrm{vol} = -\int_M f \cdot (\mathrm{div}X)\, d\mathrm{vol}.$$

To see this we observe

$$\begin{aligned}
\mathrm{div}\left(f \cdot X\right) &= g\left(\nabla f, X\right) + f \cdot (\mathrm{div}X) \\
&= g\left(df, \theta_X\right) + f \cdot (\mathrm{div}X).
\end{aligned}$$

Thus

$$\int_M \mathrm{div}\left(f \cdot X\right) d\mathrm{vol} = \int_M g(df, \theta_X)d\mathrm{vol} + \int_M f \cdot (\mathrm{div}X)\, d\mathrm{vol}.$$

On the other hand

$$\begin{aligned}
\mathrm{div}\left(f \cdot X\right) \cdot\, d\mathrm{vol} &= L_{f \cdot X}d\mathrm{vol} \\
&= di_{f \cdot X}d\mathrm{vol}.
\end{aligned}$$

So Stokes' theorem tells us that the integral on the left vanishes. ☐

This result also shows that, up to sign, the Laplacian on functions is the same as our old definition, i.e.,

$$\mathrm{div}\nabla = -\delta d.$$

The other result we need is

PROPOSITION 32. *Suppose X and θ_X are as in the previous proposition. Then $v \to \nabla_v X$ is symmetric iff $d\theta_X = 0$.*

PROOF. Recall that

$$d\theta_X(V, W) + (L_X g)(V, W) = 2g\left(\nabla_V X, W\right).$$

Since $L_X g$ is symmetric and $d\theta_X$ is skew-symmetric the result immediately follows. ☐

Therefore, if $\omega = \theta_X$ is harmonic, then we have that $\mathrm{div}X = 0$ and ∇X is a symmetric $(1, 1)$-tensor. Using this we can now prove the following generalized fundamental equations.

PROPOSITION 33. *Let X be a vector field so that ∇X is symmetric (i.e. corresponding 1-form is closed). If*

$$f = \frac{1}{2}|X|^2$$

and X is the gradient of u near p, then

(1)
$$\nabla f = \nabla_X X.$$

(2)
$$\begin{aligned}
\text{Hess} f\,(V,V) &= \text{Hess}^2 u\,(V,V) + (\nabla_X \text{Hess} u)\,(V,V) + R\,(V,X,X,V) \\
&= g\,(\nabla_V X, \nabla_V X) + g\,(\nabla^2_{X,V} X, V) + R\,(V,X,X,V)
\end{aligned}$$

(3)
$$\begin{aligned}
\Delta f &= |\text{Hess} u|^2 + D_X \Delta u + \text{Ric}\,(X,X) \\
&= |\nabla X|^2 + D_X \text{div} X + \text{Ric}\,(X,X)
\end{aligned}$$

PROOF. For (1) just observe that

$$\begin{aligned}
g(\nabla f, V) &= D_V \frac{1}{2}|X|^2 \\
&= g(\nabla_V X, X) \\
&= g(\nabla_X X, V).
\end{aligned}$$

For (2) we first observe that

$$\text{Hess} u\,(U,V) = g\,(\nabla_U X, V).$$

Thus

$$\begin{aligned}
\text{Hess}^2 u\,(V,V) &= g\,(\nabla_V X, \nabla_V X) \\
&= g\,(\nabla_{\nabla_V X} X, V)
\end{aligned}$$

and

$$\begin{aligned}
(\nabla_X \text{Hess} u)\,(V,V) &= \nabla_X g\,(\nabla_V X, V) - g\,(\nabla_{\nabla_X V} X, V) - g\,(\nabla_V X, \nabla_X V) \\
&= g\,(\nabla_X \nabla_V X, V) - g\,(\nabla_{\nabla_X V} X, V) \\
&= g\,(\nabla^2_{X,V} X, V).
\end{aligned}$$

This shows that

$$\begin{aligned}
\text{Hess} f\,(V,V) &= g\,(\nabla_V \nabla_X X, V) \\
&= g\,(R(V,X)X + \nabla_X \nabla_V X + \nabla_{[V,X]} X, V) \\
&= R\,(V,X,X,V) + g\,(\nabla_X \nabla_V X, V) \\
&\quad - g\,(\nabla_{\nabla_X V} X, V) + g\,(\nabla_{\nabla_V X} X, V) \\
&= R\,(V,X,X,V) + (\nabla_X \text{Hess} u)\,(V,V) + \text{Hess}^2 u\,(V,V)
\end{aligned}$$

For (3) we take traces in (2). We know from our calculations with Killing fields that this gives us the first and third terms. The second term, however, is new. To handle that term we observe that either $X|_p = 0$ or we can choose the orthonormal frame E_i to be parallel in the direction of X. In both cases we have

$$\begin{aligned}
\sum (\nabla_X \text{Hess} u)\,(E_i, E_i) &= D_X \sum \text{Hess} u\,(E_i, E_i) \\
&= D_X \Delta u.
\end{aligned}$$

$\square$

We can now easily show the other Bochner theorem.

THEOREM 48. (Bochner, 1948) *If (M, g) is compact, oriented, and has* Ric ≥ 0, *then every harmonic 1-form is parallel.*

PROOF. Suppose ω is a harmonic 1-form and X the dual vector field. Then

$$\Delta \left(\frac{1}{2} |X|^2 \right) = |\nabla X|^2 + \mathrm{Ric}(X, X),$$

since div$X = \Delta u = 0$. Thus Stokes' theorem together with the condition Ric ≥ 0 implies

$$
\begin{aligned}
0 &= \int_M \left(|\nabla X|^2 + \mathrm{Ric}(X, X) \right) \cdot d\mathrm{vol} \\
&\geq \int_M |\nabla X|^2 \cdot d\mathrm{vol} \\
&\geq 0.
\end{aligned}
$$

We can therefore conclude that $|\nabla X| = 0$. □

COROLLARY 18. *If (M, g) is as before and furthermore has positive Ricci curvature at one point, then all harmonic 1-forms vanish everywhere.*

PROOF. Since we just proved $\mathrm{Ric}(X, X) \equiv 0$, we must have that $X|_p = 0$ if the Ricci tensor is positive on $T_p M$. But then $X \equiv 0$, since X is parallel. □

COROLLARY 19. *If (M, g) is compact, orientable, and satisfies* Ric ≥ 0, *then $b_1(M) \leq n = \dim M$, with equality holding iff (M, g) is a flat torus.*

PROOF. We know from Hodge theory that $b_1(M) = \dim \mathcal{H}^1(M)$. Now, all harmonic 1-forms are parallel, so the linear map: $\mathcal{H}^1(M) \to T_p^* M$ that evaluates ω at p is injective. In particular, $\dim \mathcal{H}^1(M) \leq n$.

If equality holds, we obviously have n linearly independent parallel fields E_i, $i = 1, \ldots, n$. This clearly implies that (M, g) is flat. Thus the universal covering is $(\mathbb{R}^n, \mathrm{can})$ with $\Gamma = \pi_1(M)$ acting by isometries. Now pull the vector fields E_i, $i = 1, \ldots, n$, back to $\tilde{E}_i$, $i = 1, \ldots, n$, on $\mathbb{R}^n$. These vector fields are again parallel and are therefore constant vector fields. This means that we can think of them as the usual Cartesian coordinate vector fields ∂_i. In addition, they are invariant under the action of Γ, i.e., for each $\gamma \in \Gamma$ we have $D\gamma (\partial_i|_p) = \partial_i|_{\gamma(p)}$, $i = 1, \ldots, n$. All of the the coordinate fields taken together are, however, only invariant under translations. Thus, Γ consists entirely of translations. This means that Γ is finitely generated, Abelian, and torsion free. Hence $\Gamma = \mathbb{Z}^q$ for some q. To see that M is a torus, we need only show that $q = n$. If $q < n$, then $\mathbb{Z}^q$ generates a subspace V of $\mathbb{R}^n$ with dimension $< n$. Let W denote the orthogonal complement to V in $\mathbb{R}^n$. Then

$$M = \mathbb{R}^n / \mathbb{Z}^q = (V \oplus W) / \mathbb{Z}^q = (V / \mathbb{Z}^q) \oplus W,$$

which is not compact. Thus, we must have that $\Gamma = \mathbb{Z}^n$ generates $\mathbb{R}^n$. □

3.2. The Bochner Technique in General. The Bochner technique actually works in a much more general setting. Suppose we have a vector bundle $E \to M$ that is endowed with an inner product structure $\langle \cdot, \cdot \rangle$ and a connection that is compatible with the metric. To be more precise, let $\Gamma(E)$ denote the sections $s : M \to E$. The connection on E is a map

$$\nabla \quad : \quad \Gamma(E) \to \Gamma(\operatorname{Hom}(TM, E)),$$
$$s \quad \to \quad \nabla s,$$

and $\nabla s : TM \to E$. We assume, it is linear in s, tensorial in X, and compatible with the metric

$$D_X \langle s_1, s_2 \rangle = \langle \nabla_X s_1, s_2 \rangle + \langle s_1, \nabla_X s_2 \rangle.$$

If we assume that (M, g) is an oriented Riemannian manifold, then using the pointwise inner product structures on $\Gamma(E)$, $\Gamma(TM)$, and integration, we get inner product structures on $\Gamma(E)$ and $\Gamma(\operatorname{Hom}(TM, E))$ via the formulae

$$
\begin{aligned}
(s_1, s_2) &= \int_M \langle s_1, s_2 \rangle, \\
(S_1, S_2) &= \int_M \langle S_1, S_2 \rangle \\
&= \int_M \operatorname{tr}(S_1^* S_2),
\end{aligned}
$$

where $S_1^* \in \Gamma(\operatorname{Hom}(E, TM))$ is the pointwise adjoint to S_1. In case M is not compact we use compactly supported sections to make sense of this. Since the connection is a linear map

$$\nabla : \Gamma(E) \to \Gamma(\operatorname{Hom}(TM, E)),$$

we get an adjoint

$$\nabla^* : \Gamma(\operatorname{Hom}(TM, E)) \to \Gamma(E)$$

defined implicitly by

$$\int_M \langle \nabla^* S, s \rangle = \int_M \langle S, \nabla s \rangle.$$

The *connection Laplacian* of a section is defined as $\nabla^* \nabla s$. We do not call this $\triangle$, since even for forms it does not equal our previous choice for the Laplacian. In fact,

$$\int_M \langle \nabla^* \nabla s, s \rangle = \int_M |\nabla s|^2.$$

Thus, the only sections which are "harmonic" with respect to this Laplacian are the parallel sections.

There is a different way of defining the connection Laplacian. Namely, consider the second covariant derivative $\nabla_{X,Y}^2 s$ and take the trace $\sum_{i=1}^n \nabla_{E_i, E_i}^2 s$ with respect to some orthonormal frame. This is easily seen to be invariantly defined. We shall use the notation

$$
\begin{aligned}
\operatorname{tr}\left(\nabla^2 s\right) &= \sum_{i=1}^n \nabla_{E_i, E_i}^2 s, \\
\operatorname{tr}\nabla^2 &= \sum_{i=1}^n \nabla_{E_i, E_i}^2.
\end{aligned}
$$

The two Laplacians are related as follows:

PROPOSITION 34. *Let (M, g) be an oriented Riemannian manifold, and $E \to M$ a vector bundle with an inner product and compatible connection, then*

$$\nabla^* \nabla s = -\mathrm{tr} \nabla^2 s$$

for all compactly supported sections of E.

PROOF. Let s_1 and s_2 be two sections which are compactly supported in the domain of an orthonormal frame E_i on M. The left-hand side of the formula can be reduced as follows:

$$
\begin{aligned}
(\nabla^* \nabla s_1, s_2) &= \int_M \langle \nabla^* \nabla s_1, s_2 \rangle \\
&= \int_M \langle \nabla s_1, \nabla s_2 \rangle \\
&= \int_M \mathrm{tr}\left((\nabla s_1)^* \nabla s_2 \right) \\
&= \sum_{i=1}^{n} \int_M g\left(E_i, \left((\nabla s_1)^* \nabla s_2 \right)(E_i) \right) \\
&= \sum_{i=1}^{n} \int_M g\left((\nabla s_1)(E_i), (\nabla s_2)(E_i) \right) \\
&= \sum_{i=1}^{n} \int_M g\left(\nabla_{E_i} s_1, \nabla_{E_i} s_2 \right).
\end{aligned}
$$

The right-hand side reduces to something similar

$$
\begin{aligned}
\sum_{i=1}^{n} \langle \nabla^2_{E_i, E_i} s_1, s_2 \rangle &= \sum_{i=1}^{n} \langle \nabla_{E_i} \nabla_{E_i} s_1, s_2 \rangle - \sum_{i=1}^{n} \langle \nabla_{\nabla_{E_i} E_i} s_1, s_2 \rangle \\
&= -\sum_{i=1}^{n} \langle \nabla_{E_i} s_1, \nabla_{E_i} s_2 \rangle + \sum_{i=1}^{n} \nabla_{E_i} \langle \nabla_{E_i} s_1, s_2 \rangle \\
&\quad - \sum_{i=1}^{n} \langle \nabla_{\nabla_{E_i} E_i} s_1, s_2 \rangle \\
&= -\langle \nabla s_1, \nabla s_2 \rangle + \mathrm{div} X,
\end{aligned}
$$

where X is defined by

$$g(X, v) = \langle \nabla_v s_1, s_2 \rangle.$$

We can then integrate and use Stokes' theorem to conclude

$$\int_M \langle \nabla^* \nabla s_1, s_2 \rangle = -\int_M \langle \mathrm{tr} \nabla^2 s_1, s_2 \rangle.$$

Thus, we must have that $\nabla^* \nabla s_1 = -\mathrm{tr} \nabla^2 s_1$ for all such sections. It is now easy to establish the result for all compactly supported sections. $\square$

With this in mind we can, as above, try to compute $\Delta \left(\frac{1}{2} |s|^2 \right)$. Initially this works as follows:

$$\Delta\left(\frac{1}{2}|s|^2\right) = \sum_{i=1}^{n} \nabla^2_{E_i,E_i} \frac{1}{2}\langle s,s\rangle$$

$$= \sum_{i=1}^{n} \nabla_{E_i}\nabla_{E_i} \frac{1}{2}\langle s,s\rangle - \sum_{i=1}^{n} \langle \nabla_{\nabla_{E_i}E_i}s,s\rangle$$

$$= \sum_{i=1}^{n}\left(\langle\nabla_{E_i}s,\nabla_{E_i}s\rangle + \langle\nabla_{E_i}\nabla_{E_i}s,s\rangle - \langle\nabla_{\nabla_{E_i}E_i}s,s\rangle\right)$$

$$= \langle\nabla s,\nabla s\rangle + \left\langle\sum_{i=1}^{n}\nabla^2_{E_i,E_i}s,s\right\rangle$$

$$= \langle\nabla s,\nabla s\rangle - \langle\nabla^*\nabla s,s\rangle.$$

The problem now lies in getting to understand the new Laplacian $\nabla^*\nabla$. This is not always possible and needs to be handled on a case-by-case basis. Later, we shall try this out in the situation where s is the curvature tensor. In the exercises various situations where s is a $(1,1)$-tensor are also discussed.

A general procedure for handling this term comes from understanding certain differential operators. Suppose we have a second-order operator $D^2 : \Gamma(E) \to \Gamma(E)$, such as the Hodge Laplacian. Then we can often get identities of the form

$$D^2 = \nabla^*\nabla + C(R_\nabla),$$

where $C(R_\nabla)$ is a trace or contraction of the curvature

$$R_\nabla : \Gamma(TM) \otimes \Gamma(TM) \otimes \Gamma(E) \to \Gamma(E)$$

defined by

$$R_\nabla(X,Y)s = \nabla^2_{X,Y}s - \nabla^2_{Y,X}s$$

$$= \nabla_X\nabla_Y s - \nabla_Y\nabla_X s - \nabla_{[X,Y]}s.$$

As an example we shall show below that on 1-forms or vector fields

$$\Delta = \nabla^*\nabla + \mathrm{Ric}.$$

Such formulae are called *Weitzenböck formulae*.

Define $H_{D^2}(E \to M)$ as the sections with $D^2s = 0$. If we are lucky enough to have an operator D^2 with a Weitzenböck formula, then this space will probably be some sort of topological invariant of $E \to M$, or at least be related to topological invariants of M. Therefore, if $C(R_\nabla) \geq 0$, then $D^2s = 0$ implies $\nabla s = 0$, which means that s is parallel. Thus we can conclude that

$$\dim H_{D^2}(E \to M) \leq \dim E_p = \text{dimension of fiber of } E \to M.$$

In general, the problem is to identify $C(R_\nabla)$. Obviously, the X,Y variables in R_∇ have to be contracted in such a way that $C(R_\nabla) : \Gamma(E) \to \Gamma(E)$.

3.3. p-Forms. The first obvious case to try this philosophy on is that of the Hodge Laplacian on k-forms as we already know that harmonic forms compute the topology of the underlying manifold. Thus we consider $E = \Lambda^k T^*M$ with the usual inner product and Riemannian covariant derivative. In the next section we shall show that there is a Weitzenböck formula for k-forms of the form

$$\Delta = \nabla^*\nabla + C(R_\nabla).$$

This was certainly known to both Bochner and Yano. It is slightly trickier to figure out what $C(R_\nabla)$ means. When $p = 1$ $C(R_\nabla)$ is essentially the Ricci tensor after type change as mentioned above. For forms of higher degree D. Meyer was the first to observe that $C(R_\nabla)$ is positive if the curvature operator is positive. In the next section we shall establish all of these facts. For now we just observe some of the nice consequences.

THEOREM 49. (D. Meyer, 1971) *If the curvature operator $\mathfrak{R} \geq 0$, then $C(R_\nabla) \geq 0$ on k-forms; and if $\mathfrak{R} > 0$, then $C(R_\nabla) > 0$.*

COROLLARY 20. *Suppose M is orientable. If $\mathfrak{R} \geq 0$, then*

$$b_k(M) \leq \binom{n}{k} = b_k(T^n),$$

and if $\mathfrak{R} > 0$ somewhere, then $b_k(M) = 0$ for $k = 1, 2, \ldots, n - 1$.

PROOF. Evidently we have that harmonic forms must be parallel. In the case of positive curvature no such forms can exist, and if the curvature is nonnegative, then the Betti number estimate follows from the fact that a parallel form is completely determined by its value at a point. Thus

$$
\begin{aligned}
b_k &= \dim H^k \\
&= \dim \mathcal{H}^k \\
&\leq \dim \Lambda^k \left(T_p^* M \right) \\
&= \frac{n!}{k! \, (n - k)!}.
\end{aligned}
$$

$\square$

We now have a pretty good understanding of manifolds with nonnegative (or positive) curvature operator. From the generalized Gauss-Bonnet theorem we know that the Euler characteristic is ≥ 0. Thus, one of the Hopf problems is settled for this class of manifolds.

H. Hopf is famous for another problem: Does $S^2 \times S^2$ admit a metric with positive sectional curvature? We already know that this space has positive Ricci curvature and also that it doesn't admit a metric with positive curvature operator, as $\chi(S^2 \times S^2) = 4$. It is also interesting to observe that $\mathbb{C}P^2$ has positive sectional curvature but doesn't admit a metric with positive curvature operator either, as $\chi(\mathbb{C}P^2) = 3$. Thus, even among 4-manifolds, there seems to be a big difference between simply connected manifolds that admit Ric > 0, sec > 0, and $\mathfrak{R} > 0$. We shall in chapter 11 describe a simply connected manifold that has Ric > 0 but doesn't even admit a metric with sec ≥ 0.

Actually, manifolds with nonnegative curvature operator can be classified (see chapter 8). From this classification it follows that there are many manifolds that have positive or nonnegative sectional curvature but admit no metric with nonnegative curvature operator.

EXAMPLE 45. *We can exhibit a metric with nonnegative sectional curvature on $\mathbb{C}P^2 \sharp \overline{\mathbb{C}}P^2$ by observing that it is an S^1 quotient of $S^2 \times S^3$. Namely, let S^1 act on the 3-sphere by the Hopf action and on the 2-sphere by rotations. If the total rotation on the 2-sphere is $2\pi k$, then the quotient is $S^2 \times S^2$ if k is even, and $\mathbb{C}P^2 \sharp \overline{\mathbb{C}}P^2$ if k is odd. In all cases O'Neill's formula tells us that the sectional curvature is* *non-negative. From the above-mentioned classification it follows, however, that the only simply connected spaces with nonnegative curvature operator are topologically equivalent to $S^2 \times S^2$, S^4, or $\mathbb{C}P^2$.*

The Bochner technique has found many generalizations. It has, for instance, proven very successful in the study of manifolds with nonnegative scalar curvature. Briefly, what happens is that *spin manifolds* (this is a condition similar to saying that a manifold is orientable) admit certain *spinor bundles*. These bundles come with a natural first-order operator called the *Dirac operator,* often denoted by $\partial\!\!\!/$ or $D\!\!\!\!/$. The square of this operator has a Weitzenböck formula of the form

$$D\!\!\!\!/^2 = \nabla^*\nabla + \frac{1}{4}\mathrm{scal}.$$

This formula was discovered and used by Lichnerowicz (as well as I. Singer, as pointed out in [**97**]) to show that a sophisticated invariant called the $\hat{A}$-genus vanishes for spin manifolds with positive scalar curvature. Using some generalizations of this formula, Gromov-Lawson showed that any metric on a torus with scal ≥ 0 is in fact flat. We just proved this for metrics with Ric ≥ 0. Dirac operators and their Weitzenböck formulae have also been of extreme importance in physics and 4-manifolds theory. Much of Witten's work (e.g., the positive mass conjecture) uses these ideas. Also, the work of Seiberg-Witten, which has had a revolutionary impact on 4-manifold geometry, is related to these ideas.

In relation to our discussion above on positively curved manifolds, we should note that there are still no known examples of simply connected manifolds that admit positive scalar curvature but not positive Ricci curvature. This despite the fact that if (M, g) is any closed Riemannian manifold, then for small enough ε the product $\left(M \times S^2, g + \varepsilon^2 ds_2^2\right)$ clearly has positive scalar curvature. This example shows that there are manifolds with positive scalar curvature that don't admit even nonnegative Ricci curvature. To see this, select your favorite surface M^2 with $b_1 > 4$. Then $b_1\left(M^2 \times S^2\right) > 4$ and therefore by Bochner's theorem can't support a metric with nonnegative Ricci curvature.

4. Clifford Multiplication on Forms

In order to give a little perspective on the proof of the Weitzenböck formula for p-forms and also to give an indication of some of the basic ideas in spin geometry, we shall develop some new structures on forms. Instead of first developing Clifford algebras in the linear algebra setting, we just go ahead and define the desired structure on a manifold.

Throughout, we fix a Riemannian manifold (M, g) of dimension n.

We shall use the *musical* isomorphisms, $\sharp$ (sharp) and $\flat$ (flat), between 1-forms and vector fields. Thus, if X is a vector field, the dual 1-form is defined as $X^\flat(v) = g(X, v)$, and conversely, if ω is a 1-form, then the vector field $\omega^\sharp$ is defined by $\omega(v) = g(\omega^\sharp, v)$. In tensor language this means that the indices get lowered or raised, hence the musical notation.

Recall that $\Omega^*(M)$ denotes the space of all forms on M, while $\Omega^p(M)$ is the space of p-forms. On $\Omega^*(M)$ we can define a product structure that is different from the wedge product. This product is called *Clifford multiplication,* and for

$\theta, \omega \in \Omega^*$ it is denoted $\theta \cdot \omega$. If $\theta \in \Omega^1(M)$ and $\omega \in \Omega^p(M)$, then

$$\theta \cdot \omega = \theta \wedge \omega - i_{\theta\sharp}\omega,$$
$$\omega \cdot \theta = (-1)^p (\theta \wedge \omega + i_{\theta\sharp}\omega).$$

By declaring the product to be bilinear and associative, we can use these properties to define the product between any two forms. Note that even when ω is a p-form, the Clifford product with a 1-form gives a mixed form. The important property of this new product structure is that for 1-forms we have

$$\theta \cdot \theta = -|\theta|^2.$$

We can polarize this formula to get

$$\theta_1 \cdot \theta_2 + \theta_2 \cdot \theta_1 = -2g(\theta_1, \theta_2).$$

Thus, orthogonal 1-forms anticommute. Also note that orthogonal forms satisfy

$$\omega_1 \cdot \omega_2 = \omega_1 \wedge \omega_2.$$

Hence, we see that Clifford multiplication not only depends on the inner product, wedge product, and interior product, but actually reproduces these three items. This is the tremendous advantage of this new structure. Namely, after one gets used to Clifford multiplication, it becomes unnecessary to work with wedge products and interior products.

There are a few more important properties, which are easily established.

PROPOSITION 35. *For $\omega_1, \omega_2 \in \Omega^*(M)$ we have*

$$g(\theta \cdot \omega_1, \omega_2) = -g(\omega_1, \theta \cdot \omega_2) \text{ for any 1-form } \theta,$$
$$g([\psi, \omega_1], \omega_2) = -g(\omega_1, [\psi, \omega_2]) \text{ for any 2-form } \psi,$$

where $[\omega_1, \omega_2] = \omega_1 \cdot \omega_2 - \omega_2 \cdot \omega_1$.

PROOF. Evidently both formulae refer to the fact that the linear maps

$$\omega \rightarrow \theta \cdot \omega,$$
$$\omega \rightarrow [\psi, \omega]$$

are skew-symmetric. To prove the identities, one therefore only needs to prove that for any p-form,

$$g(\theta \cdot \omega, \omega) = 0,$$
$$g([\psi, \omega], \omega) = 0.$$

Both of these identities follow directly from the definition of Clifford multiplication, and the fact that the two maps

$$\Omega^p \rightarrow \Omega^{p+1},$$
$$\omega \rightarrow \theta \wedge \omega,$$

$$\Omega^{p+1} \rightarrow \Omega^p,$$
$$\omega \rightarrow i_{\theta\#}\omega,$$

are adjoints to each other. Namely, Clifford multiplication is the difference between these two operations, and since they are adjoint to each other this must be a skew-symmetric operation as desired. $\square$

PROPOSITION 36. *For ω_1, $\omega_2 \in \Omega^*(M)$ and vector fields X, Y we have the derivation properties:*

$$\nabla_X (\omega_1 \cdot \omega_2) = (\nabla_X \omega_1) \cdot \omega_2 + \omega_1 \cdot (\nabla_X \omega_2),$$

$$R(X,Y)(\omega_1 \cdot \omega_2) = (R(X,Y)\omega_1) \cdot \omega_2 + \omega_1 \cdot R(X,Y)\omega_2.$$

PROOF. In case $\omega_1 = \omega_2 = \theta$ is a 1-form, we have

$$\begin{aligned}
\nabla_X (\theta \cdot \theta) &= -\nabla_X |\theta|^2 \\
&= -2g(\nabla_X \theta, \theta) \\
&= (\nabla_X \theta) \cdot \theta + \theta \cdot (\nabla_X \theta).
\end{aligned}$$

More generally, we must use the easily established Leibniz rules for interior and exterior products (see also the Appendix). In case $\omega_1 = \theta$ is a 1-form and $\omega_2 = \omega$ is a general form, we have that

$$\begin{aligned}
\nabla_X (\theta \wedge \omega) &= (\nabla_X \theta) \wedge \omega + \theta \wedge (\nabla_X \omega), \\
\nabla_X (i_{\theta\#} \omega) &= i_{\nabla_X \theta\#} \omega + i_{\theta\#} (\nabla_X \omega),
\end{aligned}$$

from which we conclude,

$$\begin{aligned}
\nabla_X (\theta \cdot \omega) &= \nabla_X (\theta \wedge \omega - i_{\theta\#} \omega) \\
&= (\nabla_X \theta) \wedge \omega + \theta \wedge (\nabla_X \omega) \\
&\quad - i_{\nabla_X \theta\#} \omega - i_{\theta\#} (\nabla_X \omega) \\
&= (\nabla_X \theta) \wedge \omega - i_{\nabla_X \theta\#} \omega \\
&\quad + \theta \wedge (\nabla_X \omega) - i_{\theta\#} (\nabla_X \omega) \\
&= (\nabla_X \theta) \cdot \omega + \theta \cdot (\nabla_X \omega).
\end{aligned}$$

One can then easily extend this to all forms. The second formula is a direct consequence of the first formula.   □

We can now define the *Dirac operator* on forms:

$$D \quad : \quad \Omega^*(M) \to \Omega^*(M),$$

$$D(\omega) = \sum_{i=1}^{n} \theta^i \cdot \nabla_{E_i} \omega,$$

where E_i is any frame and θ^i the dual coframe. The definition is clearly independent of the frame field. The Dirac operator is related to the standard exterior derivative and its adjoint (see also the Appendix):

PROPOSITION 37. *Given a frame E_i and its dual coframe θ^i, then*

$$\begin{aligned}
d\omega &= \theta^i \wedge \nabla_{E_i} \omega, \\
\delta\omega &= -i_{(\theta^i)^\sharp} \nabla_{E_i} \omega, \\
D &= d + \delta.
\end{aligned}$$

PROOF. First one sees, as usual, that the right-hand sides are invariantly defined and give operators with the usual properties. (Note, in particular, that $d = \theta^i \wedge \nabla_{E_i}$ on functions and that $\delta = -i_{(\theta^i)^\sharp} \nabla_{E_i}$ on 1-forms.) Thus, one can compute, say, $\theta^i \wedge \nabla_{E_i} \omega$ from knowing how to compute this when $\omega = \theta^j$. Then we take an orthonormal frame such that $(\theta^i)^\sharp = E_i$, and finally we assume that the frame is normal at $p \in M$ and establish the formulae at that point. However, the assumption that the frame is normal insures us that all the quantities vanish when we use $\omega = \theta^j$.

The formula $D = d + \delta$ is then a direct consequence of the definition of Clifford multiplication.   □

The square of the Dirac operator now satisfies:

$$D^2 = (d + \delta)^2 = d\delta + \delta d = \triangle.$$

Before considering the more general k-forms let us prove the promised Weitzenböck formula for 1-forms.

COROLLARY 21. *Let X be a vector and $\theta = X^\flat$ the dual 1-form, then*

$$\triangle\theta = \nabla^*\nabla\theta + \operatorname{Ric}(X)^\flat.$$

PROOF. We do all calculations at a point p where we have an orthonormal frame E_i which satisfies $(\nabla E_i)|_p = 0$. We have that

$$\delta\omega = -i_{E_i}\nabla_{E_i}\omega,$$

and the following formula for the exterior derivative whose proof can be found in the appendix:

$$d\omega(X_0, ..., X_k) = \sum_{i=0}^{k}(-1)^i(\nabla_{X_i}\omega)\left(X_0, ..., \hat{X}_i, ...X_k\right).$$

If Z is a constant linear combinations of E_i then at p we get

$$
\begin{aligned}
(\triangle\theta)(Z) &= (d\delta\theta)(Z) + (\delta d\theta)(Z) \\
&= \nabla_Z\delta\theta - \sum_{i=1}^{n}(\nabla_{E_i}d\theta)(E_i, Z) \\
&= -\sum_{i=1}^{n}\nabla_Z((\nabla_{E_i}\theta)(E_i)) - \sum_{i=1}^{n}(\nabla_{E_i}d\theta)(E_i, Z) \\
&= -\sum_{i=1}^{n}(\nabla^2_{Z,E_i}\theta)(E_i) - \nabla_{E_i}\sum_{i=1}^{n}d\theta(E_i, Z) \\
&= -\sum_{i=1}^{n}(\nabla^2_{Z,E_i}\theta)(E_i) - \nabla_{E_i}\sum_{i=1}^{n}(\nabla_{E_i}\theta)(Z) - (\nabla_Z\theta)(E_i) \\
&= \sum_{i=1}^{n}(\nabla^2_{E_i,Z}\theta - \nabla^2_{Z,E_i}\theta)(E_i) - \sum_{i=1}^{n}(\nabla^2_{E_i,E_i}\theta)(Z) \\
&= \sum_{i=1}^{n}(R(E_i, Z)\theta)(E_i) + (\nabla^*\nabla\theta)(Z).
\end{aligned}
$$

We now need to sort out the curvature term. The trick is to figure out how the curvature tensor can be evaluated on forms. One easily checks that it works as with covariant derivatives:

$$
\begin{aligned}
(R(X,Y)\theta)W &= R(X,Y)(\theta(W)) - \theta(R(X,Y)W) \\
&= -\theta(R(X,Y)W).
\end{aligned}
$$

So

$$\sum_{i=1}^{n} \left(R\left(E_i, Z\right)\theta\right)\left(E_i\right) = -\sum_{i=1}^{n}\theta\left(R\left(E_i, Z\right)E_i\right)$$

$$= \sum_{i=1}^{n}\theta\left(R\left(Z, E_i\right)E_i\right)$$

$$= \theta\left(\mathrm{Ric}\left(Z\right)\right)$$

$$= g\left(X, \mathrm{Ric}\left(Z\right)\right)$$

$$= g\left(\mathrm{Ric}\left(X\right), Z\right)$$

$$= \mathrm{Ric}\left(X\right)^{\flat}\left(Z\right).$$

$\square$

With this behind us we can now try to generalize this to forms of higher degree

PROPOSITION 38. *Given a frame E_i and its dual coframe θ^i, we have:*

$$D^2\omega = \sum_{i,j=1}^{n}\theta^i\cdot\theta^j\cdot\nabla^2_{E_i, E_j}\omega$$

$$= \sum_{i,j=1}^{n}\left(\nabla^2_{E_i, E_j}\omega\right)\cdot\theta^j\cdot\theta^i.$$

PROOF. First, recall that

$$\nabla^2_{E_i, E_j} = \nabla_{E_i}\nabla_{E_j} - \nabla_{\nabla_{E_i}E_j}$$

is tensorial in both E_i and E_j, and thus the two expressions on the right-hand side are invariantly defined. Using invariance, we need only prove the formula at a point $p \in M$, where the frame is assumed to be normal, i.e., $\left(\nabla E_i\right)|_p = 0$ and consequently also $\left(\nabla\theta^i\right)|_p = 0$. We can then compute at p,

$$D^2\omega = \theta^i\cdot\left(\nabla_{E_i}\left(\theta^j\cdot\nabla_{E_j}\omega\right)\right)$$

$$= \theta^i\cdot\left(\nabla_{E_i}\theta^j\right)\cdot\nabla_{E_j}\omega + \theta^i\cdot\theta^j\cdot\nabla_{E_i}\nabla_{E_j}\omega$$

$$= \theta^i\cdot\theta^j\cdot\nabla_{E_i}\nabla_{E_j}\omega - \theta^i\cdot\theta^j\cdot\nabla_{\nabla_{E_i}E_j}\omega$$

$$= \sum_{i,j=1}^{n}\theta^i\cdot\theta^j\cdot\nabla^2_{E_i, E_j}\omega.$$

For the second formula the easiest thing to do is to observe that for a p-form ω we have

$$\hat{D}\omega = \left(\nabla_{E_i}\omega\right)\cdot\theta^i = (-1)^p\left(d - \delta\right)\omega.$$

Thus also,

$$\hat{D}^2 = \triangle = D^2.$$

This finishes the proof.

$\square$

We can now establish the relevant Weitzenböck formula.

THEOREM 50. *Given a frame E_i and its dual coframe θ^i, we have*

$$D^2\omega = \nabla^*\nabla\omega + \frac{1}{2}\sum_{i,j=1}^n \theta^i \cdot \theta^j \cdot R(E_i, E_j)\omega$$

$$= \nabla^*\nabla\omega + \frac{1}{2}\sum_{i,j=1}^n R(E_i, E_j)\omega \cdot \theta^j \cdot \theta^i.$$

PROOF. Using the above identities for D^2, it clearly suffices to check

$$\nabla^*\nabla\omega + \frac{1}{2}\sum_{i,j=1}^n \theta^i \cdot \theta^j \cdot R(E_i, E_j)\omega = \sum_{i,j=1}^n \theta^i \cdot \theta^j \cdot \nabla^2_{E_i, E_j}\omega,$$

$$\nabla^*\nabla\omega + \frac{1}{2}\sum_{i,j=1}^n R(E_i, E_j)\omega \cdot \theta^j \cdot \theta^i = \sum_{i,j=1}^n \left(\nabla^2_{E_i, E_j}\omega\right) \cdot \theta^j \cdot \theta^i.$$

These formulae are established in the same way, so we concentrate on the first. As usual, note that everything is invariant. We can therefore pick a frame that is orthonormal and normal at $p \in M$ and compute at $p \in M$,

$$\sum_{i,j=1}^n \theta^i \cdot \theta^j \cdot \nabla^2_{E_i, E_j}\omega = -\sum_{i=1}^n \nabla^2_{E_i, E_i}\omega + \sum_{i \neq j} \theta^i \cdot \theta^j \cdot \nabla^2_{E_i, E_j}\omega$$

$$= -\sum_{i=1}^n \nabla^2_{E_i, E_i}\omega + \sum_{i<j} \theta^i \cdot \theta^j \cdot \left(\nabla^2_{E_i, E_j}\omega - \nabla^2_{E_j, E_i}\omega\right)$$

$$= -\sum_{i=1}^n \nabla^2_{E_i, E_i}\omega + \sum_{i<j} \theta^i \cdot \theta^j \cdot R(E_i, E_j)\omega$$

$$= -\sum_{i=1}^n \nabla^2_{E_i, E_i}\omega + \frac{1}{2}\sum_{i,j=1}^n \theta^i \cdot \theta^j \cdot R(E_i, E_j)\omega,$$

where we used the relations

$$\theta^i \cdot \theta^i = -1,$$
$$\theta^i \cdot \theta^j = -\theta^j \cdot \theta^i,$$

Now use that we know

$$\nabla^*\nabla = -\sum_{i=1}^n \nabla^2_{E_i, E_i}$$

to finish the proof.   □

We can now establish the desired Bochner formula for forms.

COROLLARY 22. *Given an orthonormal frame E_i and its dual coframe θ^i, we have for any harmonic form ω, i.e., $D\omega = 0$, that*

$$0 = \nabla^*\nabla\omega + \frac{1}{4}\sum_{i,j=1}^n \left[\theta^i \cdot \theta^j, R(E_i, E_j)\omega\right].$$

PROOF. First, we use that the frame is orthonormal to conclude that

$$\sum_{i,j=1}^n R(E_i, E_j)\omega \cdot \theta^j \cdot \theta^i = -\sum_{i,j=1}^n R(E_i, E_j)\omega \cdot \theta^i \cdot \theta^j.$$

Thus, we have

$$D^2\omega = \nabla^*\nabla\omega + \frac{1}{2}\sum_{i,j=1}^n \theta^i \cdot \theta^j \cdot R(E_i, E_j)\omega,$$

$$D^2\omega = \nabla^*\nabla\omega - \frac{1}{2}\sum_{i,j=1}^n R(E_i, E_j)\omega \cdot \theta^i \cdot \theta^j.$$

Adding these equations and dividing by 2 then yields

$$D^2\omega = \nabla^*\nabla\omega + \frac{1}{4}\sum_{i,j=1}^n \left[\theta^i \cdot \theta^j, R(E_i, E_j)\omega\right].$$

Therefore, if $D\omega = 0$, then

$$0 = \nabla^*\nabla\omega + \frac{1}{4}\sum_{i,j=1}^n \left[\theta^i \cdot \theta^j, R(E_i, E_j)\omega\right],$$

yielding the desired equation. □

Having identified the curvature terms in the Weitzenböck and Bochner formulae, it now remains to be seen that this term is nonnegative when the curvature operator is nonnegative. Before doing this, let us deconstruct the curvature terms in the following way:

LEMMA 29. *For an orthonormal frame E_i and dual coframe θ^i we have*

$$R(X,Y)\omega = \frac{1}{4}\sum_{i,j=1}^n g(R(X,Y)E_i, E_j)\left(\theta^i \cdot \theta^j \cdot \omega - \omega \cdot \theta^i \cdot \theta^j\right)$$

$$= \frac{1}{4}\sum_{i,j=1}^n g(R(X,Y)E_i, E_j)\left[\theta^i \cdot \theta^j, \omega\right].$$

PROOF. Needless to say, as the right-hand side is invariant, we can assume that the frame is orthonormal and normal at $p \in M$. Moreover, both sides are derivations in ω, so it suffices to check the identities for 1-forms. Finally, we can restrict attention to 1-forms of the type $\omega = \theta^k$ and then compute

$$\theta^i \cdot \theta^j \cdot \theta^k - \theta^k \cdot \theta^i \cdot \theta^j.$$

This term depends on whether $k = i$ or $k = j$ or $k \neq i, j$. We can also assume that $i \neq j$, as $\left[\theta^i \cdot \theta^i, \omega\right] = 0$. We then get

$$\theta^i \cdot \theta^j \cdot \theta^k - \theta^k \cdot \theta^i \cdot \theta^j = \begin{cases} 0, & k \neq i, j, \\ -2\theta^i, & k = j, \\ 2\theta^j, & k = i. \end{cases}$$

Using this we can now compute

$$\sum_{i,j=1}^{n} g\left(R\left(X,Y\right)E_i, E_j\right)\left[\theta^i \cdot \theta^j, \theta^k\right] = -2\sum_{i=1}^{n} g\left(R\left(X,Y\right)E_i, E_k\right)\theta^i$$

$$+2\sum_{j=1}^{n} g\left(R\left(X,Y\right)E_k, E_j\right)\theta^j$$

$$= 4\sum_{i=1}^{n} g\left(R\left(X,Y\right)E_k, E_i\right)\theta^i.$$

As in the case of the proof of the Weitzenböck formula for 1-forms we have that the last term is the 1-form $4R\left(X,Y\right)\theta^k$.   □

With this last formula we can now relate the curvature term in the Bochner formula to the curvature operator.

LEMMA 30. *For an orthonormal frame E_i and its dual coframe θ^i we have that*

$$\sum_{i,j=1}^{n} g\left(\left[\theta^i \cdot \theta^j, R\left(E_i, E_j\right)\omega\right], \omega\right) = \sum_{\alpha} \lambda_\alpha \left|\left[\Theta_\alpha, \omega\right]\right|^2,$$

where λ_α are the eigenvalues for the curvature operator and Θ_α the duals of eigenvectors for the curvature operator.

PROOF. Using the skew-symmetry of $\omega \to \left[\theta^i \cdot \theta^j, \omega\right]$ and the definition of the curvature operator, we can compute

$$\sum_{i,j=1}^{n} g\left(\left[\theta^i \cdot \theta^j, R\left(E_i, E_j\right)\omega\right], \omega\right)$$

$$= -\sum_{i,j=1}^{n} g\left(R\left(E_i, E_j\right)\omega, \left[\theta^i \cdot \theta^j, \omega\right]\right)$$

$$= -\frac{1}{4}\sum_{i,j=1}^{n}\sum_{k,l=1}^{n} g\left(R\left(E_i, E_j\right)E_k, E_l\right) g\left(\left[\theta^k \cdot \theta^l, \omega\right], \left[\theta^i \cdot \theta^j, \omega\right]\right)$$

$$= \frac{1}{4}\sum_{i,j,k,l=1}^{n} g\left(\Re\left(E_i \wedge E_j\right), E_k \wedge E_l\right) g\left(\left[\theta^k \cdot \theta^l, \omega\right], \left[\theta^i \cdot \theta^j, \omega\right]\right)$$

$$= \sum_{i<j,k<l}^{n} g\left(\Re\left(E_i \wedge E_j\right), E_k \wedge E_l\right) g\left(\left[\theta^k \cdot \theta^l, \omega\right], \left[\theta^i \cdot \theta^j, \omega\right]\right).$$

Now observe that the $E_i \wedge E_j$ form an orthonormal basis for $\Lambda^2 TM$, and the $\theta^i \cdot \theta^j$ are the dual basis for $\Omega^2\left(M\right)$. The expression we have arrived at is obviously invariant under change of orthonormal bases in $\Lambda^2 TM$. So select an orthonormal basis Ξ_α for $\Lambda^2 TM$ such that $\Re\left(\Xi_\alpha\right) = \lambda_\alpha \Xi_\alpha$. With Θ_α denoting the dual basis for $\Omega^2\left(M\right)$, we then get

$$\sum_{i,j=1}^{n} g\left(\left[\theta^i \cdot \theta^j, R\left(E_i, E_j\right)\omega\right], \omega\right) = \sum_{\alpha} \lambda_\alpha \left|\left[\Theta_\alpha, \omega\right]\right|^2$$

as desired.   □

THEOREM 51. *On a compact oriented Riemannian n-manifold with nonnegative curvature operator every harmonic form is parallel. Moreover, if the curvature operator is positive, then harmonic p-forms vanish when $p = 1, \ldots, n-1$.*

PROOF. If ω is harmonic, then we have from the previous section that

$$
\begin{aligned}
0 &= \int_M \langle \nabla^* \nabla \omega, \omega \rangle + \frac{1}{4} \int_M \sum_\alpha \lambda_\alpha \left\| [\Theta_\alpha, \omega] \right\|^2 \\
&= \int_M |\nabla \omega|^2 + \frac{1}{4} \int_M \sum_\alpha \lambda_\alpha \left\| [\Theta_\alpha, \omega] \right\|^2 .
\end{aligned}
$$

As both terms are nonnegative, they both vanish. In particular, $\nabla \omega = 0$.

We also have that $\lambda_\alpha \left\| [\Theta_\alpha, \omega] \right\|^2 = 0$. The only way this can happen, if all $\lambda_\alpha > 0$, is if $[\Theta_\alpha, \omega] = 0$ for all α. Since the Θ_α form a basis for the 2-forms, this means that $[\psi, \omega] = 0$ for all 2-forms. To see that this makes $\omega = 0$, just pick $\psi = \theta^i \cdot \theta^j$, $\omega = \theta^{i_1} \cdots \theta^{i_p}$, and compute:

$$
\left[\theta^i \cdot \theta^j, \theta^{i_1} \cdots \theta^{i_p} \right] = \begin{cases} 0, & i, j \notin \{i_1, \ldots, i_p\}, \\ 0, & i, j \in \{i_1, \ldots, i_p\}, \\ 2\theta^i \cdot \theta^j \cdot \theta^{i_1} \cdots \theta^{i_p}, & \text{otherwise.} \end{cases}
$$

In general, we can write

$$
\omega = \sum_{i_1 < \cdots < i_p} a_{i_1 \cdots i_p} \theta^{i_1} \cdots \theta^{i_p}
$$

Therefore, $\left[\theta^i \cdot \theta^j, \omega \right]$ can only vanish if $a_{i_1 \cdots i_p} = 0$ whenever $i \in \{i_1, \cdots, i_p\}$ or $j \in \{i_1, \cdots, i_p\}$ but not both i and j belong to $\{i_1, \cdots, i_p\}$. Using this in the situation where $i < j$ shows that ω must be zero unless p is 0 or n. $\square$

5. The Curvature Tensor

It is now time to apply the Bochner technique to the most natural tensor, the curvature tensor. It is by no means clear that this will yield anything. It seems both miraculous and profound that it works. We shall present results by Lichnerowicz (see [**64**, Chapter 1] and also [**65**] for an in-depth discussion on the meaning of these matters in physics), Berger, and Tachibana (see [**89**]) that combine to show that a compact Riemannian manifold with $\operatorname{div} R = 0$ and nonnegative sectional curvature, respectively nonnegative curvature operator, has parallel Ricci tensor, respectively parallel curvature tensor.

Recall that if we consider the $(1, 3)$ version of the curvature tensor R, then we can construct two $(0, 4)$-tensors: $\operatorname{div} \nabla R$ and $\nabla \operatorname{div} R$. If for our present purposes we use the notation

$$
R^\flat (X, Y, Z, W) = g(X, R(Y, Z)W),
$$

then we can take inner products of the three tensors $R^\flat$, $\operatorname{div} \nabla R$, and $\nabla \operatorname{div} R$. Note that $R^\flat$ is not the usual $(0, 4)$-tensor. This will be very important in the proof below.

THEOREM 52. (Lichnerowicz, 1958) *The curvature tensor R on a compact oriented Riemannian manifold satisfies*

$$
2 \int_M |\operatorname{div} R|^2 - 2 \int_M K = \int_M |\nabla R|^2 ,
$$

where

$$K = g\left(R^b, \operatorname{div}\nabla R - \nabla\operatorname{div}R\right).$$

PROOF. By far the most important ingredient in the proof is that we have the second Bianchi identity at our disposal. To establish the formula, we compute at a point p where we have an orthonormal frame E_i with $(\nabla E_i)|_p = 0$:

$$
\begin{aligned}
\Delta\frac{1}{2}|R|^2 &= \frac{1}{2}\sum_{i=1}^{n}\nabla_{E_i}\nabla_{E_i}|R|^2 \\
&= \frac{1}{2}\sum_{i=1}^{n}\nabla_{E_i}\nabla_{E_i}g\left(R,R\right) \\
&= \sum_{i=1}^{n}\nabla_{E_i}g\left(\left(\nabla_{E_i}R\right),R\right) \\
&= \sum_{i=1}^{n}g\left(\left(\nabla_{E_i}\left(\nabla_{E_i}R\right)\right),R\right) \\
&\quad + \sum_{i=1}^{n}g\left(\left(\nabla_{E_i}R\right),\left(\nabla_{E_i}R\right)\right) \\
&= \sum_{i=1}^{n}g\left(\left(\nabla_{E_i}\left(\nabla_{E_i}R\right)\right),R\right) \\
&\quad + |\nabla R|^2.
\end{aligned}
$$

We now claim that

$$\sum_{i=1}^{n}g\left(\left(\nabla_{E_i}\left(\nabla_{E_i}R\right)\right),R\right) = 2g\left(R^b,\operatorname{div}\nabla R\right).$$

Using that ∇R has the same symmetry properties as R, we first compute

$$
\begin{aligned}
2\operatorname{div}\nabla R\left(E_j,E_k,E_l,E_m\right) &= 2\sum_{i=1}^{n}g\left(\left(\nabla_{E_i}\left(\nabla R\right)\right)\left(E_j,E_k,E_l,E_m\right),E_i\right) \\
&= 2\sum_{i=1}^{n}g\left(\nabla_{E_i}\left(\left(\nabla R\right)\left(E_j,E_k,E_l,E_m\right)\right),E_i\right) \\
&= 2\sum_{i=1}^{n}g\left(\nabla_{E_i}\left(\left(\nabla_{E_j}R\right)\left(E_k,E_l\right)E_m\right),E_i\right) \\
&= 2\sum_{i=1}^{n}\nabla_{E_i}g\left(\left(\nabla_{E_j}R\right)\left(E_k,E_l\right)E_m,E_i\right) \\
&= 2\sum_{i=1}^{n}\nabla_{E_i}g\left(\left(\nabla_{E_j}R\right)\left(E_m,E_i\right)E_k,E_l\right) \\
&= 2\sum_{i=1}^{n}g\left(\nabla_{E_i}\left(\nabla_{E_j}R\right)\left(E_m,E_i\right)E_k,E_l\right)
\end{aligned}
$$

and then observe that

$$2g\left(R^\flat, \operatorname{div}\nabla R\right)$$

$$= 2\sum_{i,j,k,l,m=1}^{n} g\left(\nabla_{E_i}\left(\nabla_{E_j}R\right)(E_k, E_l)E_m, E_i\right) g\left(E_j, R(E_k, E_l)E_m\right)$$

$$= 2\sum_{i,j,k,l,m=1}^{n} g\left(\nabla_{E_i}\left(\nabla_{E_j}R\right)(E_k, E_l)E_m, E_i\right) g\left(R(E_k, E_l)E_m, E_j\right)$$

$$= 2\sum_{i,j,k,l,m=1}^{n} g\left(\nabla_{E_i}\left(\nabla_{E_j}R\right)(E_k, E_l)E_m, E_i\right) g\left(R(E_j, E_m)E_l, E_k\right).$$

On the other hand, using the second Bianchi identity,

$$\sum_{i=1}^{n} g\left(\nabla_{E_i}\left(\nabla_{E_i}R\right)(E_j, E_k)E_l, E_m\right)$$

$$= \sum_{i=1}^{n} \nabla_{E_i}g\left((\nabla_{E_i}R)(E_j, E_k)E_l, E_m\right)$$

$$= -\sum_{i=1}^{n} \nabla_{E_i}g\left((\nabla_{E_j}R)(E_k, E_i)E_l, E_m\right)$$

$$- \sum_{i=1}^{n} \nabla_{E_i}g\left((\nabla_{E_k}R)(E_i, E_j)E_l, E_m\right)$$

$$= -\sum_{i=1}^{n} \nabla_{E_i}g\left((\nabla_{E_j}R)(E_k, E_i)E_l, E_m\right)$$

$$+ \sum_{i=1}^{n} \nabla_{E_i}g\left((\nabla_{E_k}R)(E_j, E_i)E_l, E_m\right),$$

and so,

$$\sum_{i=1}^{n} g\left((\nabla_{E_i}(\nabla_{E_i}R)), R\right)$$

$$= \sum_{i,j,k,l,m=1}^{n} g\left(\nabla_{E_i}(\nabla_{E_i}R)(E_j, E_k)E_l, E_m\right) g\left(R(E_j, E_k)E_l, E_m\right)$$

$$= -\sum_{i,j,k,l,m=1}^{n} \nabla_{E_i}g\left((\nabla_{E_j}R)(E_k, E_i)E_l, E_m\right) g\left(R(E_j, E_k)E_l, E_m\right)$$

$$+ \sum_{i,j,k,l,m=1}^{n} \nabla_{E_i}g\left((\nabla_{E_k}R)(E_j, E_i)E_l, E_m\right) g\left(R(E_j, E_k)E_l, E_m\right)$$

$$= \sum_{i,j,k,l,m=1}^{n} \nabla_{E_i}g\left((\nabla_{E_j}R)(E_k, E_i)E_l, E_m\right) g\left(R(E_k, E_j)E_l, E_m\right)$$

$$+ \sum_{i,j,k,l,m=1}^{n} \nabla_{E_i}g\left((\nabla_{E_k}R)(E_j, E_i)E_l, E_m\right) g\left(R(E_j, E_k)E_l, E_m\right)$$

$$
= 2 \sum_{i,j,k,l,m=1}^{n} \nabla_{E_i} g\left(\left(\nabla_{E_j} R\right)\left(E_k, E_i\right) E_l, E_m\right) g\left(R\left(E_k, E_j\right) E_l, E_m\right)
$$

$$
= 2 \sum_{i,j,k,l,m=1}^{n} \nabla_{E_i} g\left(\left(\nabla_{E_j} R\right)\left(E_m, E_l\right) E_k, E_i\right) g\left(R\left(E_j, E_k\right) E_l, E_m\right)
$$

$$
= 2 \sum_{i,j,k,l,m=1}^{n} \nabla_{E_i} g\left(\left(\nabla_{E_j} R\right)\left(E_k, E_l\right) E_m, E_i\right) g\left(R\left(E_j, E_m\right) E_l, E_k\right)
$$

$$
= 2g\left(R^\flat, \mathrm{div}\nabla R\right).
$$

Using the definition of K, we then arrive at

$$
\Delta \frac{1}{2}|R|^2 = |\nabla R|^2 + 2g\left(R^\flat, \nabla \mathrm{div} R\right) + 2K.
$$

From Stokes' theorem (see also the Appendix) it follows that

$$
\int_M \Delta \frac{1}{2}|R|^2 = 0,
$$

$$
\int_M g\left(R^\flat, \nabla \mathrm{div} R\right) = -\int_M |\mathrm{div} R|^2.
$$

This gives us the desired formula. □

We are now interested in understanding when K is nonnegative. In order to analyze this better we shall go through some generalities.

For any tensor T we can consider the curvature

$$
\begin{aligned}
R(X,Y)T &= \left(\nabla_X\left(\nabla_Y T\right)\right) - \left(\nabla_Y\left(\nabla_X T\right)\right) - \left(\nabla_{[X,Y]} T\right) \\
&= \nabla^2_{X,Y} T - \nabla^2_{Y,X} T
\end{aligned}
$$

as a new tensor of the same type. This new tensor is tensorial in X and Y. Moreover, it is also tensorial in T, so we have for any function f

$$
R(X,Y)(fT) = f R(X,Y) T.
$$

More importantly, one can easily show that

$$
\begin{aligned}
(R(X,Y)T)(X_1,\ldots,X_k) &= R(X,Y)(T(X_1,\ldots,X_k)) \\
&\quad -T(R(X,Y)X_1,\ldots,X_k) \\
&\quad \vdots \\
&\quad -T(X_1,\ldots,R(X,Y)X_k)
\end{aligned}
$$

To understand this new curvature, we can therefore simply break it down to the point where we need only worry about how it acts on vector fields and 1-forms. This we already know how to deal with.

We are particularly interested in the case where T is of type $(1,k)$. In that case we can make a special contraction. Namely, if we choose an orthonormal frame E_i, then

$$
((\mathrm{div}\nabla - \nabla\mathrm{div})T)(Y, X_1,\ldots,X_k) = \sum_{i=1}^{n} g\left((R(E_i,Y)T)(X_1,\ldots,X_k), E_i\right).
$$

It therefore appears that $(\mathrm{div}\nabla - \nabla\mathrm{div})\,T$ is something like the Ricci curvature of T. This is in line with our Weitzenböck formulae, where the curvature term is some sort of contraction in the curvature. If we make the type change

$$T^{\flat}\left(Y, X_1, \ldots, X_k\right) = g\left(Y, T\left(X_1, \ldots, X_k\right)\right),$$

then we get the quadratic expression for this Ricci curvature

$$K = g\left(T^{\flat}, (\mathrm{div}\nabla - \nabla\mathrm{div})\,T\right).$$

The claim is that this quantity is nonnegative whenever the curvature operator is nonnegative and $T = R$. In order to make our argument a little more transparent, let us first show a similar but easier result.

LEMMA 31. (Berger) *Suppose T is a symmetric $(1,1)$-tensor on a Riemannian manifold (M, g) with* $\sec \geq 0$, *then*

$$K = g\left(T^{\flat}, (\mathrm{div}\nabla - \nabla\mathrm{div})\,T\right) \geq 0.$$

PROOF. We shall calculate at a point p, where an orthonormal frame has been chosen such that $T\left(E_i\right) = \lambda_i E_i$:

$$
\begin{aligned}
g\left(T^{\flat}, (\mathrm{div}\nabla - \nabla\mathrm{div})\,T\right) &= \sum_{i,j,k=1}^{n} g\left(E_j, T\left(E_k\right)\right) g\left(\left(R\left(E_i, E_j\right) T\right)\left(E_k\right), E_i\right) \\
&= \sum_{i,j,k=1}^{n} g\left(E_j, T\left(E_k\right)\right) g\left(R\left(E_i, E_j\right) T\left(E_k\right), E_i\right) \\
&\quad - \sum_{i,j,k=1}^{n} g\left(E_j, T\left(E_k\right)\right) g\left(T\left(R\left(E_i, E_j\right) E_k\right), E_i\right) \\
&= \sum_{i,j,k=1}^{n} g\left(R\left(E_i, g\left(E_j, T\left(E_k\right)\right) E_j\right) T\left(E_k\right), E_i\right) \\
&\quad - \sum_{i,j,k=1}^{n} g\left(T\left(R\left(E_i, g\left(E_j, T\left(E_k\right)\right) E_j\right) E_k\right), E_i\right) \\
&= \sum_{i,k=1}^{n} g\left(R\left(E_i, T\left(E_k\right)\right) T\left(E_k\right), E_i\right) \\
&\quad - \sum_{i,k=1}^{n} g\left(R\left(E_i, T\left(E_k\right)\right) E_k, T\left(E_i\right)\right) \\
&= \sum_{i,k=1}^{n} \lambda_k^2 \cdot g\left(R\left(E_i, E_k\right) E_k, E_i\right) \\
&\quad - \sum_{i,k=1}^{n} \lambda_k \lambda_i \cdot g\left(R\left(E_i, E_k\right) E_k, E_i\right)
\end{aligned}
$$

$$= \sum_{i,k=1}^{n} \left(\lambda_k^2 - \lambda_k \lambda_i\right) \sec\left(E_i, E_k\right)$$

$$= \sum_{i<k} \left(\lambda_k^2 - \lambda_k \lambda_i\right) \sec\left(E_i, E_k\right)$$

$$+ \sum_{i>k} \left(\lambda_k^2 - \lambda_k \lambda_i\right) \sec\left(E_i, E_k\right)$$

$$= \sum_{i<k} \left(\lambda_k^2 - \lambda_k \lambda_i\right) \sec\left(E_i, E_k\right)$$

$$+ \sum_{i<k} \left(\lambda_i^2 - \lambda_k \lambda_i\right) \sec\left(E_i, E_k\right)$$

$$= \sum_{i<k} \left(\lambda_k - \lambda_i\right)^2 \sec\left(E_i, E_k\right)$$

$$\geq \quad 0.$$

This finishes the proof.   □

Given this, one might suspect that we should be able to do something for the Ricci tensor, given that the sectional curvature is nonnegative. This is only partially true, as we don't have a Bochner formula for the Ricci tensor. Given that the manifold has divergence-free curvature tensor, one can find a Bochner formula and then get that the Ricci tensor must be parallel. The proofs are not hard and are deferred to the exercises. Note that we can't more generally hope that the Ricci tensor is parallel if it is divergence free, as all of the Berger spheres have divergence-free Ricci tensor, but only the standard sphere has parallel Ricci tensor.

We can now go over to the more complicated result we are interested in. It was first established in [**89**], and then with a modified proof in [**42**]. After that, the result seems to have fallen into oblivion. We shall present a more general version that is analogous to the above lemma, but the proof is essentially the one proposed by Tachibana.

THEOREM 53. (Tachibana, 1974) *If $\Re \geq 0$, then*

$$g\left(T^\flat, (\operatorname{div}\nabla - \nabla\operatorname{div})T\right) \geq 0$$

for any $(1,3)$-tensor T that induces a self-adjoint map $\Re : \Lambda^2 TM \to \Lambda^2 TM$.

PROOF. The fact that $T : \Lambda^2 TM \to \Lambda^2 TM$ is self-adjoint means that T enjoys the properties

$$g\left(T\left(X,Y,Z\right),W\right) = -g\left(T\left(X,Y,W\right),Z\right) = g\left(T\left(Y,X,W\right),Z\right),$$
$$g\left(T\left(X,Y,Z\right),W\right) = g\left(T\left(Z,W,X\right),Y\right).$$

Thus, we have a tensor with some of the properties of the curvature tensor. Let us first divide K into four terms:

$$K = g\left(T^\flat, (\operatorname{div}\nabla - \nabla\operatorname{div})T\right)$$

$$= \sum_{i,j,k,l,m=1}^{n} g\left(E_j, T\left(E_k, E_l, E_m\right)\right) g\left(\left(R\left(E_i, E_j\right)T\right)\left(E_k, E_l, E_m\right), E_i\right)$$

$$= \sum_{i,j,k,l,m=1}^{n} g\left(E_j, T\left(E_k, E_l, E_m\right)\right) g\left(R\left(E_i, E_j\right)\left(T\left(E_k, E_l, E_m\right)\right), E_i\right)$$

$$+ \sum_{i,j,k,l,m=1}^{n} -g\left(E_j, T\left(E_k, E_l, E_m\right)\right) g\left(T\left(R\left(E_i, E_j\right) E_k, E_l, E_m\right), E_i\right)$$

$$+ \sum_{i,j,k,l,m=1}^{n} -g\left(E_j, T\left(E_k, E_l, E_m\right)\right) g\left(T\left(E_k, R\left(E_i, E_j\right) E_l, E_m\right), E_i\right)$$

$$+ \sum_{i,j,k,l,m=1}^{n} -g\left(E_j, T\left(E_k, E_l, E_m\right)\right) g\left(T\left(E_k, E_l, R\left(E_i, E_j\right) E_m\right), E_i\right)$$

$$= A + B + C + D.$$

We now compute each of the terms $A, B, C,$ and D:

$$A = \sum_{i,j,k,l,m=1}^{n} g\left(E_j, T\left(E_k, E_l, E_m\right)\right) g\left(R\left(E_i, E_j\right)\left(T\left(E_k, E_l, E_m\right)\right), E_i\right)$$

$$= \sum_{i,j,k,l,m=1}^{n} g\left(R\left(E_i, g\left(E_j, T\left(E_k, E_l, E_m\right)\right) E_j\right)\left(T\left(E_k, E_l, E_m\right)\right), E_i\right)$$

$$= \sum_{i,k,l,m=1}^{n} g\left(R\left(E_i, T\left(E_k, E_l, E_m\right)\right)\left(T\left(E_k, E_l, E_m\right)\right), E_i\right)$$

$$= \sum_{i,k,l,m=1}^{n} g\left(\mathfrak{R}\left(E_i \wedge T\left(E_k, E_l, E_m\right)\right), E_i \wedge T\left(E_k, E_l, E_m\right)\right);$$

$$B = \sum_{i,j,k,l,m=1}^{n} -g\left(E_j, T\left(E_k, E_l, E_m\right)\right) g\left(T\left(R\left(E_i, E_j\right) E_k, E_l, E_m\right), E_i\right)$$

$$= \sum_{i,j,k,l,m=1}^{n} -g\left(E_j, T\left(E_k, E_l, E_m\right)\right) g\left(T\left(E_m, E_i, R\left(E_i, E_j\right) E_k\right), E_l\right)$$

$$= \sum_{i,j,k,l,m=1}^{n} g\left(E_j, T\left(E_k, E_l, E_m\right)\right) g\left(T\left(E_m, E_i, E_l\right), R\left(E_i, E_j\right) E_k\right)$$

$$= \sum_{i,k,l,m=1}^{n} g\left(T\left(E_m, E_i, E_l\right), R\left(E_i, T\left(E_k, E_l, E_m\right)\right) E_k\right)$$

$$= \sum_{i,k,l,m=1}^{n} g\left(R\left(E_i, T\left(E_k, E_l, E_m\right)\right) E_k, T\left(E_m, E_i, E_l\right)\right)$$

$$= -\sum_{i,k,l,m=1}^{n} g\left(R\left(E_i, T\left(E_k, E_l, E_m\right)\right) T\left(E_m, E_i, E_l\right), E_k\right)$$

$$= -\sum_{i,k,l,m=1}^{n} g\left(\mathfrak{R}\left(E_i \wedge T\left(E_k, E_l, E_m\right)\right), E_k \wedge T\left(E_m, E_i, E_l\right)\right).$$

Similarly,

$$C = \sum_{i,k,l,m=1}^{n} g\left(\Re\left(E_i \wedge T\left(E_k, E_l, E_m\right)\right), E_l \wedge T\left(E_m, E_i, E_k\right)\right).$$

Finally, we have

$$
\begin{aligned}
D &= \sum_{i,j,k,l,m=1}^{n} -g\left(E_j, T\left(E_k, E_l, E_m\right)\right) g\left(T\left(E_k, E_l, R\left(E_i, E_j\right) E_m\right), E_i\right) \\
&= \sum_{i,j,k,l,m=1}^{n} g\left(E_j, T\left(E_k, E_l, E_m\right)\right) g\left(R\left(E_i, E_j\right) E_m, T\left(E_k, E_l, E_i\right)\right) \\
&= -\sum_{i,k,l,m=1}^{n} g\left(R\left(E_i, T\left(E_k, E_l, E_m\right)\right) T\left(E_k, E_l, E_i\right), E_m\right) \\
&= -\sum_{i,k,l,m=1}^{n} g\left(\Re\left(E_i \wedge T\left(E_k, E_l, E_m\right)\right), E_m \wedge T\left(E_k, E_l, E_i\right)\right) \\
&= \sum_{i,k,l,m=1}^{n} g\left(\Re\left(E_i \wedge T\left(E_k, E_l, E_m\right)\right), E_m \wedge T\left(E_l, E_k, E_i\right)\right).
\end{aligned}
$$

Therefore, if we define elements $\theta_{iklm} \in \Lambda^2 TM$ by

$$
\begin{aligned}
\theta_{iklm} &= E_i \wedge T\left(E_k, E_l, E_m\right) \\
&\quad -E_k \wedge T\left(E_m, E_i, E_l\right) \\
&\quad +E_l \wedge T\left(E_m, E_i, E_k\right) \\
&\quad +E_m \wedge T\left(E_l, E_k, E_i\right),
\end{aligned}
$$

then one checks that

$$\sum_{i,k,l,m=1}^{n} g\left(\Re\left(\theta_{iklm}\right), \theta_{iklm}\right) = 4K$$

by observing that after multiplying out, there are 16 terms on the left-hand side, which can be collected in groups of four. After reindexing some of the sums, each of these groups consists of four equal terms that correspond to one of $A, B, C,$ or D. Since the left-hand side is assumed to be nonnegative, we have proven the desired result. $\square$

COROLLARY 23. (Tachibana, 1974) *If (M, g) is a compact oriented Riemannian manifold with $\mathrm{div}R = 0$ and $\Re \geq 0$, then $\nabla R = 0$. If in addition, $\Re > 0$, then (M, g) has constant curvature.*

PROOF. The first part is immediate from the above theorems. For the second part we have again that $K = 0$. Since $\Re$ is assumed to be positive, we must therefore have that

$$
\begin{aligned}
\theta_{iklm} &= E_i \wedge R\left(E_k, E_l\right) E_m \\
&\quad -E_k \wedge R\left(E_m, E_i\right) E_l \\
&\quad +E_l \wedge R\left(E_m, E_i\right) E_k \\
&\quad +E_m \wedge R\left(E_l, E_k\right) E_i \\
&= 0.
\end{aligned}
$$

From this one can see that the curvature must be constant. A different proof of this can be found using the material from chapter 8.   □

6. Further Study

For more general and complete accounts of the Bochner technique and spin geometry we recommend the two texts [97] and [61]. The latter book also has a complete proof of the Hodge theorem. Other sources for this particular result are [56], [81], and [92]. For more information about Killing fields and related matters we refer the reader to [59, Chapter II] and [95]. There is also a good elementary account of Killing fields in O'Neill's book [73, Chapter 9].

For other generalizations to manifolds with integral curvature bounds the reader should consult [40]. In there the reader will find a complete discussion on generalizations of the above mentioned results about Betti numbers.

7. Exercises

(1) Let $F : (M,g) \to (\mathbb{R}^k, \mathrm{can})$ be a Riemannian submersion and let (M,g) be complete. If each of the components of F has zero Hessian, then $(M,g) = (N,h) \times (\mathbb{R}^k, \mathrm{can})$.

(2) Let $\mathfrak{t} \subset \mathrm{iso}\,(M,g)$ be an Abelian subalgebra corresponding to a torus subgroup $T^k \subset \mathrm{Iso}\,(M,g)$. Define $\mathfrak{p} \subset \mathfrak{t}$ as the set of Killing fields that correspond to circle actions, i.e., actions induced by homomorphisms $S^1 \to T^k$. Show that $\mathfrak{p}$ is a vector space over the rationals with $\dim_{\mathbb{Q}}\mathfrak{p} = \dim_{\mathbb{R}}\mathfrak{t}$.

(3) Show that for any $(1,1)$-tensor S and vector field X we have

$$\mathrm{tr}\,(\nabla_X S) = \nabla_X \mathrm{tr} S.$$

(4) Given two Killing fields X and Y on a Riemannian manifold, develop a formula for $\Delta g\,(X,Y)$. Use this to give a formula for the Ricci curvature in a frame consisting of Killing fields.

(5) For a vector field X define the Lie derivative of the connection as follows:

$$\begin{aligned}(L_X\nabla)\,(U,V) &= L_X\,(\nabla_U V) - \nabla_{L_X U} V - \nabla_U L_X V \\ &= [X,\nabla_U V] - \nabla_{[X,U]} V - \nabla_U\,[X,V].\end{aligned}$$

(a) Show that $L_X \nabla$ is a $(1,2)$-tensor.

(b) We say that X is an *affine vector field* if $L_X\nabla = 0$. Show that for such a field we have

$$\nabla^2_{U,V} X = -R\,(X,U)\,V.$$

Hint: Show that:

$$R\,(W,U)\,V + \nabla^2_{U,V} W = (L_W\nabla)\,(U,V).$$

(c) Show that Killing fields are affine. Give an example of an affine field on $\mathbb{R}^n$ which is not a Killing field.

(d) Let N be a component of the zero set for a Killing field X. Show that $\nabla_V\,(\nabla X) = 0$ for vector fields V tangent to N.

(6) Let X be a vector field on a Riemannian manifold.

(a) Show that

$$|L_X g|^2 = 2\,|\nabla X|^2 + 2\mathrm{tr}\,(\nabla X)^2.$$

(b) Establish the following integral formulae on a closed oriented Riemannian manifold:

$$\int_M \left(\mathrm{Ric}\,(X,X) + \mathrm{tr}\,(\nabla X)^2 - (\mathrm{div}X)^2 \right) \;=\; 0,$$

$$\int_M \left(\mathrm{Ric}\,(X,X) + g\left(\mathrm{tr}\nabla^2 X, X\right) + \frac{1}{2}|L_X g|^2 - (\mathrm{div}X)^2 \right) \;=\; 0.$$

(c) Finally, show that X is a Killing field iff

$$\mathrm{div}X \;=\; 0,$$
$$\mathrm{tr}\nabla^2 X \;=\; -\mathrm{Ric}\,(X).$$

(7) (Yano) If X is an affine vector field show that $\mathrm{tr}\nabla^2 X = -\mathrm{Ric}\,(X)$ and that $\mathrm{div}X$ is constant. Use this together with the above characterizations of Killing fields to show that on closed manifolds affine fields are Killing fields.

(8) If K is a Killing field show that L_K and Δ commute as operators on forms. Conversely show that X is a Killing field if L_X and Δ commute on functions.

(9) Suppose (M, g) is compact and has $b_1 = k$. If $\mathrm{Ric} \geq 0$, then the universal covering splits:

$$\left(\tilde{M}, g\right) = (N, h) \times \left(\mathbb{R}^k, \mathrm{can}\right).$$

(10) Let (M, g) be a compact $2n$-manifold with positive sectional curvature. If $\dim\left(\mathfrak{h}\,(M, g)\right) \geq k$, then $\mathcal{Z}\left(\mathfrak{h}\,(M, g)\right)$ contains an element of dimension $\geq 2\,(k-1)$.

(11) Let (M, g) be an n-dimensional Riemannian manifold that is isometric to Euclidean space outside some compact subset $K \subset M$, i.e., $M - K$ is isometric to $\mathbb{R}^n - C$ for some compact set $C \subset \mathbb{R}^n$. If $\mathrm{Ric}_g \geq 0$, show that $M = \mathbb{R}^n$. Hint: Find a metric on the n-torus that is isometric to a neighborhood of $K \subset M$ somewhere and otherwise flat. Alternatively, show that any parallel 1-form on $\mathbb{R}^n - C$ extends to a harmonic 1-form on M. Then apply Bochner's formula to show that it must in fact be parallel when $\mathrm{Ric}_g \geq 0$, and use this to conclude that the manifold is flat.

(12) Given two vector fields X and Y on (M, g) such that ∇X and ∇Y are symmetric, develop Bochner formulae for $\nabla^2 \frac{1}{2} g\,(X, Y)$ and $\Delta \frac{1}{2} g\,(X, Y)$.

(13) For general sections s_1 and s_2 of an appropriate bundle show in analogy with the formula

$$\Delta \frac{1}{2}|s|^2 = |\nabla s|^2 + \left\langle \mathrm{tr}\nabla^2 s, s \right\rangle$$

that:

$$\Delta \left\langle s_1, s_2 \right\rangle = 2\left\langle \nabla s_1, \nabla s_2 \right\rangle + \left\langle \mathrm{tr}\nabla^2 s_1, s_2 \right\rangle + \left\langle s_1, \mathrm{tr}\nabla^2 s_2 \right\rangle.$$

Use this on forms to develop Bochner formulae from the Weitzenböck formulae for inner products of such sections.

More generally we can consider the 1-form defined by $\omega\left(v\right) = \langle\nabla_v s_1, s_2\rangle$ which represents half of the differential of $\langle s_1, s_2\rangle$. Show that

$$
\begin{aligned}
-\delta\omega &= \langle\nabla s_1, \nabla s_2\rangle + \langle\mathrm{tr}\nabla^2 s_1, s_2\rangle \\
&= \langle(\nabla^*\nabla + \mathrm{tr}\nabla^2)\, s_1, s_2\rangle \\
d\omega\left(X, Y\right) &= \langle R\left(X, Y\right) s_1, s_2\rangle - \langle\nabla_X s_1, \nabla_Y s_2\rangle + \langle\nabla_Y s_1, \nabla_X s_2\rangle.
\end{aligned}
$$

(14) Show that in dimension 2,

$$
K = g\left(R^{\flat}, (\mathrm{div}\nabla - \nabla\mathrm{div})\, R\right) = 0.
$$

(15) *(Simons)* Let (M, g) be a Riemannian manifold with a $(1,1)$-tensor field T that is symmetric and whose covariant derivative is symmetric

$$
(\nabla_X T)\left(Y\right) = (\nabla_Y T)\left(X\right).
$$

Show that,

$$
\Delta\frac{1}{2}\left|T\right|^2 = \left|\nabla T\right|^2 + g\left(T^{\flat}, \nabla\mathrm{div}T\right) + g\left(T^{\flat}, (\mathrm{div}\nabla - \nabla\mathrm{div})\, T\right).
$$

When M is compact and oriented conclude that if $\sec \geq 0$ and $\mathrm{div}T = 0$, then $\nabla T = 0$. Moreover, if $\sec > 0$, then $T = c \cdot I$ for some constant c. In case T is not symmetric establish a Bochner formula that can be used to arrive at the above results.

(16) *(Berger)* On a closed Riemannian manifold (M, g) show that if $\mathrm{div}R = 0$ and $\sec \geq 0$, then $\nabla\mathrm{Ric} = 0$. (Hint: use an exercise from chapter 2 to get the symmetry for $\nabla\mathrm{Ric}$ and also the formula $2\mathrm{div}\,(\mathrm{Ric}) = d\,(\mathrm{scal})$ to conclude that $\mathrm{div}\,(\mathrm{Ric}) = 0$.)

(17) Let $(M^n, g) \hookrightarrow \mathbb{R}^{n+1}$ be an isometric immersion of an oriented manifold.

(a) Using the Codazzi equations, show that

$$
\Delta\frac{1}{2}\left|S\right|^2 = \left|\nabla S\right|^2 + g\left(S^{\flat}, \nabla\mathrm{div}S\right) + K,
$$

where S is the shape operator and K is as usual defined by

$$
K = g\left(S^{\flat}, (\mathrm{div}\nabla - \nabla\mathrm{div})\, S\right).
$$

(b) Assuming that M is compact, show that

$$
\int\left|\nabla S\right|^2 = \int\left|d\,(\mathrm{tr}S)\right|^2 - \int K.
$$

(Recall that we proved in the exercises to chapter 4 that $\mathrm{div}S = d\,(\mathrm{tr}S)$.)

(c) Show Liebmann's theorem: If (M, g) has constant mean curvature ($\mathrm{tr}S = $ constant) and nonnegative shape operator, then (M, g) is a constant-curvature sphere. Hint: Using chapter 4, find out something about the curvature from the positivity of S; then use

$$
K = \sum_{i < j}\left(\lambda_j - \lambda_i\right)^2 \cdot \sec\left(E_i, E_j\right).
$$

In case $M = S^2$, H. Hopf showed that one can prove this theorem without using the nonnegativity of the shape operator. This is not too hard to believe, as we know that

$$K(p) = (\lambda_2 - \lambda_1)^2 \cdot \sec(p),$$
$$\int \sec(p) \, d\text{vol} = 4\pi,$$

indicating that $\int K$ should be nonnegative. On the other hand, Wente has exhibited immersed tori with constant mean curvature (see Wente's article in [**45**]).

(18) Show that if one defines the divergence of a p-form by

$$\text{div}\omega(X_2, \ldots, X_p) = \sum_{i=1}^n (\nabla_{E_i}\omega)(E_i, X_2, \ldots, X_p)$$
$$= \sum_{i=1}^n i_{E_i}(\nabla_{E_i}\omega)(X_2, \ldots, X_p),$$

where E_i is an orthonormal frame, then $\delta = -\text{div}$.

(19) Suppose we have a Killing field K on a closed oriented Riemannian manifold (M, g). Assume that ω is a harmonic form.

(a) Show that $L_K\omega = 0$. Hint: Show that $L_K\omega$ is also harmonic.

(b) Show that $i_K\omega$ is closed, but not necessarily harmonic.

(20) Let (M, g) be a closed Kähler manifold with Kähler form ω. Show using the exercises from chapter 2 that

$$\omega^k = \underbrace{\omega \wedge \cdots \wedge \omega}_{k \text{ times}}$$

is closed but not exact by showing that $\omega^{\frac{\dim M}{2}}$ is proportional to the volume form. Conclude that none of the even homology groups vanish.

(21) Let $E \to M$ be a vector bundle with connection ∇.

(a) Show that ∇ induces a natural connection on $\text{Hom}(E, E)$ that we also denote ∇.

(b) Let $\Omega^p(M, E)$ denote the alternating p-linear maps from TM to E (note that $\Omega^0(M, E) = \Gamma(E)$.) Show that $\Omega^*(M)$ acts in a natural way from both left and right on $\Omega^*(M, E)$ by wedge product. Show also that there is a natural wedge product

$$\Omega^p(M, \text{Hom}(E, E)) \times \Omega^q(M, E) \to \Omega^{p+q}(M, E).$$

(c) Show that there is a connection dependent exterior derivative

$$d^\nabla : \Omega^p(M, E) \to \Omega^{p+1}(M, E)$$

with the property that it satisfies the exterior derivative version of Leibniz's rule with respect to the above defined wedge products, and such that for $s \in \Gamma(E)$ we have: $d^\nabla s = \nabla s$.

(d) If we think of $R(X, Y) s \in \Omega^2(M, \text{Hom}(E, E))$. Show that:

$$\left(d^\nabla \circ d^\nabla\right)(s) = R \wedge s$$

for any $s \in \Omega^p(M, E)$ and that Bianchi's second identity can be stated as $d^\nabla R = 0$.

(22) If we let $E = TM$ in the previous exercise, then

$$\Omega^1(M, TM) = \text{Hom}(TM, TM)$$

will just consist of all $(1, 1)$-tensors.

(a) Show that in this case $d^\nabla s = 0$ iff s is a Codazzi tensor.

(b) The entire chapter seems to indicate that whenever we have a tensor bundle $E = \mathbb{R}$, TM, $\Lambda^2 M$ etc. and an element $s \in \Omega^p(M, E)$ with $d^\nabla s = 0$, then there is a Bochner type formula for s. Moreover, when in addition s is "divergence free" and some sort of curvature is nonnegative, then s should be parallel. Can you develop a theory in this generality?

(c) Show that if X is a vector field, then ∇X is a Codazzi tensor iff $R(\cdot, \cdot) X = 0$. Give an example of a vector field such that ∇X is Codazzi but X itself is not parallel. Is it possible to establish a Bochner type formula for exact tensors like $\nabla X = d^\nabla X$ even if they are not closed?

(23) *(Thomas)* Show that in dimensions $n > 3$ the Gauss equations $(\mathfrak{R} = S \wedge S)$ imply the Codazzi equations $(d^\nabla S = 0)$ provided $\det S \neq 0$. Hint: use the second Bianchi identity and be very careful with how things are defined. It will also be useful to study the linear map

$$\text{Hom}(\Lambda^2 V, V) \rightarrow \text{Hom}(\Lambda^3 V, \Lambda^2 V),$$
$$T \rightarrow T \wedge S$$

for a linear map $S : V \rightarrow V$. In particular, one can see that this map is injective only when the rank of S is ≥ 4.

(24) In dimensions $4n$ we have that the Hodge $* : H^{2n}(M) \rightarrow H^{2n}(M)$ satisfies $** = I$. The difference in the dimensions of the eigenspaces for ± 1 is called the *signature* of M:

$$\tau(M) = \sigma(M) = \dim(\ker(* - I) - \ker(* + I)).$$

One can show that this does not depend on the metric used to define $*$, by observing that it is the index of the symmetric bilinear map

$$H^{2n}(M) \times H^{2n}(M) \rightarrow \mathbb{R},$$
$$(\omega_1, \omega_2) \rightarrow \int \omega_1 \wedge \omega_2.$$

Recall that the index of a symmetric bilinear map is the difference between positive and negative diagonal elements when it has been put into diagonal form. In dimension 4 one can show that

$$\sigma(M) = \frac{1}{12\pi^2} \int_M \left(|W^+|^2 - |W^-|^2 \right).$$

Using the exercises from chapter 4, show that for an Einstein metric in dimension 4 we have

$$\chi(M) \geq \frac{3}{2} \sigma(M),$$

with equality holding iff the metric is Ricci flat and $W^- = 0$. Conclude that not all four manifolds admit Einstein metrics. In higher dimensions there are no known obstructions to the existence of Einstein metrics. Hint: consider connected sums of $\mathbb{C}P^2$ with itself k times.

(25) Recall the curvature forms defined using an orthonormal frame E_i:

$$\Omega_i^j\left(X,Y\right)E_j = R\left(X,Y\right)E_i.$$

They yield a skew-symmetric matrix of 2-forms:

$$\Omega = \left(\Omega_i^j\right).$$

From linear algebra we know that there are various invariant polynomials that depend on the entries of matrices, e.g., the trace and determinant. We can define similar objects in this case as follows:

$$p_{2l}\left(\Omega\right) = \sum \Omega_{i_1}^{i_2} \wedge \Omega_{i_2}^{i_3} \wedge \cdots \wedge \Omega_{i_l}^{i_1}.$$

These are known as the *Pontryagin forms*. Show that they yield globally defined forms that are closed (you need to look at the exercises in chapter 2 and also understand what the second Bianchi identity has to do with $d\Omega$). Show, that they are zero when l is odd. Thus, they generate homology classes $p_l \in H^{4l}$, which are known as the Pontryagin classes of the manifold. It can be shown that these classes do not depend on the metric.

Show that the Pontryagin classes are zero on a manifold with constant curvature. Hint: Use that we know what the curvature tensor looks like. Thus even in the case where $4l = n = \dim M$, we do not necessarily have that p_l is the Euler class.

Try to compute $p_1 \in H^4$ for some of the standard 4-manifolds.

(26) In case the manifold has even dimension $n = 2m$, we can construct the *Euler form*:

$$
\begin{aligned}
e\left(\Omega\right) &= \varepsilon^{i_1\cdots i_n}\cdot\Omega_{i_2}^{i_1}\wedge\cdots\wedge\Omega_{i_n}^{i_{n-1}},\\
\varepsilon^{i_1\cdots i_n} &= \text{sign of the permutation }\left(i_1\cdots i_n\right),
\end{aligned}
$$

which modulo a factor generates the Euler class, or characteristic, of the manifold. Show that this form also yields a globally defined closed form. Note that this is essentially the square root of the determinant of Ω. However, as this determinant is a $2n$ form, it is always zero and therefore doesn't yield anything interesting. The cohomology class of $e\left(\Omega\right)$ can also be seen to be independent of the metric. Moreover, as discussed in chapter 4, it is proportional to the Euler characteristic.

CHAPTER 8

Symmetric Spaces and Holonomy

In this chapter we shall give a brief overview of (locally) symmetric spaces and holonomy. Only the simplest proofs will be presented. Thus, we will have to be sketchy in places. Still, most of the standard results are proved or at least mentioned. We give some explicit examples, including the complex projective space, in order to show how one can compute curvatures on symmetric spaces relatively easily. There is a brief introduction to holonomy and the de Rham decomposition theorem. We give a few interesting consequences of this theorem and then proceed to discuss how holonomy and symmetric spaces are related. Finally, we classify all compact manifolds with nonnegative curvature operator. We shall in a few places use results from chapter 9. They will therefore have to be taken for granted at this point.

As we have already seen, Riemann showed that locally there is only one constant curvature geometry. After Lie's work on "continuous" groups it became clear that one had many more interesting models for geometries. Next to constant curvature spaces, the most natural type of geometry to try to understand is that of (locally) symmetric spaces. One person managed to take all the glory for classifying symmetric spaces; Elie Cartan. He started out in his thesis with cleaning up and correcting Killing's classification of simple complex Lie algebras. Using this he later classified all the simple real Lie algebras. With the help of this and many of his different characterizations of symmetric spaces, Cartan, by the mid 1920s had managed to give a complete (local) classification of all symmetric spaces. This was an astonishing achievement even by today's deconstructionist standards, not least because Cartan also had to classify the real simple Lie algebras. This in itself takes so much work that most books on Lie algebras give up after having settled the complex case.

After Cartan's work, a few people worked on getting a better conceptual understanding of some of these new geometries and also on giving a more global classification. Still, not much happened until the 1950s, when people realized a interesting connection between symmetric spaces and holonomy: The de Rham decomposition theorem and Berger's classification of holonomy groups. It then became clear that almost all holonomy groups occurred for symmetric spaces and therefore gave good approximating geometries to most holonomy groups. An even more interesting question also came out of this, namely, what about those few holonomy groups that do not occur for symmetric spaces? This is related to the study of Kähler manifolds and some exotic geometries in dimensions 7 and 8. The Kähler case seems to be quite well understood by now, not least because of Yau's work on the Calabi conjecture. The exotic geometries have only very recently become better understood with D. Joyce's work.

1. Symmetric Spaces

There are many ways of representing symmetric spaces. Below we shall see how they can be described as homogeneous spaces, via Lie algebras, and finally, by their curvature tensor.

1.1. The Homogeneous Description.

We say that a Riemannian manifold (M, g) is a *symmetric space* if for each $p \in M$ the isotropy group Iso_p contains an isometry A_p such that $DA_p : T_p M \to T_p M$ is the antipodal map $-I$. Since isometries preserve geodesics, we immediately see that for any geodesic $\gamma(t)$ such that $\gamma(0) = p$ we have: $A_p \circ \gamma(t) = \gamma(-t)$. Using this, it is easy to show that symmetric spaces are homogeneous and complete. Namely, if two points are joined by a geodesic, then the symmetry in the midpoint between these points on the geodesic is an isometry that maps these points to each other. Thus, any two points that can be joined by a broken sequence of geodesics can be mapped to each other by an isometry. This shows that the space is homogeneous. It is then easy to show that the space is complete.

Given a homogeneous space $G/H = \mathrm{Iso}/\mathrm{Iso}_p$, we see that it is symmetric provided that the symmetry A_p exists for just one p. The symmetry A_q can then be constructed by selecting an isometry g that takes p to q and then observing that

$$g \circ A_p \circ g^{-1}$$

has the correct differential at q. This means, in particular, that any Lie group G with bi-invariant metric is a symmetric space, as $g \to g^{-1}$ is the desired symmetry around the identity element. Let us list some of the important families of homogeneous spaces that are symmetric. They come in pairs of compact and noncompact spaces. Below we list just a few families of examples. There are many more families and several exceptional examples as well.

Lie groups with bi-invariant metrics

group	rank	dim
$SU(n+1)$	n	$n(n+2)$
$SO(2n+1)$	n	$n(2n+1)$
$Sp(n)$	n	$n(2n+1)$
$SO(2n)$	n	$n(2n-1)$

Noncompact analogues of bi-invariant metrics

(complexified group)/group	rank	dim
$SL(n+1, \mathbb{C})/SU(n+1)$	n	$n(n+2)$
$SO(2n+1, \mathbb{C})/SO(2n+1)$	n	$n(2n+1)$
$Sp(n, \mathbb{C})/Sp(n)$	n	$n(2n+1)$
$SO(2n, \mathbb{C})/SO(2n)$	n	$n(2n-1)$

Compact homogeneous examples

Iso	Iso_p	dim	rank	description
$SO(n+1)$	$SO(n)$	n	1	Sphere
$O(n+1)$	$O(n) \times \{1, -1\}$	n	1	$\mathbb{R}P^n$
$U(n+1)$	$U(n) \times U(1)$	$2n$	1	$\mathbb{C}P^n$
$Sp(n+1)$	$Sp(n) \times Sp(1)$	$4n$	1	$\mathbb{H}P^n$
F_4	$Spin(9)$	16	1	$\mathbb{O}P^2$
$SO(p+q)$	$SO(p) \times SO(q)$	pq	$\min(p,q)$	real Grassmannian
$SU(p+q)$	$S(U(p) \times U(q))$	$2pq$	$\min(p,q)$	complex Grassmannian

Noncompact homogeneous examples

Iso	Iso_p	dim	rank	description
$SO(n,1)$	$SO(n)$	n	1	Hyperbolic space
$O(n,1)$	$O(n) \times \{1, -1\}$	n	1	Hyperbolic $\mathbb{R}P^n$
$U(n,1)$	$U(n) \times U(1)$	$2n$	1	Hyperbolic $\mathbb{C}P^n$
$Sp(n,1)$	$Sp(n) \times Sp(1)$	$4n$	1	Hyperbolic $\mathbb{H}P^n$
F_4^{-20}	$Spin(9)$	16	1	Hyperbolic $\mathbb{O}P^2$
$SO(p,q)$	$SO(p) \times SO(q)$	pq	$\min(p,q)$	Hyperbolic Grassmannian
$SU(p,q)$	$S(U(p) \times U(q))$	$2pq$	$\min(p,q)$	Complex hyperbolic Grassmannian

Recall that $Spin(n)$ is the universal double covering of $SO(n)$ for $n > 2$. We also have the following special identities for low dimensions:

$$
\begin{aligned}
SO(2) &= U(1), \\
Spin(3) &= SU(2) = Sp(1), \\
Spin(4) &= Spin(3) \times Spin(3).
\end{aligned}
$$

Note that all of the compact examples have sec ≥ 0 by O'Neill's formula. It also follows from this formula that all the projective spaces (compact and noncompact) have quarter pinched metrics, i.e., the ratio between the smallest and largest curvatures is $\frac{1}{4}$. This was all proven in chapter 3. Below we shall do some different calculations to justify these remarks.

In the above list of examples there is a column called *rank*. This is related to the rank of a Lie group as discussed in chapter 7. Here, however, we need a rank concept for more general spaces. The *rank* of a geodesic $\gamma : \mathbb{R} \to M$ is simply the dimension of parallel fields E along γ such that

$$
R(E(t), \dot{\gamma}(t))\dot{\gamma}(t) = 0
$$

for all t. The rank of a geodesic is therefore always ≥ 1. The rank of a Riemannian manifold is now defined as the minimum rank over all of the geodesics in M. For symmetric spaces the rank can be computed from knowledge of Abelian subgroups in Lie groups and is therefore more or less algebraic. For a general manifold there might of course be metrics with different ranks, but this is actually not so obvious. Is it, for example, possible to find a metric on the sphere of rank > 1? A general remark is that any Cartesian product has rank ≥ 2, and also many symmetric spaces have rank ≥ 2. In general, it is unclear to what extent other manifolds can also have rank ≥ 2. However, see below for the case of nonpositive curvature and nonnegative curvature operators. Note that there are five compact rank one symmetric spaces (CROSS) in the above lists. These are the only simply connected compact rank 1 symmetric spaces.

1.2. Isometries and Parallel Curvature. Another interesting property for symmetric spaces is that they have parallel curvature tensor. This is because the symmetries A_p leave the curvature tensor and its covariant derivative invariant. In particular, we have

$$DA_p\left(\left(\nabla_X R\right)\left(Y, Z, W\right)\right) = \left(\nabla_{DA_p X} R\right)\left(DA_p Y, DA_p Z, DA_p W\right),$$

which at p implies

$$
\begin{aligned}
-\left(\nabla_X R\right)\left(Y, Z, W\right) &= \left(\nabla_{-X} R\right)\left(-Y, -Z, -W\right) \\
&= \left(\nabla_X R\right)\left(Y, Z, W\right).
\end{aligned}
$$

Thus, $\nabla R = 0$. This almost characterizes symmetric spaces.

THEOREM 54. (E. Cartan) *If (M, g) is a Riemannian manifold with parallel curvature tensor, then for each $p \in M$ there is an isometry A_p defined in a neighborhood of p with $DA_p = -I$ on $T_p M$. Moreover, if (M, g) is simply connected and complete, then the symmetry is defined on all of M, and in particular, the space is symmetric.*

PROOF. The global statement follows from the local one using an analytic continuation argument and the next Theorem below. Note that for the local statement we already have a candidate for a map. Namely, if ε is so small that $\exp_p : B(0, \varepsilon) \to B(p, \varepsilon)$ is a diffeomorphism, then we can just define $A_p(x) = -x$ in these coordinates. It now remains to see why this is an isometry when we have parallel curvature tensor. This means that in these coordinates the metric is the same at x and $-x$. Switching to polar coordinates, we have the fundamental equations relating curvature and the metric. So the claim follows if we can prove that the curvature tensor is the same when we go in opposite directions. To check this, first observe

$$R\left(\cdot, v\right) v = R\left(\cdot, -v\right)\left(-v\right).$$

So the curvatures start out being the same. If ∂_r is the radial field, we also have

$$\left(\nabla_{\partial_r} R\right) = 0.$$

Thus, the curvature tensors not only start out being equal, but also satisfy the same simple first-order equation. Consequently they must remain the same as we go equal distance in opposite directions. $\square$

A Riemannian manifold with parallel curvature tensor is called a *locally symmetric space.*

It is worth mentioning that there are left-invariant metrics that are not locally symmetric. Namely, in the exercises to chapter 3 it is shown that the Berger spheres ($\varepsilon \neq 1$) and the Heisenberg group do not have parallel curvature tensor. In fact, as they are 3 dimensional they can't even have parallel Ricci tensor.

With very little extra work we can generalize the above theorem on the existence of local symmetries. Recall that in our discussion about existence of isometries with a given differential in chapter 5 we decided that they could exist only when the spaces had the same constant curvature. However, there is a generalization to symmetric spaces. Namely, we know that any isometry preserves the curvature tensor. Thus, if we start with a linear isometry that preserves the curvatures at a point, then we should be able to extend this map in the situation where curvatures are everywhere the same. This is the content of the next theorem.

THEOREM 55. (E. Cartan) *Suppose we have a simply connected symmetric space (M, g) and a complete locally symmetric space $(N, \bar{g})$ of the same dimension. Given a linear isometry $L : T_pM \to T_qN$ such that*

$$L\left(R^g\left(x, y\right) z\right) = R^{\bar{g}}\left(Lx, Ly\right) Lz$$

for all $x, y, z \in T_pM$, there is a unique Riemannian isometry $F : M \to N$ such that $D_pF = L$.

PROOF. The proof of this is, as in the constant curvature case, by analytic continuation. So we need only find these isometries locally. Given that there is an isometry defined locally, we know that it must look like

$$F = \exp_q \circ L \circ \exp_p^{-1}.$$

To see that this indeed defines an isometry, we have to show that the metrics in exponential coordinates are the same via the identification of the tangent spaces by L. As usual the radial curvatures determine the metrics. In addition, the curvatures are parallel and therefore satisfy the same first-order equation. We assume that initially the curvatures are the same at p and q via the linear isometry. But then they must be the same in frames that are radially parallel around these points. Consequently, the spaces are locally isometric. □

This result shows that the curvature tensor completely characterizes the symmetric space. It also tells us what the isometry group must be in case the symmetric space is simply connected. We shall study this further below.

1.3. Algebraic Descriptions of Symmetric Spaces. It is worthwhile to try to get a more algebraic description of symmetric spaces. Note that there are many ways of writing homogeneous spaces as quotients G/H, e.g.,

$$S^3 = SU\left(2\right) = SO\left(4\right)/SO\left(3\right) = O\left(4\right)/O\left(3\right).$$

But only one of these, $O\left(4\right)/O\left(3\right)$, tells us directly that S^3 is a symmetric space. This is because the isometry A_p modulo conjugation lies in $O\left(3\right)$ as it is orientation reversing. We shall in this section try to get a Lie algebraic description based on Killing fields rather than the Lie group description based on isometries.

To get a more complete picture, we have to understand how the involution acts, not just on the space M, but as a map in Iso (M, g), and then in the Lie algebra iso (M, g) of Killing fields.

Let us fix a symmetric space (M, g) and a point $p \in M$. Recall from chapter 7 that the map

$$\text{iso} \quad \to \quad T_pM \times \mathfrak{so}\left(T_pM\right),$$
$$X \quad \to \quad \left(X|_p, (\nabla X)|_p\right)$$

is an injection. Since (M, g) is homogeneous, this linear map will be a surjection onto the first factor. Thus, iso can be identified with $T_pM \times \text{iso}_p$. This then induces a Lie algebra structure on $T_pM \times \text{iso}_p$ from that on iso. To understand this structure a little better, let us first observe that the decomposition $T_pM \times \text{iso}_p$ at the level of Killing fields looks like

$$X \in T_pM \text{ iff } (\nabla X)|_p = 0,$$
$$X \in \text{iso}_p \text{ iff } X|_p = 0.$$

So as not to confuse Killing fields with vectors, let us introduce the terminology

$$\mathfrak{t}_p = \{X \in \mathfrak{iso} : (\nabla X)|_p = 0\}.$$

Let us check where the Lie brackets of various combinations of Killing fields X, Y lie.

(a) If $X, Y \in \mathfrak{t}_p$ or $X, Y \in \mathfrak{iso}_p$, then

$$[X, Y]|_p = \nabla_{X|_p} Y - \nabla_{Y|_p} X = 0.$$

So we conclude that $[X, Y] \in \mathfrak{iso}_p$ in these cases. In the case where $X, Y \in \mathfrak{iso}_p$, we even have that the Lie bracket coincides, up to sign, with the Lie bracket coming from $\mathfrak{so}(T_p M)$ (see also the section on Lie derivatives in the appendix). To see this recall that

$$\nabla^2_{V,W} Y = \nabla_V \nabla_W Y - \nabla_{\nabla_V W} Y$$

is tensorial in V and W. Therefore, if $v \in T_p M$ and $X|_p = 0$, then

$$\nabla_v \nabla_X Y = \nabla_{\nabla_v X} Y.$$

In the case $X, Y \in \mathfrak{iso}_p$ and $v \in T_p M$ this implies

$$
\begin{aligned}
[(\nabla X)|_p, (\nabla Y)|_p](v) &= (\nabla X \circ \nabla Y - \nabla Y \circ \nabla X)(v) \\
&= \nabla_{\nabla_v Y} X - \nabla_{\nabla_v X} Y \\
&= \nabla_v (\nabla_Y X - \nabla_X Y) \\
&= -\nabla_v [X, Y].
\end{aligned}
$$

Hence, the element $[X, Y] \in \mathfrak{iso}$ is identified with $-[(\nabla X)|_p, (\nabla Y)|_p]$ inside $\mathfrak{so}(T_p M)$.

(b) If $X \in \mathfrak{t}_p$ and $Y \in \mathfrak{iso}_p$, then

$$[X, Y]|_p = \nabla_{X|_p} Y = (\nabla Y)(X|_p).$$

Which is simply the way the elements $Y \in \mathfrak{so}(T_p M)$ act on $T_p M$.

In conclusion, we see that the Lie algebra $\mathfrak{iso}$ can be represented as a direct sum: $\mathfrak{iso} = \mathfrak{t}_p \oplus \mathfrak{iso}_p$, where $\mathfrak{t}_p$ is a vector space with a Euclidean metric, and $\mathfrak{iso}_p$ is a subalgebra of the skew-symmetric transformations on $\mathfrak{t}_p$. Moreover, the Lie algebra structure on $\mathfrak{iso} = \mathfrak{t}_p \oplus \mathfrak{iso}_p$ is given by

$$
\begin{aligned}
[h_1, h_2] &= -(h_1 \circ h_2 - h_2 \circ h_1) \text{ if } h_1, h_2 \in \mathfrak{iso}_p, \\
-[h, x] &= [x, h] = h(x) \text{ if } h \in \mathfrak{iso}_p \text{ and } x \in \mathfrak{t}_p, \\
[x, y] &\in \mathfrak{iso}_p \text{ for } x, y \in \mathfrak{t}_p.
\end{aligned}
$$

Thus, the only Lie brackets that are not given canonically are $[x, y]$, where $x, y \in \mathfrak{t}_p$. We shall soon be able to show that as an element of $\mathfrak{so}(T_p M)$ this Lie bracket is represented by the curvature

$$R(x, y) : T_p M \to T_p M.$$

It will, however, take considerable more work to show directly that $R(x, y)$ is an element of $\mathfrak{iso}_p$ when we are on a symmetric space.

All of this, of course, works for homogeneous spaces, so we still have to see what is special in the symmetric setting? This means that we have to understand how the map A_p acts on this Lie algebra. We can guess that it should be the identity on $\mathfrak{iso}_p$ and multiplication by -1 on $\mathfrak{t}_p$. To see why this is, let us start by checking how it acts on Iso. We define it as conjugation $\sigma : \text{Iso} \to \text{Iso}$, i.e.,

$$\sigma(g) = A_p \circ g \circ A_p.$$

Thus σ is an automorphism with

$$\sigma(g) = g \text{ iff } g \in \text{Iso}_p,$$
$$\sigma \circ \sigma = id.$$

This doesn't quite characterize σ, but it does characterize the differential as acting in the way we suspected:

$$D\sigma(h) = h \text{ for all } h \in \mathfrak{iso}_p,$$
$$D\sigma(x) = -x \text{ for all } x \in \mathfrak{t}_p.$$

Since σ also fixes Iso_p, it induces a map on Iso/Iso_p whose differential is $-id$ on $T_{\text{Iso}_p}(\text{Iso}/\text{Iso}_p)$. This means that we have found a completely algebraic description of a symmetric space.

Conversely, suppose we have a Lie algebra $\mathfrak{g}$ and a Lie algebra involution $L:$ $\mathfrak{g} \to \mathfrak{g}$. Then we can try to construct a symmetric space as follows: First decompose $\mathfrak{g} = \mathfrak{t} \oplus \mathfrak{k}$ where $\mathfrak{t}$ is the -1 eigenspace for L and $\mathfrak{k}$ is the 1 eigenspace for L. Then observe that $\mathfrak{k}$ is a Lie subalgebra as

$$L[h_1, h_2] = [Lh_1, Lh_2]$$
$$= [h_1, h_2].$$

Note also that for similar reasons,

$$[\mathfrak{k}, \mathfrak{t}] \subset \mathfrak{t},$$
$$[\mathfrak{t}, \mathfrak{t}] \subset \mathfrak{k}.$$

Suppose now that there is a connected compact Lie group K with Lie algebra is $\mathfrak{k}$ such that the action of $\mathfrak{k}$ on $\mathfrak{t}$ yields an action of K on $\mathfrak{t}$. In case K is simply connected this always true. Compactness of K then allows us to choose a Euclidean metric on $\mathfrak{t}$ making the action of K isometric. Then we see that the decomposition $\mathfrak{g} = \mathfrak{t} \oplus \mathfrak{k}$ is exactly of the type described for $\mathfrak{iso}$. Next pick a bi-invariant metric on K such that $\mathfrak{g}$ gets a Euclidean metric. Finally, if we can also choose a Lie group $G \supset K$ whose Lie algebra is $\mathfrak{g}$, then we have constructed a Riemannian manifold G/K. To make it symmetric we need to be able to find an involution σ on G such that $D\sigma = L$. If G and K are chosen so that G/K is simply connected, then σ can be constructed from L. There is a long exact sequence that shows when G/K is simply connected. Assuming that G/K is connected it looks like

$$\pi_1(K) \to \pi_1(G) \to \pi_1(G/K) \to \pi_0(K) \to \pi_0(G) \to 1,$$

where π_0 denotes the set of connected components. As K and G are Lie groups these spaces are in fact groups. From this sequence we see that G/K is simply connected if $\pi_0(K) \to \pi_0(G)$ is an isomorphism and $\pi_1(K) \to \pi_1(G)$ is surjective.

Note that the algebraic approach might not immediately give us the isometry group of the symmetric space. For Euclidean space we can, aside from the standard way using $\mathfrak{g} = \mathfrak{iso}$, also simply use $\mathfrak{g} = \mathbb{R}^n$ and let the involution be multiplication by -1 on all of $\mathfrak{g}$. For $S^3 = O(4)/O(3)$, we see that the algebraic approach might give us the description $S^3 = Spin(4)/Spin(3)$.

It is important to realize that a Lie algebra $\mathfrak{g}$, in itself, does not give rise to a symmetric space. The involution is really an integral part of the construction and does not necessarily exist on a given Lie algebra. The map $-id$ can, for instance, not be used, as it does not preserve the bracket. Rather, it is an *anti-automorphism*. This is particularly interesting if $\mathfrak{g}$ comes from a Lie group G with bi-invariant metric. There the involution $A_e : G \rightarrow G$, which shows that G is symmetric, is the anti-automorphism: $g \rightarrow g^{-1}$, whose differential at e is $-id$. In fact, the algebraic description of G as a symmetric space comes from using $\mathfrak{g} \times \mathfrak{g}$ with $L(X,Y) = (Y,X)$. This will be investigated in the next section.

1.4. Curvature Description of Symmetric Spaces. Given the algebraic nature of symmetric spaces, there must, of course, be a purely algebraic way of computing the curvatures. This is the content of our next lemma. Note that the formula is similar, but identical, to the one that was developed for bi-invariant metrics in chapter 3.

LEMMA 32. *On a symmetric space we have that if* $X, Y, Z \in \mathfrak{t}_p$, *then*

$$R(X,Y)Z = [Z,[X,Y]]$$

at p.

PROOF. By assumption, we suppose that the Killing fields are globally defined and satisfy $\nabla X = \nabla Y = \nabla Z = 0$ at p. The right-hand side does lie in T_pM rather than $\mathfrak{iso}_p$, so we are on the right track. The proof follows from the fact, proved below, that if K is a Killing field on a Riemannian manifold, then

$$\nabla^2_{X,Y}K = -R(K,X)Y.$$

Using this, $\nabla X = \nabla Y = \nabla Z = 0$ at p, and Bianchi's first identity we have

$$
\begin{aligned}
R(X,Y)Z &= R(X,Z)Y - R(Y,Z)X \\
&= -\nabla_Z \nabla_Y X + \nabla_Z \nabla_X Y \\
&= \nabla_Z [X,Y] \\
&= [Z,[X,Y]],
\end{aligned}
$$

which is what we wanted to prove.   □

LEMMA 33. *If* K *is a Killing field on a Riemannian manifold* (M,g), *then*

$$\nabla^2_{X,Y}K = -R(K,X)Y.$$

PROOF. The fact that K is a Killing field is used in the sense that $Y \rightarrow \nabla^2_{X,Y}K$ is skew-adjoint. This follows from skew-symmetric of $Y \rightarrow \nabla_Y K$ in the following way

$$
\begin{aligned}
g(\nabla^2_{X,Y}K,Y) &= g(\nabla_X \nabla_Y K,Y) - g(\nabla_{\nabla_X Y}K,Y) \\
&= g(\nabla_X \nabla_Y K,Y) + g(\nabla_Y K, \nabla_X Y) \\
&= D_X g(\nabla_Y K,Y) \\
&= 0.
\end{aligned}
$$

For any vector field Z we can now compute

$$
\begin{aligned}
g\left(\nabla^2_{X,Y}K, Z\right) &= -g\left(\nabla^2_{X,Z}K, Y\right) \\
&= -g\left(\nabla^2_{Z,X}K, Y\right) - g\left(R\left(X, Z\right)K, Y\right) \\
&= g\left(\nabla^2_{Z,Y}K, X\right) - g\left(R\left(X, Z\right)K, Y\right) \\
&= g\left(\nabla^2_{Y,Z}K, X\right) + g\left(R\left(Z, Y\right)K, X\right) - g\left(R\left(X, Z\right)K, Y\right) \\
&= -g\left(\nabla^2_{Y,X}K, Z\right) + g\left(R\left(Z, Y\right)K, X\right) - g\left(R\left(X, Z\right)K, Y\right) \\
&= -g\left(\nabla^2_{X,Y}K, Z\right) - g\left(R\left(Y, X\right)K, Z\right) \\
&\quad + g\left(R\left(Z, Y\right)K, X\right) - g\left(R\left(X, Z\right)K, Y\right).
\end{aligned}
$$

Thus,

$$
2g\left(\nabla^2_{X,Y}K, Z\right) = -g\left(R\left(Y, X\right)K, Z\right) + g\left(R\left(Z, Y\right)K, X\right) - g\left(R\left(X, Z\right)K, Y\right).
$$

Bianchi's first identity, together with the other symmetry properties of the curvature tensor, now tell us that

$$
\begin{aligned}
&g\left(R\left(Z, Y\right)K, X\right) - g\left(R\left(X, Z\right)K, Y\right) - g\left(R\left(Y, X\right)K, Z\right) \\
&= -g\left(R\left(K, X\right)Y, Z\right) + g\left(R\left(Y, K\right)X, Z\right) + g\left(R\left(X, Y\right)K, Z\right) \\
&= -2g\left(R\left(K, X\right)Y, Z\right).
\end{aligned}
$$

Hence

$$
2g\left(\nabla^2_{X,Y}K, Z\right) = -2g\left(R\left(K, X\right)Y, Z\right),
$$

which yields the desired property. $\qquad\square$

Note that the curvatures now contain all the information about the Lie algebra structure that is needed for defining the brackets of vectors in $\mathfrak{t}_p$. More specifically

$$
\begin{aligned}
[\mathfrak{t}_p, \mathfrak{t}_p] &= \operatorname{span}\left\{R\left(X, Y\right)\right\} \\
&= \mathfrak{r}_p \subset \mathfrak{so}\left(T_pM\right).
\end{aligned}
$$

This can be used to give a more efficient description of a symmetric space than the one using Iso. This description is called the *curvature description*. Suppose (M, g) is a symmetric space and $p \in M$. Let $\mathfrak{r}_p \subset \mathfrak{so}\left(T_pM\right)$ be the Lie algebra generated by the skew-symmetric endomorphisms $R\left(x, y\right) : T_pM \to T_pM$. Then we get a bracket operation on $\mathfrak{c}_p = T_pM \oplus \mathfrak{r}_p$ by defining

$$
\begin{aligned}
[x, y] &= R\left(x, y\right) \in \mathfrak{r}_p \text{ for } x, y \in T_pM, \\
-[r, x] &= [x, r] = r\left(x\right) \in T_pM \text{ for } x \in T_pM \text{ and } r \in \mathfrak{r}_p, \\
[r, s] &= -\left(r \circ s - s \circ r\right) \in \mathfrak{r}_p \text{ for } r, s \in \mathfrak{r}_p.
\end{aligned}
$$

Using Bianchi's first identity for the curvature tensor, one can show that the Jacobi identity holds. Thus, this bracket operation defines a Lie algebra. Also, the linear involution L, which is the identity on $\mathfrak{r}_p$ and multiplication by -1 on T_pM, is a Lie algebra automorphism. Since this construction works on any manifold, we still have to worry about why it reconstructs the symmetric space we started with. However, we saw that $R\left(x, y\right) = [x, y] \in \mathfrak{iso}_p$ on a symmetric space. From this it follows that $(\mathfrak{c}_p, \mathfrak{r}_p) \subset (\mathfrak{iso}, \mathfrak{iso}_p)$, $\mathfrak{iso} \cap \mathfrak{c}_p = \mathfrak{r}_p$, and that L is merely the restriction of $D\sigma$ onto $\mathfrak{c}_p$. It is then easy to see that this new description gives a possibly different way of representing the symmetric space. Below we shall use a holonomy argument to show directly that $R\left(x, y\right) \in \mathfrak{iso}_p$ on a symmetric space.

Note also that given any Lie algebra description $(\mathfrak{g}, L)$ for a symmetric space, we can use this description to compute the curvature tensor.

2. Examples of Symmetric Spaces

We shall here try to explain how some of the above constructions work in the concrete case of the Grassmann manifold and its hyperbolic counterpart. We shall also look at complex Grassmannians, but there we restrict attention to the complex projective space. After these examples we give a formula for the curvature tensor on a compact Lie group with bi-invariant metric. Finally, we briefly discuss the symmetric space structure of $Sl(n)/SO(n)$. The moral of all of these examples and the above Lie algebra descriptions is that one can compute the curvature tensor algebraically without knowing the connection. Based on some general features of these examples, we shall see in the next section that the simplest symmetric spaces have either nonnegative or nonpositive curvature operator.

2.1. The Compact Grassmannian.

First consider the Grassmannian of oriented k-planes in $\mathbb{R}^{k+l}$, denoted by $M = \tilde{G}_k\left(\mathbb{R}^{k+l}\right)$. Thus, each element in M is a k-dimensional subspace of $\mathbb{R}^{k+l}$ together with an orientation. In particular, $\tilde{G}_1\left(\mathbb{R}^{n+1}\right) = S^n$. We shall assume that we have the orthogonal splitting $\mathbb{R}^{k+l} = \mathbb{R}^k \oplus \mathbb{R}^l$, where the distinguished element $p = \mathbb{R}^k$ takes up the first k coordinates in $\mathbb{R}^{k+l}$ and is endowed with its natural positive orientation.

Let us first identify M as a homogeneous space. Observe that $O(k+l)$ acts on $\mathbb{R}^{k+l}$. As such, it maps k-dimensional subspaces to k-dimensional subspaces, and does something uncertain to the orientations of these subspaces. We therefore get that $O(k+l)$ acts transitively on M. This is, however, not the isometry group as the matrix $-I \in SO(k+l)$ acts trivially if k and l are even.

The isotropy group consists of those elements that keep $\mathbb{R}^k$ fixed as well as preserving the orientation. Clearly, the correct isotropy group is then $SO(k) \times O(l) \subset O(k+l)$.

The tangent space at $p = \mathbb{R}^k$ is naturally identified with the space of $k \times l$ matrices $Mat_{k \times l}$, or equivalently, with $\mathbb{R}^k \otimes \mathbb{R}^l$. To see this, just observe that any k-dimensional subspace of $\mathbb{R}^{k+l}$ can be represented as a linear graph over $\mathbb{R}^k$ with values in the orthogonal complement $\mathbb{R}^l$. The isotropy action of $SO(k) \times O(l)$ on $Mat_{k \times l}$ now acts as follows:

$$SO(k) \times O(l) \times Mat_{k \times l} \quad \rightarrow \quad Mat_{k \times l},$$
$$(A, B, X) \quad \rightarrow \quad AXB^{-1} = AXB^t.$$

If we define X to be the matrix that is 1 in the $(1,1)$ entry and otherwise zero, then $AXB^t = A_1\left(B^1\right)^t$, where A_1 is the first column of A and B^1 is the first column of B. Thus, the orbit of X, under the isotropy action, generates a basis for $Mat_{k \times l}$ but does not cover all of the space. This is an example of an irreducible action on Euclidean space that is not transitive on the unit sphere. The representation, when seen as acting on $\mathbb{R}^k \otimes \mathbb{R}^l$, is denoted by $SO(k) \otimes O(l)$.

To see that M is a symmetric space, we have to show that the isotropy group contains the required involution. On the tangent space $T_pM = Mat_{k \times l}$ it is supposed to act by multiplication by -1. Thus, we have to find $(A, B) \in SO(k) \times O(l)$ such that for all X,

$$AXB^t = -X.$$

Clearly, we can just set

$$A = I_k,$$
$$B = -I_l.$$

Depending on k and l, other choices are possible, but they will act in the same way.

We have now exhibited M as a symmetric space, although we didn't use the isometry group of the space. Instead, we used a finite covering of the isometry group and then had some extra elements that acted trivially.

Let us now give the Lie algebra description and compute the curvature tensor. Since we actually found the isometry group modulo a finite covering, we see that

$$\mathfrak{iso} = \mathfrak{so}\,(k + l),$$
$$\mathfrak{iso}_p = \mathfrak{so}\,(k) \times \mathfrak{so}\,(l).$$

We shall use the block decomposition of matrices in $\mathfrak{so}\,(k + l)$:

$$X = \begin{pmatrix} X_1 & B \\ -B^t & X_2 \end{pmatrix},$$
$$X_1 \in \mathfrak{so}\,(k)\,, X_2 \in \mathfrak{so}\,(l)\,, B \in Mat_{k \times l}.$$

If we set

$$\mathfrak{t}_p = \left\{ \begin{pmatrix} 0 & B \\ -B^t & 0 \end{pmatrix} : B \in Mat_{k \times l} \right\},$$

then we have an orthogonal decomposition:

$$\mathfrak{so}\,(k + l) = \mathfrak{t}_p \oplus \mathfrak{so}\,(k) \oplus \mathfrak{so}\,(l)\,,$$

where we can identify $\mathfrak{t}_p = T_p M$. The inner product on $\mathfrak{t}_p$ is the standard Euclidean metric defined by

$$
\begin{aligned}
\left\langle \begin{pmatrix} 0 & A \\ -A^t & 0 \end{pmatrix}, \begin{pmatrix} 0 & B \\ -B^t & 0 \end{pmatrix} \right\rangle &= \mathrm{tr}\left(\begin{pmatrix} 0 & A \\ -A^t & 0 \end{pmatrix} \begin{pmatrix} 0 & B \\ -B^t & 0 \end{pmatrix}^t \right) \\
&= \mathrm{tr}\left(\begin{pmatrix} 0 & A \\ -A^t & 0 \end{pmatrix} \begin{pmatrix} 0 & -B \\ B^t & 0 \end{pmatrix} \right) \\
&= \mathrm{tr}\begin{pmatrix} AB^t & 0 \\ 0 & A^t B \end{pmatrix} \\
&= \mathrm{tr}\,(AB^t) + \mathrm{tr}\,(A^t B) \\
&= 2\mathrm{tr}\,(AB^t).
\end{aligned}
$$

Thus, it is twice the usual Euclidean metric on $\mathbb{R}^{k \cdot l}$ that we used above. But that, of course, does not change matters much.

We now have to compute Lie brackets of elements in $\mathfrak{t}_p$ and then see how $\mathfrak{so}\,(k) \oplus \mathfrak{so}\,(l)$ acts on $\mathfrak{t}_p$ in order to find the curvature tensor. For $A, B \in \mathfrak{t}_p$ we have

$$
\begin{aligned}
[A, B] &= \begin{pmatrix} 0 & A \\ -A^t & 0 \end{pmatrix} \begin{pmatrix} 0 & B \\ -B^t & 0 \end{pmatrix} - \begin{pmatrix} 0 & B \\ -B^t & 0 \end{pmatrix} \begin{pmatrix} 0 & A \\ -A^t & 0 \end{pmatrix} \\
&= \begin{pmatrix} -AB^t & 0 \\ 0 & -A^t B \end{pmatrix} - \begin{pmatrix} -BA^t & 0 \\ 0 & -B^t A \end{pmatrix} \\
&= \begin{pmatrix} BA^t - AB^t & 0 \\ 0 & B^t A - A^t B \end{pmatrix} \in \mathfrak{so}\,(k) \oplus \mathfrak{so}\,(l)\,.
\end{aligned}
$$

Observe that there is a basis for $\mathfrak{so}(k) \oplus \mathfrak{so}(l)$ that can be written in this way, so there will be no difference between the curvature and isometry descriptions. Now take $C \in \mathfrak{t}_p$ and compute

$$
\begin{aligned}
R(A, B) C &= [C, [A, B]] \\
&= \left[\begin{pmatrix} 0 & C \\ -C^t & 0 \end{pmatrix}, \begin{pmatrix} BA^t - AB^t & 0 \\ 0 & B^t A - A^t B \end{pmatrix} \right] \\
&= \begin{pmatrix} 0 & C(B^t A - A^t B) \\ & -(BA^t - AB^t)C \\ -(A^t B - B^t A)C^t & \\ +C^t(AB^t - BA^t) & 0 \end{pmatrix}.
\end{aligned}
$$

This does not seem very illuminating, so let us find the sectional curvatures by considering the directional curvature transformation

$$
R(A, B) B = \begin{pmatrix} 0 & BB^t A - 2BA^t B + AB^t B \\ -A^t BB^t + 2B^t AB^t - B^t BA^t & 0 \end{pmatrix}.
$$

We now have to take the inner product with A giving us

$$
\begin{aligned}
\langle R(A, B) B, A \rangle &= \operatorname{tr}\left(BB^t AA^t - 2BA^t BA^t + AB^t BA^t \right) \\
&\quad + \operatorname{tr}\left(A^t BB^t A - 2B^t AB^t A + B^t BA^t A \right) \\
&= \operatorname{tr}\left(BB^t AA^t \right) - 2\operatorname{tr}\left(BA^t BA^t \right) + \operatorname{tr}\left(AB^t BA^t \right) \\
&\quad + \operatorname{tr}\left(A^t BB^t A \right) - 2\operatorname{tr}\left(B^t AB^t A \right) + \operatorname{tr}\left(B^t BA^t A \right) \\
&= \operatorname{tr}\left(BA^t AB^t \right) - 2\operatorname{tr}\left(BA^t BA^t \right) + \operatorname{tr}\left(AB^t BA^t \right) \\
&\quad + \operatorname{tr}\left(A^t BB^t A \right) + \operatorname{tr}\left(B^t AA^t B \right) - 2\operatorname{tr}\left(B^t AB^t A \right) \\
&= \langle BA^t, BA^t \rangle - 2\langle BA^t, AB^t \rangle + \langle AB^t, AB^t \rangle \\
&\quad + \langle A^t B, A^t B \rangle - 2\langle A^t B, B^t A \rangle + \langle B^t A, B^t A \rangle \\
&= \left| BA^t - AB^t \right|^2 + \left| A^t B - B^t A \right|^2 \geq 0.
\end{aligned}
$$

Here we recklessly used Euclidean norms for matrices in various different spaces. The conclusion is that the sectional curvatures are all ≥ 0.

When $k = 1$ or $l = 1$, it is easy to see that one gets a metric of constant positive curvature. Otherwise, the metric will have some zero sectional curvatures.

2.2. The Hyperbolic Grassmannian.

Let us now turn to the hyperbolic analogue. In the Euclidean space $\mathbb{R}^{k,l}$ we use, instead of the positive definite inner product $v^t \cdot w$, the quadratic form:

$$
\begin{aligned}
v^t I_{k,l} w &= v^t \begin{pmatrix} -I_k & 0 \\ 0 & I_l \end{pmatrix} w \\
&= -\sum_{i=1}^{k} v_i w_i + \sum_{i=k+1}^{k+l} v_i w_i.
\end{aligned}
$$

The group of linear transformations that preserve this form is denoted by $O(k, l)$. These transformations are defined by the relation

$$
X \cdot I_{k,l} \cdot X^t = I_{k,l}.
$$

Note that if $k, l > 0$, then $O(k, l)$ is not compact. But it clearly contains the (maximal) compact subgroup $O(k) \times O(l)$.

The Lie algebra $\mathfrak{so}(k, l)$ of $O(k, l)$ consists of the matrices satisfying

$$Y \cdot I_{k,l} + I_{k,l} \cdot Y^t = 0.$$

If we use the same block decomposition for Y as we did for $I_{k,l}$ above, then we have that it looks like

$$Y = \begin{pmatrix} A & B \\ B^t & C \end{pmatrix},$$

$$A \in \mathfrak{so}(k),$$

$$C \in \mathfrak{so}(l).$$

$$B \in Mat_{k \times l}$$

We now consider only those (oriented) k-dimensional subspaces of $\mathbb{R}^{k,l}$ on which this quadratic form generates a positive definite inner product. This space is the hyperbolic Grassmannian $M = \tilde{G}_k(\mathbb{R}^{k,l})$. Our selected point is as before $p = \mathbb{R}^k$. One can easily see that topologically: $\tilde{G}_k(\mathbb{R}^{k,l})$ is an open subset of $\tilde{G}_k(\mathbb{R}^{k+l})$. The metric on this space is another story, however. Clearly, $O(k, l)$ acts transitively on M, and those elements that fix p are of the form $SO(k) \times O(l)$. One can, as before, find the desired involution, and thus exhibit M as a symmetric space. Again some of these elements act trivially, but at the Lie algebra level this makes no difference. Thus, we have

$$\mathfrak{iso} = \mathfrak{so}(k, l),$$

$$\mathfrak{iso}_p = \mathfrak{so}(k) \times \mathfrak{so}(l),$$

$$\mathfrak{t}_p = \left\{ \begin{pmatrix} 0 & A \\ A^t & 0 \end{pmatrix} : A \in Mat_{k \times l} \right\}.$$

On $\mathfrak{t}_p$ we use the Euclidean metric

$$\left\langle \begin{pmatrix} 0 & A \\ A^t & 0 \end{pmatrix}, \begin{pmatrix} 0 & B \\ B^t & 0 \end{pmatrix} \right\rangle = \mathrm{tr}\left(\begin{pmatrix} 0 & A \\ A^t & 0 \end{pmatrix} \cdot \begin{pmatrix} 0 & B \\ B^t & 0 \end{pmatrix}^t \right)$$

$$= \mathrm{tr}\left(\begin{pmatrix} 0 & A \\ A^t & 0 \end{pmatrix} \cdot \begin{pmatrix} 0 & B \\ B^t & 0 \end{pmatrix} \right)$$

$$= \mathrm{tr}\left(\begin{pmatrix} AB^t & 0 \\ 0 & A^t B \end{pmatrix} \right)$$

$$= \mathrm{tr}\left(AB^t \right) + \mathrm{tr}\left(A^t B \right)$$

$$= 2\mathrm{tr}\left(AB^t \right).$$

So while $\mathfrak{t}_p$ looks different, we seem to use the same metric.

On $\mathfrak{t}_p$ we have the Lie bracket

$$\left[\begin{pmatrix} 0 & A \\ A^t & 0 \end{pmatrix}, \begin{pmatrix} 0 & B \\ B^t & 0 \end{pmatrix}\right]$$

$$= \begin{pmatrix} 0 & A \\ A^t & 0 \end{pmatrix}\begin{pmatrix} 0 & B \\ B^t & 0 \end{pmatrix} - \begin{pmatrix} 0 & B \\ B^t & 0 \end{pmatrix}\begin{pmatrix} 0 & A \\ A^t & 0 \end{pmatrix}$$

$$= \begin{pmatrix} AB^t & 0 \\ 0 & A^tB \end{pmatrix} - \begin{pmatrix} BA^t & 0 \\ 0 & B^tA \end{pmatrix}$$

$$= \begin{pmatrix} AB^t - BA^t & 0 \\ 0 & A^tB - B^tA \end{pmatrix} \in \mathfrak{so}(k) \oplus \mathfrak{so}(l).$$

This is the negative of what we had before. We can now compute the curvature tensor:

$$R(A,B)C = [C,[A,B]]$$

$$= \left[\begin{pmatrix} 0 & C \\ C^t & 0 \end{pmatrix}, \begin{pmatrix} AB^t - BA^t & 0 \\ 0 & A^tB - B^tA \end{pmatrix}\right]$$

$$= \begin{pmatrix} 0 & C(A^tB - B^tA) \\ & -(AB^t - BA^t)C \\ C^t(AB^t - BA^t) & 0 \\ -(A^tB - B^tA)C^t & \end{pmatrix}.$$

If we let $C = B$ and compute the sectional curvature as before, we arrive at

$$\langle R(A,B)B, A\rangle$$

$$= \operatorname{tr}\left(\begin{pmatrix} 0 & 2BA^tB \\ & -BB^tA - AB^tB \\ 2B^tAB^t & 0 \\ -B^tBA^t - A^tBB^t & \end{pmatrix}\cdot\begin{pmatrix} 0 & A \\ A^t & 0 \end{pmatrix}\right)$$

$$= \operatorname{tr}\begin{pmatrix} 2BA^tBA^t & 0 \\ -BB^tAA^t - AB^tBA^t & \\ 0 & 2B^tAB^tA \\ & -B^tBA^tA - A^tBB^tA \end{pmatrix}.$$

This is exactly the negative of the expression we got in the compact case. Hence, the hyperbolic Grassmannians have nonpositive curvature. When $k = 1$, we have reconstructed the hyperbolic space together with its isometry group.

2.3. Complex Projective Space Revisited. We shall view complex projective space as a complex Grassmannian. Namely, let $M = \mathbb{C}P^n = G_1(\mathbb{C}^{n+1})$, i.e., the complex lines in $\mathbb{C}^{n+1}$. More generally we can consider $G_k(\mathbb{C}^{k+l})$ and of course the hyperbolic counterparts $G_k(\mathbb{C}^{k,l})$, but we leave this to the reader.

The group $U(n+1) \subset SO(2n+2)$ consists of those orthogonal transformations that also preserve the complex structure. If we use complex coordinates, then the Hermitian metric on $\mathbb{C}^{n+1}$ can be written as

$$z^*w = \sum \bar{z}_i w_i,$$

where as usual, $A^* = \bar{A}^t$ is the conjugate transpose. Thus, the elements of $U(n+1)$ satisfy

$$A^{-1} = A^*.$$

As with the Grassmannian, $U(n+1)$ acts on M, but this time, all of the transformations of the form aI, where $a\bar{a} = 1$, act trivially. Thus, we restrict attention to $SU(n+1)$, which still acts transitively, but now with a finite kernel consisting of those aI such that $a^{n+1} = 1$.

If we let $p = \mathbb{C}$ be the first coordinate axis, then the isotropy group is given by $S(U(1) \times U(n))$, i.e., the matrices in $U(1) \times U(n)$ of determinant 1. This group is naturally isomorphic to $U(n)$ via the map

$$A \to \begin{pmatrix} \det A^{-1} & 0 \\ 0 & A \end{pmatrix}.$$

The involution that makes M symmetric is then given by

$$\begin{pmatrix} (-1)^n & 0 \\ 0 & -I_n \end{pmatrix}.$$

Let us now pass to the Lie algebra level in order to compute the curvature tensor. From above, we have

$$\begin{aligned} \mathfrak{iso} &= \mathfrak{su}(n+1) = \{A : A = -A^*, \operatorname{tr} A = 0\}, \\ \mathfrak{iso}_p &= \mathfrak{u}(n) = \{B : B = -B^*\}. \end{aligned}$$

The inclusion looks like

$$B \to \begin{pmatrix} -\operatorname{tr} B & 0 \\ 0 & B \end{pmatrix}.$$

Thus we should write elements of $\mathfrak{su}(n+1)$ in the form

$$\begin{pmatrix} -\operatorname{tr} B & -z^* \\ z & B \end{pmatrix},$$

and then identify

$$\mathfrak{t}_p = \left\{ \begin{pmatrix} 0 & -z^* \\ z & 0 \end{pmatrix} : z \in \mathbb{C}^n \right\}$$

and use the inner product

$$\begin{aligned} & \left\langle \begin{pmatrix} 0 & -z^* \\ z & 0 \end{pmatrix}, \begin{pmatrix} 0 & -w^* \\ w & 0 \end{pmatrix} \right\rangle \\ &= \frac{1}{2} \operatorname{tr} \left(\begin{pmatrix} 0 & -z^* \\ z & 0 \end{pmatrix} \cdot \begin{pmatrix} 0 & -w^* \\ w & 0 \end{pmatrix}^* \right) \\ &= \frac{1}{2} \operatorname{tr} \begin{pmatrix} 0 & -z^* \\ z & 0 \end{pmatrix} \cdot \begin{pmatrix} 0 & w^* \\ -w & 0 \end{pmatrix} \\ &= \frac{1}{2} \operatorname{tr} \begin{pmatrix} z^* w & 0 \\ 0 & zw^* \end{pmatrix} \\ &= \frac{1}{2} \left(z^* w + \operatorname{tr}(zw^*) \right) \\ &= \frac{1}{2} \left(z^* w + w^* z \right) \\ &= \operatorname{Re} \langle z, w \rangle. \end{aligned}$$

Here $\langle z, w \rangle$ is the usual Hermitian inner product on $\mathbb{C}^n$, which is conjugate linear in the w variable.

For the curvature tensor we first compute the Lie bracket on $\mathfrak{t}_p$:

$$\left[\begin{pmatrix} 0 & -z^* \\ z & 0 \end{pmatrix}, \begin{pmatrix} 0 & -w^* \\ w & 0 \end{pmatrix} \right]$$

$$= \begin{pmatrix} 0 & -z^* \\ z & 0 \end{pmatrix} \begin{pmatrix} 0 & -w^* \\ w & 0 \end{pmatrix} - \begin{pmatrix} 0 & -w^* \\ w & 0 \end{pmatrix} \begin{pmatrix} 0 & -z^* \\ z & 0 \end{pmatrix}$$

$$= \begin{pmatrix} -z^*w & 0 \\ 0 & -zw^* \end{pmatrix} - \begin{pmatrix} -w^*z & 0 \\ 0 & -wz^* \end{pmatrix}$$

$$= \begin{pmatrix} w^*z - z^*w & 0 \\ 0 & wz^* - zw^* \end{pmatrix}.$$

Then, we get

$$R(z,w)\,w = \left[\begin{pmatrix} 0 & -w^* \\ w & 0 \end{pmatrix}, \begin{pmatrix} w^*z - z^*w & 0 \\ 0 & wz^* - zw^* \end{pmatrix} \right]$$

$$= \begin{pmatrix} 0 & -w^* \\ w & 0 \end{pmatrix} \begin{pmatrix} w^*z - z^*w & 0 \\ 0 & wz^* - zw^* \end{pmatrix}$$

$$- \begin{pmatrix} w^*z - z^*w & 0 \\ 0 & wz^* - zw^* \end{pmatrix} \begin{pmatrix} 0 & -w^* \\ w & 0 \end{pmatrix}$$

$$= \begin{pmatrix} 0 & w^*(zw^* - wz^*) \\ & + (w^*z - z^*w)w^* \\ w(w^*z - z^*w) & 0 \\ + (zw^* - wz^*)w & \end{pmatrix}.$$

Now identify $\mathfrak{t}_p$ with $\mathbb{C}^n$ and observe that

$$R(z,w)\,w = w(w^*z - z^*w) + (zw^* - wz^*)\,w.$$

To compute the sectional curvatures we need to pick an orthonormal basis z, w for a plane. This means that $|z|^2 = |w|^2 = 1$ and $\mathrm{Re}\,\langle z, w \rangle = 0$. The sectional curvature of the plane spanned by z, w is therefore

$$\begin{aligned} \sec(z,w) &= \mathrm{Re}\,\langle w(w^*z - z^*w) + (zw^* - wz^*)\,w, z \rangle \\ &= \mathrm{Re}z^*w(w^*z - z^*w) + \mathrm{Re}z^*(zw^* - wz^*)\,w \\ &= |\langle w, z \rangle|^2 - 2\mathrm{Re}\left(\langle w, z \rangle^2\right) + 1 \\ &= 1 + 3\,|\mathrm{Im}\,\langle w, z \rangle|^2. \end{aligned}$$

Thus, if z, w are orthogonal with respect to the Hermitian metric, i.e., $\langle z, w \rangle = 0$, then $\sec(z,w) = 1$, while if, e.g., $w = iz$, then we get that the sectional curvature of a complex line is $\sec(z, iz) = 4$. Since $|\mathrm{Im}\,\langle w, z \rangle| \le |z|\,|w| = 1$, all other curvatures lie between these two values. Thus we have shown that the complex projective space is quarter pinched. This should be compared to our discussion in chapter 3 where we established a similar formula for the sectional curvature using O'Neill's formula.

2.4. Lie Groups with Bi-Invariant Metrics. In a more abstract vein, let us see how Lie groups with bi-invariant metrics behave when considered as symmetric spaces. To this end, suppose we have a compact Lie group G with a bi-invariant metric. As usual, the Lie algebra $\mathfrak{g}$ of G is identified with T_eG and is also the set of left-invariant vector fields on G. The object is then to find an appropriate Lie algebra description.

The claim is that a Lie algebra description is $(\mathfrak{g} \oplus \mathfrak{g}, L)$, where $L(X,Y) = (Y,X)$. Clearly, the diagonal $\mathfrak{g}^\Delta = \{(X,X) : X \in \mathfrak{g}\}$ is the 1-eigenspace, while the complement $\mathfrak{g}^\perp = \{(X,-X) : X \in \mathfrak{g}\}$ is the -1-eigenspace. Thus, we should identify

$$\mathfrak{k} = \mathfrak{g}^\Delta \cong \mathfrak{g},$$
$$\mathfrak{t} = \mathfrak{g}^\perp.$$

We already know that $\mathfrak{g}$ corresponds to the compact Lie group G, so we are simply saying that

$$G = (G \times G) / G^\Delta.$$

On $\mathfrak{t}$, the Lie bracket looks like

$$
\begin{aligned}
[(X,-X),(Y,-Y)] &= ([X,Y],[-X,-Y]) \\
&= ([X,Y],[X,Y]) \in \mathfrak{k}.
\end{aligned}
$$

Thus, the curvature tensor can be computed as follows:

$$
\begin{aligned}
R(X,Y)Z &= R((X,-X),(Y,-Y))(Z,-Z) \\
&= [(Z,-Z),([X,Y],[X,Y])] \\
&= ([Z,[X,Y]],-[Z,[X,Y]]) \in \mathfrak{t}.
\end{aligned}
$$

Hence, we arrive at that the formula

$$R(X,Y)Z = [Z,[X,Y]]$$

for the curvature tensor on a compact Lie group with bi-invariant metric. This formula looks exactly like the one for the curvature of a symmetric space, but it is interpreted differently. Another curious feature is that if one computes the curvature tensor in the standard way using a bi-invariant metric, then the formula has a factor $\frac{1}{4}$ on it (see chapter 3). The reason for this discrepancy is that left- or right-invariant vector fields do not lie in $\mathfrak{t}$ unless they are parallel. Conversely, a Killing field from $\mathfrak{t}$ is left- or right-invariant only when it is parallel.

2.5. $Sl(n)/SO(n)$. The manifold is the quotient space of the $n \times n$ matrices with determinant 1 by the orthogonal matrices. The Lie algebra of $Sl(n)$ is

$$\mathfrak{sl}(n) = \{X \in Mat_{n \times n} : \mathrm{tr} X = 0\}.$$

This Lie algebra is naturally divided up into symmetric and skew-symmetric matrices

$$\mathfrak{sl}(n) = \mathfrak{t} \oplus \mathfrak{so}(n),$$

where $\mathfrak{t}$ consists of the symmetric matrices. On $\mathfrak{t}$ we can use the usual Euclidean metric. The involution is obviously given by $-id$ on $\mathfrak{t}$ and id on $\mathfrak{so}(n)$. Holistically, this is the map

$$L(X) = -X^t.$$

Let us now grind out the curvature tensor:

$$
\begin{aligned}
R(X,Y)Z &= [Z,[X,Y]] \\
&= Z[X,Y] - [X,Y]Z \\
&= ZXY - ZYX - XYZ + YXZ
\end{aligned}
$$

This yields

$$
\begin{aligned}
g\left(R\left(X,Y\right)Z,W\right) &= \operatorname{tr}\left([Z,[X,Y]]W^{t}\right) \\
&= \operatorname{tr}\left([Z,[X,Y]]W\right) \\
&= \operatorname{tr}\left(Z\left[X,Y\right]W-\left[X,Y\right]ZW\right) \\
&= \operatorname{tr}\left(WZ\left[X,Y\right]-\left[X,Y\right]ZW\right) \\
&= \operatorname{tr}\left(\left[X,Y\right]\left[W,Z\right]\right) \\
&= -\operatorname{tr}\left(\left[X,Y\right]\left[W,Z\right]^{t}\right) \\
&= -g\left(\left[X,Y\right],\left[W,Z\right]\right).
\end{aligned}
$$

In particular, the sectional curvatures must be nonpositive.

3. Holonomy

First we discuss holonomy for general manifolds and the de Rham decomposition theorem. We then use holonomy to give a brief discussion of how symmetric spaces can be classified according to whether they are compact or not.

3.1. The Holonomy Group. Let (M,g) be a Riemannian n-manifold. If $c:[a,b]\to M$ is a smooth curve, then

$$
P_{c(a)}^{c(b)}:T_{c(a)}M\to T_{c(b)}M
$$

denotes the effect of parallel translating a vector in $T_{c(a)}M$ along c to $T_{c(b)}M$. This property will in general depend not only on the endpoints of the curve, but also on the actual curve. We can generalize this to work for piecewise smooth curves by breaking up the process at the breakpoints in the curve.

Suppose now the curve is a loop, i.e., $c(a)=c(b)=p$. Then parallel translation gives an isometry on T_pM. The set of all such isometries is called the *holonomy group* at p and is denoted by $\operatorname{Hol}_p=\operatorname{Hol}_p(M,g)$. One can easily see that this forms a subgroup of $O(T_pM)=O(n)$. Moreover, it is actually a Lie group, which is often a closed subgroup of $O(n)$. We also have the *restricted holonomy group* $\operatorname{Hol}_p^0=\operatorname{Hol}_p^0(M,g)$, which is the connected normal subgroup that comes from using only contractible loops. It can be shown that the restricted holonomy group is always compact. Here are some elementary properties that are easy to establish:

(a) $\operatorname{Hol}_p(\mathbb{R}^n)=\{1\}$.

(b) $\operatorname{Hol}_p(S^n(r))=SO(n)$.

(c) $\operatorname{Hol}_p(H^n)=SO(n)$.

(d) $\operatorname{Hol}_p(M,g)\subset SO(n)$ iff M is orientable.

(e) $\operatorname{Hol}_p\left(\tilde{M},\tilde{g}\right)=\operatorname{Hol}_p^0\left(\tilde{M},\tilde{g}\right)=\operatorname{Hol}_p^0(M,g)$, where $\tilde{M}$ is the universal covering of M.

(f) $\operatorname{Hol}_{(p,q)}(M_1\times M_2,g_1+g_2)=\operatorname{Hol}_p(M_1,g_1)\times\operatorname{Hol}_q(M_2,g_2)$.

(g) $\operatorname{Hol}_p(M,g)$ is conjugate to $\operatorname{Hol}_q(M,g)$ via parallel translation along any curve from p to q.

(h) A tensor on (M,g) is parallel iff it is invariant under the restricted holonomy group; e.g., if ω is a 2-form, then $\nabla\omega=0$ iff $\omega(Pv,Pw)=\omega(v,w)$ for all $P\in\operatorname{Hol}_p^0(M,g)$ and $v,w\in T_pM$.

We are now ready to study how the Riemannian manifold decomposes according to the holonomy. Guided by (f) we see that Cartesian products are reflected in a product structure at the level of the holonomy. Furthermore, (g) shows that if the holonomy decomposes at just one point, then it decomposes everywhere.

To make things more precise, let us consider the action of Hol_p^0 on T_pM. If $E \subset T_pM$ is an invariant subspace, i.e., $\mathrm{Hol}_p^0(E) \subset E$, then the orthogonal complement is also preserved, i.e., $\mathrm{Hol}_p^0(E^\perp) \subset E^\perp$. Thus, T_pM decomposes into irreducible invariant subspaces:

$$T_pM = E_1 \oplus \cdots \oplus E_k.$$

Here, irreducible means that there are no nontrivial invariant subspaces inside E_i. Since parallel translation around loops at p preserves this decomposition, we see that parallel translation along any curve from p to q preserves this decomposition. Thus, we get a global decomposition of the tangent bundle into distributions, each of which is invariant under parallel translation:

$$TM = \eta_1 \oplus \cdots \oplus \eta_k.$$

With this we can prove de Rham's decomposition theorem.

THEOREM 56. (de Rham, 1952) *If we decompose the tangent bundle of a Riemannian manifold (M, g) into irreducible components according to the holonomy*

$$TM = \eta_1 \oplus \cdots \oplus \eta_k,$$

then around each point $p \in M$ there is a neighborhood U that has a product structure of the form

$$
\begin{aligned}
(U, g) &= (U_1 \times \cdots \times U_k, g_1 + \cdots + g_k), \\
TU_i &= \eta_{i|U_i}.
\end{aligned}
$$

Moreover, if (M, g) is simply connected and complete, then there is a global splitting

$$
\begin{aligned}
(M, g) &= (M_1 \times \cdots \times M_k, g_1 + \cdots + g_k), \\
TM_i &= \eta_i.
\end{aligned}
$$

PROOF. Given the decomposition into parallel distributions, we first observe that each of the distributions must be integrable. Thus, we do get a local splitting into submanifolds at the manifold level. To see that the metric splits as well, just observe that the submanifolds are totally geodesic, as their tangent spaces are invariant under parallel translation. This gives the local splitting. The global result is not just a trivial analytic continuation argument. Apparently, one must understand how simple connectivity forces the maximal integral submanifolds to be embedded submanifolds. Instead of going that route, let M_i be the maximal integral submanifolds through some fixed $p \in M$, and define the abstract Riemannian manifold

$$(M_1 \times \cdots \times M_k, g_1 + \cdots + g_k).$$

Locally, (M, g) and

$$(M_1 \times \cdots \times M_k, g_1 + \cdots + g_k)$$

are isometric to each other. Given that (M, g) is complete and the M_is are totally geodesic we see that also

$$(M_1 \times \cdots \times M_k, g_1 + \cdots + g_k)$$

is complete. Therefore, if M is simply connected, we can find an isometric embedding

$$(M, g) \to (M_1 \times \cdots \times M_k, g_1 + \cdots + g_k).$$

Completeness insures us that the map is onto and in fact a Riemannian covering map. We have then shown that M is isometric to the universal covering of

$$(M_1 \times \cdots \times M_k, g_1 + \cdots + g_k),$$

which is the product manifold

$$\left(\tilde{M}_1 \times \cdots \times \tilde{M}_k, \tilde{g}_1 + \cdots + \tilde{g}_k\right)$$

with the induced pull-back metric.   □

Given this decomposition it is reasonable, when studying classification problems for Riemannian manifolds, to study only those Riemannian manifolds that are *irreducible*, i.e., those where the holonomy has no invariant subspaces. Guided by this we have some nice characterizations of Einstein manifolds.

THEOREM 57. *If (M, g) is an irreducible Riemannian manifold with parallel Ricci tensor, then (M, g) is Einstein. In particular, irreducible symmetric spaces are Einstein.*

PROOF. The fact that $\nabla \mathrm{Ric} = 0$ means that the Ricci tensor is invariant under parallel translation. Now decompose

$$T_p M = E_1 \oplus \cdots \oplus E_k$$

into the eigenspaces for $\mathrm{Ric} : T_p M \to T_p M$ with respect to distinct eigenvalues $\lambda_1 < \cdots < \lambda_k$. As above, we can now parallel translate these eigenspaces to get a global decomposition

$$TM = \eta_1 \oplus \cdots \oplus \eta_k$$

into parallel distributions, with the property that

$$\mathrm{Ric}_{|\eta_i} = \lambda_i \cdot I$$

But then the decomposition theorem tells us that (M, g) is reducible unless there is only one eigenvalue.   □

3.2. Rough Classification of Symmetric Spaces. Guided by our examples and the results on holonomy, we can now try to classify irreducible symmetric spaces. They seem to come in three groups.

Compact Type: If the Einstein constant is positive, then it follows from Myers' diameter bound (chapter 6) that the space is compact. In this case one can show that the curvature operator is nonnegative.

Flat Type: If the space is Ricci flat, then it follows that it must be flat. In case the space is compact, this is immediate from Bochner's theorem, while if the space is noncompact and complete a little more work is needed. Thus, the only Ricci flat irreducible examples are S^1 and $\mathbb{R}^1$.

Noncompact Type: If the Einstein constant is negative, then it follows from Bochner's theorem on Killing fields that the space is noncompact. In this case, one can show that the curvature operator is nonpositive.

We won't give a complete list of all irreducible symmetric spaces, but one interesting feature is that they come in compact/noncompact dual pairs, as described in the above lists. Also, there is a further subdivision. Among the compact types there are Lie groups with bi-invariant metrics and then all the others. Similarly, in the noncompact regime there are the duals to the bi-invariant metrics and then the rest. This gives us the following division:

Type I: Compact irreducible symmetric spaces of the form G/K where G is a compact simple real Lie group and K a maximal compact subgroup, e.g.,

$$SO\,(k+l)\,/\,(SO\,(k) \times SO\,(l))\,.$$

Type II: Compact irreducible symmetric spaces G, where G is a compact simple real Lie group with a bi-invariant metric, e.g.,

$$SO\,(n)\,.$$

Type III: Non-compact symmetric spaces G/K, where G is a noncompact simple real Lie group and K a maximal compact subgroup, e.g.,

$$SO\,(k,l)\,/\,(SO\,(k) \times SO\,(l)) \ \text{ or } \ Sl\,(n)\,/SO\,(n)\,.$$

Type IV: Noncompact symmetric spaces G/K, where K is a compact simple real Lie group and G its complexification, e.g.,

$$SO\,(n,\mathbb{C})\,/SO\,(n)\,.$$

The algebraic difference between compact and noncompact can be seen by looking at the examples above. There we saw that in the compact case $\mathfrak{t}$ consists of skew-symmetric matrices, while in the noncompact case $\mathfrak{t}$ consists of symmetric matrices. Thus, the metric looks like

$$g\,(X,Y) = \mp\mathrm{tr}\,(XY)\,,$$

where the minus is for the compact case and the plus for the noncompact case. It is this difference that ultimately gave us the different sign for the curvatures. Before getting to the curvature we see that for $X, Y \in \mathfrak{t}$ and $K \in \mathfrak{k}$,

$$
\begin{aligned}
g\,([X,K]\,,Y) &= \mp\mathrm{tr}\,((XK - KX)\,Y) \\
&= \mp\,(\mathrm{tr}XKY - \mathrm{tr}KXY) \\
&= \mp\,(\mathrm{tr}KYX - \mathrm{tr}KXY) \\
&= \pm\mathrm{tr}\,(K\,(XY - YX)) \\
&= \pm\mathrm{tr}\,(K\,[X,Y]) \\
&= \mp\,\langle K,[X,Y]\rangle\,,
\end{aligned}
$$

where for elements of $\mathfrak{k}$ we use that they are always skew-symmetric, and therefore their inner product is given by

$$\langle K_1, K_2\rangle = -\mathrm{tr}\,(K_1 K_2)\,.$$

Using this, one can see that

$$
\begin{aligned}
g\,(R\,(X,Y)\,Z,W) &= g\,([Z,[X,Y]]\,,W) \\
&= \mp\,\langle [X,Y]\,,[Z,W]\rangle\,.
\end{aligned}
$$

With this information we can compute the diagonal terms for the curvature operator

$$
\begin{aligned}
g\left(\mathfrak{R}\left(\sum X_i \wedge Y_i\right),\left(\sum X_i \wedge Y_i\right)\right) &= \sum g\,(R\,(X_i,Y_i)\,Y_j,X_j) \\
&= \mp\sum \langle [X_i,Y_i]\,,[Y_j,X_j]\rangle \\
&= \pm\left|\sum [X_i,Y_i]\right|^2
\end{aligned}
$$

and conclude that it is either nonnegative or nonpositive according to type. This seems to have been noticed for the first time in the literature in [**42**]. It also means

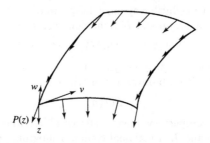

Figure 8.1

that Tachibana's result from the last chapter is not vacuous. In fact, it gives a characterization of symmetric spaces of compact type.

Note that since compact type symmetric spaces have nonnegative curvature operator, it becomes relatively easy to compute their cohomology. The Bochner technique tells us that all harmonic forms are parallel. Now, a parallel form is necessarily invariant under the holonomy. Thus, we are left with a classical invariance problem. Namely, determine all forms on a Euclidean space that are invariant under a given group action on the space. It is particularly important to know the cohomology of the real and complex Grassmannians, as one can use that information to define Pontryagin and Chern classes for vector bundles. We refer the reader to [**87**, vol. 5] and [**69**] for more on this.

4. Curvature and Holonomy

To get a better understanding of holonomy and how it relates to symmetric spaces, we need to figure out how it can be computed from the curvature tensor.

We denote the Lie algebra of $\mathrm{Hol}_p^0 \subset SO(n)$ by $\mathfrak{hol}_p \subset \mathfrak{so}(n)$. This Lie algebra is therefore an algebra of skew-symmetric transformations on T_pM. We have on T_pM several other skew-symmetric transformations. Namely, for each pair of vectors $v, w \in T_pM$ there is the curvature tensor $R(v, w) : T_pM \to T_pM$ that maps x to $R(v, w) x$. We can show that this transformation lies in $\mathfrak{hol}_p$. To see this select a coordinate system x^i so that $\partial_1|_p = v$ and $\partial_2|_p = w$. For each $t > 0$ consider the loop c_t at p which in the x^1, x^2 coordinates corresponds to the square with side lengths $\sqrt{t}$, i.e., it is obtained by first following the flow of ∂_1 for time $\sqrt{t}$, then the flow of ∂_2 for time $\sqrt{t}$, then the flow of $-\partial_1$ for time $\sqrt{t}$, and finally the flow of $-\partial_2$ for time $\sqrt{t}$. Now let P_t be parallel translation along this loop (see Figure 8.1). Using our variational characterization of the curvature tensor from chapter 6 one can prove Cartan's characterization of the curvature:

$$R(v, w) = \lim_{t \to 0} \frac{P_t - I}{t}.$$

To completely determine $\mathfrak{hol}_p$, it is of course necessary to look at all contractible loops, not just the short ones. However, each contractible loop can be decomposed into lassos, that is, loops that consist of a curve emanating from p and ending at some q, and then at q we have a very small loop (see Figure 8.2). Thus, any element of $\mathfrak{hol}_p$ is the composition of elements of the form

$$P^{-1} \circ R(P(v), P(w)) \circ P : T_pM \to T_pM,$$

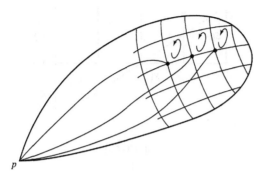

Figure 8.2

where $P : T_pM \rightarrow T_qM$ denotes parallel translation along some curve from p to q. This characterization of holonomy was first proved by Ambrose and Singer, the proof we have indicated is due to Nijenhuis. For the complete proofs the reader is referred to [**11**].

It is therefore possible, in principle, to compute holonomy from knowledge of the curvature tensor at all points. In reality this is not so useful, but for locally symmetric spaces we know that the curvature tensor is invariant under parallel translation, so we have

THEOREM 58. *For a locally symmetric space the holonomy Lie algebra* $\mathfrak{hol}_p$ *is generated by curvature transformations* $R(v, w)$*, where* $v, w \in T_pM$*, i.e.,* $\mathfrak{r}_p = \mathfrak{hol}_p$*. Moreover,* $\mathfrak{hol}_p \subset \mathfrak{iso}_p$*.*

PROOF. We have already indicated the first part. For the second we could use our curvature description of the symmetric space. Instead, we give a more geometric proof, which also establishes that $R(x, y) \in \mathfrak{iso}_p$ as a by-product.

Observe that not only do isometries map geodesics to geodesics, but also parallel fields to parallel fields. Therefore, if we have a geodesic $\gamma : [0, 1] \rightarrow M$ and a parallel field E along γ, then we could apply the involution $A_{\gamma(\frac{1}{2})}$ to E (this involution exists if the geodesic is sufficiently short.) This involution reverses γ and at the same time changes the sign of E. Thus we have $DA_{\gamma(\frac{1}{2})}(E(0)) = -E(1)$, or in other words (see also Figure 8.3)

$$P_{\gamma(0)}^{\gamma(1)} = -DA_{\gamma(\frac{1}{2})}$$

Now use that any curve can be approximated by a broken geodesic to conclude that parallel translation along any curve can be approximated by a successive composition of differentials of isometries. For a loop that is also a broken geodesic, we see that the composition of these isometries must belong to Iso_p. Hence, we have shown the stronger statement that

$$\text{Hol}_p \subset \text{Iso}_p$$

In Figure 8.4 we have sketched how one can parallel translate along a broken geodesic from p to q in a symmetric space. This finishes the proof. $\square$

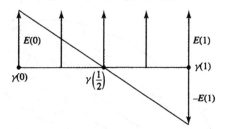

Figure 8.3

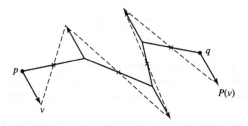

Figure 8.4

Armed with this information, it is possible for us to determine the holonomy of irreducible symmetric spaces.

COROLLARY 24. *For an irreducible symmetric space*

$$\mathfrak{hol}_p = \mathfrak{iso}_p.$$

PROOF. We know that $\mathfrak{hol}_p \subset \mathfrak{iso}_p$ and that $\mathfrak{iso}_p$ acts effectively on T_pM. By assumption we have that $\mathfrak{hol}_p$ acts irreducibly on T_pM. It is then a question of using that an irreducible symmetric space has a unique Lie algebra description to finish this proof. Proving this uniqueness result is a little beyond what we wish to do here. □

Note that irreducibility is important for this theorem since Euclidean space has trivial holonomy. Also, Iso_p might contain orientation-reversing elements, so we cannot show equality at the Lie group level.

We shall now mention, without any indication of proof whatsoever, the classification of connected irreducible holonomy groups. Berger classified all possible holonomies. Simons also deserves to be mentioned as he gave a direct proof of the fact that spaces with non-transitive holonomy must be locally symmetric, i.e., he did not use Berger's classification of holonomy groups.

THEOREM 59. (Berger, 1955 and Simons, 1962) *Let* (M, g) *be a simply connected irreducible Riemannian n-manifold. The holonomy* Hol_p *either acts transitively on the unit sphere in* T_pM *or* (M, g) *is a symmetric space of rank* ≥ 2. *Moreover, in the first case the holonomy is one of the following groups:*

dim = n	Hol$_p$	Properties
n	$SO(n)$	Generic case
$n = 2m$	$U(m)$	Kähler metric
$n = 2m$	$SU(m)$	Kähler metric and Ric = 0
$n = 4m$	$Sp(1) \cdot Sp(m)$	Quaternionic-Kähler and Einstein
$n = 4m$	$Sp(m)$	Hyper-Kähler and Ric = 0
$n = 16$	$Spin(9)$	Symmetric and Einstein
$n = 8$	$Spin(7)$	Ric = 0
$n = 7$	G_2	Ric = 0

It is curious that all but the two largest irreducible holonomy groups, $SO(n)$ and $U(m)$, force the metric to be Einstein and in some cases even Ricci flat. Looking at the relationship between curvature and holonomy, it is clear that having small holonomy forces the curvature tensor to have some special properties. One can, using a case-by-case check, see that various traces of the curvature tensor must be zero, thus forcing the metric to be either Einstein or Ricci flat (see [11] for details.) Note that Kähler metrics do not have to be Einstein (see the exercises to chapter 3), and quaternionic Kähler manifolds are not necessarily Kähler, as $Sp(1) \cdot Sp(m)$ is not contained in $U(2m)$. Using a little bit of the theory of Kähler manifolds, it is not hard to see that metrics with holonomy $SU(n)$ are Ricci flat. Since $Sp(m) \subset SU(2m)$, we then get that also hyper-Kähler manifolds are Ricci flat. One can see that the last two holonomies occur only for Ricci flat manifolds. In particular, they never occur as the holonomy of a symmetric space. With the exception of the four types of Ricci flat holonomies all other holonomies occur for symmetric spaces. This follows from the above classification and the fact that the rank one symmetric spaces have holonomy $SO(n)$, $U(m)$, $Sp(1) \cdot Sp(m)$, or $Spin(9)$.

This leads to another profound question. Are there compact simply connected Ricci flat spaces with holonomy $SU(m)$, $Sp(m)$, G_2, or $Spin(7)$? The answer is yes. But it is a highly nontrivial yes. Yau got the Fields medal, in part, for establishing the $SU(m)$ case. Actually, he solved the Calabi conjecture, and the holonomy question was a by-product (see, e.g., [11] for more information on the Calabi conjecture). Note that we have the Eguchi-Hanson metric which is a complete Ricci flat Kähler metric and therefore has $SU(2)$ as holonomy group. Recently, D. Joyce solved the cases of $Spin(7)$ and G_2 by methods similar to those employed by Yau. An even more intriguing question is whether there are compact simply connected manifolds that are Ricci flat but have $SO(n)$ as a holonomy group. Note that the Schwarzschild metric is complete, Ricci flat and has $SO(4)$ as holonomy group. For more in-depth information on these issues we refer the reader to [11].

A general remark about how special ($\neq SO(n)$) holonomies occur: It seems that they are all related to the existence of parallel forms. In the Kähler case, for example, the Kähler form is a parallel nondegenerate 2-form. Correspondingly, one has a parallel 4-form for quaternionic-Kähler manifolds and a parallel 8-form for Cayley-Kähler manifolds (which are all known to be locally symmetric). This is studied in more detail in the proof of the classification of manifolds with nonnegative curvature operator below. For the last two exceptional holonomies there are also some special 4-forms that do the job.

From the classification of holonomy groups we immediately get an interesting corollary.

COROLLARY 25. *If a Riemannian manifold has the property that the holonomy doesn't act transitively on the unit sphere, then it is either reducible or a locally symmetric space of rank ≥ 2. In particular, the rank must be ≥ 2.*

It is unclear to what extent the converse fails for general manifolds. For non-positive curvature, however, there is the famous higher-rank rigidity result proved independently by W. Ballmann and K. Burns-R. Spatzier (see [6] and [17]).

THEOREM 60. *A compact Riemannian manifold of non-positive curvature of rank ≥ 2 does not have transitive holonomy. In particular, it must be either reducible or locally symmetric.*

It is worthwhile mentioning that in [8] it was shown that the rank of a compact non-positively curved manifold can be computed from the fundamental group. Thus, a good deal of geometric information is automatically encoded into the topology. The rank rigidity theorem is proved by dynamical systems methods. The idea is to look at the geodesic flow on the unit sphere bundle, i.e., the flow that takes a unit vector and moves it time t along the unit speed geodesic in the direction of the unit vector. This flow has particularly nice properties on non-positively curved manifolds, which we won't go into. The idea is to use the flat parallel fields to show that the holonomy can't be transitive. The Berger-Simons result then shows that the manifold has to be locally symmetric if it is irreducible. Nice as this method of proof is, it would also be very pleasing to have a proof that goes more along the lines of the Bochner technique. In nonpositive curvature this method is a bit different. It usually centers on studying harmonic maps into the space rather than harmonic forms on the space (for more on this see [97]).

In nonnegative curvature, on the other hand, it is possible to find irreducible spaces that are not symmetric and have rank ≥ 2. On $S^2 \times S^2$ we have a product metric that is reducible and has rank 3. But if we take another metric on this space that comes as a quotient of $S^2 \times S^3$ by an action of S^1 (acting by rotations on the first factor and the Hopf action on the second), then we get a metric which has rank 2. The only way in which a rank 2 metric can split off a de Rham factor is if it splits off something 1-dimensional, but that is topologically impossible in this case. So in conclusion, the holonomy must be transitive and irreducible.

By assuming the stronger condition that the curvature operator is nonnegative, one can almost classify all such manifolds. This was first done in [42] and in more generality in Chen's article in [45]. This classification allows us to conclude that higher rank gives rigidity. The theorem and proof are a nice synthesis of everything we have learned in this and the previous chapter. In particular, the proof uses the Bochner technique in the two most non-trivial cases we have covered: for forms and the curvature tensor.

THEOREM 61. (S. Gallot and D. Meyer, 1975) *If (M, g) is a compact Riemannian n-manifold with nonnegative curvature operator, then one of the following cases must occur:*

(a) (M, g) has rank > 1 and is either reducible or locally symmetric.

(b) $\mathrm{Hol}^0(M, g) = SO(n)$ and the universal covering is homeomorphic to a sphere.

(c) $\mathrm{Hol}^0(M, g) = U\left(\frac{n}{2}\right)$ and the universal covering is biholomorphic to $\mathbb{C}P^{\frac{n}{2}}$.

(d) $\mathrm{Hol}^0(M, g) = Sp(1) Sp\left(\frac{n}{4}\right)$ and the universal covering is up to scaling isometric to $\mathbb{H}P^{\frac{n}{4}}$.

(e) $\mathrm{Hol}^0\left(M, g\right) = Spin\left(9\right)$ *and the universal covering is up to scaling isometric to* $\mathbb{O}P^2$.

PROOF. First we use a result from chapter (The splitting theorem) which shows that a finite covering of M is isometric to a product $N \times T^n$. So if we assume that (M, g) is irreducible, then the fundamental group is finite. Therefore, we can assume that we work with a simply connected manifold M. Now we observe, that either all of the homology groups $H^p\left(M, \mathbb{R}\right) = 0$ for $p = 1, \ldots, n-1$, in which case the space is a homology sphere, or some homology group $H^p\left(M, \mathbb{R}\right) \neq 0$ for some $p \neq 0, n$. In the latter case, we then have a harmonic p-form by the Hodge theorem. The Bochner technique now tells us that this form must be parallel, since the curvature operator is nonnegative.

The proof is now based on the following observation: A Riemannian n-manifold with holonomy $SO\left(n\right)$ cannot admit any nontrivial parallel p-forms for $0 < p < n$. Note that the volume form is always parallel, so it is clearly necessary to use the condition $p \neq 0, n$. We are also allowed to assume that $p \leq \frac{n}{2}$, since the Hodge star $*\omega$ of ω is parallel iff ω is parallel. The observation is proved by contradiction, so suppose that ω is a parallel p-form, where $0 < p < n$.

First suppose $p = 1$. Then the dual of the 1-form is a parallel vector field. This means that the manifold splits locally. In particular, it must be reducible and have special holonomy.

More generally, when $p \leq n - 2$ is odd, we can for $v_1, \ldots, v_p \in T_p M$ find an element of $P \in SO\left(n\right)$ such that $P\left(v_i\right) = -v_i$. Therefore, if the holonomy is $SO\left(n\right)$ and ω is invariant under parallel translation, we must have,

$$
\begin{aligned}
\omega\left(v_1, \ldots, v_p\right) &= \omega\left(Pv_1, \ldots, Pv_p\right) \\
&= \omega\left(-v_1, \ldots, -v_p\right) \\
&= -\omega\left(v_1, \ldots, v_p\right).
\end{aligned}
$$

This shows that $\omega = 0$. In case n is odd, we can then conclude using the Hodge star operator that no parallel forms exist when the holonomy is $SO\left(n\right)$.

We can then assume that we have an even dimensional manifold and that ω is a parallel p-form with $p \leq \frac{n}{2}$ even. We claim again that if the holonomy is $SO\left(n\right)$, then $\omega = 0$. First select vectors $v_1, \ldots, v_p \in T_p M$, then find orthonormal vectors $e_1, \ldots, e_p \in T_p M$ such that

$$
\mathrm{span}\left\{v_1, \ldots, v_p\right\} = \mathrm{span}\left\{e_1, \ldots, e_p\right\}.
$$

Then we know that $\omega\left(e_1, \ldots, e_p\right)$ is zero iff $\omega\left(v_1, \ldots, v_p\right)$ is zero. Now use that $p \leq \frac{n}{2}$ to find $P \in SO\left(n\right)$ such that

$$
\begin{aligned}
P\left(e_1\right) &= e_2, \\
P\left(e_2\right) &= e_1, \\
P\left(e_i\right) &= e_i \text{ for } i = 3, \ldots, p.
\end{aligned}
$$

Using invariance of ω under P then yields

$$
\begin{aligned}
\omega\left(e_1, \ldots, e_p\right) &= \omega\left(Pe_1, \ldots, Pe_p\right) \\
&= \omega\left(e_2, e_1, e_3, \ldots, e_p\right) \\
&= -\omega\left(e_1, \ldots, e_p\right).
\end{aligned}
$$

In summary, we have shown that any Riemannian manifold with nonnegative curvature operator has holonomy $SO(n)$ only if all homology groups vanish. Supposing that the manifold is irreducible and has transitive holonomy, we can then use the above classification to see what holonomy groups are potentially allowed. The Ricci flat cases are, however, not allowed, as the nonnegative curvature would then make the manifold flat. Thus, we have only the three possibilities $U\left(\frac{n}{2}\right)$, $Sp(1)Sp\left(\frac{n}{4}\right)$, or $Spin(9)$. In the latter two cases one can show from holonomy considerations that the manifold must be Einstein. Thus, Tachibana's result from chapter 7 shows that the metric is locally symmetric. From the classification of symmetric spaces it then follows that the space is isometric to either $\mathbb{H}P^{\frac{n}{4}}$ or $\mathbb{O}P^2$. This leaves us with the Kähler case. In this situation we can show that the cohomology ring must be the same as that of $\mathbb{C}P^{\frac{n}{2}}$, i.e., there is a homology class $\omega \in H^2(M,\mathbb{R})$ such that any homology class is proportional to some power of $\omega : \omega^k = \omega \wedge \ldots \wedge \omega$. This can be seen as follows. When the holonomy is $U\left(\frac{n}{2}\right)$ observe that there must be an almost complex structure on the tangent spaces that is invariant under parallel translation. After type change this gives us a Kähler form ω. This 2-form is necessarily parallel. If ω^k doesn't generate H^{2k}, then each form not proportional to ω^k will by the above arguments reduce the holonomy to a proper subgroup of $U\left(\frac{n}{2}\right)$. As H^n is also generated by the volume form we see that $\omega^{\frac{n}{2}} \neq 0$. In particular, none of the forms ω^k, $k = 1, \ldots, n/2$ are closed. This shows that M has the cohomology ring of $\mathbb{C}P^{\frac{n}{2}}$.

To get the stronger conclusions on the topological type one must use more profound results from [66] and [71]. □

There are two questions left over in this classification. Namely, for the sphere and complex projective space we get only topological rigidity. For the sphere one can clearly perturb the standard metric and still have positive curvature operator, so one couldn't expect more there. On $\mathbb{C}P^2$, say, we know that the curvature operator has exactly two zero eigenvalues. These two zero eigenvalues and eigenvectors are actually forced on us by the fact that the metric is Kähler. Therefore; if we perturb the standard metric, while keeping the same Kähler structure, then these two zero eigenvalues will persist and the positive eigenvalues will stay positive. Thus, the curvature operator stays nonnegative.

Given that there is such a big difference between the classes of manifolds with nonnegative curvature operator and nonnegative sectional curvature, one might think the same is true for nonpositive curvature. However, the above rank rigidity theorem tells us that in fact nonpositive sectional curvature is much more rigid than nonnegative sectional curvature. Nevertheless, there is a recent example of Aravinda and Farrell showing that there are nonpositively curved manifolds that do not admit metrics with nonpositive curvature operator (see [4]).

5. Further Study

We have eliminated many important topics about symmetric spaces. For more in-depth information we recommend the texts by Besse, Helgason, and Jost (see [11, Chapters 7,10], [12, Chapter 3], [53], and [56, Chapter 6]). Another very good text which covers the theory of Lie groups and symmetric spaces is [55]. O'Neill's book [73, Chapter 8] also has a nice elementary account of symmetric spaces.

6. Exercises

(1) (a) Show that the holonomy of $\mathbb{C}P^n$ is $U(n)$.

(b) Show that the holonomy of a Riemannian manifold is contained in $U(m)$ iff it has a Kähler structure.

(2) Assume that M has nonpositive or nonnegative sectional curvature. Let γ be a geodesic and E a parallel field along γ. Show that the following conditions are equivalent.

(a) $g\left(R\left(E,\dot{\gamma}\right)\dot{\gamma},E\right)=0$ everywhere.

(b) $R\left(E,\dot{\gamma}\right)\dot{\gamma}=0$ everywhere.

(c) E is a Jacobi field.

(3) Show that $SO\left(n,\mathbb{C}\right)/SO\left(n\right)$ and $Sl\left(n,\mathbb{C}\right)/SU\left(n\right)$ are symmetric spaces with nonpositive curvature operator.

(4) The *quaternionic projective space* is defined as being the quaternionic lines in $\mathbb{H}^{n+1}$. Here the quaternions $\mathbb{H}$ are the complex matrices

$$\begin{pmatrix} z & w \\ -\bar{w} & \bar{z} \end{pmatrix}.$$

If we identify $\mathbb{H}$ with $\mathbb{R}^4$, then we usually write elements as $x_1 + ix_2 + jx_3 + kx_4$. Multiplication is done using

$$\begin{aligned} i^2 &= j^2 = k^2 = -1, \\ ij &= k = -ji, \\ jk &= i = -kj, \\ ki &= j = -ik. \end{aligned}$$

(a) Show that if we define

$$i = \begin{pmatrix} i & 0 \\ 0 & -i \end{pmatrix},$$

$$j = \begin{pmatrix} 0 & 1 \\ -1 & 0 \end{pmatrix},$$

$$k = \begin{pmatrix} 0 & i \\ i & 0 \end{pmatrix},$$

then these two descriptions are equivalent.

(b) The *symplectic group* $Sp\left(n\right) \subset SU\left(2n\right) \subset SO\left(4n\right)$ consists of those orthogonal matrices that commute with the three complex structures generated by i, j, k on $\mathbb{R}^{4n}$. A better way of looking at this group is by considering $n \times n$ matrices A with quaternionic entries such that

$$A^{-1} = A^*.$$

Here the conjugate of a quaternion is

$$\overline{x_1 + ix_2 + jx_3 + kx_4} = x_1 - ix_2 - jx_3 - kx_4,$$

so we have as usual that

$$|q|^2 = q\bar{q}.$$

Now show that

$$\mathbb{H}P^n = Sp\left(n+1\right)/\left(Sp\left(1\right) \times Sp\left(n\right)\right).$$

Use this to exhibit $\mathbb{H}P^n$ as a symmetric space. Show that the holonomy is $Sp(1) \times Sp(n)$ and that the space is quarter pinched.

(5) Construct the hyperbolic analogues of the complex and quaternionic projective spaces. Show that they have negative curvature and are quarter pinched.

(6) Show that any locally symmetric space (not necessarily complete) is locally isometric to a symmetric space. Conclude that a simply connected locally symmetric space admits a monodromy map into a unique symmetric space. Show that if the locally symmetric space is complete, then the monodromy map is bijective.

(7) Let (M, g) be a compact Riemannian n-manifold that is irreducible and with $\mathfrak{R} \geq 0$. Show that the following are equivalent:
   (a) $\chi(M) > 0$.
   (b) The odd Betti numbers are zero.
   (c) $-I \in \mathrm{Hol}_p^0$.
   (d) The dimension n is even.
   Use this to show that any compact manifold with $\mathfrak{R} \geq 0$ has $\chi(M) \geq 0$.

(8) Show that if an irreducible symmetric space has strictly positive or negative curvature operator, then it has constant curvature.

(9) Using the skew-symmetric linear maps
$$x \wedge y \quad : \quad T_pM \to T_pM,$$
$$x \wedge y (v) \quad = \quad g(x, v) y - g(y, v) x,$$
show that $\Lambda^2 T_pM = \mathfrak{so}(T_pM)$. Using this identification, show that the image of the curvature operator $\mathfrak{R}(\Lambda^2 T_pM) \subset \mathfrak{hol}_p$, with equality for symmetric spaces. Use this to conclude that the holonomy is $SO(n)$ if the curvature operator is positive or negative.

(10) Let M be a symmetric space. If $X \in \mathfrak{t}_p$ and $Y \in \mathfrak{iso}_p$, then $[X, Y] \in \mathfrak{t}_p$.

(11) Let M be a symmetric space and $X, Y, Z \in \mathfrak{t}_p$. Show that
$$R(X, Y) Z \quad = \quad [L_X, L_Y] Z,$$
$$\mathrm{Ric}(X, Y) \quad = \quad -\mathrm{tr}([L_X, L_Y]).$$

CHAPTER 9

Ricci Curvature Comparison

In this chapter we shall prove some of the fundamental results for manifolds with lower Ricci curvature bounds. Two important techniques will be developed: Relative volume comparison and weak upper bounds for the Laplacian of distance functions. With these techniques we shall show numerous results on restrictions of fundamental groups of such spaces and also present a different proof of the estimate for the first Betti number by Bochner.

We have already seen how variational calculus can be used to obtain Myers' diameter bounds and also how the Bochner technique can be used. In the 50s Calabi discovered that one has weak upper bounds for the Laplacian of distance function given lower Ricci curvature bounds even at points where this function isn't smooth. However, it wasn't until around 1970, when Cheeger and Gromoll proved their splitting theorem, that this was fully appreciated. Around 1980, Gromov exposed the world to his view of how volume comparison can be used. The relative volume comparison theorem was actually first proved by Bishop in [**13**]. At the time, however, one only considered balls of radius less than the injectivity radius. Later, Gromov observed that the result holds for all balls and immediately put it to use in many situations. In particular, he showed how one could generalize the Betti number estimate from Bochner's theorem using only topological methods and volume comparison. Anderson then refined this to get information about fundamental groups. One's intuition about Ricci curvature has generally been borrowed from experience with sectional curvature. This has led to many naive conjectures that haven proven to be false through the construction of several interesting examples of manifolds with nonnegative Ricci curvature. On the other hand, much good work has also come out of this, as we shall see. The reason for treating Ricci curvature before the more advanced results on sectional curvature is that we want to break the link between the two. The techniques for dealing with these two subjects, while similar, are not the same.

1. Volume Comparison

1.1. The Fundamental Equations. Throughout this section, assume that we have a complete Riemannian manifold (M, g) of dimension n. Furthermore, we are given a point $p \in M$ and with that the distance function $r(x) = d(x, p)$. We know that this distance function is smooth on the image of the interior of the segment domain. In analogy with the fundamental equations for the metric:

(1) $\quad L_{\partial_r} g = 2\mathrm{Hess}r,$
(2) $\quad (\nabla_{\partial_r} \mathrm{Hess}r)(X, Y) + \mathrm{Hess}^2 r(X, Y) = -R(X, \partial_r, \partial_r, Y),$

we also have a similar set of equations for the volume form.

PROPOSITION 39. *The volume form dvol and Laplacian Δr of r are related by:*

(tr1) $L_{\partial_r} \text{dvol} = \Delta r \text{dvol},$

(tr2) $\partial_r \Delta r + \dfrac{(\Delta r)^2}{n-1} \le \partial_r \Delta r + |\text{Hess} r|^2 = -\text{Ric}(\partial_r, \partial_r).$

PROOF. The way to establish the first equation is by first selecting orthonormal 1-forms θ^i. The volume form is then given by

$$\text{dvol} = \theta^1 \wedge \cdots \wedge \theta^n.$$

As with the metric g, we also have that dvol is parallel. Next observe that

$$
\begin{aligned}
\left(L_{\partial_r} \theta^i\right)(X) &= \partial_r \left(\theta^i(X)\right) - \theta^i \left(L_{\partial_r} X\right) \\
&= \partial_r \left(\theta^i(X)\right) - \theta^i \left(\nabla_{\partial_r} X\right) + \theta^i \left(\nabla_X \partial_r\right) \\
&= \left(\nabla_{\partial_r} \theta^i\right)(X) + \theta^i \left(\nabla_X \partial_r\right).
\end{aligned}
$$

This shows that

$$
\begin{aligned}
L_{\partial_r} \text{dvol} &= L_{\partial_r} \left(\theta^1 \wedge \cdots \wedge \theta^n\right) \\
&= \sum \theta^1 \wedge \cdots \wedge L_{\partial_r} \theta^i \wedge \cdots \wedge \theta^n \\
&= \sum \theta^1 \wedge \cdots \wedge \nabla_{\partial_r} \theta^i \wedge \cdots \wedge \theta^n \\
&\quad + \sum \theta^1 \wedge \cdots \wedge \theta^i \circ \nabla. \partial_r \wedge \cdots \wedge \theta^n \\
&= \nabla_{\partial_r} \left(\theta^1 \wedge \cdots \wedge \theta^n\right) + \text{tr}\left(\nabla. \partial_r\right) \theta^1 \wedge \cdots \wedge \theta^n \\
&= \nabla_{\partial_r} \text{dvol} + \text{tr}\left(\nabla. \partial_r\right) \text{dvol} \\
&= \Delta r \text{dvol}.
\end{aligned}
$$

To establish the second equation we take traces in (2). Thus we select an orthonormal frame E_i, set $X = Y = E_i$ and sum over i. We can in addition assume that $\nabla_{\partial_r} E_i = 0$. We already know that

$$\sum_{i=1}^n R\left(E_i, \partial_r, \partial_r, E_i\right) = \text{Ric}(\partial_r, \partial_r).$$

On the left hand side we get

$$
\begin{aligned}
\sum_{i=1}^n \left(\nabla_{\partial_r} \text{Hess} r\right)(E_i, E_i) &= \sum_{i=1}^n \partial_r \text{Hess} r\left(E_i, E_i\right) \\
&= \partial_r \Delta r
\end{aligned}
$$

and

$$
\begin{aligned}
\sum_{i=1}^n \text{Hess}^2 r\left(E_i, E_i\right) &= \sum_{i=1}^n g\left(\nabla_{E_i} \partial_r, \nabla_{E_i} \partial_r\right) \\
&= \sum_{i,j=1}^n g\left(\nabla_{E_i} \partial_r, g\left(\nabla_{E_i} \partial_r, E_j\right) E_j\right) \\
&= \sum_{i,j=1}^n g\left(\nabla_{E_i} \partial_r, E_j\right) g\left(\nabla_{E_i} \partial_r, E_j\right) \\
&= |\text{Hess} r|^2
\end{aligned}
$$

Finally we need to show that

$$\frac{(\Delta r)^2}{n-1} \leq |\text{Hess} r|^2.$$

To this end we also assume that $E_1 = \partial_r$. Then

$$
\begin{aligned}
|\text{Hess} r|^2 &= \sum_{i,j=1}^{n} \left(g\left(\nabla_{E_i}\partial_r, E_j\right)\right)^2 \\
&= \sum_{i,j=2}^{n} \left(g\left(\nabla_{E_i}\partial_r, E_j\right)\right)^2 \\
&\leq \frac{1}{n-1} \left(\sum_{i=2}^{n} g\left(\nabla_{E_i}\partial_r, E_i\right)\right)^2 \\
&= \frac{1}{n-1}(\Delta r)^2.
\end{aligned}
$$

The inequality

$$|A|^2 \leq \frac{1}{k}|\text{tr}(A)|^2$$

for a $k \times k$ matrix A is a direct consequence of the Cauchy-Schwarz inequality

$$
\begin{aligned}
|(A, I_k)|^2 &\leq |A|^2 |I_k|^2 \\
&= |A|^2 k,
\end{aligned}
$$

where I_k is the identity $k \times k$ matrix. $\qquad\square$

If we use the polar coordinate decomposition $g = dr^2 + g_r$ and let $d\text{vol}_{n-1}$ be the standard volume form on $S^{n-1}(1)$, then we have that

$$d\text{vol} = \lambda(r, \theta)\, dr \wedge d\text{vol}_{n-1},$$

where θ indicates a coordinate on S^{n-1}. If we apply (tr1) to this version of the volume form we get

$$
\begin{aligned}
L_{\partial_r} d\text{vol} &= L_{\partial_r}\left(\lambda(r, \theta)\, dr \wedge d\text{vol}_{n-1}\right) \\
&= \partial_r(\lambda)\, dr \wedge d\text{vol}_{n-1}
\end{aligned}
$$

as both $L_{\partial_r} dr = 0$ and $L_{\partial_r} d\text{vol}_{n-1} = 0$. We can therefore simplify (tr1) to

$$\partial_r \lambda = \lambda \Delta r.$$

In constant curvature k we know that

$$g_k = dr^2 + \text{sn}_k^2(r)\, ds_{n-1}^2,$$

thus the volume form is

$$
\begin{aligned}
d\text{vol}_k &= \lambda_k(r)\, dr \wedge d\text{vol}_{n-1} \\
&= \text{sn}_k^{n-1}(r)\, dr \wedge d\text{vol}_{n-1},
\end{aligned}
$$

this conforms with the fact that

$$\Delta r = (n-1)\frac{\text{sn}_k'(r)}{\text{sn}_k(r)},$$

$$\partial_r\left(\text{sn}_k^{n-1}(r)\right) = (n-1)\frac{\text{sn}_k'(r)}{\text{sn}_k(r)}\text{sn}_k^{n-1}(r).$$

1.2. Volume Estimation. With the above information we can prove the estimates that are analogous to our basic comparison estimates for the metric and Hessian of r assuming lower sectional curvature bounds (see chapter 6).

LEMMA 34. (Ricci Comparison Result) *Suppose that (M, g) has $\mathrm{Ric} \geq (n-1) \cdot k$ for some $k \in \mathbb{R}$. Then*

$$\Delta r \ \leq \ (n-1) \frac{\mathrm{sn}_k'(r)}{\mathrm{sn}_k(r)},$$

$$d\mathrm{vol} \ \leq \ d\mathrm{vol}_k,$$

where $d\mathrm{vol}_k$ is the volume form in constant sectional curvature k.

PROOF. Notice that the right-hand sides of the inequalities correspond exactly to what one would get in constant curvature k.

For the first inequality, we use that

$$\partial_r \Delta r + \frac{(\Delta r)^2}{n-1} \leq -(n-1) \cdot k$$

dividing by $n-1$ and using λ_k this gives

$$\partial_r \left(\frac{\Delta r}{n-1} \right) + \left(\frac{\Delta r}{n-1} \right)^2 \leq -k = \partial_r (\lambda_k) + (\lambda_k)^2$$

Separation of variables then yields:

$$\frac{\partial_r \frac{\Delta r}{n-1}}{k + \left(\frac{\Delta r}{n-1} \right)^2} \leq \frac{\partial_r \lambda_k}{k + (\lambda_k)^2}.$$

Thus

$$F(\lambda(r)) \leq F(\lambda_k(r)),$$

where F is the antiderivative of $\frac{1}{\lambda^2 + k}$ satisfying $\lim_{\lambda \to \infty} F(\lambda) = 0$. Since F has positive derivative we can conclude that $\lambda(r) \leq \lambda_k(r)$.

For the second inequality we now know that

$$\partial_r \lambda \leq (n-1) \frac{\mathrm{sn}_k'(r)}{\mathrm{sn}_k(r)} \lambda$$

while

$$\partial_r \lambda_k = (n-1) \frac{\mathrm{sn}_k'(r)}{\mathrm{sn}_k(r)} \lambda_k.$$

In addition the metrics g and g_k agree at p. Thus also the volume forms agree at p. This means that

$$\lim_{r \to 0} (\lambda - \lambda_k) \ = \ 0,$$

$$\partial_r (\lambda - \lambda_k) \ \leq \ (n-1) \frac{\mathrm{sn}_k'(r)}{\mathrm{sn}_k(r)} (\lambda - \lambda_k).$$

Whence the volume form inequality follows. $\square$

Our first volume comparison gives the obvious upper volume bound coming from our upper bound on the volume density.

LEMMA 35. *If* (M, g) *has* $\mathrm{Ric} \geq (n-1) \cdot k$, *then*

$$\mathrm{vol} B(p, r) \leq v(n, k, r),$$

where $v(n, k, r)$ *denotes the volume of a ball of radius* r *in the constant-curvature space form* S_k^n.

PROOF. Above, we showed that in polar coordinates around p we have

$$d\mathrm{vol} \leq d\mathrm{vol}_k.$$

Thus

$$
\begin{aligned}
\mathrm{vol} B(p, r) &= \int_{\mathrm{seg}_p \cap B(0,r)} d\mathrm{vol} \\
&\leq \int_{\mathrm{seg}_p \cap B(0,r)} d\mathrm{vol}_k \\
&\leq \int_{B(0,r)} d\mathrm{vol}_k \\
&= v(n, k, r).
\end{aligned}
$$

$\square$

With a little more technical work, the above absolute volume comparison result can be improved in a rather interesting direction. The result one obtains is referred to as the relative volume comparison estimate. It will prove invaluable in many situations throughout the rest of the text.

LEMMA 36. (Relative Volume Comparison, Bishop-Cheeger-Gromov, 1964-1980) *Suppose* (M, g) *is a complete Riemannian manifold with* $\mathrm{Ric} \geq (n-1) \cdot k$. *Then*

$$r \rightarrow \frac{\mathrm{vol} B(p, r)}{v(n, k, r)}$$

is a nonincreasing function whose limit is 1 as $r \rightarrow 0$.

PROOF. We will use exponential polar coordinates. The volume form $\lambda(r, \theta) dr \wedge d\theta$ for (M, g) is initially defined only on some star-shaped subset of

$$T_p M = \mathbb{R}^n = (0, \infty) \times S^{n-1},$$

but we can just set $\lambda = 0$ outside this set. The comparison density λ_k is defined on all of $\mathbb{R}^n$ for $k \leq 0$ and on $B\left(0, \pi/\sqrt{k}\right)$ for $k > 0$. We can likewise extend $\lambda_k = 0$ outside $B\left(0, \pi/\sqrt{k}\right)$. Myers' theorem says that $\lambda = 0$ on $\mathbb{R}^n - B\left(0, \pi/\sqrt{k}\right)$ in this case. So we might as well just consider $r < \pi/\sqrt{k}$ when $k > 0$.

The ratio of the volumes is

$$\frac{\mathrm{vol} B(p, R)}{v(n, k, R)} = \frac{\int_0^R \int_{S^{n-1}} \lambda dr \wedge d\theta}{\int_0^R \int_{S^{n-1}} \lambda_k dr \wedge d\theta},$$

and we know that

$$0 \leq \lambda(r, \theta) \leq \lambda_k(r, \theta) = \mathrm{sn}_k^{n-1}(r)$$

everywhere.

Differentiation of this quotient with respect to R yields

$$\frac{d}{dR}\left(\frac{\text{vol}B(p,R)}{v(n,k,R)}\right)$$

$$= \frac{\left(\int_{S^{n-1}}\lambda\left(R,\theta\right)d\theta\right)\left(\int_0^R\int_{S^{n-1}}\lambda_k\left(r,\theta\right)dr\wedge d\theta\right)}{(v(n,k,R))^2}$$

$$- \frac{\left(\int_{S^{n-1}}\lambda_k\left(R,\theta\right)d\theta\right)\left(\int_0^R\int_{S^{n-1}}\lambda\left(r,\theta\right)dr\wedge d\theta\right)}{(v(n,k,R))^2}$$

$$= (v(n,k,R))^{-2}\cdot\int_0^R\left[\left(\int_{S^{n-1}}\lambda\left(R,\theta\right)d\theta\right)\cdot\left(\int_{S^{n-1}}\lambda_k\left(r,\theta_{n-1}\right)d\theta\right)\right.$$

$$\left. - \left(\int_{S^{n-1}}\lambda_k\left(R,\theta\right)d\theta\right)\left(\int_{S^{n-1}}\lambda\left(r,\theta\right)d\theta\right)\right]dr.$$

So to see that

$$R\to\frac{\text{vol}B(p,R)}{v(n,k,R)}$$

is nonincreasing, it suffices to check that

$$\frac{\int_{S^{n-1}}\lambda\left(r,\theta\right)d\theta}{\int_{S^{n-1}}\lambda_k\left(r,\theta\right)d\theta} = \frac{1}{\omega_{n-1}}\int_{S^{n-1}}\frac{\lambda\left(r,\theta\right)}{\lambda_k\left(r,\theta\right)}d\theta$$

is nonincreasing. This follows from

$$\partial_r\left(\frac{\lambda\left(r,\theta\right)}{\lambda_k\left(r,\theta\right)}\right) = \frac{\lambda_k\partial_r\lambda - \lambda\partial_r\lambda_k}{\lambda_k^2}$$

$$\leq \frac{\lambda_k\left(n-1\right)\frac{\text{sn}_k'(r)}{\text{sn}_k(r)}\lambda - \lambda\left(n-1\right)\frac{\text{sn}_k'(r)}{\text{sn}_k(r)}\lambda_k}{\lambda_k^2}$$

$$= 0.$$

$\square$

1.3. Maximal Diameter Rigidity.
Given Myers' diameter estimate, it is natural to ask what happens if the diameter attains it maximal value. The next result shows that only the sphere has this property.

THEOREM 62. (S. Y. Cheng, 1975) *If (M,g) is a complete Riemannian manifold with* $\text{Ric}\geq(n-1)k>0$ *and* $\text{diam}=\pi/\sqrt{k}$, *then (M,g) is isometric to S_k^n.*

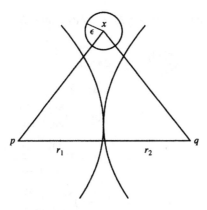

Figure 9.1

PROOF. Fix $p, q \in M$ such that $d(p, q) = \pi/\sqrt{k}$. Define $r(x) = d(x, p)$, $\tilde{r}(x) = d(x, q)$. We will show that

(1) $r + \tilde{r} = d(p, x) + d(x, q) = d(p, q) = \pi/\sqrt{k}$, $x \in M$.

(2) $r, \tilde{r}$ are smooth on $M - \{p, q\}$.

(3) $\mathrm{Hess} r = (\mathrm{sn}_k'/\mathrm{sn}_k) \, ds_{n-1}^2$ on $M - \{p, q\}$.

(4) $g = dr^2 + \mathrm{sn}_k^2 ds_{n-1}^2$.

We know that (3) implies (4) and that (4) implies M must be S_k^n.

Proof of (1): The triangle inequality shows that

$$d(p, x) + d(x, q) \geq \pi/\sqrt{k},$$

so if (1) does not hold, we can find $\varepsilon > 0$ such that (see Figure 9.1)

$$d(p, x) + d(x, q) = 2 \cdot \varepsilon + \frac{\pi}{\sqrt{k}} = 2 \cdot \varepsilon + d(p, q).$$

Then the metric balls $B(p, r_1)$, $B(q, r_2)$, and $B(x, \varepsilon)$ are pairwise disjoint, when $r_1 \leq d(p, x)$, $r_2 \leq d(q, x)$ and $r_1 + r_2 = \pi/\sqrt{k}$. Thus,

$$
\begin{aligned}
1 &= \frac{\mathrm{vol} M}{\mathrm{vol} M} \geq \frac{\mathrm{vol} B(x, \varepsilon) + \mathrm{vol} B(p, r_1) + \mathrm{vol} B(q, r_2)}{\mathrm{vol} M} \\
&\geq \frac{v(n, k, \varepsilon)}{v\left(n, k, \frac{\pi}{\sqrt{k}}\right)} + \frac{v(n, k, r_1)}{v\left(n, k, \frac{\pi}{\sqrt{k}}\right)} + \frac{v(n, k, r_2)}{v\left(n, k, \frac{\pi}{\sqrt{k}}\right)} \\
&= \frac{v(n, k, \varepsilon)}{v\left(n, k, \frac{\pi}{\sqrt{k}}\right)} + 1,
\end{aligned}
$$

which is a contradiction.

Proof of (2): If $x \in M - \{q, p\}$, then x can be joined to both p and q by segments σ_1, σ_2. The previous statement says that if we put these two segments together, then we get a segment from p to q through x. Such a segment must be smooth, and thus σ_1 and σ_2 are both subsegments of a larger segment. This implies from our characterization of when distance functions are smooth that both r and $\tilde{r}$ are smooth at $x \in M - \{p, q\}$.

Proof of (3): We have $r(x) + \tilde{r}(x) = \pi/\sqrt{k}$, thus $\Delta r = -\Delta\tilde{r}$. On the other hand,

$$
\begin{aligned}
(n-1)\frac{\mathrm{sn}_k'(r(x))}{\mathrm{sn}_k(r(x))} &\geq \Delta r(x) \\
&= -\Delta\tilde{r}(x) \\
&\geq -(n-1)\frac{\mathrm{sn}_k'(\tilde{r}(x))}{\mathrm{sn}_k(\tilde{r}(x))} \\
&= -(n-1)\frac{\mathrm{sn}_k'\left(\frac{\pi}{\sqrt{k}} - r(x)\right)}{\mathrm{sn}_k\left(\frac{\pi}{\sqrt{k}} - r(x)\right)} \\
&= (n-1)\frac{\mathrm{sn}_k'(r(x))}{\mathrm{sn}_k(r(x))}.
\end{aligned}
$$

This implies,

$$
\Delta r = (n-1)\frac{\mathrm{sn}_k'}{\mathrm{sn}_k}
$$

and

$$
\begin{aligned}
-(n-1)k &= \partial_r(\Delta r) + \frac{(\Delta r)^2}{n-1} \\
&\leq \partial_r(\Delta r) + |\mathrm{Hess}r|^2 \\
&\leq -\mathrm{Ric}(\partial_r, \partial_r) \\
&\leq -(n-1)k.
\end{aligned}
$$

Hence, all inequalities are equalities, and in particular

$$
(\Delta r)^2 = (n-1)|\mathrm{Hess}r|^2.
$$

Recall that this gives us equality in the Cauchy-Schwarz inequality $|A|^2 \leq k\,(\mathrm{tr}A)^2$. Thus $A = \frac{\mathrm{tr}A}{k}I_k$. In our case we have restricted $\mathrm{Hess}r$ to the $(n-1)$ dimensional space orthogonal to ∂_r so on this space we obtain:

$$
\begin{aligned}
\mathrm{Hess}r &= \frac{\Delta r}{n-1}g_r \\
&= \frac{\mathrm{sn}_k'}{\mathrm{sn}_k}g_r.
\end{aligned}
$$

$\square$

We have now proved that any complete manifold with $\mathrm{Ric} \geq (n-1)\cdot k > 0$ has diameter $\leq \pi/\sqrt{k}$, where equality holds only when the space is S_k^n. A natural perturbation question is therefore: Do manifolds with $\mathrm{Ric} \geq (n-1)\cdot k > 0$ and diam $\approx \pi/\sqrt{k}$, have to be homeomorphic or diffeomorphic to a sphere?

For $n = 2, 3$ this is true, when $n \geq 4$, however, there are counterexamples. The case $n = 2$ will be settled later, while $n = 3$ goes beyond the scope of this book (see [**85**]). The examples for $n \geq 4$ are divided into two cases: $n = 4$ and $n \geq 5$.

EXAMPLE 46. *(Anderson, 1990) For $n = 4$ consider metrics on $I \times S^3$ of the form*

$$
dr^2 + \varphi^2\sigma_1^2 + \psi^2(\sigma_2^2 + \sigma_3^2).
$$

If we define

$$\varphi(r) = \begin{cases} \frac{\sin(ar)}{a} & r \leq r_0, \\ c_1 \sin(r + \delta) & r \geq r_0, \end{cases}$$

$$\psi(r) = \begin{cases} br^2 + c & r \leq r_0, \\ c_2 \sin(r + \delta) & r \geq r_0, \end{cases}$$

and then reflect these function in $r = \pi/2 - \delta$, we get a metric on $\mathbb{C}P^2 \# \overline{\mathbb{C}}P^2$. For any small $r_0 > 0$ we can now adjust the parameters so that φ and ψ become C^1 and generate a metric with $\mathrm{Ric} \geq (n-1)$. For smaller and smaller choices of r_0 we see that $\delta \to 0$, so the interval $I \to [0, \pi]$ as $r_0 \to 0$. This means that the diameters converge to π.

EXAMPLE 47. *(Otsu, 1991) For $n \geq 5$ we only need to consider standard doubly warped products:*

$$dr^2 + \varphi^2 \cdot ds_2^2 + \psi^2 ds_{n-3}^2$$

on $I \times S^2 \times S^{n-3}$. Similar choices for φ and ψ will yield metrics on $S^2 \times S^{n-2}$ with $\mathrm{Ric} \geq n - 1$ and diameter $\to \pi$.

In both of the above examples we actually only constructed C^1 functions φ, ψ and therefore only C^1 metrics. The functions are, however, concave and can easily be smoothed near the break points so as to stay concave. This will not change the values or first derivatives much and only increase the second derivative in absolute value. Thus the lower curvature bound still holds.

2. Fundamental Groups and Ricci Curvature

We shall now attempt to generalize the estimate on the first Betti number we obtained using the Bochner technique to the situation where one has more general Ricci curvature bounds. This requires some knowledge about how fundamental groups are tied in with the geometry.

2.1. The First Betti Number. Suppose M is a compact Riemannian manifold of dimension n and $\tilde{M}$ its universal covering space. The fundamental group $\pi_1(M)$ acts by isometries on $\tilde{M}$. Recall from algebraic topology that

$$H_1(M, \mathbb{Z}) = \pi_1(M) / [\pi_1(M), \pi_1(M)],$$

where $[\pi_1(M), \pi_1(M)]$ is the commutator subgroup. Thus, $H_1(M, \mathbb{Z})$ acts by deck transformations on the covering space

$$\tilde{M} / [\pi_1(M), \pi_1(M)]$$

with quotient M. Since $H_1(M, \mathbb{Z})$ is a finitely generated Abelian group, we know that the set of torsion elements T is a finite normal subgroup. We can then consider $\Gamma = H_1(M, \mathbb{Z}) / T$ as acting by deck transformations on

$$\bar{M} = \tilde{M} / [\pi_1(M), \pi_1(M)] / T.$$

Thus, we have a covering $\pi : \bar{M} \to M$ with a torsion free and Abelian Galois group of deck transformations. The rank of the torsion-free group Γ is clearly equal to

$$b_1(M) = \dim H_1(M, \mathbb{R}).$$

Next recall that any finite-index subgroup of Γ has the same rank as Γ. So if we can find a finite-index subgroup that is generated by elements that can be geometrically controlled, then we might be able to bound b_1. To this end we have a very interesting result.

LEMMA 37. (M. Gromov, 1980) *For fixed* $x \in \bar{M}$ *there exists a finite-index subgroup* $\Gamma' \subset \Gamma$ *that is generated by elements* $\gamma_1, \ldots, \gamma_m$ *such that*

$$d\left(x, \gamma_i\left(x\right)\right) \leq 2 \cdot \operatorname{diam}\left(M\right).$$

Furthermore, for all $\gamma \in \Gamma' - \{1\}$ *we have*

$$d\left(x, \gamma\left(x\right)\right) > \operatorname{diam}\left(M\right).$$

PROOF. First we find a finite-index subgroup that can be generated by elements satisfying the first condition. Then we modify this group so that it also satisfies the second condition.

For each $\varepsilon > 0$ let Γ_ε be the group generated by

$$\left\{\gamma \in \Gamma : d\left(x, \gamma\left(x\right)\right) < 2\operatorname{diam}\left(M\right) + \varepsilon\right\},$$

and let $\pi_\varepsilon : \bar{M} \to \bar{M}/\Gamma_\varepsilon$ denote the covering projection. We claim that for each $z \in \bar{M}$ we have

$$d\left(\pi_\varepsilon\left(z\right), \pi_\varepsilon\left(x\right)\right) < \operatorname{diam}\left(M\right) + \varepsilon.$$

Otherwise, we could find $z \in \bar{M}$ such that

$$d\left(x, z\right) = d\left(\pi_\varepsilon\left(z\right), \pi_\varepsilon\left(x\right)\right) = \operatorname{diam}\left(M\right) + \varepsilon.$$

Now, we can find $\gamma \in \Gamma$ such that $d\left(\gamma\left(x\right), z\right) \leq \operatorname{diam}\left(M\right)$, but then we would have

$$\begin{aligned} d\left(\pi_\varepsilon\left(\gamma\left(x\right)\right), \pi_\varepsilon\left(x\right)\right) &\geq d\left(\pi_\varepsilon\left(z\right), \pi_\varepsilon\left(x\right)\right) - d\left(\pi_\varepsilon\left(z\right), \pi_\varepsilon\left(\gamma\left(x\right)\right)\right) \geq \varepsilon, \\ d\left(x, \gamma\left(x\right)\right) &\leq d\left(x, z\right) + d\left(z, \gamma\left(x\right)\right) \leq 2\operatorname{diam}\left(M\right) + \varepsilon. \end{aligned}$$

Here we have a contradiction, as the first line says that $\gamma \notin \Gamma_\varepsilon$, while the second line says $\gamma \in \Gamma_\varepsilon$.

Note that compactness of $\bar{M}/\Gamma_\varepsilon$ shows that $\Gamma_\varepsilon \subset \Gamma$ has finite index.

Now observe that there are at most finitely many elements in the set

$$\left\{\gamma \in \Gamma : d\left(x, \gamma\left(x\right)\right) < 3\operatorname{diam}\left(M\right)\right\},$$

as Γ acts discretely on $\bar{M}$. Hence, there must be a sufficiently small $\varepsilon > 0$ such that

$$\left\{\gamma \in \Gamma : d\left(x, \gamma\left(x\right)\right) < 2\operatorname{diam}\left(M\right) + \varepsilon\right\} = \left\{\gamma \in \Gamma : d\left(x, \gamma\left(x\right)\right) \leq 2\operatorname{diam}\left(M\right)\right\}.$$

Then we have a finite-index subgroup Γ_ε of Γ generated by

$$\left\{\gamma \in \Gamma : d\left(x, \gamma\left(x\right)\right) \leq 2\operatorname{diam}\left(M\right)\right\} = \left\{\gamma_1, \ldots, \gamma_m\right\}.$$

We shall now modify these generators until we get the desired group Γ'.

First, observe that as the rank of Γ_ε is b_1, we can assume that $\left\{\gamma_1, \ldots, \gamma_{b_1}\right\}$ are linearly independent and generate a subgroup $\Gamma'' \subset \Gamma_\varepsilon$ of finite index. Next, we recall that only finitely many elements γ in Γ'' lie in

$$\left\{\gamma \in \Gamma : d\left(x, \gamma\left(x\right)\right) \leq 2\operatorname{diam}\left(M\right)\right\}.$$

We can therefore choose

$$\left\{\tilde{\gamma}_1, \ldots, \tilde{\gamma}_{b_1}\right\} \subset \left\{\gamma \in \Gamma : d\left(x, \gamma\left(x\right)\right) \leq 2\operatorname{diam}\left(M\right)\right\}$$

with the following properties (we use additive notation here, as it is easier to read):

(1) span $\left\{\tilde{\gamma}_1, \ldots, \tilde{\gamma}_k\right\} \subset$ span $\left\{\gamma_1, \ldots, \gamma_k\right\}$ has finite index for all $k = 1, \ldots, b_1$.

(2) $\tilde{\gamma}_k = l_{1k} \cdot \gamma_1 + \cdots + l_{kk} \cdot \gamma_k$ is chosen such that l_{kk} is maximal in absolute value among all elements in

$$\Gamma'' \cap \{\gamma \in \Gamma : d(x, \gamma(x)) \le 2\text{diam}(M)\}.$$

The group Γ' generated by $\{\tilde{\gamma}_1, \ldots, \tilde{\gamma}_{b_1}\}$ clearly has finite index in Γ'' and hence also in Γ. The generators lie in

$$\{\gamma \in \Gamma : d(x, \gamma(x)) \le 2\text{diam}(M)\},$$

as demanded by the first property. It only remains to show that the second property is also satisfied. The see this, let

$$\gamma = m_1 \cdot \tilde{\gamma}_1 + \cdots + m_k \cdot \tilde{\gamma}_k$$

be chosen such that $m_k \ne 0$. If $d(x, \gamma(x)) \le \text{diam}(M)$, then we also have that

$$
\begin{aligned}
d\left(x, \gamma^2(x)\right) &\le d(x, \gamma(x)) + d\left(\gamma(x), \gamma^2(x)\right) \\
&= 2d(x, \gamma(x)) \\
&\le 2\text{diam}(M).
\end{aligned}
$$

Thus,

$$\gamma^2 \in \Gamma'' \cap \{\gamma \in \Gamma : d(x, \gamma(x)) \le 2\text{diam}(M)\},$$

and also,

$$
\begin{aligned}
\gamma^2 &= 2m_1 \cdot \tilde{\gamma}_1 + \cdots + 2m_k \cdot \tilde{\gamma}_k \\
&= \sum_{i=1}^{k-1} n_i \cdot \gamma_i + 2m_k \cdot l_{kk} \cdot \gamma_k.
\end{aligned}
$$

But this violates the maximality of l_{kk}, as we assumed $m_k \ne 0$. □

With this lemma we can now give Gromov's proof of

THEOREM 63. (S. Gallot and M. Gromov, 1980) *If M is a Riemannian manifold of dimension n such that $\text{Ric} \ge (n-1)k$ and $\text{diam}(M) \le D$, then there is a function $C(n, k \cdot D^2)$ such that*

$$b_1(M) \le C\left(n, k \cdot D^2\right).$$

Moreover, $\lim_{\varepsilon \to 0} C(n, \varepsilon) = n$. In particular, there is $\varepsilon(n) > 0$ such that if $k \cdot D^2 \ge -\varepsilon(n)$, then $b_1(M) \le n$.

PROOF. First observe that for $k > 0$ there is nothing to prove, as we know that $b_1 = 0$ from Myers' theorem.

Suppose we have chosen a covering $\bar{M}$ of M with torsion-free Abelian Galois group of deck transformations $\Gamma = \langle \gamma_1, \ldots, \gamma_{b_1} \rangle$ such that for some $x \in \bar{M}$ we have

$$
\begin{aligned}
d(x, \gamma_i(x)) &\le 2\text{diam}(M), \\
d(x, \gamma(x)) &> \text{diam}(M), \gamma \ne 1.
\end{aligned}
$$

Then we clearly have that all of the balls $B\left(\gamma(x), \frac{\text{diam}(M)}{2}\right)$ are disjoint. Now set

$$I_r = \left\{\gamma \in \Gamma : \gamma = l_1 \cdot \gamma_1 + \cdots + l_{b_1} \cdot \gamma_{b_1}, |l_1| + \cdots + |l_{b_1}| \le r\right\}.$$

Note that for $\gamma \in I_r$ we have

$$B\left(\gamma(x), \frac{\text{diam}(M)}{2}\right) \subset B\left(x, r \cdot 2\text{diam}(M) + \frac{\text{diam}(M)}{2}\right).$$

All of these balls are disjoint and have the same volume, as γ acts isometrically. We can therefore use the relative volume comparison theorem to conclude that the cardinality of I_r is bounded from above by

$$\frac{\mathrm{vol}B\left(x, r \cdot 2\mathrm{diam}\,(M) + \frac{\mathrm{diam}(M)}{2}\right)}{\mathrm{vol}B\left(x, \frac{\mathrm{diam}(M)}{2}\right)} \leq \frac{v\left(n, k, r \cdot 2\mathrm{diam}\,(M) + \frac{\mathrm{diam}(M)}{2}\right)}{v\left(n, k, \frac{\mathrm{diam}(M)}{2}\right)}.$$

This shows that

$$\begin{aligned} b_1 &\leq |I_1| \\ &\leq \frac{v\left(n, k, 2\mathrm{diam}\,(M) + \frac{\mathrm{diam}(M)}{2}\right)}{v\left(n, k, \frac{\mathrm{diam}(M)}{2}\right)}, \end{aligned}$$

which gives us a general bound for b_1. To get a more refined bound we have to use I_r for larger r. If r is an integer, then

$$|I_r| = (2r + 1)^{b_1}.$$

The upper bound for $|I_r|$ can be reduced to

$$\begin{aligned} \frac{v\left(n, k, r \cdot 2\mathrm{diam}\,(M) + \frac{\mathrm{diam}(M)}{2}\right)}{v\left(n, k, \frac{\mathrm{diam}(M)}{2}\right)} &\leq \frac{v\left(n, k, \left(r \cdot 2 + \frac{1}{2}\right) D\right)}{v\left(n, k, \frac{D}{2}\right)} \\ &= \frac{\int_0^{(r \cdot 2 + \frac{1}{2})D} \left(\frac{\sinh(\sqrt{-k}t)}{\sqrt{-k}}\right)^{n-1} dt}{\int_0^{\frac{1}{2}D} \left(\frac{\sinh(\sqrt{-k}t)}{\sqrt{-k}}\right)^{n-1} dt} \\ &= \frac{\int_0^{(r \cdot 2 + \frac{1}{2})D\sqrt{-k}} \sinh^{n-1}(t)\, dt}{\int_0^{\frac{1}{2}D\sqrt{-k}} \sinh^{n-1}(t)\, dt} \\ &= 2^n \left(r \cdot 2 + \frac{1}{2}\right)^n + \cdots \leq 5^n \cdot r^n, \end{aligned}$$

where in the last step we assume that $D\sqrt{-k}$ is very small relative to r. If $b_1 \geq n+1$, this cannot be larger than $|I_r| = (2r + 1)^{b_1}$ when $r = 5^n$. Thus select $r = 5^n$ and the assume $D\sqrt{-k}$ is small enough that

$$\frac{\int_0^{(r \cdot 2 + \frac{1}{2})D\sqrt{-k}} \sinh^{n-1}(t)\, dt}{\int_0^{\frac{1}{2}D\sqrt{-k}} \sinh^{n-1}(t)\, dt} \leq 5^n \cdot r^n$$

in order to force $b_1 \leq n$.   □

Gallot's proof of the above theorem uses techniques that are sophisticated generalizations of the Bochner technique.

2.2. Finiteness of Fundamental Groups. One can get even more information from these volume comparison techniques. Instead of considering just the first homology group, we can actually get some information about fundamental groups as well.

For our next result we need a different kind of understanding of how fundamental groups can be represented.

LEMMA 38. (M. Gromov, 1980) *For a Riemannian manifold M and $\tilde{x} \in \tilde{M}$, we can always find generators $\{\gamma_1, \ldots \gamma_m\}$ for the fundamental group $\Gamma = \pi_1(M)$ such that $d(x, \gamma_i(x)) \leq 2\mathrm{diam}(M)$ and such that all relations for Γ in these generators are of the form $\gamma_i \cdot \gamma_j \cdot \gamma_k^{-1} = 1$.*

PROOF. For any $\varepsilon \in (0, \mathrm{inj}(M))$ choose a triangulation of M such that adjacent vertices in this triangulation are joined by a curve of length less that ε. Let $\{x_1, \ldots, x_k\}$ denote the set of vertices and $\{e_{ij}\}$ the edges joining adjacent vertices (thus, e_{ij} is not necessarily defined for all i, j). If x is the projection of $\tilde{x} \in \tilde{M}$, then join x and x_i by a segment σ_i for all $i = 1, \ldots, k$ and construct the loops

$$\sigma_{ij} = \sigma_i e_{ij} \sigma_j^{-1}$$

for adjacent vertices. Now, any loop in M based at x is homotopic to a loop in the 1-skeleton of the triangulation, i.e., a loop that is constructed out of juxtaposing edges e_{ij}. Since $e_{ij} e_{jk} = e_{ij}\sigma_j^{-1}\sigma_j e_{jk}$ such loops are the product of loops of the form σ_{ij}. Therefore Γ is generated by σ_{ij}.

Now observe that if three vertices x_i, x_j, x_k are adjacent to each other, then they span a 2-simplex $\triangle_{ijk}$. Thus, we have that the loop $\sigma_{ij}\sigma_{jk}\sigma_{ki} = \sigma_{ij}\sigma_{jk}\sigma_{ik}^{-1}$ is homotopically trivial. We claim that these are the only relations needed to describe Γ. To see this, let σ be any loop in the 1-skeleton that is homotopically trivial. Now use that σ in fact contracts in the 2-skeleton. Thus, a homotopy corresponds to a collection of 2-simplices $\triangle_{ijk}$. In this way we can represent the relation $\sigma = 1$ as a product of elementary relations of the form $\sigma_{ij}\sigma_{jk}\sigma_{ik}^{-1} = 1$.

Finally, use discreteness of Γ to get rid of ε as in the above case. $\qquad\square$

A simple example might be instructive here.

EXAMPLE 48. *Consider $M_k = S^3/\mathbb{Z}_k$; the constant-curvature 3-sphere divided out by the cyclic group of order k. As $k \to \infty$ the volume of these manifolds goes to zero, while the curvature is 1 and the diameter $\frac{\pi}{2}$. Thus, the fundamental groups can only get bigger at the expense of having small volume. If we insist on writing the cyclic group $\mathbb{Z}_k$ in the above manner, then the number of generators needed goes to infinity as $k \to \infty$. This is also justified by the next theorem.*

For numbers $n \in \mathbb{N}$, $k \in \mathbb{R}$, and $v, D \in (0, \infty)$, let $\mathfrak{M}(n, k, v, D)$ denote the class of compact Riemannian n-manifolds with

$$\begin{aligned} \mathrm{Ric} &\geq (n-1)k, \\ \mathrm{vol} &\geq v, \\ \mathrm{diam} &\leq D. \end{aligned}$$

We can now prove:

THEOREM 64. (M. Anderson, 1990) *There are only finitely many fundamental groups among the manifolds in $\mathfrak{M}(n, k, v, D)$ for fixed n, k, v, D.*

PROOF. Choose generators $\{\gamma_1, \ldots, \gamma_m\}$ as in the lemma. Since the number of possible relations is bounded by 2^{m^3}, we have reduced the problem to showing that m is bounded. We have that $d(x, \gamma_i(x)) \leq 2D$. Fix a fundamental domain $F \subset \tilde{M}$

that contains x, i.e., a closed set such that $\pi : F \to M$ is onto and $\mathrm{vol}F = \mathrm{vol}M$. One could, for example, choose the Dirichlet domain

$$F = \left\{ z \in \tilde{M} : d\left(x, z\right) \leq d\left(\gamma\left(x\right), z\right) \text{ for all } \gamma \in \pi_1\left(M\right) \right\}.$$

Then we have that the sets $\gamma_i\left(F\right)$ are disjoint up to sets of measure 0, all have the same volume, and all lie in the ball $B\left(x, 4D\right)$. Thus,

$$m \leq \frac{\mathrm{vol}B\left(x, 4D\right)}{\mathrm{vol}F} \leq \frac{v\left(n, k, 4D\right)}{v}.$$

In other words, we have bounded the number of generators in terms of n, D, v, k alone. $\qquad\square$

Another related result shows that groups generated by short loops must in fact be finite.

LEMMA 39. (M. Anderson, 1990) *For fixed numbers $n \in \mathbb{N}$, $k \in \mathbb{R}$, and $v, D \in (0, \infty)$ we can find $L = L\left(n, k, v, D\right)$ and $N = N\left(n, k, v, D\right)$ such that if $M \in \mathfrak{M}\left(n, k, v, D\right)$, then any subgroup of $\pi_1\left(M\right)$ that is generated by loops of length $\leq L$ must have order $\leq N$.*

PROOF. Let $\Gamma \subset \pi_1\left(M\right)$ be a group generated by loops $\left\{\gamma_1, \ldots, \gamma_k\right\}$ of length $\leq L$. Consider the universal covering $\pi : \tilde{M} \to M$ and let $x \in \tilde{M}$ be chosen such that the loops are based at $\pi\left(x\right)$. Then select a fundamental domain $F \subset \tilde{M}$ as above with $x \in F$. Thus for any $\gamma_1, \gamma_2 \in \pi_1\left(M\right)$, either $\gamma_1 = \gamma_2$ or $\gamma_1\left(F\right) \cap \gamma_2\left(F\right)$ has measure 0.

Now define $U\left(r\right)$ as the set of $\gamma \in \Gamma$ such that γ can be written as a product of at most r elements from $\left\{\gamma_1, \ldots, \gamma_k\right\}$. We assumed that $d\left(x, \gamma_i\left(x\right)\right) \leq L$ for all i, and thus $d\left(x, \gamma\left(x\right)\right) \leq r \cdot L$ for all $\gamma \in U\left(r\right)$. This means that $\gamma\left(F\right) \subset B\left(x, r \cdot L + D\right)$. As the sets $\gamma\left(F\right)$ are disjoint up to sets of measure zero, we obtain

$$\begin{aligned} \left|U\left(r\right)\right| &\leq \frac{\mathrm{vol}B\left(x, r \cdot L + D\right)}{\mathrm{vol}F} \\ &\leq \frac{v\left(n, k, r \cdot L + D\right)}{v}. \end{aligned}$$

Now define

$$N = \frac{v\left(n, k, 2D\right)}{v} + 1,$$

$$L = \frac{D}{N}.$$

If Γ has more than N elements we get a contradiction by using $r = N$ as we would have

$$\begin{aligned} \frac{v\left(n, k, 2D\right)}{v} + 1 &= N \\ &\leq \left|U\left(N\right)\right| \\ &\leq \frac{v\left(n, k, 2D\right)}{v}. \end{aligned}$$

$\qquad\square$

3. Manifolds of Nonnegative Ricci Curvature

In this section we shall prove the splitting theorem of Cheeger-Gromoll. This theorem is analogous to the maximal diameter theorem in many ways. It also has far-reaching consequences for compact manifolds with nonnegative Ricci curvature. For instance, we shall see that $S^3 \times S^1$ does not admit any complete metrics with zero Ricci curvature. One of the critical ingredients in the proof of the splitting theorem is the maximum principle for continuous functions. These analytical matters will be taken care of in the first subsection.

3.1. The Maximum Principle. We shall try to understand how one can assign second derivatives to (distance) functions at points where the function is not smooth. In chapter 11 we shall also discuss generalized gradients, but this theory is completely different and works only for Lipschitz functions.

The key observation for our development of generalized Hessians and Laplacians is

LEMMA 40. *If $f, h : (M, g) \to \mathbb{R}$ are C^2 functions such that $f(p) = h(p)$ and $f(x) \geq h(x)$ for all x near p, then*

$$
\begin{aligned}
\nabla f(p) &= \nabla h(p), \\
\mathrm{Hess} f|_p &\geq \mathrm{Hess} h|_p, \\
\Delta f(p) &\geq \Delta h(p).
\end{aligned}
$$

PROOF. If $(M, g) \subset (\mathbb{R}, \mathrm{can})$, then the theorem is simple calculus. In general, We can take $\gamma : (-\varepsilon, \varepsilon) \to M$ to be a geodesic with $\gamma(0) = p$, then use this observation on $f \circ \gamma$, $h \circ \gamma$ to see that

$$
\begin{aligned}
df(\dot{\gamma}(0)) &= dh(\dot{\gamma}(0)), \\
\mathrm{Hess} f(\dot{\gamma}(0), \dot{\gamma}(0)) &\geq \mathrm{Hess} h(\dot{\gamma}(0), \dot{\gamma}(0)).
\end{aligned}
$$

This clearly implies the lemma if we let $v = \dot{\gamma}(0)$ run over all $v \in T_p M$. $\qquad\square$

This lemma implies that a C^2 function $f : M \to \mathbb{R}$ has $\mathrm{Hess} f|_p \geq B$, where B is a symmetric bilinear map on $T_p M$ (or $\Delta f(p) \geq a \in \mathbb{R}$), iff for every $\varepsilon > 0$ there exists a function $f_\varepsilon(x)$ defined in a neighborhood of p such that
   (1) $f_\varepsilon(p) = f(p)$.
   (2) $f(x) \geq f_\varepsilon(x)$ in some neighborhood of p.
   (3) $\mathrm{Hess} f_\varepsilon|_p \geq B - \varepsilon \cdot g|_p$ (or $\Delta f_\varepsilon(p) \geq a - \varepsilon$).
Such functions f_ε are called *support functions from below*. One can analogously use *support functions from above* to find upper bounds for $\mathrm{Hess} f$ and Δf. Support functions are also known as barrier functions in PDE theory.

For a continuous function $f : (M, g) \to \mathbb{R}$ we say that: $\mathrm{Hess} f|_p \geq B$ (or $\Delta f(p) \geq a$) iff there exist smooth support functions f_ε satisfying (1)-(3). One also says that $\mathrm{Hess} f|_p \geq B$ (or $\Delta f(p) \geq a$) hold in the support or barrier sense. In PDE theory there are other important ways of defining weak derivatives. The notion used here is guided by what we can obtain from geometry.

One can easily check that if $(M, g) \subset (\mathbb{R}, \mathrm{can})$, then f is convex if $\mathrm{Hess} f \geq 0$ everywhere. Thus, $f : (M, g) \to \mathbb{R}$ is convex if $\mathrm{Hess} f \geq 0$ everywhere. Using this, one can easily prove

THEOREM 65. *If $f : (M, g) \to \mathbb{R}$ is continuous with* Hess$f \geq 0$ *everywhere, then f is constant near any local maximum. In particular, f cannot have a global maximum unless f is constant.*

We shall need a more general version of this theorem called the maximum principle. As stated below, it was first proved for smooth functions by E. Hopf in 1927 and then later for continuous functions by Calabi in 1958 using the idea of support functions. A continuous function $f : (M, g) \to \mathbb{R}$ with $\Delta f \geq 0$ everywhere is said to be *subharmonic*. If $\Delta f \leq 0$, then f is *superharmonic*.

THEOREM 66. (The Strong Maximum Principle) *If $f : (M, g) \to \mathbb{R}$ is continuous and subharmonic, then f is constant in a neighborhood of every local maximum. In particular, if f has a global maximum, then f is constant.*

PROOF. First, suppose that $\Delta f > 0$ everywhere. Then f can't have any local maxima at all. For if f has a local maximum at $p \in M$, then there would exist a smooth support function $f_\varepsilon(x)$ with

 (1) $f_\varepsilon(p) = f(p)$,
 (2) $f_\varepsilon(x) \leq f(x)$ for all x near p,
 (3) $\Delta f_\varepsilon(p) > 0$.

Here (1) and (2) imply that f_ε must also have a local maximum at p. But this implies that Hess$f_\varepsilon(p) \leq 0$, which contradicts (3).

Next just assume that $\Delta f \geq 0$ and let $p \in M$ be a local maximum for f. For sufficiently small $r < \text{inj}(p)$ we therefore have a function $f : (B(p, r), g) \to \mathbb{R}$ with $\Delta f \geq 0$ and a global maximum at p. If f is constant on $B(p, r)$, then we are done, otherwise, we can assume (by possibly decreasing r) that $f(x) \neq f(p)$ for some

$$x \in S(p, r) = \{x \in M : d(p, x) = r\}.$$

Then define

$$V = \{x \in S(p, r) : f(x) = f(p)\}.$$

Our goal is to construct a smooth function $h = e^{\alpha\varphi} - 1$ such that

$$
\begin{aligned}
h &< 0 \text{ on } V, \\
h(p) &= 0, \\
\Delta h &> 0 \text{ on } \bar{B}(p, r).
\end{aligned}
$$

This function is found by first selecting an open disc $U \subset S(p, r)$ that contains V. We can then find φ such that

$$
\begin{aligned}
\varphi(p) &= 0, \\
\varphi &< 0 \text{ on } U, \\
\nabla\varphi &\neq 0 \text{ on } \bar{B}(p, r).
\end{aligned}
$$

In an appropriate coordinate system $(x^1, \dots, x^n)$ we can simply assume that U lies in the lower half-plane: $x^1 < 0$ and then let $\varphi = x^1$ (see also Figure 9.2). Lastly, choose α so big that

$$\Delta h = \alpha e^{\alpha\varphi}(\alpha|\nabla\varphi|^2 + \Delta\varphi) > 0 \text{ on } \overline{B}(p, r).$$

Now consider the function $\bar{f} = f + \delta h$ on $\overline{B}(p, r)$. Provided δ is very small, this function has a local maximum in the interior $B(p, r)$, since

$$
\begin{aligned}
\bar{f}(p) &= f(p) \\
&> \max\{f(x) + \delta h(x) = \bar{f}(x) : x \in \partial B(p, r)\}.
\end{aligned}
$$

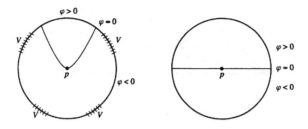

Figure 9.2

On the other hand, we can also show that $\bar{f}$ has positive Laplacian, thus giving a contradiction with the first part of the proof. To see that the Laplacian is positive, select f_ε as a support function from below for f at $q \in B(p, r)$. Then $f_\varepsilon + \delta h$ is a support function from below for $\bar{f}$ at q. The Laplacian of this support function is estimated by

$$\Delta (f_\varepsilon + \delta h)(p) \geq -\varepsilon + \delta \Delta h(p),$$

which for given δ must become positive as $\varepsilon \to 0$.   □

A continuous function $f : (M, g) \to \mathbb{R}$ is said to be *linear* if $\operatorname{Hess} f \equiv 0$ (i.e., both of the inequalities $\operatorname{Hess} f \geq 0$, $\operatorname{Hess} f \leq 0$ hold everywhere). One can easily prove that this implies that

$$(f \circ \gamma)(t) = f(\gamma(0)) + \alpha t$$

for each geodesic γ. This implies that

$$f \circ \exp_p(x) = f(p) + g(v_p, x)$$

for each $p \in M$ and some $v_p \in T_pM$. In particular f is C^∞ with $\nabla f|_p = v_p$.

More generally, we have the concept of a harmonic function. This is a continuous function $f : (M, g) \to \mathbb{R}$ with $\Delta f = 0$. The maximum principle shows that if M is closed, then all harmonic functions are constant. On incomplete or complete open manifolds, however, there are often many harmonic functions. This is in contrast to the existence of linear functions, where ∇f is necessary parallel and therefore splits the manifold locally into a product where one factor is an interval. It is an important fact that any harmonic function is C^∞ if the metric is C^∞. Using the above maximum principle we can reduce this to a standard result in PDE theory (see also chapter 10).

THEOREM 67. (Regularity of harmonic functions) *If* $f : (M, g) \to \mathbb{R}$ *is continuous and harmonic in the weak sense, then* f *is smooth.*

PROOF. We fix $p \in M$ and a neighborhood Ω around p with smooth boundary. We can in addition assume that Ω is contained in a coordinate neighborhood. It is now a standard fact from PDE theory that the following Dirichlet boundary value problem has a solution:

$$\Delta u = 0,$$
$$u|_{\partial\Omega} = f|_{\partial\Omega}.$$

Moreover, such a solution u is smooth on the interior of Ω. Now consider the two functions $u - f$ and $f - u$ on Ω. If they are both nonpositive, then they must vanish and hence $f = u$ is smooth near p. Otherwise one of these functions must be

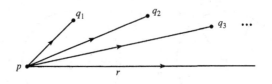

Figure 9.3

positive somewhere. However, as it vanishes on the boundary and is subharmonic this implies that it has an interior global maximum. The maximum principle then shows that the function is constant, but this is only possible if it vanishes. □

3.2. Rays and Lines. We will work only with complete and noncompact manifolds in this section. A *ray* $r(t) : [0, \infty) \to (M, g)$ is a unit speed geodesic such that

$$d(r(t),\ r(s)) = |t - s| \text{ for all } t, s \geq 0.$$

One can think of a ray as a semi-infinite segment or as a segment from $r(0)$ to infinity. A *line* $\ell(t) : \mathbb{R} \to (M, g)$ is a unit speed geodesic such that

$$d(\gamma(t), \gamma(s)) = |t - s| \text{ for all } t, s \in \mathbb{R}.$$

LEMMA 41. *If* $p \in (M, g)$, *then there is always a ray emanating from* p. *If* M *is disconnected at infinity then* (M, g) *contains a line.*

PROOF. Let $p \in M$ and consider a sequence $q_i \to \infty$. Find unit vectors $v_i \in T_p M$ such that:

$$\sigma_i(t) = \exp_p(tv_i),\ t \in [0, d(p, q_i)]$$

is a segment from p to q_i. By possibly passing to a subsequence, we can assume that $v_i \to v \in T_p M$ (see Figure 9.3). Now

$$\sigma(t) = \exp_p(tv),\ t \in [0, \infty),$$

becomes a segment. This is because σ_i converges pointwise to σ by continuity of $\exp_p$, and thus

$$d(\sigma(s),\ \sigma(t)) = \lim d(\sigma_i(s),\ \sigma_i(t)) = |s - t|.$$

A complete manifold is *connected at infinity* if for every compact set $K \subset M$ there is a compact set $C \supset K$ such that any two points in $M - C$ can be joined by a curve in $M - K$. If M is not connected at infinity, we say that M is *disconnected at infinity*.

If M is disconnected at infinity, we can obviously find a compact set K and sequences of points $p_i \to \infty$, $q_i \to \infty$ such that any curve from p_i to q_i must pass through K. If we join these points by segments $\sigma_i : (-a_i, b_i) \to M$ such that $a_i, b_i \to \infty$, $\sigma_i(0) \in K$, then the sequence will subconverge to a line (see Figure 9.4). □

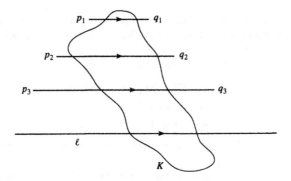

Figure 9.4

EXAMPLE 49. *Surfaces of revolution $dr^2 + \varphi^2(r)ds^2_{n-1}$, where $\varphi : [0,\infty) \to [0,\infty)$ and $\dot{\varphi}(t) < 1$, $\ddot{\varphi}(t) < 0$, $t > 0$, cannot contain any lines. These manifolds look like paraboloids.*

EXAMPLE 50. *Any complete metric on $S^{n-1} \times \mathbb{R}$ must contain a line, since the manifold is disconnected at infinity.*

EXAMPLE 51. *The Schwarzschild metric on $S^2 \times \mathbb{R}^2$ does not contain any lines. This will also follow from our main result in this section.*

THEOREM 68. (The Splitting Theorem, Cheeger-Gromoll, 1971): *If (M, g) contains a line and has $\mathrm{Ric} \geq 0$, then (M, g) is isometric to a product $(H \times \mathbb{R}, g_0 + dt^2)$.*

The proof is quite involved and will require several constructions. The main idea is to find a distance function $r : M \to \mathbb{R}$ (i.e. $|\nabla r| \equiv 1$) that is linear (i.e. $\mathrm{Hess}\, r \equiv 0$). Having found such a function, one can easily see that $M = U_0 \times \mathbb{R}$, where $U_0 = \{r = 0\}$ and $g = dt^2 + g_0$. The maximum principle will play a key role in showing that r, when it has been constructed, is both smooth and linear. Recall that in the proof of the maximal diameter theorem we used two distance functions $r, \tilde{r}$ placed at maximal distance from each other and then proceeded to show that $r + \tilde{r} = $ constant. This implied that $r, \tilde{r}$ were smooth, except at the two chosen points, and that Δr is exactly what it is in constant curvature. We then used the rigidity part of the Cauchy-Schwarz inequality to compute $\mathrm{Hess}\, r$. In the construction of our linear distance function we shall do something similar. In this situation the two ends of the line play the role of the points at maximal distance. Using this line we will construct two distance functions $b_\pm$ from infinity that are continuous, satisfy $b_+ + b_- \geq 0$ (from the triangle inequality), $\Delta b_\pm \leq 0$, and $b_+ + b_- = 0$ on the line. Thus, $b_+ + b_-$ is superharmonic and has a global minimum. The minimum principle will therefore show that $b_+ + b_- \equiv 0$. Thus, $b_+ = -b_-$ and

$$0 \geq \Delta b_+ = -\Delta b_- \geq 0,$$

which shows that both of $b_\pm$ are harmonic and therefore C^∞. We then show that they are actually distance functions (i.e., $|\nabla b_\pm| \equiv 1$). Finally we can conclude that

$$
\begin{aligned}
0 &= \nabla b_\pm(\Delta b_\pm) + \frac{(\Delta b_\pm)^2}{n-1} \\
&\leq \nabla b_\pm(\Delta b_\pm) + |\mathrm{Hess} b_\pm|^2 \\
&= |\mathrm{Hess} b_\pm|^2 \\
&\leq -\mathrm{Ric}(\nabla b_\pm, \nabla b_\pm) \leq 0.
\end{aligned}
$$

This establishes that $|\mathrm{Hess} b_\pm|^2 = 0$, so that we have two linear distance functions $b_\pm$ as desired.

The proof proceeds through several results some of which we will need later.

3.3. Laplacian Comparison.

LEMMA 42. (E. Calabi, 1958) *Let* $r(x) = d(x,p)$, $p \in (M,g)$. *If* $\mathrm{Ric}(M,g) \geq 0$, *then*

$$
\Delta r(x) \leq \frac{n-1}{r(x)} \text{ for all } x \in M.
$$

PROOF. We know that the result is true whenever r is smooth. For any other $q \in M$, choose a unit speed segment $\sigma : [0, \ell] \to M$ with $\sigma(0) = p$, $\sigma(\ell) = q$. Then the triangle inequality implies that $r_\varepsilon(x) = \varepsilon + d(\sigma(\varepsilon), x)$ is a support function from above for r at q. If all these support functions are smooth at q, then

$$
\begin{aligned}
\Delta r_\varepsilon(q) &\leq \frac{n-1}{r_\varepsilon(q) - \varepsilon} \\
&= \frac{n-1}{r(q) - \varepsilon} \\
&\leq \frac{n-1}{r(q)} + \varepsilon \cdot \frac{2(n-1)}{(r(q))^2}
\end{aligned}
$$

for small ε, and hence $\Delta r(q) \leq \frac{n-1}{r(q)}$ in the support sense.

Now for the smoothness. Fix $\varepsilon > 0$ and suppose r_ε is not smooth at q. Then we know that either
(1) there are two segments from $\sigma(\varepsilon)$ to q, or
(2) q is a critical value for $\exp_{\sigma(\varepsilon)} : \mathrm{seg}(\sigma(\varepsilon)) \to M$.

Case (1) would give us a nonsmooth curve of length ℓ from p to q, which we know is impossible. Thus, case (2) must hold. To get a contradiction out of this, we show that this implies that $\exp_q$ has $\sigma(\varepsilon)$ as a critical value.

Using that q is critical for $\exp_{\sigma(\varepsilon)}$, we find a Jacobi field $J(t) : [\varepsilon, \ell] \to TM$ along $\sigma|_{[\varepsilon, \ell]}$ such that $J(\varepsilon) = 0$, $\dot{J}(\varepsilon) \neq 0$ and $J(\ell) = 0$ (see chapter 6). Then also $\dot{J}(\ell) \neq 0$ as it solves a linear second order equation. Running backwards from q to $\sigma(\varepsilon)$ then shows that $\exp_q$ is critical at $\sigma(\varepsilon)$. This however contradicts that $\sigma : [0, \ell] \to M$ is a segment. $\square$

By a similar analysis, we can prove

LEMMA 43. *If* (M,g) *is complete and* $\mathrm{Ric}(M,g) \geq (n-1)k$, *then any distance function* $r(x) = d(x,p)$ *satisfies:*

$$
\Delta r(x) \leq (n-1)\frac{\mathrm{sn}_k'(r(x))}{\mathrm{sn}_k(r(x))}.
$$

This lemma together with the maximum principle allows us to eliminate the use of relative volume comparison in the proof of Cheng's diameter theorem.

As in the other proof, consider $\tilde{r}(x) = d(x, q)$, $r(x) = d(x, p)$, where $d(p, q) = \pi/\sqrt{k}$. Then we have $r + \tilde{r} \geq \pi/\sqrt{k}$, and equality will hold for any $x \in M - \{p, q\}$ that lies on a segment joining p and q. On the other hand, the above lemma tells us that

$$
\begin{aligned}
\Delta(r + \tilde{r}) &\leq \Delta r + \Delta \tilde{r} \\
&\leq (n-1)\sqrt{k}\cot(\sqrt{k}r(x)) + (n-1)\sqrt{k} \cdot (\sqrt{k}\tilde{r}(x)) \\
&\leq (n-1)\sqrt{k}\cot(\sqrt{k}r(x)) + (n-1)\sqrt{k}\cot\left(\sqrt{k}\left(\frac{\pi}{\sqrt{k}} - r(x)\right)\right) \\
&= (n-1)\sqrt{k}(\cot(\sqrt{k}r(x)) + \cot(\pi - \sqrt{k}r(x))) = 0.
\end{aligned}
$$

So $r + \tilde{r}$ is superharmonic on $M - \{p, q\}$ and has a global minimum on this set. Thus, the minimum principle tells us that $r + \tilde{r} = \pi/\sqrt{k}$ on M. The proof can now be completed as before.

3.4. Busemann Functions. For the rest of this section we fix a complete noncompact Riemannian manifold (M, g) with nonnegative Ricci curvature. Let $\gamma : [0, \infty) \to (M, g)$ be a unit speed ray, and define

$$
b_t(x) = d(x, \gamma(t)) - t.
$$

PROPOSITION 40. *(1) For fixed x, the function $t \to b_t(x)$ is decreasing and bounded in absolute value by $d(x, \gamma(0))$.*
(2) $|b_t(x) - b_t(y)| \leq d(x, y)$.
(3) $\Delta b_t(x) \leq \frac{n-1}{b_t+t}$ everywhere.

PROOF. (2) and (3) are obvious, since $b_t(x) + t$ is a distance function from $\gamma(t)$. For (1), first observe that the triangle inequality implies

$$
|b_t(x)| = |d(x, \gamma(t) - t| = |d(x, \gamma(t)) - d(\gamma(0), \gamma(t))| \leq d(x, \gamma(0)).
$$

Second, if $s < t$ then

$$
\begin{aligned}
b_t(x) - b_s(x) &= d(x, \gamma(t)) - t - d(x, \gamma(s)) + s \\
&= d(x, \gamma(t)) - d(x, \gamma(s)) - d(\gamma(t), \gamma(s)) \\
&\leq d(\gamma(t), \gamma(s)) - d(\gamma(t), \gamma(s)) = 0.
\end{aligned}
$$

$\square$

This proposition shows that the family of functions $\{b_t\}_{t \geq 0}$ forms a pointwise bounded equicontinuous family that is also pointwise decreasing. Thus, b_t must converge to a distance-decreasing function b_γ satisfying

$$
\begin{aligned}
|b_\gamma(x) - b_\gamma(y)| &\leq d(x, y), \\
|b_\gamma(x)| &\leq d(x, \gamma(0)),
\end{aligned}
$$

and

$$
b_\gamma(\gamma(r)) = \lim b_t(\gamma(r)) = \lim(d(\gamma(r), \gamma(t)) - t) = -r.
$$

This function b_γ is called the *Busemann function* for γ and should be interpreted as a distance function from "$\gamma(\infty)$."

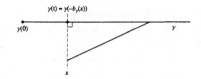

Figure 9.5

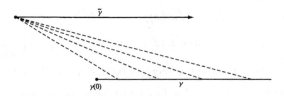

Figure 9.6

EXAMPLE 52. *If $M = (\mathbb{R}^n, \mathrm{can})$, then all Busemann functions are of the form*

$$b_\gamma(x) = \gamma(0) - \dot{\gamma}(0) \cdot x$$

(see Figure 9.5).

The level sets $b_\gamma^{-1}(t)$ are called *horospheres*. In $(\mathbb{R}^n, \mathrm{can})$ these are obviously hyperplanes.

Given our ray γ, as before, and $p \in M$, consider a family of unit speed segments $\sigma_t : [0, \ell_t] \to (M, g)$ from p to $\gamma(t)$. As when we constructed rays, this family must subconverge to some ray $\tilde{\gamma} : [0, \infty) \to M$, with $\tilde{\gamma}(0) = p$. A ray coming from such a construction is called an *asymptote* for γ from p (see Figure 9.6). Such asymptotes from p need not be unique.

PROPOSITION 41. *(1) $b_\gamma(x) \leq b_\gamma(p) + b_{\tilde{\gamma}}(x)$.*
(2) $b_\gamma(\tilde{\gamma}(t)) = b_\gamma(p) + b_{\tilde{\gamma}}(\tilde{\gamma}(t)) = b_\gamma(p) - t$.

PROOF. Let $\sigma_i : [0, \ell_i] \to (M, g)$ be the segments converging to $\tilde{\gamma}$. To check (1), observe that

$$
\begin{aligned}
d(x, \gamma(s)) - s \quad &\leq \quad d(x, \tilde{\gamma}(t)) + d(\tilde{\gamma}(t), \gamma(s)) - s \\
&= \quad d(x, \tilde{\gamma}(t)) - t + d(p, \tilde{\gamma}(t)) + d(\tilde{\gamma}(t), \gamma(s)) - s \\
&\to \quad d(x, \tilde{\gamma}(t)) - t + d(p, \tilde{\gamma}(t)) + b_\gamma(\tilde{\gamma}(t)) \quad \text{as} \quad s \to \infty.
\end{aligned}
$$

Thus, we see that (1) is true provided that (2) is true. To establish (2), we notice that

$$d(p, \gamma(t_i)) = d(p, \sigma_i(s)) + d(\sigma_i(s), \gamma(t_i))$$

for some sequence $t_i \to \infty$. Now, $\sigma_i(s) \to \tilde{\gamma}(s)$, so we obtain

$$
\begin{aligned}
b_\gamma(p) \quad &= \quad \lim(d(p, \gamma(t_i)) - t_i) \\
&= \quad \lim(d(p, \tilde{\gamma}(s)) + d(\tilde{\gamma}(s), \gamma(t_i)) - t_i) \\
&= \quad d(p, \tilde{\gamma}(s)) + \lim(d(\tilde{\gamma}(s), \gamma(t_i)) - t_i) \\
&= \quad s + b_\gamma(\tilde{\gamma}(s)) \\
&= \quad -b_{\tilde{\gamma}}(\tilde{\gamma}(s)) + b_\gamma(\tilde{\gamma}(s)).
\end{aligned}
$$

$\square$

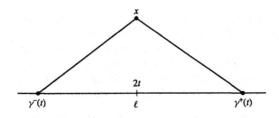

Figure 9.7

We have now shown that b_γ has $b_\gamma(p) + b_{\tilde{\gamma}}$ as support function from above at $p \in M$.

LEMMA 44. *If* $\mathrm{Ric}(M, g) \geq 0$, *then* $\Delta b_\gamma \leq 0$ *everywhere.*

PROOF. Since $b_\gamma(p) + b_{\tilde{\gamma}}$ is a support function from above at p, we only need to check that $\Delta b_{\tilde{\gamma}}(p) \leq 0$. To see this, observe that the functions

$$b_t(x) = d(x, \tilde{\gamma}(t)) - t$$

are actually support functions from above for $b_{\tilde{\gamma}}$ at p. Furthermore, these functions are smooth at p with

$$\Delta b_t(p) \leq \frac{n-1}{t} \to 0 \text{ as } t \to \infty.$$

$\square$

Now suppose (M, g) has $\mathrm{Ric} \geq 0$ and contains a line $\gamma(t) : \mathbb{R} \to M$. Let b^+ be the Busemann function for $\gamma : [0, \infty) \to M$, and b^- the Busemann function for $\gamma : (-\infty, 0] \to M$. Thus,

$$b^+(x) = \lim_{t \to +\infty} (d(x, \gamma(t)) - t),$$
$$b^-(x) = \lim_{t \to +\infty} (d(x, \gamma(-t)) - t).$$

Clearly,

$$b^+(x) + b^-(x) = \lim_{t \to +\infty} (d(x, \gamma(t)) + d(x, \gamma(-t)) - 2t),$$

so by the triangle inequality

$$(b^+ + b^-)(x) \geq 0 \text{ for all } x.$$

Moreover,

$$(b^+ + b^-)(\gamma(t)) = 0$$

since γ is a line (see Figure 9.7).

This gives us a function $b^+ + b^-$ with $\Delta(b^+ + b^-) \leq 0$ and a global minimum at $\gamma(t)$. The minimum principle then shows that $b^+ + b^- = 0$ everywhere. In particular, $b^+ = -b^-$ and $\Delta b^+ = \Delta b^- = 0$ everywhere.

To finish the proof of the splitting theorem, we still need to show that $b^\pm$ are distance functions, i.e. $|\nabla b^\pm| \equiv 1$. To see this, let $p \in M$ and construct asymptotes $\tilde{\gamma}^\pm$ for $\gamma^\pm$ from p. Then consider

$$b_t^\pm(x) = d(x, \tilde{\gamma}^\pm(t)) - t,$$

and observe:

$$b_t^+(x) \geq b^+(x) - b^+(p) = -b^-(x) + b^-(p) \geq -b_t^-(x)$$

with equality holding for $x = p$. Since both $b_t^{\pm}$ are smooth at p with unit gradient, we must therefore have that $\nabla b_t^+(p) = -\nabla b_t^-(p)$. Then also, $b^{\pm}$ must be differentiable at p with unit gradient. We have therefore shown (without using that $b^{\pm}$ are smooth from $\Delta b^{\pm} = 0$) that $b^{\pm}$ are everywhere differentiable with unit gradient. The result that harmonic functions are smooth can now be invoked and the proof is finished as explained in the beginning of the section.

3.5. Structure Results in Nonnegative Ricci Curvature. The splitting theorem gives several nice structure results for compact manifolds with nonnegative Ricci curvature.

COROLLARY 26. $S^p \times S^1$ does not admit any Ricci flat metrics when $p = 2, 3$.

PROOF. The universal covering is $S^p \times \mathbb{R}$, As this space is disconnected at infinity any metric with nonnegative Ricci curvature must split. If the original metric is Ricci flat, then after the splitting, we will get a Ricci flat metric on S^p. If $p \leq 3$, such a metric must also be flat. But we know that S^p, $p = 2, 3$ do not admit any flat metrics. $\square$

When $p \geq 4$ it is not known whether S^p admits a Ricci flat metric.

THEOREM 69. (Structure Theorem for Nonnegative Ricci Curvature, Cheeger-Gromoll, 1971) *Suppose* (M, g) *is a compact Riemannian manifold with* Ric ≥ 0. *Then the universal cover* $(\widetilde{M}, \tilde{g})$ *splits isometrically as a product* $N \times \mathbb{R}^p$, *where* N *is a compact manifold.*

PROOF. By the splitting theorem, we can write $\widetilde{M} = N \times \mathbb{R}^p$, where N does not contain any lines. Observe that if

$$\gamma(t) = (\gamma_1(t), \gamma_2(t)) \in N \times \mathbb{R}^p$$

is a geodesic, then both γ_i are geodesics, and if γ is a line, then both γ_i are also lines unless they are constant. Thus, all lines in $\widetilde{M}$ must be of the form $\gamma(t) = (x, \sigma(t))$, where $x \in N$ and σ is a line in $\mathbb{R}^p$.

If N is not compact, then it must contain a ray $\gamma(t) : [0, \infty) \to N$. If $\pi : \widetilde{M} \to M$ is the covering map, then we can consider $c(t) = \pi \circ (\gamma(t), 0)$ in M. This is of course a geodesic in M, and since M is compact, there must be a sequence $t_i \to \infty$ such that $\dot{c}(t_i) \to v \in T_x M$ for some $x \in M$, $v \in T_x M$. Choose $\tilde{x} \in \widetilde{M}$ such that $\pi(\tilde{x}) = x$, and consider lifts $\gamma_i(t) : [-t_i, \infty) \to \widetilde{M}$ of $c(t + t_i)$, where $D\pi(\dot{\gamma}_i(0)) = \dot{c}(t_i)$ and $\gamma_i(0) \to \tilde{x}$. On the one hand, these geodesics converge to a geodesic $\hat{\gamma} : (-\infty, \infty) \to \widetilde{M}$ with $\hat{\gamma}(0) = \tilde{x}$. On the other hand, since $D\pi(\dot{\gamma}(t_i)) = \dot{c}(t_i)$, there must be deck transformations $g_i \in \pi_1(M)$ such that $g_i \circ \gamma(t + t_i) = \gamma_i(t)$. Thus, the γ_is are rays and must converge to a line. From our earlier observations, this line must be in $\mathbb{R}^p$. The deck transformations g_i therefore map $\dot{\gamma}(t + t_i)$, which are tangent to N, to vectors that are almost perpendicular to N. This, however, contradicts the following property for isometries on $\widetilde{M}$.

Let $F : \widetilde{M} \to \widetilde{M}$ be an isometry, e.g., $F = g_i$. If $\ell(t)$ is a line in $\widetilde{M}$, then $F \circ \ell$ must also be a line in $\widetilde{M}$. Since all lines in $\widetilde{M}$ lie in $\mathbb{R}^p$ and every vector tangent to $\mathbb{R}^p$ is the velocity of some line, we see that for each $c \in N$ we can find $F_1(c) \in N$ such that

$$F : \{c\} \times \mathbb{R}^p \to \{F_1(c)\} \times \mathbb{R}^p.$$

This implies that F must be of the form $F = (F_1, F_2)$, where $F_1 : N \to N$ is an isometry and $F_2 : N \times \mathbb{R}^p \to \mathbb{R}^p$. In particular, the tangent bundles TN and $T\mathbb{R}^p$ are preserved by DF.   □

This theorem also gives a strong structure for $\pi_1(M)$. Consider the group G of isometries on N that are split off by the action of $\pi_1(M)$ on $\widetilde{M} = N \times \mathbb{R}^p$. Since N is compact and G acts discretely on N we see that it is finite. The kernel of the homomorphism $\pi_1(M) \to G$ is then a finite index subgroup that acts discretely and cocompactly on $\mathbb{R}^p$. Such groups are known as *crystallographic groups* and are fairly well understood. It is a theorem of Bieberbach that any group of isometries $\Gamma \subset \mathrm{Iso}(\mathbb{R}^p)$ that is discrete and cocompact must contain a rank p Abelian group $\mathbb{Z}^p$ of finite index. This structure comes from the exact sequence

$$1 \to \mathbb{R}^p \to \mathrm{Iso}(\mathbb{R}^p) \to O(p) \to 1,$$

where the map $\mathrm{Iso}(\mathbb{R}^p) \to O(p)$ is the assignment that takes the isometry $Ox + v$ to O. If we restrict this short exact sequence to Γ we see that the kernel is an Abelian subgroup of $\mathbb{R}^p$ which acts discretely. This shows that it must be of the form $\mathbb{Z}^q$ where $q \le p$. If $q < p$, then the action of $\mathbb{Z}^q$ leaves the q-dimensional subspace $V = \mathrm{span}\{\mathbb{Z}^q\}$ invariant and therefore fixes the orthogonal complement. This shows that the action can't be cocompact. Finally we also note that the image in $O(p)$ is discrete and hence finite. (For more details see also [**34**], [**96**]). Note that there are non-discrete actions of $\mathbb{Z}^n$ on $\mathbb{R}$ for any $n \ge 1$. To see this simply take real numbers $\alpha_1, ..., \alpha_n$ that are linearly independent over $\mathbb{Q}$ and use these as a basis for the action. Note, however, that all orbits of this action are dense so it is not a discrete action.

We can now prove some further results about the structure of compact manifolds with nonnegative Ricci curvature.

COROLLARY 27. *Suppose (M, g) is a complete, compact Riemannian manifold with $\mathrm{Ric} \ge 0$. If M is $K(\pi, 1)$, i.e., the universal cover is contractible, then the universal covering is Euclidean space and (M, g) is a flat manifold.*

PROOF. We know that $\widetilde{M} = \mathbb{R}^p \times C$, where C is compact. The only way in which this space can be contractible is if C is contractible. But the only compact manifold that is contractible is the one-point space.   □

COROLLARY 28. *If (M, g) is compact with $\mathrm{Ric} \ge 0$ and has $\mathrm{Ric} > 0$ on some tangent space $T_p M$, then $\pi_1(M)$ is finite.*

PROOF. Since $\mathrm{Ric} > 0$ on an entire tangent space, the universal cover cannot split into a product $\mathbb{R}^p \times C$, where $p \ge 1$. Thus, the universal covering is compact.
                                                                                  □

COROLLARY 29. *If (M, g) is compact and has $\mathrm{Ric} \ge 0$, then $b_1(M) \le \dim M = n$, with equality holding iff (M, g) is a flat torus.*

PROOF. We always have a surjection

$$h : \pi_1(M) \to H_1(M, \mathbb{Z}),$$

that maps loops to cycles. The above mentioned structure result for the fundamental groups shows that we have a finite index subgroup $\mathbb{Z}^p \subset \pi_1(M)$ with $p \le n$. The image $h(\mathbb{Z}^p) \subset H_1(M, \mathbb{Z})$ is therefore also of finite index. This shows that the rank of the torsion free part of $H_1(M, \mathbb{Z})$ must be $\leq p$. In case $b_1 = n$, we must have that $p = n$ as $h(\mathbb{Z}^p)$ otherwise couldn't have finite index. This shows in addition that

$$h|_{\mathbb{Z}^n} : \mathbb{Z}^n \to H_1(M, \mathbb{Z})$$

has trivial kernel as the image otherwise couldn't have finite index. Thus $\widetilde{M} = \mathbb{R}^n$ as p was the dimension of the Euclidean factor. Consequently M is flat. We now observe that the kernel of

$$h : \pi_1(M) \to H_1(M, \mathbb{Z})$$

has to be a finite subgroup as it does not intersect the finite index subgroup $\mathbb{Z}^n \subset \pi_1(M)$. Since all isometries on $\mathbb{R}^n$ of finite order have a fixed point we have shown that the inclusion $\mathbb{Z}^n \subset \pi_1(M)$ is an isomorphism. This shows that M is a torus. $\square$

The penultimate result is a bit stronger than simply showing that $H^1(M, \mathbb{R}) = 0$ as we did using the Bochner technique. The last result is equivalent to Bochner's theorem, but the proof is quite a bit different.

4. Further Study

The adventurous reader could consult [47] for further discussions. Anderson's article [2] contains the finiteness results for fundamental groups mentioned here and also some interesting examples of manifolds with nonnegative Ricci curvature. For the examples with almost maximal diameter we refer the reader to [3] and [74]. It is also worthwhile to consult the original paper on the splitting theorem [27] and the elementary proof of it in [37]. We already mentioned in chapter 7 Gallot's contributions to Betti number bounds, and the reference [40] works here as well. The reader should also consult the articles by Colding, Perel'man, and Zhu in [50] to get an idea of how rapidly this subject has grown in the past few years.

5. Exercises

(1) With notation as in the first section:
$$d\mathrm{vol} = \lambda dr \wedge d\mathrm{vol}_{n-1}.$$

Show that $\mu = \lambda^{\frac{1}{n-1}}$ satisfies

$$\partial_r^2 \mu \leq -\frac{\mu}{n-1} \mathrm{Ric}(\partial_r, \partial_r),$$
$$\mu(0, \theta) = 0,$$
$$\lim_{r \to 0} \partial_r \mu(r, \theta) = 1.$$

This can be used to show the desired estimates for the volume form as well.

(2) Assume the distance function $r = d(\cdot, p)$ is smooth on $B(p, R)$. If in our usual polar coordinates

$$\mathrm{Hess}\, r = \frac{\mathrm{sn}'_k(r)}{\mathrm{sn}_k(r)} g_r,$$

then all sectional curvatures on $B(p, R)$ are equal to k.

(3) Show that if (M, g) has $\mathrm{Ric} \geq (n-1)k$ and for some $p \in M$ we have $\mathrm{vol} B(p, R) = v(n, k, R)$, then the metric has constant curvature k on $B(p, R)$.

(4) Let X be a vector field on a Riemannian manifold and consider $F_t(p) = \exp_p(tX|_p)$.

   (a) For $v \in T_pM$ show that $J(t) = DF_t(v)$ is a Jacobi field along $t \to \gamma(t) = \exp(tX)$ with the initial conditions $J(0) = v$, $\dot{J}(0) = \nabla_v X$.

   (b) Select an orthonormal basis e_i for T_pM and let $J_i(t) = DF_t(e_i)$. Show that

$$(\det(DF_t))^2 = \det(g(J_i(t), J_j(t))).$$

   (c) Show that as long as $\det(DF_t) \neq 0$ it satisfies

$$\frac{d^2(\det(DF_t))^{\frac{1}{n}}}{dt^2} \leq -\frac{(\det(DF_t))^{\frac{1}{n}}}{n} \operatorname{Ric}(\dot{\gamma}, \dot{\gamma}).$$

   Hint: Use that any $n \times n$ matrix satisfies $(\operatorname{tr}(A))^2 \leq n\operatorname{tr}(A^*A)$.

(5) Show that a complete manifold (M, g) with the property that

$$\operatorname{Ric} \geq 0,$$

$$\lim_{r \to \infty} \frac{\operatorname{vol}B(p, r)}{\omega_n r^n} = 1,$$

for some $p \in M$, must be isometric to Euclidean space.

(6) *(Cheeger)* The relative volume comparison estimate can be generalized as follows: Suppose (M, g) has $\operatorname{Ric} \geq (n-1)k$ and dimension n.

   (a) Select points $p_1, \ldots, p_k \in M$. Then the function

$$r \to \frac{\operatorname{vol}\left(\bigcup_{i=1}^k B(p_i, r)\right)}{v(n, k, r)}$$

   is nonincreasing and converges to k as $r \to 0$.

   (b) If $A \subset M$, then

$$r \to \frac{\operatorname{vol}\left(\bigcup_{p \in A} B(p, r)\right)}{v(n, k, r)}$$

   is nonincreasing. To prove this, use the above with the finite collection of points taken to be very dense in A.

(7) The absolute volume comparison can also be slightly generalized. Namely, for $p \in M$ and a subset $\Gamma \subset T_pM$ of unit vectors, consider the cones defined in polar coordinates:

$$B^\Gamma(p, r) = \{(t, \theta) \in M : t \leq r \text{ and } \theta \in \Gamma\}.$$

If $\operatorname{Ric} M \geq (n-1)k$, show that

$$\operatorname{vol}B^\Gamma(p, r) \leq \operatorname{vol}\Gamma \cdot \int_0^r (\operatorname{sn}_k(t))^{n-1} dt.$$

(8) Let G be a compact connected Lie group with a bi-invariant metric. Use the results from this chapter to prove

   (a) If G has finite center, then G has finite fundamental group.

   (b) A finite covering of G looks like $G' \times T^k$, where G' is compact simply connected, and T^k is a torus.

   (c) If G has finite fundamental group, then the center is finite.

(9) Show that a compact Riemannian manifold with irreducible restricted holonomy and $\operatorname{Ric} \geq 0$ has finite fundamental group.

(10) Let (M, g) be an n-dimensional Riemannian manifold that is isometric to Euclidean space outside some compact subset $K \subset M$, i.e., $M - K$ is isometric to $\mathbb{R}^n - C$ for some compact set $C \subset \mathbb{R}^n$. If $\mathrm{Ric}_g \geq 0$, show that $M = \mathbb{R}^n$. (In chapter 7 we gave two different hints for this problem, here is a third. Use the splitting theorem.)

(11) Show that if $\mathrm{Ric} \geq n - 1$, then $\mathrm{diam} \leq \pi$, by showing that if $d(p, q) > \pi$, then
$$e_{p,q}(x) = d(p, x) + d(x, q) - d(p, q)$$
has negative Laplacian at a local minimum.

CHAPTER 10

Convergence

In this chapter we will give an introduction to several of the convergence ideas for Riemannian manifolds. The goal is to understand what it means for a sequence of Riemannian manifolds, or more generally metric spaces, to converge to a space. In the first section we develop the weakest convergence concept: Gromov-Hausdorff convergence. We then go on to explain some of the elliptic regularity theory we need for some of the later developments that use stronger types of convergence. In section 3 we develop the idea of norms of Riemannian manifolds. This is a concept developed by the author in the hope that it will make it easier to understand convergence theory as a parallel to the easier Hölder theory for functions (as is explained in section 2.) At the same time, we also feel that it has made some parts of the theory more concise. In this section we examine some stronger convergence ideas that were developed by Cheeger and Gromov and study their relation to the norms of manifolds. These preliminary discussions will enable us in subsequent sections to establish the convergence theorem of Riemannian geometry and its generalizations by Anderson and others. These convergence theorems contain the Cheeger finiteness theorem stating that certain very general classes of Riemannian manifolds contain only finitely many diffeomorphism types.

The idea of measuring the distance between subspaces of a given space goes back to Hausdorff and was extensively studied in the Polish and Russian schools of topology. The more abstract versions we use here seem to begin with Shikata's proof of the differentiable sphere theorem. In Cheeger's thesis, the idea that abstract manifolds can converge to each other is also evident. In fact, as we shall see below, he proved his finiteness theorem by showing that certain classes of manifolds are precompact in various topologies. After these two early forays into convergence theory it wasn't until Gromov bombarded the mathematical community with his highly original approaches to geometry that the theory developed further. He introduced a very weak kind of convergence that is simply an abstract version of Hausdorff distance. The first use of this new idea was to prove a group-theoretic question about the nilpotency of groups with polynomial growth. Soon after the introduction of this weak convergence, the earlier ideas on strong convergence by Cheeger resurfaced. There are various conflicting accounts on who did what and when. Certainly, the Russian school, notably Nikolaev and Berestovskii, deserve a lot of credit for their work on synthetic geometry, which could and should have been used in the convergence context. It appears that they were concerned primarily with studying generalized metrics in their own right. By contrast, the western school studied convergence and thereby developed an appreciation for studying Riemannian manifolds with little regularity, and even metric spaces.

1. Gromov-Hausdorff Convergence

1.1. Hausdorff Versus Gromov Convergence. At the beginning of the twentieth century, Hausdorff introduced what we call the *Hausdorff distance* between subsets of a metric space. If (X,d) is the metric space and $A, B \subset X$, then we define

$$
\begin{aligned}
d(A,B) &= \inf\{d(a,b) : a \in A, b \in B\}, \\
B(A,\varepsilon) &= \{x \in X : d(x,A) < \varepsilon\}, \\
d_H(A,B) &= \inf\{\varepsilon : A \subset B(B,\varepsilon), B \subset B(A,\varepsilon)\}.
\end{aligned}
$$

Thus, $d(A,B)$ is small if some points in these sets are close, while the *Hausdorff distance* $d_H(A,B)$ is small iff every point of A is close to a point in B and vice versa. One can easily see that the Hausdorff distance defines a metric on the closed subsets of X and that this collection is compact when X is compact.

We shall concern ourselves only with compact metric spaces and *proper* metric spaces. The latter have by definition proper distance functions, i.e., all closed balls are compact. This implies, in particular, that the spaces are separable, complete, and locally compact.

Around 1980, Gromov extended this concept to a distance between abstract metric spaces. If X and Y are metric spaces, then an *admissible* metric on the disjoint union $X \amalg Y$ is a metric that extends the given metrics on X and Y. With this we can define the *Gromov-Hausdorff distance* as

$$
d_{G-H}(X,Y) = \inf\{d_H(X,Y) : \text{admissible metrics on } X \amalg Y\}.
$$

Thus, we try to put a metric on $X \amalg Y$ such that X and Y are as close as possible in the Hausdorff distance, with the constraint that the extended metric restricts to the given metrics on X and Y. In other words, we are trying to define distances between points in X and Y without violating the triangle inequality.

EXAMPLE 53. *If Y is the one-point space, then*

$$
\begin{aligned}
d_{G-H}(X,Y) &\leq \operatorname{rad}X \\
&= \inf_{y \in X} \sup_{x \in X} d(x,y) \\
&= \text{radius of smallest ball covering } X.
\end{aligned}
$$

EXAMPLE 54. *By defining $d(x,y) = D/2$, where $\operatorname{diam}X, \operatorname{diam}Y \leq D$ and $x \in X$, $y \in Y$ we see that*

$$
d_{G-H}(X,Y) \leq D/2.
$$

Let $(\mathcal{M}, d_{G-H})$ denote the collection of compact metric spaces. We shall study this class as a metric space in its own right. To justify this we must show that only isometric spaces are within distance zero of each other.

PROPOSITION 42. *If X and Y are compact metric spaces with $d_{G-H}(X,Y) = 0$, then X and Y are isometric.*

PROOF. Choose a sequence of metrics d_i on $X \amalg Y$ such that the Hausdorff distance between X and Y in this metric is $< i^{-1}$. Then we can find (possibly discontinuous) maps

$$
\begin{aligned}
I_i &: \quad X \to Y, \text{ where } d_i(x, I_i(x)) \leq i^{-1}, \\
J_i &: \quad Y \to X, \text{ where } d_i(y, J_i(y)) \leq i^{-1}.
\end{aligned}
$$

Using the triangle inequality and that d_i restricted to either X or Y is the given metric d on these spaces yields

$$\begin{aligned}
d\left(I_i\left(x_1\right), I_i\left(x_2\right)\right) &\leq 2i^{-1} + d\left(x_1, x_2\right), \\
d\left(J_i\left(y_1\right), J_i\left(y_2\right)\right) &\leq 2i^{-1} + d\left(y_1, y_2\right), \\
d\left(x, J_i \circ I_i\left(x\right)\right) &\leq 2i^{-1}, \\
d\left(y, I_i \circ J_i\left(y\right)\right) &\leq 2i^{-1}.
\end{aligned}$$

We construct $I : X \to Y$ and $J : Y \to X$ as limits of these maps in the same way the Arzela-Ascoli lemma is proved. For each x the sequence $\left(I_i\left(x\right)\right)$ in Y has an accumulation point since Y is compact. As in the Arzela-Ascoli lemma select a dense countable set $A \subset X$. Using a diagonal argument select a subsequence I_{i_j} such that $I_{i_j}\left(a\right) \to I\left(a\right)$ for all $a \in A$. The first inequality now shows that I is distance decreasing on A. In particular, it is uniformly continuous and therefore has a unique extension to a map $I : X \to Y$, which is also distance decreasing. In a similar fashion we also get a distance decreasing map $J : Y \to X$.

The last two inequalities imply that I and J are inverses to each other. It then follows that both I and J are isometries. □

Both symmetry and the triangle inequality are easily established for d_{G-H}. Thus, $\left(\mathcal{M}, d_{G-H}\right)$ is a pseudometric space, and if we consider equivalence classes of isometric spaces it becomes a metric space. In fact, as we shall see, this metric space is both complete and separable. First we show how spaces can be approximated by finite metric spaces.

EXAMPLE 55. *Let X be compact and $A \subset X$ a finite subset such that every point in X is within distance ε of some element in A, i.e., $d_H\left(A, X\right) \leq \varepsilon$. Such sets A are called ε-dense in X. It is then clear that if we use the metric on A induced by X, then also $d_{G-H}\left(X, A\right) \leq \varepsilon$. The importance of this remark is that for any $\varepsilon > 0$ we can in fact find such finite subsets of X, since X is compact.*

EXAMPLE 56. *Suppose we have ε-dense subsets*

$$\begin{aligned}
A &= \left\{x_1, \ldots, x_k\right\} \subset X, \\
B &= \left\{y_1, \ldots, y_k\right\} \subset Y,
\end{aligned}$$

with the further property that

$$\left|d\left(x_i, x_j\right) - d\left(y_i, y_j\right)\right| \leq \varepsilon, \ 1 \leq i, j \leq k.$$

Then $d_{G-H}\left(X, Y\right) \leq 3\varepsilon$. We already have that the finite subsets are ε-close to the spaces, so by the triangle inequality it suffices to show that $d_{G-H}\left(A, B\right) \leq \varepsilon$. For this we must exhibit a metric d on $A \amalg B$ that makes A and B ε-Hausdorff close. Define

$$\begin{aligned}
d\left(x_i, y_i\right) &= \varepsilon, \\
d\left(x_i, y_j\right) &= \min_k \left\{d\left(x_i, x_k\right) + \varepsilon + d\left(y_j, y_k\right)\right\}.
\end{aligned}$$

Thus, we have extended the given metrics on A and B in such a way that no points from A and B get identified, and in addition the potential metric is symmetric. It *then remains to check the triangle inequality. Here we must show*

$$d(x_i, y_j) \leq d(x_i, z) + d(y_j, z),$$
$$d(x_i, x_j) \leq d(y_k, x_i) + d(y_k, x_j),$$
$$d(y_i, y_j) \leq d(x_k, y_i) + d(x_k, y_j).$$

It suffices to check the first two cases as the third is similar to the second. In the first one we can assume that $z = x_k$. Then we can find l such that

$$d(y_j, x_k) = \varepsilon + d(y_j, y_l) + d(x_l, x_k).$$

Hence,

$$
\begin{aligned}
d(x_i, x_k) + d(y_j, x_k) &= d(x_i, x_k) + \varepsilon + d(y_j, y_l) + d(x_l, x_k) \\
&\geq d(x_i, x_l) + \varepsilon + d(y_j, y_l) \\
&\geq d(x_i, y_j).
\end{aligned}
$$

For the second case select l, m with

$$d(y_k, x_i) = d(y_k, y_l) + \varepsilon + d(x_l, x_i),$$
$$d(y_k, x_j) = d(y_k, y_m) + \varepsilon + d(x_m, x_j).$$

Then, using our assumption about the comparability of the metrics on A and B, we have

$$
\begin{aligned}
d(y_k, x_i) + d(y_k, x_j) &= d(y_k, y_l) + \varepsilon + d(x_l, x_i) + d(y_k, y_m) + \varepsilon + d(x_m, x_j) \\
&\geq d(x_k, x_l) + d(x_l, x_i) + d(x_k, x_m) + d(x_m, x_j) \\
&\geq d(x_i, x_j).
\end{aligned}
$$

EXAMPLE 57. *Suppose $M_k = S^3/\mathbb{Z}_k$ with the usual metric induced from $S^3(1)$. Then we have a Riemannian submersion $M_k \to S^2(1/2)$ whose fibers have diameter $2\pi/k \to 0$ as $k \to \infty$. Using the previous example, we can therefore easily check that $M_k \to S^2(1/2)$ in the Gromov-Hausdorff topology.*

One can similarly see that the Berger metrics $(S^3, g_\varepsilon) \to S^2(1/2)$ as $\varepsilon \to 0$. Notice that in both cases the volume goes to zero, but the curvatures and diameters are uniformly bounded. In the second case the manifolds are even simply connected. It should also be noted that the topology changes rather drastically from the sequence to the limit, and in the first case the elements of the sequence even have mutually different fundamental groups.

PROPOSITION 43. *The "metric space" $(\mathcal{M}, d_{G-H})$ is separable and complete.*

PROOF. To see that it is separable, first observe that the collection of all finite metric spaces is dense in this collection. Now take the countable collection of all finite metric spaces that in addition have the property that all distances are rational. Clearly, this collection is dense as well.

To show completeness, select a Cauchy sequence $\{X_n\}$. To show convergence of this sequence, it suffices to check that some subsequence is convergent. Select a subsequence $\{X_i\}$ such that $d_{G-H}(X_i, X_{i+1}) < 2^{-i}$ for all i. Then select metrics $d_{i,i+1}$ on $X_i \amalg X_{i+1}$ making these spaces 2^{-i}-Hausdorff close. Now define a metric $d_{i,i+j}$ on $X_i \amalg X_{i+j}$ by

$$d_{i,i+j}(x_i, x_{i+j}) = \min_{\{x_{i+k} \in X_{i+k}\}} \left\{ \sum_{k=0}^{j-1} d(x_{i+k}, x_{i+k+1}) \right\}.$$

We have then defined a metric d on $Y = \amalg_i X_i$ with the property that in this metric $d_H(X_i, X_{i+j}) \leq 2^{-i+1}$. This metric space is not complete, but the "boundary" of the completion is exactly our desired limit space. To define it, first consider

$$\hat{X} = \{\{x_i\} : x_i \in X_i \text{ and } d(x_i, x_j) \to 0 \text{ as } i, j \to \infty\}.$$

This space has a pseudometric defined by

$$d(\{x_i\}, \{y_i\}) = \lim_{i \to \infty} d(x_i, y_i).$$

Given that we are only considering Cauchy sequences $\{x_i\}$, this must yield a metric on the quotient space X, obtained by the equivalence relation

$$\{x_i\} \sim \{y_i\} \text{ iff } d(\{x_i\}, \{y_i\}) = 0.$$

Now we can extend the metric on Y to one on $X \amalg Y$ by declaring

$$d(x_k, \{x_i\}) = \lim_{i \to \infty} d(x_k, x_i).$$

Using that $d_H(X_j, X_{j+1}) \leq 2^{-j}$, we can for any $x_i \in X_i$ find a sequence $\{x_{i+j}\} \in \hat{X}$ such that $x_{i+0} = x_i$ and $d(x_{i+j}, x_{i+j+1}) \leq 2^{-j}$. Then we must have $d(x_i, \{x_{i+j}\}) \leq 2^{-i+1}$. Thus, every X_i is 2^{-i+1}-close to the limit space X. Conversely, for any given sequence $\{x_i\}$ we can find an equivalent sequence $\{y_i\}$ with the property that $d(y_k, \{y_i\}) \leq 2^{-k+1}$ for all k. Thus, X is 2^{-i+1}-close to X_i. $\square$

From the proof of this theorem we get the useful information that Gromov-Hausdorff convergence can always be thought of as Hausdorff convergence. In other words, if we know that $X_i \to X$ in the Gromov-Hausdorff sense, then after possibly passing to a subsequence, we can assume that there is a metric on $X \amalg (\amalg_i X_i)$ in which X_i Hausdorff converges to X. With such a selection of a metric, it then makes sense to say that $x_i \to x$, where $x_i \in X_i$ and $x \in X$. We shall often use this without explicitly mentioning a choice of ambient metric on $X \amalg (\amalg_i X_i)$.

There is an equivalent way of picturing convergence. For a compact metric space X, let $C(X)$ denote the continuous functions on X, and $L^\infty(X)$ the bounded measurable functions with the sup-norm (not the essential sup-norm). We know that $L^\infty(X)$ is a Banach space. When X is bounded, we construct a map $X \to L^\infty(X)$, by sending x to the continuous function $d(x, \cdot)$. This is usually called the *Kuratowski embedding* when we consider it as a map into $C(X)$. From the triangle inequality, we can easily see that this is in fact a distance-preserving map. Thus, any compact metric space is isometric to a subset of some Banach space $L^\infty(X)$. The important observation now is that two such spaces $L^\infty(X)$ and $L^\infty(Y)$ are isometric if the spaces X and Y are Borel equivalent (there exists a measurable bijection). Also, if $X \subset Y$, then $L^\infty(X)$ sits isometrically as a linear subspace of $L^\infty(Y)$. Now recall that any compact metric space is Borel equivalent to some subset of $[0, 1]$. Thus all compact metric spaces X are isometric to some subset of $L^\infty([0, 1])$. We can then define

$$d_{G-H}(X, Y) = \inf d_H(i(X), j(Y)),$$

where $i : X \to L^\infty([0, 1])$ and $j : Y \to L^\infty([0, 1])$ are distance-preserving maps.

1.2. Pointed Convergence. So far, we haven't really dealt with noncompact spaces. There is, of course, nothing wrong with defining the Gromov-Hausdorff distance between unbounded spaces, but it will almost never be finite. In order to change this, we should have in mind what is done for convergence of functions on unbounded domains. There, one usually speaks about convergence on compact subsets. To do something similar, we first define the pointed Gromov-Hausdorff distance

$$d_{G-H}\left((X,x),(Y,y)\right) = \inf\left\{d_H\left(X,Y\right) + d\left(x,y\right)\right\}.$$

Here we take as usual the infimum over all Hausdorff distances and in addition require the selected points to be close. The above results are still true for this modified distance. We can then introduce the Gromov-Hausdorff topology on the collection of proper pointed metric spaces $\mathcal{M}_* = \{(X,x,d)\}$ in the following way: We say that

$$(X_i, x_i, d_i) \to (X, x, d)$$

in the *pointed Gromov-Hausdorff topology* if for all R, the closed metric balls

$$\left(\bar{B}\left(x_i, R\right), x_i, d_i\right) \to \left(\bar{B}\left(x, R\right), x, d\right)$$

converge with respect to the pointed Gromov-Hausdorff metric.

1.3. Convergence of Maps. We shall also have recourse to speak about *convergence of maps*. Suppose we have

$$\begin{aligned} f_k &: & X_k \to Y_k, \\ X_k &\to& X, \\ Y_k &\to& Y. \end{aligned}$$

Then we say that f_k converges to $f : X \to Y$ if for every sequence $x_k \in X_k$ converging to $x \in X$ we have that $f_k\left(x_k\right) \to f\left(x\right)$. This definition obviously depends in some sort of way on having the spaces converge in the Hausdorff sense, but we shall ignore this. It is also a very strong kind of convergence for if we assume that $X_k = X$, $Y_k = Y$, and $f_k = f$, then f can converge to itself only if it is continuous.

Note also that convergence of functions preserves such properties as being distance preserving or submetries.

Another useful observation is that we can regard the sequence of maps f_k as one continuous map

$$F : (\amalg_i X_i) \to Y \amalg (\amalg_i Y_i).$$

The sequence converges iff this map has an extension

$$X \amalg (\amalg_i X_i) \to Y \amalg (\amalg_i Y_i),$$

in which case the limit map is the restriction to X. Thus, a sequence is convergent iff the map

$$F : (\amalg_i X_i) \to Y \amalg (\amalg_i Y_i)$$

is uniformly continuous.

A sequence of functions as above is called *equicontinuous,* if for every $\varepsilon > 0$ there is an $\delta > 0$ such that

$$f_k\left(B\left(x_k, \delta\right)\right) \subset B\left(f_k\left(x_k\right), \varepsilon\right)$$

for all k and $x_k \in X_k$. A sequence is therefore equicontinuous if, for example, all the functions are Lipschitz continuous with the same Lipschitz constant. As for standard equicontinuous sequences, we have the Arzela-Ascoli lemma:

LEMMA 45. *An equicontinuous family* $f_k : X_k \to Y_k$, *where* $X_k \to X$, *and* $Y_k \to Y$ *in the (pointed) Gromov-Hausdorff topology, has a convergent subsequence. When the spaces are not compact, we also assume that* f_k *preserves the base point.*

PROOF. The standard proof carries over without much change. Namely, first choose dense subsets

$$A_i = \{a_1^i, a_2^i, \ldots\} \subset X_i$$

such that the sequences

$$\{a_j^i\} \to a_j \in X.$$

Then also, $A = \{a_j\} \subset X$ is dense. Next, use a diagonal argument to find a subsequence of functions that converge on the above sequences. Finally, show that this sequence converges as promised. $\square$

1.4. Compactness of Classes of Metric Spaces.
We now turn our attention to conditions that ensure convergence of spaces. More precisely we want some good criteria for when a collection of (pointed) spaces is precompact (i.e., closure is compact).

For a compact metric space X, define the capacity and covering as follows

$$\text{Cap}\,(\varepsilon) \;=\; \text{Cap}_X\,(\varepsilon) = \text{maximum number of disjoint } \frac{\varepsilon}{2}\text{-balls in } X,$$

$$\text{Cov}\,(\varepsilon) \;=\; \text{Cov}_X\,(\varepsilon) = \text{minimum number of } \varepsilon\text{-balls it takes to cover } X.$$

First, we observe that $\text{Cov}\,(\varepsilon) \leq \text{Cap}\,(\varepsilon)$. To see this select disjoint balls $B\left(x_i, \frac{\varepsilon}{2}\right)$, then consider the collection $B\left(x_i, \varepsilon\right)$. In case the latter do not cover X there exists $x \in X - \cup B\left(x_i, \varepsilon\right)$. This would imply that $B\left(x, \frac{\varepsilon}{2}\right)$ is disjoint from all of the balls $B\left(x_i, \frac{\varepsilon}{2}\right)$. Thus showing that the former balls do not form a maximal disjoint family.

Another important observation is that if two compact metric spaces X and Y satisfy $d_{G-H}\,(X, Y) < \delta$, then it follows from the triangle inequality that:

$$\text{Cov}_X\,(\varepsilon + 2\delta) \;\leq\; \text{Cov}_Y\,(\varepsilon),$$
$$\text{Cap}_X\,(\varepsilon) \;\geq\; \text{Cap}_Y\,(\varepsilon + 2\delta).$$

With this information we can now characterize precompact classes of compact metric spaces.

PROPOSITION 44. (M. Gromov, 1980) *For a class* $\mathcal{C} \subset (\mathcal{M}, d_{G-H})$, *the following statements are equivalent:*

(1) $\mathcal{C}$ *is precompact, i.e., every sequence in* $\mathcal{C}$ *has a subsequence that is convergent in* $(\mathcal{M}, d_{G-H})$.

(2) There is a function $N_1\,(\varepsilon) : (0, \alpha) \to (0, \infty)$ *such that* $\text{Cap}_X\,(\varepsilon) \leq N_1\,(\varepsilon)$ *for all* $X \in \mathcal{C}$.

(3) There is a function $N_2\,(\varepsilon) : (0, \alpha) \to (0, \infty)$ *such that* $\text{Cov}_X\,(\varepsilon) \leq N_2\,(\varepsilon)$ *for all* $X \in \mathcal{C}$.

PROOF. (1) $\Rightarrow$ (2): If $\mathcal{C}$ is precompact, then for every $\varepsilon > 0$ we can find $X_1, \ldots, X_k \in \mathcal{C}$ such that for any $X \in \mathcal{C}$ we have that $d_{G-H}(X, X_i) < \frac{\varepsilon}{4}$ for some i. Then

$$\text{Cap}_X(\varepsilon) \le \text{Cap}_{X_i}\left(\frac{\varepsilon}{2}\right) \le \max_i \text{Cap}_{X_i}\left(\frac{\varepsilon}{2}\right).$$

This gives a bound for $\text{Cap}_X(\varepsilon)$ for each $\varepsilon > 0$.

(2) $\Rightarrow$ (3) Use $N_2 = N_1$.

(3) $\Rightarrow$ (1): It suffices to show that $\mathcal{C}$ is totally bounded, i.e., for each $\varepsilon > 0$ we can find finitely many metric spaces $X_1, \ldots, X_k \in \mathcal{M}$ such that any metric space in $\mathcal{C}$ is within ε of some X_i in the Gromov-Hausdorff metric. Since

$$\text{Cov}_X\left(\frac{\varepsilon}{2}\right) \le N\left(\frac{\varepsilon}{2}\right),$$

we know that any $X \in \mathcal{C}$ is within $\frac{\varepsilon}{2}$ of a finite subset with at most $N\left(\frac{\varepsilon}{2}\right)$ elements in it. Using the induced metric we think of these finite subsets as finite metric spaces. Next, observe that

$$\text{diam}X \le 2\delta\text{Cov}_X(\delta)$$

for any fixed δ. This means that these finite metric spaces have no distances that are bigger than $\varepsilon N\left(\frac{\varepsilon}{2}\right)$. The metric on such a finite metric space then consists of a matrix (d_{ij}), $1 \le i, j \le N\left(\frac{\varepsilon}{2}\right)$, where each entry satisfies $d_{ij} \in \left[0, \varepsilon N\left(\frac{\varepsilon}{2}\right)\right]$. From among all such finite metric spaces it is then possible to select a finite number of them such that any of the matrices (d_{ij}) is within $\frac{\varepsilon}{2}$ of one matrix from the finite selection of matrices. This means that the spaces are within $\frac{\varepsilon}{2}$ of each other. We have then found the desired finite collection of metric spaces. $\square$

As a corollary we can also get a precompactness theorem in the pointed category.

COROLLARY 30. *A collection $\mathcal{C} \subset \mathcal{M}_*$ is precompact iff for each $R > 0$ the collection*

$$\{B(x, R) : B(x, R) \subset (X, x) \in \mathcal{C}\} \subset (\mathcal{M}, d_{G-H})$$

is precompact.

Using the relative volume comparison theorem we can now show

COROLLARY 31. *For any integer $n \ge 2$, $k \in \mathbb{R}$, and $D > 0$ we have that the following classes are precompact:*

(1) The collection of closed Riemannian n-manifolds with $\text{Ric} \ge (n-1)k$ and diam $\le D$.

(2) The collection of pointed complete Riemannian n-manifolds with $\text{Ric} \ge (n-1)k$.

PROOF. It suffices to prove (2). Fix $R > 0$. We have to show that there can't be too many disjoint balls inside $B(x, R) \subset M$. To see this, suppose $B(x_1, \varepsilon), \ldots,$ $B(x_\ell, \varepsilon) \subset B(x, R)$ are disjoint. If $B(x_i, \varepsilon)$ is the ball with the smallest volume, we have

$$\ell \le \frac{\text{vol}B(x, R)}{\text{vol}B(x_i, \varepsilon)} \le \frac{\text{vol}B(x_i, 2R)}{\text{vol}B(x_i, \varepsilon)} \le \frac{v(n, k, 2R)}{v(n, k, \varepsilon)}.$$

This gives the desired bound. $\square$

It seems intuitively clear that an n-dimensional space should have $\mathrm{Cov}\left(\varepsilon\right) \sim$ ε^{-n} as $\varepsilon \to 0$. In fact, the Minkowski dimension of a metric space is defined as

$$\dim X = \limsup_{\varepsilon \to 0} \frac{\log \mathrm{Cov}\left(\varepsilon\right)}{-\log \varepsilon}.$$

This definition will in fact give the right answer for Riemannian manifolds. Some fractal spaces might, however, have nonintegral dimension. Now observe that

$$\frac{v\left(n, k, 2R\right)}{v\left(n, k, \varepsilon\right)} \sim \varepsilon^{-n}.$$

Therefore, if we can show that covering functions carry over to limit spaces, then we will have shown that manifolds with lower curvature bounds can only collapse in dimension.

LEMMA 46. *Let* $\mathcal{C}\left(N\left(\varepsilon\right)\right)$ *be the collection of metric spaces with* $\mathrm{Cov}\left(\varepsilon\right) \leq$ $N\left(\varepsilon\right)$. *Suppose* N *is continuous. Then* $\mathcal{C}\left(N\left(\varepsilon\right)\right)$ *is compact.*

PROOF. We already know that this class is precompact. So we only have to show that if $X_i \to X$ and $\mathrm{Cov}_{X_i}\left(\varepsilon\right) \leq N\left(\varepsilon\right)$, then also $\mathrm{Cov}_X\left(\varepsilon\right) \leq N\left(\varepsilon\right)$. This follows easily from

$$\mathrm{Cov}_X\left(\varepsilon\right) \leq \mathrm{Cov}_{X_i}\left(\varepsilon - 2d_{G-H}\left(X, X_i\right)\right) \leq N\left(\varepsilon - 2d_{G-H}\left(X, X_i\right)\right),$$

and

$$N\left(\varepsilon - 2d_{G-H}\left(X, X_i\right)\right) \to N\left(\varepsilon\right) \text{ as } i \to \infty.$$

$\square$

2. Hölder Spaces and Schauder Estimates

First, we shall define the Hölder norms and Hölder spaces. We will then briefly discuss the necessary estimates we need for elliptic operators for later applications. The standard reference for all the material here is the classic book by Courant and Hilbert [**30**], especially chapter IV, and the thorough text [**44**], especially chapters 1-6. A more modern text that also explains how PDE's are used in geometry, including some of the facts we need, is [**90**], especially vol. III.

2.1. Hölder Spaces. Let us fix a bounded domain $\Omega \subset \mathbb{R}^n$. The bounded continuous functions from Ω to $\mathbb{R}^k$ are denoted by $C^0\left(\Omega, \mathbb{R}^k\right)$, and we use the sup-norm, denoted by

$$\|u\|_{C^0} = \sup_{x \in \Omega} |u\left(x\right)|,$$

on this space. This makes $C^0\left(\Omega, \mathbb{R}^k\right)$ into a Banach space. We wish to generalize this so that we still have a Banach space, but in addition also take into account derivatives of the functions. The first natural thing to do is to define $C^m\left(\Omega, \mathbb{R}^k\right)$ as the functions with m continuous partial derivatives. Using multi-index notation, we define

$$\partial^i u = \partial_1^{i_1} \cdots \partial_n^{i_n} u = \frac{\partial^l u}{\partial\left(x^1\right)^{i_1} \cdots \partial\left(x^n\right)^{i_n}},$$

where $i = \left(i_1, \ldots, i_n\right)$ and $l = |i| = i_1 + \cdots + i_n$. Then the C^m-norm is

$$\|u\|_{C^m} = \sup_{x \in \Omega} |u\left(x\right)| + \sum_{1 \leq |i| \leq m} \sup_{\Omega} \left|\partial^i u\right|.$$

This norm does result in a Banach space, but the inclusions

$$C^m\left(\Omega,\mathbb{R}^k\right)\subset C^{m-1}\left(\Omega,\mathbb{R}^k\right)$$

do not yield closed subspaces. For instance, $f\left(x\right)=|x|$ is in the closure of

$$C^1\left([-1,1]\,,\mathbb{R}\right)\subset C^0\left([-1,1]\,,\mathbb{R}\right).$$

To accommodate this problem, we define for each $\alpha\in(0,1]$ the C^α-pseudonorm of $u:\Omega\to\mathbb{R}^k$ as

$$\|u\|_\alpha=\sup_{x,y\in\Omega}\frac{|u\left(x\right)-u\left(y\right)|}{|x-y|^\alpha}.$$

When $\alpha=1$, this gives the best Lipschitz constant for u.

Define the *Hölder space* $C^{m,\alpha}\left(\Omega,\mathbb{R}^k\right)$ as being the functions in $C^m\left(\Omega,\mathbb{R}^k\right)$ such that all mth-order partial derivatives have finite C^α-pseudonorm. On this space we use the norm

$$\|u\|_{C^{m,\alpha}}=\|u\|_{C^m}+\sum_{|i|=m}\|\partial^i u\|_\alpha.$$

If we wish to be specific about the domain, then we write

$$\|u\|_{C^{m,\alpha},\Omega}.$$

We can now show

LEMMA 47. $C^{m,\alpha}\left(\Omega,\mathbb{R}^k\right)$ *is a Banach space with the $C^{m,\alpha}$-norm. Furthermore, the inclusion*

$$C^{m,\alpha}\left(\Omega,\mathbb{R}^k\right)\subset C^{m,\beta}\left(\Omega,\mathbb{R}^k\right),$$

where $\beta<\alpha$ is always compact, i.e., it maps closed bounded sets to compact sets.

PROOF. We only need to show this in the case where $m=0$; the more general case is then a fairly immediate consequence.

First, we must show that any Cauchy sequence $\{u_i\}$ in $C^\alpha\left(\Omega,\mathbb{R}^k\right)$ converges. Since it is also a Cauchy sequence in $C^0\left(\Omega,\mathbb{R}^k\right)$ we have that $u_i\to u\in C^0$ in the C^0-norm. For fixed $x\neq y$ observe that

$$\frac{|u_i\left(x\right)-u_i\left(y\right)|}{|x-y|^\alpha}\to\frac{|u\left(x\right)-u\left(y\right)|}{|x-y|^\alpha}.$$

As the left-hand side is uniformly bounded, we also get that the right-hand side is bounded, thus showing that $u\in C^\alpha$.

Finally select $\varepsilon>0$ and N so that for $i,j\geq N$ and $x\neq y$

$$\frac{|\left(u_i\left(x\right)-u_j\left(x\right)\right)-\left(u_i\left(y\right)-u_j\left(y\right)\right)|}{|x-y|^\alpha}\leq\varepsilon.$$

If we let $j\to\infty$, this shows that

$$\frac{|\left(u_i\left(x\right)-u\left(x\right)\right)-\left(u_i\left(y\right)-u\left(y\right)\right)|}{|x-y|^\alpha}\leq\varepsilon.$$

Hence $u_i\to u$ in the C^α-topology.

Now for the last statement. A bounded sequence in $C^\alpha\left(\Omega,\mathbb{R}^k\right)$ is equicontinuous so the inclusion

$$C^\alpha\left(\Omega,\mathbb{R}^k\right)\subset C^0\left(\Omega,\mathbb{R}^k\right)$$

is compact. We then use

$$\frac{|u(x) - u(y)|}{|x - y|^{\beta}} = \left(\frac{|u(x) - u(y)|}{|x - y|^{\alpha}}\right)^{\beta/\alpha} \cdot |u(x) - u(y)|^{1 - \beta/\alpha}$$

to conclude that

$$\|u\|_{\beta} \leq (\|u\|_{\alpha})^{\beta/\alpha} \cdot (2 \cdot \|u\|_{C^0})^{1 - \beta/\alpha}.$$

Therefore, a sequence that converges in C^0 and is bounded in C^{α}, also converges in C^{β}, as long as $\beta < \alpha \leq 1$. $\square$

2.2. Elliptic Estimates. We now turn our attention to *elliptic operators*. We shall consider equations of the form

$$Lu = a^{ij}\partial_i\partial_j u + b^i\partial_i u = f,$$

where $a^{ij} = a^{ji}$. The operator is called *elliptic* if the matrix (a^{ij}) is positive definite. Throughout we assume that all eigenvalues for (a^{ij}) lie in some interval $[\lambda, \lambda^{-1}]$, $\lambda > 0$, and that the coefficients are bounded

$$\begin{aligned}\|a^{ij}\|_{\alpha} &\leq \lambda^{-1}, \\ \|b^i\|_{\alpha} &\leq \lambda^{-1}.\end{aligned}$$

Let us state without proof the a priori estimates, usually called the *Schauder estimates*, or *elliptic estimates*, that we shall need.

THEOREM 70. *Let $\Omega \subset \mathbb{R}^n$ be an open domain of diameter $\leq D$ and $K \subset \Omega$ a subdomain such that $d(K, \partial\Omega) \geq \delta$. Moreover assume $\alpha \in (0, 1)$, then there is a constant $C = C(n, \alpha, \lambda, \delta, D)$ such that*

$$\|u\|_{C^{2,\alpha}, K} \leq C\left(\|Lu\|_{C^{\alpha}, \Omega} + \|u\|_{C^{\alpha}, \Omega}\right),$$

$$\|u\|_{C^{1,\alpha}, K} \leq C\left(\|Lu\|_{C^0, \Omega} + \|u\|_{C^{\alpha}, \Omega}\right).$$

Furthermore, if Ω has smooth boundary and $u = \varphi$ on $\partial\Omega$, then there is a constant $C = C(n, \alpha, \lambda, D)$ such that on all of Ω we have

$$\|u\|_{C^{2,\alpha}, \Omega} \leq C\left(\|Lu\|_{C^{\alpha}, \Omega} + \|\varphi\|_{C^{2,\alpha}, \partial\Omega}\right).$$

One way of proving these results is to establish them first for the simplest operator:

$$Lu = \Delta u = \delta^{ij}\partial_i\partial_j u.$$

Then observe that a linear change of coordinates shows that we can handle operators with constant coefficients:

$$Lu = \Delta u = a^{ij}\partial_i\partial_j u.$$

Finally, Schauder's trick is that the assumptions about the functions a^{ij} imply that they are almost constant locally. A partition of unity type argument then finishes the analysis.

The first-order term doesn't cause much trouble and can even be swept under the rug in the case where the operator is in divergence form:

$$Lu = a^{ij}\partial_i\partial_j u + b^i\partial_i u = \partial_i\left(a^{ij}\partial_j u\right).$$

Such operators are particularly nice when one wishes to use integration by parts, as we have

$$\int_{\Omega} \left(\partial_i\left(a^{ij}\partial_j u\right)\right) h = -\int_{\Omega} a^{ij}\partial_j u\partial_i h$$

when $h = 0$ on $\partial\Omega$. This is interesting in the context of geometric operators, as the Laplacian on manifolds in local coordinates looks like

$$
\begin{aligned}
Lu &= \Delta_g u \\
&= \frac{1}{\sqrt{\det g_{ij}}} \partial_i \left(\sqrt{\det g_{ij}} \cdot g^{ij} \cdot \partial_j u \right).
\end{aligned}
$$

The above theorem has an almost immediate corollary.

COROLLARY 32. *If in addition we assume that* $\left\| a^{ij} \right\|_{C^{m,\alpha}}$, $\left\| b^i \right\|_{C^{m,\alpha}} \le \lambda^{-1}$, *then there is a constant* $C = C(n, m, \alpha, \lambda, \delta, D)$ *such that*

$$
\|u\|_{C^{m+2,\alpha}, K} \le C \left(\|Lu\|_{C^{m,\alpha}, \Omega} + \|u\|_{C^\alpha, \Omega} \right).
$$

And on a domain with smooth boundary,

$$
\|u\|_{C^{m+2,\alpha}, \Omega} \le C \left(\|Lu\|_{C^{m,\alpha}, \Omega} + \|\varphi\|_{C^{m+2,\alpha}, \partial\Omega} \right).
$$

The Schauder estimates can be used to show that the Dirichlet problem always has a unique solution.

THEOREM 71. *Suppose* $\Omega \subset \mathbb{R}^n$ *is a bounded domain with smooth boundary, then the Dirichlet problem*

$$
\begin{aligned}
Lu &= f, \\
u|_{\partial\Omega} &= \varphi
\end{aligned}
$$

always has a unique solution $u \in C^{2,\alpha}(\Omega)$ *if* $f \in C^\alpha(\Omega)$ *and* $\varphi \in C^{2,\alpha}(\partial\Omega)$.

Observe that uniqueness is an immediate consequence of the maximum principle. The existence part requires a bit more work.

2.3. Harmonic Coordinates. The above theorem makes it possible to introduce *harmonic coordinates* on Riemannian manifolds.

LEMMA 48. *If* (M, g) *is an* n-*dimensional Riemannian manifold and* $p \in M$, *then there is a neighborhood* $U \ni p$ *on which we can find a harmonic coordinate system*

$$
x = \left(x^1, \dots, x^n \right) : U \to \mathbb{R}^n,
$$

i.e., a coordinate system such that the functions x^i *are harmonic with respect to the Laplacian on* (M, g).

PROOF. First select a coordinate system $y = (y^1, \dots, y^n)$ on a neighborhood around p such that $y(p) = 0$. We can then think of M as being an open subset of $\mathbb{R}^n$ and $p = 0$. The metric g is written as

$$
g = g_{ij} = g(\partial_i, \partial_j) = g \left(\frac{\partial}{\partial y^i}, \frac{\partial}{\partial y^j} \right)
$$

in the standard Cartesian coordinates $(y^1, \dots, y^n)$. We must then find a coordinate transformation $y \to x$ such that

$$
\Delta x^k = \frac{1}{\sqrt{\det g_{ij}}} \partial_i \left(\sqrt{\det g_{ij}} \cdot g^{ij} \cdot \partial_j x^k \right) = 0
$$

To find these coordinates, fix a small ball $B(0,\varepsilon)$ and solve the Dirichlet problem

$$\begin{aligned}
\Delta x^k &= 0 \\
x^k &= y^k \text{ on } \partial B(0,\varepsilon)
\end{aligned}$$

We have then found n harmonic functions that should be close to the original coordinates. The only problem is that we don't know if they actually are coordinates. The Schauder estimates tell us that

$$\begin{aligned}
\|x - y\|_{C^{2,\alpha}, B(0,\varepsilon)} &\leq C \left(\|\Delta(x-y)\|_{C^{\alpha}, B(0,\varepsilon)} + \left\| (x-y)_{|\partial B(0,\varepsilon)} \right\|_{C^{2,\alpha}, \partial B(0,\varepsilon)} \right) \\
&= C \|\Delta y\|_{C^{\alpha}, B(0,\varepsilon)}.
\end{aligned}$$

If matters were arranged such that

$$\|\Delta y\|_{C^{\alpha}, B(0,\varepsilon)} \to 0 \text{ as } \varepsilon \to 0,$$

then we could conclude that Dx and Dy are close for small ε. Since y does form a coordinates system, we would then also be able to conclude that x formed a coordinate system.

Now we just observe that if y were chosen as exponential Cartesian coordinates, then we would have that $\partial_k g_{ij} = 0$ at p. The formula for Δy then shows that $\Delta y = 0$ at p. Hence, we have

$$\|\Delta y\|_{C^{\alpha}, B(0,\varepsilon)} \to 0 \text{ as } \varepsilon \to 0.$$

Finally recall that the constant C depends only on an upper bound for the diameter of the domain aside from α, n, λ. Thus,

$$\|x - y\|_{C^{2,\alpha}, B(0,\varepsilon)} \to 0 \text{ as } \varepsilon \to 0.$$

$\square$

One reason for using harmonic coordinates on Riemannian manifolds is that both the Laplacian and Ricci curvature tensor have particularly nice formulae in such coordinates.

LEMMA 49. *Let (M,g) be an n-dimensional Riemannian manifold and suppose we have a harmonic coordinate system $x : U \to \mathbb{R}^n$. Then*

(1) $\Delta u = \frac{1}{\sqrt{\det g_{st}}} \partial_i \left(\sqrt{\det g_{st}} \cdot g^{ij} \cdot \partial_j u \right) = g^{ij} \partial_i \partial_j u$.

(2) $\frac{1}{2}\Delta g_{ij} + Q(g, \partial g) = -\mathrm{Ric}_{ij} = -\mathrm{Ric}(\partial_i, \partial_j)$. Here Q is some universal analytic expression that is polynomial in the matrix g, quadratic in ∂g, and a denominator term depending on $\sqrt{\det g_{ij}}$.

PROOF. (1) By definition, we have that

$$
\begin{aligned}
0 &= \Delta x^k \\
&= \frac{1}{\sqrt{\det g_{st}}} \partial_i \left(\sqrt{\det g_{st}} \cdot g^{ij} \cdot \partial_j x^k \right) \\
&= g^{ij} \partial_i \partial_j x^k + \frac{1}{\sqrt{\det g_{st}}} \partial_i \left(\sqrt{\det g_{st}} \cdot g^{ij} \right) \cdot \partial_j x^k \\
&= g^{ij} \partial_i \delta_j^k + \frac{1}{\sqrt{\det g_{st}}} \partial_i \left(\sqrt{\det g_{st}} \cdot g^{ij} \right) \cdot \delta_j^k \\
&= 0 + \frac{1}{\sqrt{\det g_{st}}} \partial_i \left(\sqrt{\det g_{st}} \cdot g^{ik} \right) \\
&= \frac{1}{\sqrt{\det g_{st}}} \partial_i \left(\sqrt{\det g_{st}} \cdot g^{ik} \right).
\end{aligned}
$$

Thus, it follows that

$$
\begin{aligned}
\Delta u &= \frac{1}{\sqrt{\det g_{st}}} \partial_i \left(\sqrt{\det g_{st}} \cdot g^{ij} \cdot \partial_j u \right) \\
&= g^{ij} \partial_i \partial_j u + \frac{1}{\sqrt{\det g_{st}}} \partial_i \left(\sqrt{\det g_{st}} \cdot g^{ij} \right) \cdot \partial_j u \\
&= g^{ij} \partial_i \partial_j u.
\end{aligned}
$$

(2) Recall that if u is harmonic, then the Bochner formula for ∇u is

$$
\Delta \left(\frac{1}{2} |\nabla u|^2 \right) = |\mathrm{Hess}\, u|^2 + \mathrm{Ric} \left(\nabla u, \nabla u \right).
$$

Here the term $|\mathrm{Hess}\, u|^2$ can be computed explicitly and depends only on the metric and its first derivatives. In particular,

$$
\frac{1}{2} \Delta g \left(\nabla x^k, \nabla x^k \right) - \left| \mathrm{Hess}\, x^k \right|^2 = \mathrm{Ric} \left(\nabla x^k, \nabla x^k \right).
$$

Polarizing this quadratic expression gives us an identity of the form

$$
\frac{1}{2} \Delta g \left(\nabla x^i, \nabla x^j \right) - g \left(\mathrm{Hess}\, x^i, \mathrm{Hess}\, x^j \right) = \mathrm{Ric} \left(\nabla x^i, \nabla x^j \right).
$$

Now use that

$$
\nabla x^k = g^{ij} \partial_j x^k \partial_i = g^{ik} \partial_i
$$

to see that $g \left(\nabla x^i, \nabla x^j \right) = g^{ij}$. We then have

$$
\frac{1}{2} \Delta g^{ij} - g \left(\mathrm{Hess}\, x^i, \mathrm{Hess}\, x^j \right) = \mathrm{Ric} \left(\nabla x^i, \nabla x^j \right),
$$

which in matrix form looks like

$$
\frac{1}{2} \left[\Delta g^{ij} \right] - \left[g \left(\mathrm{Hess}\, x^i, \mathrm{Hess}\, x^j \right) \right] = \left[g^{ik} \right] \cdot \left[\mathrm{Ric} \left(\partial_k, \partial_l \right) \right] \cdot \left[g^{lj} \right].
$$

This is, of course, not the promised formula. Instead, it is a similar formula for the inverse of (g_{ij}). One can now use the matrix equation $[g_{ik}] \cdot [g^{kj}] = \left[\delta_i^j \right]$ to conclude that

$$
\begin{aligned}
0 &= \Delta\left([g_{ik}]\cdot[g^{kj}]\right) \\
&= [\Delta g_{ik}]\cdot[g^{kj}] + 2\left[\sum_k g\left(\nabla g_{ik}, \nabla g^{kj}\right)\right] + [g_{ik}]\cdot[\Delta g^{kj}] \\
&= [\Delta g_{ik}]\cdot[g^{kj}] + 2[\nabla g_{ik}]\cdot[\nabla g^{kj}] + [g_{ik}]\cdot[\Delta g^{kj}]
\end{aligned}
$$

Inserting this in the above equation yields

$$
\begin{aligned}
[\Delta g_{ij}] &= -2[\nabla g_{ik}]\cdot[\nabla g^{kl}]\cdot[g_{lj}] - [g_{ik}]\cdot[\Delta g^{kl}]\cdot[g_{lj}] \\
&= -2[\nabla g_{ik}]\cdot[\nabla g^{kl}]\cdot[g_{lj}] \\
&\quad -2[g_{ik}]\cdot[g\left(\mathrm{Hess}x^k, \mathrm{Hess}x^l\right)]\cdot[g_{lj}] \\
&\quad -2[g_{ik}]\cdot[g^{ks}]\cdot[\mathrm{Ric}\left(\partial_s, \partial_t\right)]\cdot[g^{tl}]\cdot[g_{lj}] \\
&= -2[\nabla g_{ik}]\cdot[\nabla g^{kl}]\cdot[g_{lj}] - 2[g_{ik}]\cdot[g\left(\mathrm{Hess}x^k, \mathrm{Hess}x^l\right)]\cdot[g_{lj}] \\
&\quad -2[\mathrm{Ric}\left(\partial_i, \partial_j\right)] .
\end{aligned}
$$

Each entry in these matrices then satisfies

$$
\begin{aligned}
\frac{1}{2}\Delta g_{ij} + Q_{ij}\left(g, \partial g\right) &= -\mathrm{Ric}_{ij}, \\
Q_{ij} &= -2\sum_{k,l} g\left(\nabla g_{ik}, \nabla g^{kl}\right)\cdot g_{lj} \\
&\quad -2\sum_{k,l} g_{ik}\cdot g\left(\mathrm{Hess}x^k, \mathrm{Hess}x^l\right)\cdot g_{lj}.
\end{aligned}
$$

$\square$

It is interesting to apply this formula to the case of an Einstein metric, where $\mathrm{Ric}_{ij} = (n-1)\,kg_{ij}$. In this case, it reads

$$
\frac{1}{2}\Delta g_{ij} = -(n-1)\,kg_{ij} - Q\left(g, \partial g\right).
$$

This formula makes sense even when g_{ij} is only $C^{1,\alpha}$. Namely, multiply by some test function, integrate, and use integration by parts to obtain a formula that uses only first derivatives of g_{ij}. If now g_{ij} is $C^{1,\alpha}$, then the left-hand side lies in C^α; but then our elliptic estimates show that g_{ij} must be in $C^{2,\alpha}$. This can be continued until we have that the metric is C^∞. In fact, one can even show that it is analytic. We can therefore conclude that any metric which in harmonic coordinates is a weak solution to the Einstein equation must in fact be smooth. We have obviously left out a few details about weak solutions. A detailed account can be found in [90, vol. III].

3. Norms and Convergence of Manifolds

We shall now explain how the $C^{m,\alpha}$ norm and convergence concepts for functions generalize to Riemannian manifolds. We shall also see how these ideas can be used to prove various compactness and finiteness theorems for classes of Riemannian manifolds.

3.1. Norms of Riemannian Manifolds. Before defining norms for manifolds, let us discuss which spaces should have norm zero. Clearly Euclidean space is a candidate. But what about open subsets of Euclidean space and other flat manifolds? If we agree that all open subsets of Euclidean space also have norm zero, then any flat manifold becomes a union of manifolds with norm zero and should therefore also have norm zero. In order to create a useful theory, it is often best to have only one space with zero norm. Thus we must agree that subsets of Euclidean space cannot have norm zero. To accommodate this problem, we define a family of norms of a Riemannian manifold, i.e., we use a function $N : (0, \infty) \to (0, \infty)$ rather than just a number. The number $N(r)$ then measures the degree of flatness on the scale of r, where the standard measure of flatness on the scale of r is the Euclidean ball $B(0, r)$. For small r, all flat manifolds then have norm zero; but as r increases we see that the space looks less and less like $B(0, r)$, and therefore the norm will become positive unless the space is Euclidean space.

For the precise definition, suppose A is a subset of a Riemannian n-manifold (M, g). We say that the $C^{m,\alpha}$-*norm on the scale of* r of $A \subset (M, g)$:

$$\|A \subset (M, g)\|_{C^{m,\alpha}, r} \le Q,$$

if we can find charts

$$\varphi_s : B(0, r) \subset \mathbb{R}^n \longleftrightarrow U_s \subset M$$

such that
- (n1) Every ball $B\left(p, \frac{1}{10} e^{-Q} r\right), p \in A$ is contained in some U_s.
- (n2) $|D\varphi_s| \le e^Q$ on $B(0, r)$ and $\left|D\varphi_s^{-1}\right| \le e^Q$ on U_s.
- (n3) $r^{|j|+\alpha} \left\|D^j g_{s\cdot\cdot}\right\|_\alpha \le Q$ for all multi indices j with $0 \le |j| \le m$.
- (n4) $\left\|\varphi_s^{-1} \circ \varphi_t\right\|_{C^{m+1,\alpha}} \le (10 + r) e^Q$.

Here $g_{s\cdot\cdot}$ is the matrix of functions of metric coefficients in the φ_s coordinates regarded as a matrix on $B(0, r)$.

First, observe that we think of the charts as maps from the fixed space $B(0, r)$ into the manifold. This is in order to have domains for the functions which do not refer to M itself. This simplifies some technical issues and makes it more clear that we are trying to measure how different the manifolds are from the standard objects, namely, Euclidean balls. The first condition says that we have a Lebesgue number for the covering of A. The second condition tells us that in the chosen coordinates the metric coefficients are bounded from below and above (in particular, we have uniform ellipticity for the Laplacian). The third condition gives us bounds on the derivatives of the metric. The fourth condition is included to ensure that the bounds for the metric in individual coordinates don't vary drastically in places where coordinates overlap. This last condition can be eliminated in many cases. We shall give another norm concept below that does this.

It will be necessary on occasion to work with Riemannian manifolds that are not smooth. The above definition clearly only requires that the metric be $C^{m,\alpha}$ in the coordinates we use, and so there is no reason to assume more about the metric. Some of the basic constructions, like exponential maps, then come into question, and indeed, if $m \le 1$ these items might not be well-defined. We shall therefore have to be a little careful in some situations.

When it is clear from the context where A is, we shall merely write $\|A\|_{C^{m,\alpha}, r}$, or for the whole space, $\|(M, g)\|_{C^{m,\alpha}, r}$ or $\|M\|_{C^{m,\alpha}, r}$. If A is precompact in M, then it is clear that the norm is bounded for all r. For unbounded domains or manifolds the norm might not be finite.

EXAMPLE 58. *Suppose (M, g) is a complete flat manifold. Then $\|(M, g)\|_{C^{m,\alpha}, r}$ $= 0$ for all $r \leq \text{inj}(M, g)$. In particular, $\|(\mathbb{R}^n, \text{can})\|_{C^{m,\alpha}, r} = 0$ for all r. We shall later see that these properties characterize flat manifolds and Euclidean space.*

3.2. Convergence of Riemannian Manifolds. Now for the convergence concept that relates to this new norm. As we can't subtract manifolds, we have to resort to a different method for defining this. If we fix a closed manifold M, or more generally a precompact subset $A \subset M$, then we say that a sequence of functions on A converges in $C^{m,\alpha}$, if they converge in the charts for some fixed finite covering of coordinate patches. This definition is clearly independent of the finite covering we choose. We can then more generally say that a sequence of tensors converges in $C^{m,\alpha}$ if the components of the tensors converge in these patches. This then makes it possible to speak about convergence of Riemannian metrics on compact subsets of a fixed manifold.

A sequence of pointed complete Riemannian manifolds is said to *converge in the pointed $C^{m,\alpha}$ topology* $(M_i, p_i, g_i) \to (M, p, g)$ if for every $R > 0$ we can find a domain $\Omega \supset B(p, R) \subset M$ and embeddings $F_i : \Omega \to M_i$ for large i such that $F_i(\Omega) \supset B(p_i, R)$ and $F_i^* g_i \to g$ on Ω in the $C^{m,\alpha}$ topology. It is easy to see that this type of convergence implies pointed Gromov-Hausdorff convergence. When all manifolds in question are closed, then we have that the maps F_i are diffeomorphisms. This means that for closed manifolds we can speak about unpointed convergence. In this case, convergence can therefore only happen if all the manifolds in the tail end of the sequence are diffeomorphic. In particular, we have that classes of closed Riemannian manifolds that are precompact in some $C^{m,\alpha}$ topology contain at most finitely many diffeomorphism types.

A warning about this kind of convergence is in order here. Suppose we have a sequence of metrics g_i on a fixed manifold M. It is possible that these metrics might converge in the sense just defined, without converging in the traditional sense of converging in some fixed coordinate systems. To be more specific, let g be the standard metric on $M = S^2$. Now define diffeomorphisms F_t coming from the flow corresponding to the vector field that is 0 at the north and south poles and otherwise points in the direction of the south pole. As t increases, the diffeomorphisms will try to map the whole sphere down to a small neighborhood of the south pole. The metrics $F_t^* g$ will therefore in some fixed coordinates converge to 0 (except at the poles). They can therefore not converge in the classical sense. If, however, we pull these metrics back by the diffeomorphisms F_{-t}, then we just get back to g. Thus the sequence (M, g_t), from the new point of view we are considering, is a constant sequence. This is really the right way to think about this as the spaces $(S^2, F_t^* g)$ are all isometric as abstract metric spaces.

3.3. Properties of the Norm. Let us now consider some of the elementary properties of norms and their relation to convergence.

PROPOSITION 45. *If $A \subset (M, g)$ is precompact, then*

(1) $\|A \subset (M, g)\|_{C^{m,\alpha}, r} = \|A \subset (M, \lambda^2 g)\|_{C^{m,\alpha}, \lambda r}$ for all $\lambda > 0$.

(2) The function $r \to \|A \subset (M, g)\|_{C^{m,\alpha}, r}$ is continuous and converges to 0 as $r \to 0$.

(3) Suppose $(M_i, p_i, g_i) \to (M, p, g)$ in $C^{m,\alpha}$. Then for a precompact domain $A \subset M$ we can find precompact domains $A_i \subset M_i$ such that

$$\|A_i\|_{C^{m,\alpha},r} \to \|A\|_{C^{m,\alpha},r} \quad \text{for all } r > 0$$

When all the manifolds are closed, we can let $A = M$ and $A_i = M_i$.

PROOF. (1) If we change the metric g to $\lambda^2 g$, then we can change the charts $\varphi_s : B(0, r) \to M$ to

$$\varphi_s^\lambda(x) = \varphi_s(\lambda^{-1}x) : B(0, \lambda r) \to M.$$

Since we scale the metric at the same time, the conditions n1-n4 will still hold with the same Q.

(2) Suppose, as above, we change the charts

$$\varphi_s : B(0, r) \to M$$

to

$$\varphi_s^\lambda(x) = \varphi_s(\lambda^{-1}x) : B(0, \lambda r) \to M,$$

without changing the metric g. If we assume that

$$\|A \subset (M, g)\|_{C^{m,\alpha},r} < Q,$$

then

$$\|A \subset (M, g)\|_{C^{m,\alpha},\lambda r} \le \max\left\{Q + |\log \lambda|, Q \cdot \lambda^2\right\}.$$

Denoting

$$N(r) = \|A \subset (M, g)\|_{C^{m,\alpha},r},$$

we therefore obtain

$$N(\lambda r) \le \max\left\{N(r) + |\log \lambda|, N(r) \cdot \lambda^2\right\}.$$

By letting $\lambda = \frac{r_i}{r}$, where $r_i \to r$, we see that this implies

$$\limsup N(r_i) \le N(r).$$

Conversely, we have that

$$
\begin{aligned}
N(r) &= N\left(\frac{r}{r_i} r_i\right) \\
&\le \max\left\{N(r_i) + \left|\log \frac{r}{r_i}\right|, N(r_i) \cdot \left(\frac{r}{r_i}\right)^2\right\}.
\end{aligned}
$$

So

$$
\begin{aligned}
N(r) &\le \liminf N(r_i) \\
&= \liminf \max\left\{N(r_i) + \left|\log \frac{r}{r_i}\right|, N(r_i) \cdot \left(\frac{r}{r_i}\right)^2\right\}.
\end{aligned}
$$

This shows that $N(r)$ is continuous. To see that $N(r) \to 0$ as $r \to 0$, just observe that any coordinate system around a point $p \in M$ can, after a linear change, be assumed to have the property that the metric $g_{ij} = \delta_{ij}$ at p. In particular $|D\varphi|_p| = |D\varphi^{-1}|_p| = 1$. Using these coordinates on sufficiently small balls will therefore give the desired charts.

(3) We fix $r > 0$ in the definition of $\|A \subset (M, g)\|_{C^{m,\alpha},r}$. For the given $A \subset M$, pick a domain $\Omega \supset A$ such that for large i we have embeddings $F_i : \Omega \to M_i$ with the property that: $F_i^* g_i \to g$ in $C^{m,\alpha}$ on Ω. Define $A_i = F_i(A)$.

For $Q > \|A \subset (M,g)\|_{C^{m,\alpha},r}$, choose appropriate charts $\varphi_s : B(0,r) \to M$ covering A, with the properties n1-n4. Then define charts in M_i by

$$\varphi_{i,s} = F_i \circ \varphi_s : B(0,r) \to M_i.$$

Condition n1 will hold just because we have Gromov-Hausdorff convergence. Condition n4 is trivial. Conditions n2 and n3 will hold for constants $Q_i \to Q$, since $F_i^* g_i \to g$ in $C^{m,\alpha}$. We can therefore conclude that

$$\limsup \|A_i\|_{C^{m,\alpha},r} \leq \|A\|_{C^{m,\alpha},r}.$$

On the other hand, for large i and $Q > \|A_i\|_{C^{m,\alpha},r}$, we can take charts $\varphi_{i,s} : B(0,r) \to M_i$ and then pull them back to M by defining $\varphi_s = F_i^{-1} \circ \varphi_{i,s}$. As before, we then have

$$\|A\|_{C^{m,\alpha},r} \leq Q_i,$$

where $Q_i \to Q$. This implies

$$\liminf \|A_i\|_{C^{m,\alpha},r} \geq \|A\|_{C^{m,\alpha},r},$$

and hence the desired result. $\qquad\square$

3.4. Compact Classes of Riemannian Manifolds. We are now ready to prove the result that is our manifold equivalent of the Arzela-Ascoli lemma. This theorem is essentially due to J. Cheeger, although our use of norms makes the statement look different.

THEOREM 72. (Fundamental Theorem of Convergence Theory) *For given $Q > 0$, $n \geq 2$, $m \geq 0$, $\alpha \in (0,1]$, and $r > 0$ consider the class $\mathcal{M}^{m,\alpha}(n,Q,r)$ of complete, pointed Riemannian n-manifolds (M,p,g) with $\|(M,g)\|_{C^{m,\alpha},r} \leq Q$. $\mathcal{M}^{m,\alpha}(n,Q,r)$ is compact in the pointed $C^{m,\beta}$ topology for all $\beta < \alpha$.*

PROOF. We proceed in stages. First, we make some general comments about the charts we use. We then show that $\mathcal{M} = \mathcal{M}^{m,\alpha}(n,Q,r)$ is pre-compact in the pointed Gromov-Hausdorff topology. Next we prove that $\mathcal{M}$ is closed in the Gromov-Hausdorff topology. The last and longest part is then devoted to the compactness statement.

Setup: First fix $K > Q$. Whenever we select an $M \in \mathcal{M}$, we shall assume that it comes equipped with an atlas of charts satisfying n1-n4 with K in place of Q. Thus we implicitly assume that all charts under consideration belong to these atlases. We will consequently only prove that limit spaces (M,p,g) satisfy $\|(M,g)\|_{C^{m,\alpha},r} \leq K$. But as K was arbitrary, we still get that $(M,p,g) \in \mathcal{M}$.

(1) Every chart $\varphi : B(0,r) \to U \subset M \in \mathcal{M}$ satisfies
   (a) $d(\varphi(x_1), \varphi(x_2)) \leq e^K |x_1 - x_2|$
   (b) $d(\varphi(x_1), \varphi(x_2)) \geq \min\{e^{-K}|x_1 - x_2|, \ e^{-K}(2r - |x_1| - |x_2|)\}$.

Here, d is distance measured in M, and $|\cdot|$ is the usual Euclidean norm.

The condition $|D\varphi| \leq e^K$, together with convexity of $B(0,r)$, immediately implies the first inequality. For the other, first observe that if any segment from x_1 to x_2 lies in U, then $|D\varphi^{-1}| \leq e^K$ implies, that

$$d(\varphi(x_1), \ \varphi(x_2)) \geq e^{-K}|x_1 - x_2|.$$

So we may assume that $\varphi(x_1)$ and $\varphi(x_2)$ are joined by a segment $\sigma : [0,1] \to M$ that leaves U. Split σ into $\sigma : [0, t_1) \to U$ and $\sigma : (t_2, 1) \to U$ such that $\sigma(t_i) \notin U$.

Then we clearly have

$$
\begin{aligned}
d(\varphi(x_1), \varphi(x_2)) \;=\;& L(\sigma) \geq L(\sigma|_{[0,t_1)}) + L(\sigma|_{(t_2,1]}) \\
\geq\;& e^{-K}(L(\varphi^{-1} \circ \sigma|_{[0,t_1)}) + L(\varphi^{-1} \circ \sigma|_{(t_2,1]})) \\
\geq\;& e^{-K}(2r - |x_1| - |x_2|).
\end{aligned}
$$

The last inequality follows from the fact that $\varphi^{-1} \circ \sigma(0) = x_1$ and $\varphi^{-1} \circ \sigma(1) = x_2$, and that $\varphi^{-1} \circ \sigma(t)$ approaches the boundary of $B(0,r)$ as $t \nearrow t_1$ or $t \searrow t_2$.

(2) Every chart

$$
\varphi : B(0,r) \to U \subset M \in \mathcal{M},
$$

and hence any δ-ball $\delta = \frac{1}{10} e^{-K} r$ in M can be covered by at most N balls of radius $\delta/4$. Here, N depends only on n, K, r.

Clearly, there exists an $N(n, K, r)$ such that $B(0,r)$ can be covered by at most N balls of radius $e^{-K} \cdot \delta/4$. Since $\varphi : B(0,r) \to U$ is a Lipschitz map with Lipschitz constant $\leq e^K$, we get the desired covering property.

(3) Every ball $B(x, \ell \cdot \delta/2) \subset M$ can be covered by $\leq N^\ell$ balls of radius $\delta/4$.

For $\ell = 1$ we just proved this. Suppose we know that $B(x, \ell \cdot \delta/2)$ is covered by $B(x_1, \delta/4), \ldots, B(x_{N^\ell}, \delta/4)$. Then

$$
B(x, \ell \cdot \delta/2 + \delta/2) \subset \cup B(x_i, \delta).
$$

Now each $B(x_i, \delta)$ can be covered by $\leq N$ balls of radius $\delta/4$, and hence $B(x, (\ell + 1)\delta/2)$ can be covered by $\leq N \cdot N^\ell = N^{\ell+1}$ balls of radius $\delta/4$.

(4) $\mathcal{M}$ is precompact in the pointed Gromov-Hausdorff topology.

This is equivalent to asserting, that for each $R > 0$ the family of metric balls

$$
B(p, R) \subset (M, p, g) \in \mathcal{M}
$$

is precompact in the Gromov-Hausdorff topology. This claim is equivalent to showing that we can find a function $N(\varepsilon) = N(\varepsilon, R, K, r, n)$ such that each $B(p, R)$ can contain at most $N(\varepsilon)$ disjoint ε-balls. To check this, let $B(x_1, \varepsilon), \ldots, B(x_s, \varepsilon)$ be a collection of disjoint balls in $B(p, R)$. Suppose that

$$
\ell \cdot \delta/2 < R \leq (\ell + 1)\delta/2.
$$

Then

$$
\begin{aligned}
\mathrm{vol} B(p, R) \;\leq\;& (N^{(\ell+1)}) \cdot (\text{maximal volume of } \tfrac{\delta}{4}\text{-ball}) \\
\leq\;& (N^{(\ell+1)}) \cdot (\text{maximal volume of chart}) \\
\leq\;& N^{(\ell+1)} \cdot e^{nK} \cdot \mathrm{vol} B(0, r) \\
\leq\;& V(R) = V(R, n, K, r).
\end{aligned}
$$

As long as $\varepsilon < r$ each $B(x_i, \varepsilon)$ lies in some chart $\varphi : B(0, r) \to U \subset M$ whose preimage in $B(0, r)$ contains an $e^{-K} \cdot \varepsilon$-ball. Thus

$$
\mathrm{vol} B(p_i, \varepsilon) \geq e^{-2nK} \mathrm{vol} B(0, \varepsilon).
$$

All in all, we get

$$
\begin{aligned}
V(R) \;\geq\;& \mathrm{vol} B(p, R) \\
\geq\;& \sum \mathrm{vol} B(p_i, \varepsilon) \\
\geq\;& s \cdot e^{-2nK} \cdot \mathrm{vol} B(0, \varepsilon).
\end{aligned}
$$

Thus,

$$s \leq N(\varepsilon) = V(R) \cdot e^{2nK} \cdot (\text{vol}B(0,\varepsilon))^{-1}.$$

Now select a sequence (M_i, g_i, p_i) in $\mathcal{M}$. From the previous considerations we can assume that $(M_i, g_i, p_i) \to (X, d, p)$ converge to some metric space in the Gromov-Hausdorff topology. It will be necessary in many places to pass to subsequences of (M_i, g_i, p_i) using various diagonal processes. Whenever this happens, we shall not reindex the family, but merely assume that the sequence was chosen to have the desired properties from the beginning. For each (M_i, p_i, g_i) choose charts

$$\varphi_{is} : B(0, r) \to U_{is} \subset M_i$$

satisfying n1-n4. We can furthermore assume that the index set $\{s\} = \{1, 2, 3, 4, \cdots\}$ is the same for all M_i, that $p_i \in U_{i1}$, and that the balls $B(p_i, \ell \cdot \delta/2)$ are covered by the first N^ℓ charts. Note that these N^ℓ charts will then be contained in $\bar{B}(p_i, \ell \cdot \delta/2 + [e^K + 1]\delta)$. Finally, for each ℓ the sequence $\bar{B}(p_i, \ell \cdot \delta/2)$ converges to $\bar{B}(p, \ell \cdot \delta/2) \subset X$, so we can choose a metric on the disjoint union

$$Y_\ell = \left(\bar{B}(p, \ell \cdot \delta/2) \coprod \left(\coprod_{i=1}^{\infty} \bar{B}(p_i, \ell \cdot \delta/2) \right) \right)$$

such that

$$\begin{aligned} p_i &\to p, \\ \bar{B}(p_i, \ell \cdot \delta/2) &\to \bar{B}(p, \ell \cdot \delta/2) \end{aligned}$$

in the Hausdorff distance inside this metric space.

(5) (X, d, p) is a Riemannian manifold of class $C^{m,\alpha}$ with norm $\leq K$.

Obviously, we need to find bijections

$$\varphi_s : B(0, r) \to U_s \subset X$$

satisfying n1-n4. For each s, consider the maps

$$\varphi_{is} : B(0, r) \to U_{is} \subset Y_{\ell'}$$

for some fixed $\ell' \gg \ell$. From 1 we have that this is a family of equicontinuous maps into the compact space $Y_{\ell'}$. The Arzela-Ascoli lemma shows that this sequence must subconverge (in the C^0 topology) to a map

$$\varphi_s : B(0, r) \subset Y_{\ell'}$$

that also has Lipschitz constant e^K. Furthermore, the inequality

$$d(\varphi(x_1), \varphi(x_2)) \geq \min\{e^{-K}|x_1 - x_2|, \ e^{-K}(2r - |x_1| - |x_2|)\}$$

will also hold for this map, as it holds for all the φ_{is} maps. In particular, φ_s is one-to-one. Finally, since $U_{is} \subset \bar{B}(p_i, \ell')$ and $\bar{B}(p_i, \ell')$ Hausdorff converges to $\bar{B}(p, \ell') \subset X$, we see that

$$\varphi_s(B(0, r)) = U_s \subset X.$$

A simple diagonal argument yields that we can pass to a subsequence of (M_i, g_i, p_i) having the property that $\varphi_{is} \to \varphi_s$ for all s. In this way, we have constructed (topological) charts

$$\varphi_s : B(0, r) \to U_s \subset X,$$

and we can easily check that they satisfy n1. Since the φ_s also satisfy 1(a) and 1(b), they would also satisfy n2 if they were differentiable (equivalent to saying that the transition functions are C^1). Now the transition functions $\varphi_{is}^{-1} \circ \varphi_{it}$ approach

$\varphi_s^{-1} \circ \varphi_t$, because $\varphi_{is} \to \varphi_s$. Note that these transition functions are not defined on the same domains, but we do know that the domain for $\varphi_s^{-1} \circ \varphi_t$ is the limit of the domains for $\varphi_{is}^{-1} \circ \varphi_{it}$, so the convergence makes sense on all compact subsets of the domain of $\varphi_s^{-1} \circ \varphi_t$. Now,

$$\|\varphi_{is}^{-1} \circ \varphi_{it}\|_{C^{m+1,\alpha}} \le (10 + r) e^K,$$

so a further application (and subsequent passage to subsequences) of Arzela-Ascoli tells us that

$$\|\varphi_s^{-1} \circ \varphi_t\|_{C^{m+1,\alpha}} \le (10 + r) e^K,$$

and that we can assume $\varphi_{is}^{-1} \circ \varphi_{it} \to \varphi_s^{-1} \circ \varphi_t$ in the $C^{m+1,\beta}$ topology. This then establishes n4. We now construct a compatible Riemannian metric on X that satisfies n2 and n3. For each s, consider the metric $g_{is} = g_{is..}$ written out in its components on $B(0,r)$ with respect to the chart φ_{is}. Since all of the $g_{is..}$ satisfy n2 and n3, we can again use Arzela-Ascoli to insure that also $g_{is..} \to g_{s..}$ on $B(0,r)$ in the $C^{m,\beta}$ topology to functions $g_{s..}$ that also satisfy n2 and n3. The local "tensors" $g_{s..}$ satisfy the right change of variables formulae to make them into a global tensor on X. This is because all the $g_{is..}$ satisfy these properties, and everything we want to converge, to carry these properties through to the limit, also converges. Recall that the rephrasing of n2 gives the necessary C^0 bounds and also shows that $g_{s..}$ is positive definite. We have now exhibited a Riemannian structure on X such that the

$$\varphi_s : B(0,r) \to U_s \subset X$$

satisfy n1-n4 with respect to this structure. This, however, does not guarantee that the metric generated by this structure is identical to the metric we got from X being the pointed Gromov-Hausdorff limit of (M_i, p_i, g_i). However, since Gromov-Hausdorff convergence implies that distances converge, and we know at the same time that the Riemannian metric converges locally in coordinates, it follows that the limit Riemannian structure must generate the "correct" metric, at least locally, and therefore also globally.

(6) $(M_i, p_i, g_i) \to (X, p, d) = (X, p, g)$ in the pointed $C^{m,\beta}$ topology.

We assume that the setup is as in 5, where charts φ_{is}, transitions $\varphi_{is}^{-1} \circ \varphi_{it}$, and metrics $g_{is..}$ converge to the same items in the limit space. First, let us agree that two maps F_1, F_2 between subsets in M_i and X are $C^{m+1,\beta}$ close if all the coordinate compositions $\varphi_s^{-1} \circ F_1 \circ \varphi_{it}$, $\varphi_s^{-1} \circ F_2 \circ \varphi_{it}$ are $C^{m+1,\beta}$ close. Thus, we have a well-defined $C^{m+1,\beta}$ topology on maps from M_i to X. Our first observation is that

$$\begin{aligned} f_{is} &= \varphi_{is} \circ \varphi_s^{-1} : U_s \to U_{is}, \\ f_{it} &= \varphi_{it} \circ \varphi_t^{-1} : U_t \to U_{it} \end{aligned}$$

"converge to each other" in the $C^{m+1,\beta}$ topology. Furthermore,

$$(f_{is})^* g_i|_{U_{is}} \to g|_{U_s}$$

in the $C^{m,\beta}$ topology. These are just restatements of what we already know. In order to finish the proof, we construct maps

$$F_{i\ell} : \Omega_\ell = \bigcup_{s=1}^\ell U_s \to \Omega_{i\ell} = \bigcup_{s=1}^\ell U_{is}$$

that are closer and closer to the f_{is}, $s = 1, \ldots, \ell$ maps (and therefore all f_{is}) as $i \to \infty$. We will construct $F_{i\ell}$ by induction on ℓ and large i depending on ℓ. For this purpose we shall need a partition of unity (λ_s) on X subordinate to (U_s). We can find such a partition, since the covering (U_s) is locally finite by choice, and we can furthermore assume that λ_s is $C^{m+1,\beta}$.

For $\ell = 1$ simply define $F_{i1} = f_{i1}$.

Suppose we have $F_{i\ell} : \Omega_\ell \to \Omega_{i\ell}$ for large i that are arbitrarily close to f_{is}, $s = 1, \ldots, \ell$ as $i \to \infty$. If $U_{\ell+1} \cap \Omega_\ell = \emptyset$, then we just define $F_{i\ell+1} = F_{i\ell}$ on $\Omega_{i\ell}$, and $F_{i\ell+1} = f_{i\ell+1}$ on $U_{\ell+1}$. In case $U_{\ell+1} \subset \Omega_\ell$, we simply let $F_{i\ell+1} = F_{i\ell}$. Otherwise, we know that $F_{i\ell}$ and $f_{i\ell+1}$ are as close as we like in the $C^{m+1,\beta}$ topology as $i \to \infty$. So the natural thing to do is to average them on $U_{\ell+1}$. Define $F_{i\ell+1}$ on $U_{\ell+1}$ by

$$
\begin{aligned}
F_{i\ell+1}(x) & \\
= \ \ \varphi_{i\ell+1} & \circ \left(\left(\sum_{s=\ell+1}^{\infty} \lambda_s(x) \right) \cdot \varphi_{i\ell+1}^{-1} \circ f_{i\ell+1}(x) + \left(\sum_{s=1}^{\ell} \lambda_s(x) \right) \cdot \varphi_{i\ell+1}^{-1} \circ F_{i\ell}(x) \right) \\
= \ \ \varphi_{i\ell+1} & \circ (\mu_1(x) \cdot \varphi_{i\ell+1}^{-1} \circ f_{i\ell+1}(x) + \mu_2(x) \cdot \varphi_{i\ell+1}^{-1} \circ F_{i\ell}(x)).
\end{aligned}
$$

This map is clearly well-defined on $U_{\ell+1}$, since $\mu_2(x) = 0$ on $U_{\ell+1} - \Omega_\ell$. Moreover, as $\mu_1(x) = 0$ on Ω_ℓ it is a smooth $C^{m+1,\beta}$ extension of $F_{i\ell}$. Now consider this map in coordinates

$$
\begin{aligned}
\varphi_{i\ell+1}^{-1} \circ F_{i\ell+1} \circ \varphi_{\ell+1}(y) &= \left(\mu_1 \circ \varphi_{\ell+1}(y) \right) \cdot \varphi_{\ell+1}^{-1} \circ f_{i\ell+1} \circ \varphi_{\ell+1}(y) \\
&\quad + \left(\mu_2 \circ \varphi_{\ell+1}(y) \right) \cdot \varphi_{i\ell+1}^{-1} \circ F_{i\ell} \circ \varphi_{\ell+1}(y) \\
&= \tilde{\mu}_1(y) F_1(y) + \tilde{\mu}_2(y) F_2(y).
\end{aligned}
$$

Then

$$
\begin{aligned}
\|\tilde{\mu}_1 F_1 + \tilde{\mu}_2 F_2 - F_1 \|_{C^{m+1,\beta}} &= \|\tilde{\mu}_1 (F_1 - F_1) + \tilde{\mu}_2 (F_2 - F_1) \|_{C^{m+1,\beta}} \\
&\leq \|\tilde{\mu}_2\|_{k+1+\beta} \cdot \|F_2 - F_1\|_{C^{m+1,\beta}}.
\end{aligned}
$$

This inequality is valid on all of $B(0, r)$, despite the fact that F_2 is not defined on all of $B(0, r)$, since

$$
\tilde{\mu}_1 \cdot F_1 + \tilde{\mu}_2 \cdot F_2 = F_1
$$

on the region where F_2 is undefined. By assumption

$$
\|F_2 - F_1\|_{C^{m+1,\beta}} \to 0 \text{ as } i \to \infty,
$$

so $F_{i\ell+1}$ is $C^{m+1,\beta}$-close to f_{is}, $s = 1, \ldots, \ell+1$ as $i \to \infty$.

Finally we see that the closeness of $F_{i\ell}$ to the coordinate charts shows that it is an embedding on all compact subsets of the domain. $\qquad \square$

COROLLARY 33. *The subclasses of $\mathcal{M}^{m,\alpha}(n, Q, r)$, where the elements in addition satisfy* diam $\leq D$, *respectively* vol $\leq V$, *are compact in the $C^{m,\beta}$ topology. In particular, they contain only finitely many diffeomorphism types.*

PROOF. We use notation as in the fundamental theorem. If $\text{diam}(M, g, p) \leq D$, then clearly $M \subset B(p, k \cdot \delta/2)$ for $k > D \cdot 2/\delta$. Hence, each element in $\mathcal{M}^{m,\alpha}(n, Q, r)$ can be covered by $\leq N^k$ charts. Thus, $C^{m,\beta}$-convergence is actually in the unpointed topology, as desired.

If instead, $\text{vol} M \leq V$, then we can use part 4 in the proof to see that we can never have more than

$$
k = V \cdot e^{2nK} \cdot (\text{vol} B(0, \varepsilon))^{-1}
$$

disjoint ε-balls. In particular, diam $\leq 2\varepsilon \cdot k$, and we can use the above argument.

Finally, compactness in any $C^{m,\beta}$ topology implies that the class cannot contain infinitely many diffeomorphism types. □

COROLLARY 34. *The norm* $\|A \subset (M,g)\|_{C^{m,\alpha},r}$ *for compact* A *is always realized by some charts* $\varphi_s : B(0,r) \to U_s$ *satisfying n1-n4, with* $\|(M,g)\|_{C^{m,\alpha},r}$ *in place of* Q.

PROOF. Choose appropriate charts

$$\varphi_s^Q : B(0,r) \to U_s^Q \subset M$$

for each $Q > \|(M,g)\|_{C^{m,\alpha},r}$, and let $Q \to \|(M,g)\|_{C^{m,\alpha},r}$. If the charts are chosen to conform with the proof of the fundamental theorem, we will obviously get some limit charts with the desired properties. □

COROLLARY 35. M *is a flat manifold if* $\|(M,g)\|_{C^{m,\alpha},r} = 0$ *for some* r, *and* M *is Euclidean space with the canonical metric if* $\|(M,g)\|_{C^{m,\alpha},r} = 0$ *for all* $r > 0$.

PROOF. The proof works even if $m = \alpha = 0$. As in the previous corollary and part (1) of the theorem, M can be covered by charts $\varphi : B(0,r) \to U \subset M$ satisfying

(a) $d(\varphi(x_1), \varphi(x_2)) \leq e^Q |x_1 - x_2|$
(b) $d(\varphi(x_1), \varphi(x_2)) \geq \min\{e^{-Q}|x_1 - x_2|,\ e^{-Q}(2r - |x_1| - |x_2|)\}$.

for each $Q > 0$. By letting $Q \to 0$, we can then use Arzela-Ascoli to find a covering of charts such that

(a) $d(\varphi(x_1), \varphi(x_2)) \leq |x_1 - x_2|$
(b) $d(\varphi(x_1), \varphi(x_2)) \geq \min\{|x_1 - x_2|,\ (2r - |x_1| - |x_2|)\}$.

This shows that the maps φ are locally distance preserving and injective. Hence they are distance preserving maps. This shows that they are also Riemannian isometries. This finishes the proof. □

3.5. Alternative Norms. Finally, we should mention that all properties of this norm concept would not change if we changed n1-n4 to say

(n1') U_s has Lebesgue number $f_1(n,Q,r)$.
(n2') $|D\varphi_s|, |D\varphi_s^{-1}| \leq f_2(n,Q)$.
(n3') $r^{|j|+\alpha} \cdot \|\partial^j g_{s..}\|_\alpha \leq f_3(n,Q)$, $0 \leq |j| \leq m$.
(n4') $\|\varphi_s^{-1} \circ \varphi_t\|_{C^{m+1,\alpha}} \leq f_4(n,Q,r)$.

As long as the f_is are all continuous, $f_1(n,0,r) = 0$, and $f_2(n,0) = 1$. The key properties we want to preserve are continuity of $\|(M,g)\|$ with respect to r, the fundamental theorem, and the characterization of flat manifolds and Euclidean space.

Another interesting thing happens if in the definition of $\|(M,g)\|_{C^{m,\alpha},r}$ we let $m = \alpha = 0$. Then n3 no longer makes sense, because $\alpha = 0$, but aside from that, we still have a C^0-norm concept. Note also that n4 is an immediate consequence of n2 in this case. The class $\mathcal{M}^0(n,Q,r)$ is now only precompact in the pointed Gromov-Hausdorff topology, but the characterization of flat manifolds is still valid. The subclasses with bounded diameter, or volume, are also only precompact with respect to the Gromov-Hausdorff topology, and the finiteness of diffeomorphism types apparently fails. It is, however, possible to say more. If we investigate the proof of the fundamental theorem, we see that the problem lies in constructing the maps $F_{ik} : \Omega_k \to \Omega_{ik}$, because we now have convergence of the coordinates only in the C^0 (actually $C^\alpha, \alpha < 1$) topology, and so the averaging process fails as it is described. We can, however, use a deep theorem from topology about local contractibility of homeomorphism groups (see [**35**]) to conclude that two C^0-close topological embeddings can be "glued" together in some way without altering them too much in the C^0 topology. This makes it possible to exhibit topological embeddings $F_{ik} : \Omega \hookrightarrow M_i$ such that the pullback metrics (not Riemannian metrics) converge. As a consequence, we see that the classes with bounded diameter or volume contain only finitely many homeomorphism types. This is exactly the content of the original version of Cheeger's finiteness theorem, including the proof as we have outlined it. But, as we have pointed out earlier, Cheeger also considered the easier to prove finiteness theorem for diffeomorphism types given better bounds on the coordinates.

Notice that we cannot easily use the fact that the charts converge in $C^\alpha (\alpha < 1)$. But it is possible to do something interesting along these lines. There is an even weaker norm concept called the *Reifenberg norm* which is related to the Gromov-Hausdorff distance. For a metric space (X, d) we define the n-dimensional norm on the scale of r as

$$\|(X, d)\|_r^n = \frac{1}{r} \sup_{p \in X} d_{G-H} (B(p, r), B(0, r)),$$

where $B(0, R) \subset \mathbb{R}^n$. The the r^{-1} factor insures that we don't have small distance between $B(p, r)$ and $B(0, r)$ just because r is small. Note also that if $(X_i, d_i) \to (X, d)$ in the Gromov-Hausdorff topology then

$$\|(X_i, d_i)\|_r^n \to \|(X, d)\|_r^n$$

for fixed n, r.

For an n-dimensional Riemannian manifold one sees immediately that

$$\lim_{r \to 0} \|(M, g)\|_r^n \to 0 = 0.$$

Cheeger and Colding have proven a converse to this (see [**25**]). There is an $\varepsilon(n) > 0$ such that if $\|(X, d)\|_r^n \le \varepsilon(n)$ for all small r, then X is in a weak sense an n-dimensional Riemannian manifold. Among other things, they show that for small r the α-Hölder distance between $B(p, r)$ and $B(0, r)$ is small. Here the α-Hölder distance $d_\alpha(X, Y)$ between metric spaces is defined as the infimum of

$$\log \max \left\{ \sup_{x_1 \ne x_2} \frac{d(F(x_1), F(x_2))}{(d(x_1, x_2))^\alpha}, \sup_{y_1 \ne y_2} \frac{d(F^{-1}(y_1), F^{-1}(y_2))}{(d(y_1, y_2))^\alpha} \right\},$$

where $F : X \to Y$ runs over all homeomorphisms. They also show that if $(M_i, g_i) \to (X, d)$ in the Gromov-Hausdorff distance and $\|(M_i, g_i)\|_r^n \le \varepsilon(n)$ for all i and small r, then $(M_i, g_i) \to (X, d)$ in the Hölder distance. In particular, all of the M_is have to be homeomorphic (and in fact diffeomorphic) to X for large i.

This is enhanced by an earlier result of Colding (see [**29**]) stating that for a Riemannian manifold (M, g) with Ric $\ge (n - 1) k$ we have that $\|(M, g)\|_r^n$ is small iff and only if

$$\text{vol} B(p, r) \ge (1 - \delta) \text{vol} B(0, r)$$

for some small δ. Relative volume comparison tells us that the volume condition holds for all small r if it holds for just one r. Thus the smallness condition for the norm holds for all small r provided we have the volume condition for just some r.

4. Geometric Applications

We shall now study the relationship between volume, injectivity radius, sectional curvature, and the norm.

First let us see what exponential coordinates can do for us. Let (M, g) be a Riemannian manifold with $|\sec M| \leq K$ and $\operatorname{inj} M \geq i_0$. On $B(0, i_0)$ we have from chapter 6 that

$$\max\left\{\left|D\exp_p\right|, \left|D\exp_p^{-1}\right|\right\} \leq \exp\left(f\left(n, K, i_0\right)\right)$$

for some function $f(n, K, i_0)$ that depends only on the dimension, K, and i_0. Moreover, as $K \to 0$ we have that $f(n, K, i_0) \to 0$. This implies

THEOREM 73. *For every $Q > 0$ there exists $r > 0$ depending only on i_0 and K such that any complete (M, g) with $|\sec M| \leq K$, $\operatorname{inj} M \geq i_0$ has $\|(M, g)\|_{C^0, r} \leq Q$. Furthermore, if (M_i, p_i, g_i) satisfy $\operatorname{inj} M_i \geq i_0$ and $|\sec M_i| \leq K_i \to 0$, then a subsequence will converge in the pointed Gromov-Hausdorff topology to a flat manifold with $\operatorname{inj} \geq i_0$.*

The proof follows immediately from our previous constructions.

This theorem does not seem very satisfactory, because even though we have assumed a C^2 bound on the Riemannian metric, we get only a C^0 bound. To get better bounds under the same circumstances, we must look for different coordinates. Our first choice for alternative coordinates uses distance functions, i.e., distance coordinates.

LEMMA 50. *Given a Riemannian manifold (M, g) with $\operatorname{inj} \geq i_0$, $|\sec| \leq K$, and $p \in M$, then the distance function $d(x) = d(x, p)$ is smooth on $B(p, i_0)$, and the Hessian is bounded in absolute value on the annulus $B(p, i_0) - B(p, i_0/2)$ by a function $F(n, K, i_0)$.*

PROOF. From chapter 6 we know that in polar coordinates

$$\sqrt{K}\cot\left(\sqrt{K}r\right)g_r \leq \operatorname{Hess}d \leq \sqrt{K}\coth\left(\sqrt{K}r\right)g_r.$$

Thus, we get the desired estimate as long as $r \in (i_0/2, i_0)$. $\square$

Now fix (M, g), $p \in M$, as in the lemma, and choose an orthonormal basis $e_1, \ldots, e_n$ for $T_p M$. Then consider the geodesics $\gamma_i(t)$ with $\gamma_i(0) = p$, $\dot{\gamma}_i(0) = e_i$, and together with those, the distance functions

$$d_i(x) = d\left(x, \gamma_i\left(i_0 \cdot \left(4\sqrt{K}\right)^{-1}\right)\right).$$

These distance functions will then have uniformly bounded Hessians on $B(p, \delta)$, $\delta = i_0 \cdot \left(8\sqrt{K}\right)^{-1}$. Define

$$\varphi(x) = (d_1(x), \ldots, d_n(x))$$

and recall that $g^{ij} = g\left(\nabla d_i, \nabla d_j\right)$.

THEOREM 74. (The Convergence Theorem of Riemannian Geometry) *Given i_0, $K > 0$, there exist $Q, r > 0$ such that any (M, g) with*

$$\operatorname{inj} \geq i_0,$$
$$|\sec| \leq K$$

has $\|(M,g)\|_{C^1,r} \leq Q$. *In particular, this class is compact in the pointed C^α topology for all $\alpha < 1$.*

PROOF. The inverse of φ is our potential chart. First, observe that $g_{ij}(p) = \delta_{ij}$, so the uniform Hessian estimate shows that $|D\varphi_p| \leq e^Q$ on $B(p,\varepsilon)$ and $\left|(D\varphi_p)^{-1}\right| \leq e^Q$ on $B(0,\varepsilon)$, where Q,ε depend only on i_0, K. The proof of the inverse function theorem then tells us that there is an $\hat{\varepsilon} > 0$ depending only on Q, n such that $\varphi : B(0,\hat{\varepsilon}) \to \mathbb{R}^n$ is one-to-one. We can then easily find r such that

$$\varphi^{-1} : B(0,r) \to U_p \subset B(p,\varepsilon)$$

satisfies n2. The conditions n3 and n4 now immediately follow from the Hessian estimates, except, we might have to increase Q somewhat. Finally, n1 holds since we have coordinates centered at every $p \in M$. $\qquad\square$

Notice that Q cannot be chosen arbitrarily small, as our Hessian estimates cannot be improved by going to smaller balls. This will be taken care of in the next section by using a different set of coordinates. This convergence result, as stated, was first proven by M. Gromov. The reader should be aware that what Gromov refers to as a $C^{1,1}$-manifold is in our terminology a manifold with $\|(M,h)\|_{C^{0,1},r} < \infty$, i.e., $C^{0,1}$-bounds on the Riemannian metric.

Using the diameter bound in positive curvature and Klingenberg's estimate for the injectivity radius from chapter 6 we get

COROLLARY 36. (J. Cheeger, 1967) *For given $n \geq 1$ and $k > 0$, the class of Riemannian $2n$-manifolds with $k \leq \sec \leq 1$ is compact in the C^α topology and consequently contains only finitely many diffeomorphism types.*

A similar result was also proven by A. Weinstein at the same time. The hypotheses are the same, but Weinstein only showed that the class contained finitely many homotopy types.

Our next result shows that one can bound the injectivity radius provided that one has lower volume bounds and bounded curvature. This result is usually referred to as Cheeger's lemma. With a little extra work one can actually prove this lemma for complete manifolds. This requires that we work with pointed spaces and also to some extent incomplete manifolds as it isn't clear from the beginning that the complete manifolds in question have global lower bounds for the injectivity radius.

LEMMA 51. (J. Cheeger, 1967) *Given $n \geq 2$ and $v, K \in (0,\infty)$ and a compact n-manifold (M,g) with*

$$|\sec| \leq K,$$
$$\mathrm{vol}B(p,1) \geq v,$$

for all $p \in M$, then $\mathrm{inj}M \geq i_0$, where i_0 depends only on n, K, and v.

PROOF. The proof goes by contradiction using the previous theorem. So assume we have (M_i, g_i) with $\mathrm{inj}M_i \to 0$ and satisfying the assumptions of the lemma. Find $p_i \in M_i$ such that $\mathrm{inj}_{p_i} = \mathrm{inj}(M_i, g_i)$, and consider the pointed sequence $(M_i, p_i, \bar{g}_i)$, where $\bar{g}_i = (\mathrm{inj}M_i)^{-2}g_i$ is rescaled so that

$$\mathrm{inj}(M_i, \bar{g}_i) = 1,$$
$$|\sec(M_i, \bar{g}_i)| \leq (\mathrm{inj}(M_i, g_i))^2 \cdot K = K_i \to 0.$$

The two previous theorems, together with the fundamental theorem, then implies that some subsequence of $(M_i, p_i, \bar{g}_i)$ will converge in the pointed $C^\alpha, \alpha < 1$, topology to a flat manifold (M, p, g).

The first observation about (M, p, g) is that $\operatorname{inj}(p) \le 1$. This follows because the conjugate radius for $(M_i, \bar{g}_i) \ge \pi/\sqrt{K_i} \to \infty$, so Klingenberg's estimate for the injectivity radius implies that there must be a geodesic loop of length 2 at $p_i \in M_i$. Since $(M_i, p_i, \bar{g}_i) \to (M, p, g)$ in the pointed C^α topology, the geodesic loops must converge to a geodesic loop in M based at p of length 2. Hence, $\operatorname{inj}(M) \le 1$.

The other contradictory observation is that $(M, g) = (\mathbb{R}^n, \operatorname{can})$. Recall that $\operatorname{vol}B(p_i, 1) \ge v$ in (M_i, g_i), so relative volume comparison shows that there is a $v'(n, K, v)$ such that $\operatorname{vol}B(p_i, r) \ge v' \cdot r^n$, for $r \le 1$. The rescaled manifold $(M_i, \bar{g}_i)$ therefore satisfies $\operatorname{vol}B(p_i, r) \ge v' \cdot r^n$, for $r \le (\operatorname{inj}(M_i, g_i))^{-1}$. Using again that $(M_i, p_i, \bar{g}_i) \to (M, p, g)$ in the pointed C^α topology, we get $\operatorname{vol}B(p, r) \ge v' \cdot r^n$ for all r. Since (M, g) is flat, this shows that it must be Euclidean space.

This last statement requires some justification. Let M be a complete flat manifold. As the elements of the fundamental group act by isometries on Euclidean space, we know that they must have infinite order (any isometry of finite order is a rotation around a point and therefore has a fixed point). Therefore, if M is not simply connected, then there is an intermediate covering $\hat{M}$:

$$\mathbb{R}^n \to \hat{M} \to M,$$

where $\pi_1\left(\hat{M}\right) = \mathbb{Z}$. This means that $\hat{M}$ looks like a cylinder. Hence, for any $p \in \hat{M}$ we must have

$$\lim_{r \to \infty} \frac{\operatorname{vol}B(p, r)}{r^{n-1}} < \infty.$$

The same must then also hold for M itself, contradicting our volume growth assumption. $\qquad\square$

This lemma was proved with a more direct method by Cheeger. We have included this, perhaps more convoluted, proof in order to show how our convergence theory can be used. The lemma also shows that the convergence theorem of Riemannian geometry remains true if the injectivity radius bound is replaced by a lower bound on the volume of 1-balls. The following result is now immediate.

COROLLARY 37. (J. Cheeger, 1967) *Let $n \ge 2$, $\Lambda, D, v \in (0, \infty)$ be given. The class of closed Riemannian n-manifolds with*

$$
\begin{aligned}
|\sec| &\le \Lambda, \\
\operatorname{diam} &\le D, \\
\operatorname{vol} &\ge v
\end{aligned}
$$

is precompact in the C^α topology for any $\alpha \in (0, 1)$ and in particular, contains only finitely many diffeomorphism types.

This convergence theorem can be generalized in another interesting direction, as observed by S.-h. Zhu.

THEOREM 75. *Given $i_0, k > 0$, there exist Q, r depending on i_0, k such that any manifold (M, g) with*

$$
\begin{aligned}
\sec &\ge -k^2, \\
\operatorname{inj} &\ge i_0
\end{aligned}
$$

satisfies $\|(M,g)\|_{C^1,r} \leq Q$.

PROOF. It suffices to get a Hessian estimate for distance functions $d(x) = d(x,p)$. We have, as before, that

$$\text{Hess} d(x) \leq k \cdot \coth(k \cdot d(x)) g_r.$$

Conversely, if $d(x_0) < i_0$, then $d(x)$ is supported from below by $f(x) = i_0 - d(x,y_0)$, where $y_0 = \gamma(i_0)$ and γ is the unique unit speed geodesic that minimizes the distance from p to x_0. Thus, $\text{Hess} d(x) \geq \text{Hess} f$ at x_0. But

$$\text{Hess} f \geq -k \cdot \coth(d(x_0,y_0) \cdot k) g_r = -k \cdot \coth(k(i_0 - r(x_0))) g_r$$

at x_0. Hence, we have two-sided bounds for $\text{Hess} d(x)$ on appropriate sets. The proof can then be finished as before. $\qquad\square$

This theorem is interestingly enough optimal. Consider rotationally symmetric metrics $dr^2 + f_\varepsilon^2(r) d\theta^2$, where f_ε is concave and satisfies

$$f_\varepsilon(r) = \begin{cases} r & \text{for } 0 \leq r \leq 1 - \varepsilon, \\ \frac{3}{4}r & \text{for } 1 + \varepsilon \leq r. \end{cases}$$

These metrics have $\sec \geq 0$ and $\text{inj} \geq 1$. As $\varepsilon \to 0$, we get a $C^{1,1}$ manifold with a $C^{0,1}$ Riemannian metric (M,g). In particular, $\|(M,g)\|_{C^{0,1},r} < \infty$ for all r. Limit spaces of sequences with $\text{inj} \geq i_0$, $\sec \geq k$ can therefore not in general be assumed to be smoother than the above example.

With a more careful construction, we can also find g_ε with

$$g_\varepsilon(r) = \begin{cases} \sin r & \text{for } 0 \leq r \leq \frac{\pi}{2} - \varepsilon, \\ 1 & \text{for } \frac{\pi}{2} \leq r. \end{cases}$$

Then the metric $dr^2 + g_\varepsilon^2(r) d\theta^2$ satisfies $|\sec| \leq 4$ and $\text{inj} \geq \frac{1}{4}$. As $\varepsilon \to 0$, we get a limit metric that is $C^{1,1}$. So while we may suspect (this is still unknown) that limit metrics from the convergence theorem are $C^{1,1}$, we prove only that they are $C^{0,1}$. In the next section we shall show that they are in fact $C^{1,\alpha}$ for all $\alpha < 1$.

5. Harmonic Norms and Ricci curvature

To get better estimates on the norms, we must use some more analysis. The idea of using harmonic coordinates for similar purposes goes back to [33]. In [57] it was shown that manifolds with bounded sectional curvature and lower bounds for the injectivity radius admit harmonic coordinates on balls of an a priori size. This result was immediately seized by the geometry community and put to use in improving the theorems from the previous section. At the same time, Nikolaev developed a different, more synthetic approach to these ideas. For the whole story we refer the reader to Greene's survey in [45]. Here we shall develop these ideas from a different point of view initiated by Anderson.

5.1. The Harmonic Norm. We shall now define another norm, called the *harmonic norm* and denoted

$$\|A \subset (M,g)\|_{C^{m,\alpha},r}^{harm}.$$

The only change in our previous definition is that condition n4 is replaced by the requirement that $\varphi_s^{-1} : U_s \to \mathbb{R}^n$ be harmonic with respect to the Riemannian metric g on M. Recall that this is equivalent to saying that for each j

$$\frac{1}{\sqrt{\det g_{st}}} \partial_i \left(\sqrt{\det g_{st}} \cdot g^{ij} \right) = 0$$

We can use the elliptic estimates to compare this norm with our old norm. Namely, recall that in harmonic coordinates $\Delta = g^{ij} \partial_i \partial_j$, conditions n2 and n3 insure that these coefficients are bounded in the required way. Therefore, if $u : U \to \mathbb{R}$ is any harmonic function, then we get that on compact subsets $K \subset U \cap U_s$,

$$\|u\|_{C^{m+1,\alpha},K} \leq C \|u\|_{C^\alpha,U}.$$

Using a coordinate function φ_t^{-1} as u then shows that we can get bounds for the transition functions on compact subsets of their domains. Changing the scale will then allow us to conclude that for each $r_1 < r_2$, there is a constant $C = C(n,m,\alpha,r_1,r_2)$ such that

$$\|A \subset (M,g)\|_{C^{m,\alpha},r_1} \leq C \|A \subset (M,g)\|_{C^{m,\alpha},r_2}^{harm}.$$

We can then show the harmonic analogue to the fundamental theorem.

COROLLARY 38. *For given* $Q > 0$, $n \geq 2$, $m \geq 0, \alpha \in (0,1]$, *and* $r > 0$ *consider the class of complete, pointed Riemannian* n-*manifolds* (M,p,g) *with* $\|(M,g)\|_{C^{m,\alpha},r}^{harm} \leq Q$. *This class is closed in the pointed* $C^{m,\alpha}$ *topology and compact in the pointed* $C^{m,\beta}$ *topology for all* $\beta < \alpha$.

The only issue to worry about is whether it is really true that limit spaces have $\|(M,g)\|_{C^{m,\alpha},r}^{harm} \leq Q$. But one can easily see that harmonic charts converge to harmonic charts. This is also discussed in the next proposition.

PROPOSITION 46. (M. Anderson, 1990) *If* $A \subset (M,g)$ *is precompact, then:*

(1) $\|A \subset (M,g)\|_{C^{m,\alpha},r}^{harm} = \|A \subset (M,\lambda^2 g)\|_{C^{m,\alpha},\lambda r}^{harm}$ *for all* $\lambda > 0$.

(2) The function $r \to \|A \subset (M,g)\|_{C^{m,\alpha},r}^{harm}$ *is continuous. Moreover, when* $m \geq 1$, *it converges to* 0 *as* $r \to 0$.

(3) Suppose $(M_i, p_i, g_i) \to (M,p,g)$ *in* $C^{m,\alpha}$ *and in addition that* $m \geq 1$. *Then for* $A \subset M$ *we can find precompact domains* $A_i \subset M_i$ *such that*

$$\|A_i\|_{C^{m,\alpha},r}^{harm} \to \|A\|_{C^{m,\alpha},r}^{harm}$$

for all $r > 0$. *When all the manifolds are closed, we can let* $A = M$ *and* $A_i = M_i$.

(4) $\|A \subset (M,g)\|_{C^{m,\alpha},r}^{harm} = \sup_{p \in A} \|\{p\} \subset (M,g)\|_{C^{m,\alpha},r}^{harm}$.

PROOF. Properties (1) and (2) are proved as for the regular norm. For the statement that the norm goes to zero as the scale decreases, just solve the Dirichlet problem as we did when existence of harmonic coordinates was established. Here it was necessary to have coordinates around every point $p \in M$ such that in these coordinates the metric satisfies $g_{ij} = \delta_{ij}$ and $\partial_k g_{ij} = 0$ at p. If $m \geq 1$, then it is easy to show that any coordinates system around p can be changed in such a way that the metric has the desired properties.

(3) The proof of this statement is necessarily somewhat different, as we must use and produce harmonic coordinates. Let the set-up be as before. First we show the easy part:

$$\liminf \|A_i\|_{C^{m,\alpha},r}^{harm} \geq \|A\|_{C^{m,\alpha},r}^{harm}.$$

To this end, select $Q > \liminf \|A_i\|^{harm}_{C^{m,\alpha},r}$. For large i we can then select charts $\varphi_{i,s} : B(0,r) \to M_i$ with the requisite properties. After passing to a subsequence, we can make these charts converge to charts

$$\varphi_s = \lim F_i^{-1} \circ \varphi_{i,s} : B(0,r) \to M.$$

Since the metrics converge in $C^{m,\alpha}$, the Laplacians of the inverse functions must also converge. Hence, the limit charts are harmonic as well. We can then conclude that $\|A\|^{harm}_{C^{m,\alpha},r} \leq Q$.

For the reverse inequality

$$\limsup \|A_i\|^{harm}_{C^{m,\alpha},r} \leq \|A\|^{harm}_{C^{m,\alpha},r},$$

select $Q > \|A\|^{harm}_{C^{m,\alpha},r}$. Then, from the continuity of the norm we can find $\varepsilon > 0$ such that also $\|A\|^{harm}_{C^{m,\alpha},r+\varepsilon} < Q$. For this scale, select charts

$$\varphi_s : B(0, r+\varepsilon) \to U_s \subset M$$

satisfying the usual conditions. Now define

$$U_{i,s} = F_i\left(\varphi_s\left(B\left(0, r+\varepsilon/2\right)\right)\right) \subset M_i.$$

This is clearly a closed disc with smooth boundary

$$\partial U_{i,s} = F_i\left(\varphi_s\left(\partial B\left(0, r+\varepsilon/2\right)\right)\right).$$

On each $U_{i,s}$ solve the Dirichlet problem

$$\psi_{i,s} \quad : \quad U_{i,s} \to \mathbb{R}^n,$$
$$\Delta_{g_i}\psi_{i,s} \quad = \quad 0,$$
$$\psi_{i,s} \quad = \quad \varphi_s^{-1} \circ F_i^{-1} \text{ on } \partial U_{i,s}.$$

The inverse of $\psi_{i,s}$, if it exists, will then be a coordinate map $B(0,r) \to U_{i,s}$. On the set $B(0, r+\varepsilon/2)$ we can now compare $\psi_{i,s} \circ F_i \circ \varphi_s$ with the identity map I. Note that these maps agree on the boundary of $B(0, r+\varepsilon/2)$. We know that $F_i^*g_i \to g$ in the fixed coordinate system φ_s. Now pull these metrics back to $B\left(0, r+\frac{\varepsilon}{2}\right)$ and refer to them as $g\left(= \varphi_s^* g\right)$ and $g_i\left(= \varphi_s^* F_i^* g_i\right)$. In this way the harmonicity conditions read $\Delta_g I = 0$ and $\Delta_{g_i}\psi_{i,s} \circ F_i \circ \varphi_s = 0$. In these coordinates we have the correct bounds for the operator

$$\Delta_{g_i} = g_i^{kl}\partial_k\partial_l + \frac{1}{\sqrt{\det g_i}}\partial_k\left(\sqrt{\det g_i} \cdot g_i^{kl}\right)\partial_l$$

to use the elliptic estimates for domains with smooth boundary. Note that this is where the condition $m \geq 1$ becomes important, so that we can bound

$$\frac{1}{\sqrt{\det g_i}}\partial_k\left(\sqrt{\det g_i} \cdot g_i^{kl}\right)$$

in C^α. The estimates then imply

$$\|I - \psi_{i,s} \circ F_i \circ \varphi_s\|_{C^{m+1,\alpha}} \quad \leq \quad C\|\Delta_{g_i}\left(I - \psi_{i,s} \circ F_i \circ \varphi_s\right)\|_{C^{m-1,\alpha}}$$
$$= \quad C\|\Delta_{g_i}I\|_{C^{m-1,\alpha}}.$$

However, we have that

$$
\begin{aligned}
\left\| \Delta_{g_i} I \right\|_{C^{m-1,\alpha}} &= \left\| \frac{1}{\sqrt{\det g_i}} \partial_k \left(\sqrt{\det g_i} \cdot g_i^{kl} \right) \right\|_{C^{m-1,\alpha}} \\
&\rightarrow \left\| \frac{1}{\sqrt{\det g}} \partial_k \left(\sqrt{\det g} \cdot g^{kl} \right) \right\|_{C^{m-1,\alpha}} \\
&= \left\| \Delta_g I \right\|_{C^{m-1,\alpha}} = 0.
\end{aligned}
$$

In particular, we must have

$$
\left\| I - \psi_{i,s} \circ F_i \circ \varphi_s \right\|_{C^{m+1,\alpha}} \rightarrow 0.
$$

It is now evident that $\psi_{i,s}$ must become coordinates for large i. Also, these coordinates will show that $\|A_i\|_{C^{m,\alpha},r}^{harm} < Q$ for large i.

(4) Since there is no transition function condition to be satisfied in the definition of $\|A\|_{C^{m,\alpha},r}^{harm}$, it is obvious that

$$
\|A \cup B\|_{C^{m,\alpha},r}^{harm} = \max \left\{ \|A\|_{C^{m,\alpha},r}^{harm}, \|B\|_{C^{m,\alpha},r}^{harm} \right\}.
$$

This shows that the norm is always realized locally. $\square$

5.2. Ricci Curvature and the Harmonic Norm. The most important feature about harmonic coordinates is that the metric is apparently controlled by the Ricci curvature. This is exploited in the next lemma, where we show how one can bound the harmonic $C^{1,\alpha}$ norm in terms of the harmonic C^1 norm and Ricci curvature.

LEMMA 52. (M. Anderson, 1990) *Suppose that a Riemannian manifold (M, g) has bounded Ricci curvature $|\mathrm{Ric}| \leq \Lambda$. For any $r_1 < r_2$, $K \geq \|A \subset (M, g)\|_{C^1, r_2}^{harm}$, and $\alpha \in (0,1)$ we can find $C(n, \alpha, K, r_1, r_2, \Lambda)$ such that*

$$
\|A \subset (M, g)\|_{C^{1,\alpha}, r_1}^{harm} \leq C(n, \alpha, K, r_1, r_2, \Lambda).
$$

Moreover, if g is an Einstein metric $\mathrm{Ric} = kg$, then for each integer m we can find a constant $C(n, \alpha, K, r_1, r_2, k, m)$ such that

$$
\|A \subset (M, g)\|_{C^{m+1,\alpha}, r_1}^{harm} \leq C(n, \alpha, K, r_1, r_2, k, m).
$$

PROOF. We just need to bound the metric components g_{ij} in some fixed harmonic coordinates. In these coordinates we have that $\Delta = g^{ij} \partial_i \partial_j$. Given that $\|A \subset (M, g)\|_{C^1, r_2}^{harm} \leq K$, we can conclude that we have the necessary conditions on the coefficients of $\Delta = g^{ij} \partial_i \partial_j$ to use the elliptic estimate

$$
\|g_{ij}\|_{C^{1,\alpha}, B(0,r_1)} \leq C(n, \alpha, K, r_1, r_2) \left(\|\Delta g_{ij}\|_{C^0, B(0,r_2)} + \|g_{ij}\|_{C^\alpha, B(0,r_2)} \right).
$$

Now use that

$$
\Delta g_{ij} = -2\mathrm{Ric}_{ij} - 2Q(g, \partial g)
$$

to conclude that

$$
\|\Delta g_{ij}\|_{C^0, B(0,r_2)} \leq 2\Lambda \|g_{ij}\|_{C^0, B(0,r_2)} + \hat{C} \|g_{ij}\|_{C^1, B(0,r_2)}.
$$

Using this we then have

$$
\begin{aligned}
\|g_{ij}\|_{C^{1,\alpha}, B(0,r_1)} &\leq C(n, \alpha, K, r_1, r_2) \left(\|\Delta g_{ij}\|_{C^0, B(0,r_2)} + \|g_{ij}\|_{C^\alpha, B(0,r_2)} \right) \\
&\leq C(n, \alpha, K, r_1, r_2) \left(2\Lambda + \hat{C} + 1 \right) \|g_{ij}\|_{C^1, B(0,r_2)}.
\end{aligned}
$$

For the Einstein case we can use a bootstrap method, as we get $C^{1,\alpha}$ bounds on the Ricci tensor from the Einstein equation $\text{Ric} = kg$. Thus, we have that Δg_{ij} is bounded in C^α rather than just C^0. Hence,

$$
\begin{aligned}
\|g_{ij}\|_{C^{2,\alpha},B(0,r_1)} &\leq C(n,\alpha,K,r_1,r_2)\left(\|\Delta g_{ij}\|_{C^\alpha,B(0,r_2)} + \|g_{ij}\|_{C^\alpha,B(0,r_2)}\right) \\
&\leq C(n,\alpha,K,r_1,r_2,k)\cdot C\cdot\|g_{ij}\|_{C^{1,\alpha},B(0,r_2)}.
\end{aligned}
$$

This gives $C^{2,\alpha}$ bounds on the metric. Then, of course, Δg_{ij} is bounded in $C^{1,\alpha}$, and thus the metric will be bounded in $C^{3,\alpha}$. Clearly, one can iterate this until one gets $C^{m+1,\alpha}$ bounds on the metric. □

Combining this with the fundamental theorem gives a very interesting compactness result.

COROLLARY 39. *For given* $n \geq 2, Q, r, \Lambda \in (0,\infty)$ *consider the class of Riemannian n-manifolds with*

$$
\begin{aligned}
\|(M,g)\|_{C^1,r}^{harm} &\leq Q, \\
|\text{Ric}| &\leq \Lambda.
\end{aligned}
$$

This class is precompact in the pointed $C^{1,\alpha}$ *topology for any* $\alpha \in (0,1)$. *Moreover, if we take the subclass of Einstein manifolds, then this class is compact in the* $C^{m,\alpha}$ *topology for any* $m \geq 0$ *and* $\alpha \in (0,1)$.

We can now prove our generalizations of the convergence theorems from the last section.

THEOREM 76. (M. Anderson, 1990) *Given* $n \geq 2$ *and* $\alpha \in (0,1)$, Λ, $i_0 > 0$, *one can for each* $Q > 0$ *find* $r(n,\alpha,\Lambda,i_0) > 0$ *such that any complete Riemannian n-manifold* (M,g) *with*

$$
\begin{aligned}
|\text{Ric}| &\leq \Lambda, \\
\text{inj} &\geq i_0
\end{aligned}
$$

satisfies $\|(M,g)\|_{C^{1,\alpha},r}^{harm} \leq Q$.

PROOF. The proof goes by contradiction. So suppose that there is a $Q > 0$ such that for each $i \geq 1$ there is a Riemannian manifold (M_i, g_i) with

$$
\begin{aligned}
|\text{Ric}| &\leq \Lambda, \\
\text{inj} &\geq i_0, \\
\|(M_i,g_i)\|_{C^{1,\alpha},i^{-1}}^{harm} &> Q.
\end{aligned}
$$

Using that the norm goes to zero as the scale goes to zero, and that it is continuous as a function of the scale, we can for each i find $r_i \in (0,i^{-1})$ such that $\|(M_i,g_i)\|_{C^{1,\alpha},r_i}^{harm} = Q$. Now rescale these manifolds: $\bar{g}_i = r_i^{-2}g_i$. Then we have that $(M_i,\bar{g}_i)$ satisfies

$$
\begin{aligned}
|\text{Ric}| &\leq r_i^2\Lambda, \\
\text{inj} &\geq r_i^{-1}i_0, \\
\|(M_i,\bar{g}_i)\|_{C^{1,\alpha},1}^{harm} &= Q.
\end{aligned}
$$

We can then select $p_i \in M_i$ such that

$$\|p_i \in (M_i, \bar{g}_i)\|_{C^{1,\alpha},1}^{harm} \in \left[\frac{Q}{2}, Q\right].$$

The first important step is now to use the bounded Ricci curvature of $(M_i, \bar{g}_i)$ to conclude that in fact the $C^{1,\gamma}$ norm must be bounded for any $\gamma \in (\alpha, 1)$. Then we can assume by the fundamental theorem that the sequence $(M_i, p_i, \bar{g}_i)$ converges in the pointed $C^{1,\alpha}$ topology, to a Riemannian manifold (M, p, g) of class $C^{1,\gamma}$. Since the $C^{1,\alpha}$ norm is continuous in the $C^{1,\alpha}$ topology we can conclude that

$$\|p \in (M, g)\|_{C^{1,\alpha},1}^{harm} \in \left[\frac{Q}{2}, Q\right].$$

The second thing we can prove is that $(M, g) = (\mathbb{R}^n, \text{can})$. This clearly violates what we just established about the norm of the limit space. To see that the limit space is Euclidean space, recall that the manifolds in the sequence $(M_i, \bar{g}_i)$ are covered by harmonic coordinates that converge to harmonic coordinates in the limit space. In these harmonic coordinates the metric components satisfy

$$\frac{1}{2}\Delta \bar{g}_{kl} + Q(\bar{g}, \partial \bar{g}) = -\text{Ric}_{kl}.$$

But we know that

$$|-\text{Ric}| \le r_i^{-2}\Lambda \bar{g}_i$$

and that the $\bar{g}_{kl}$ converge in the $C^{1,\alpha}$ topology to the metric coefficients g_{kl} for the limit metric. We can therefore conclude that the limit manifold is covered by harmonic coordinates and that in these coordinates the metric satisfies:

$$\frac{1}{2}\Delta g_{kl} + Q(g, \partial g) = 0.$$

The limit metric is therefore a weak solution to the Einstein equation $\text{Ric} = 0$ and must therefore be a smooth Ricci flat Riemannian manifold. It is now time to use that: $\text{inj}(M_i, \bar{g}_i) \to \infty$. In the limit space we have that any geodesic is a limit of geodesics from the sequence $(M_i, \bar{g}_i)$, since the Riemannian metrics converge in the $C^{1,\alpha}$ topology. If a geodesic in the limit is a limit of segments, then it must itself be a segment. We can then conclude that as $\text{inj}(M_i, \bar{g}_i) \to \infty$ any finite length geodesic must be a segment. This, however, implies that $\text{inj}(M, g) = \infty$. The splitting theorem then shows that the limit space is Euclidean space.   □

From this theorem we immediately get

COROLLARY 40. (M. Anderson, 1990) *Let* $n \ge 2$ *and* Λ, D, $i \in (0, \infty)$ *be given. The class of closed Riemannian n-manifolds satisfying*

$$\begin{aligned}|\text{Ric}| &\le \Lambda, \\ \text{diam} &\le D, \\ \text{inj} &\ge i\end{aligned}$$

is precompact in the $C^{1,\alpha}$ *topology for any* $\alpha \in (0, 1)$ *and in particular contains only finitely many diffeomorphism types.*

Notice how the above theorem depended on the characterization of Euclidean space we obtained from the splitting theorem. There are other similar characterizations of Euclidean space. One of the most interesting ones uses volume pinching.

5.3. Volume Pinching. The idea is to use the relative volume comparison theorem rather than the splitting theorem. We know from the exercises to chapter 9 that Euclidean space is the only space with

$$\text{Ric} \geq 0,$$

$$\lim_{r \to \infty} \frac{\text{vol}B\,(p, r)}{\omega_n r^n} = 1,$$

where $\omega_n r^n$ is the volume of a Euclidean ball of radius r. This result has a very interesting gap phenomenon associated with it, when one assumes the stronger hypothesis that the space is Ricci flat.

LEMMA 53. (M. Anderson, 1990) *For each $n \geq 2$ there is an $\varepsilon(n) > 0$ such that any complete Ricci flat manifold (M, g) that satisfies*

$$\text{vol}B\,(p, r) \geq (\omega_n - \varepsilon)\, r^n$$

for some $p \in M$ is isometric to Euclidean space.

PROOF. First observe that on any complete Riemannian manifold with $\text{Ric} \geq 0$, relative volume comparison can be used to show that

$$\text{vol}B\,(p, r) \geq (1 - \varepsilon)\, \omega_n r^n$$

as long as

$$\lim_{r \to \infty} \frac{\text{vol}B\,(p, r)}{\omega_n r^n} \geq (1 - \varepsilon).$$

It is then easy to see that if this holds for one p, then it must hold for all p. Moreover, if we scale the metric to $(M, \lambda^2 g)$, then the same volume comparison still holds, as the lower curvature bound $\text{Ric} \geq 0$ can't be changed by scaling.

If our assertion were not true, then we could for each integer i find Ricci flat manifolds (M_i, g_i) with

$$\lim_{r \to \infty} \frac{\text{vol}B\,(p_i, r)}{\omega_n r^n} \geq (1 - i^{-1}),$$

$$\|(M_i, g_i)\|_{C^{1,\alpha}, r}^{harm} \neq 0 \text{ for all } r > 0.$$

By scaling these metrics suitably, it is then possible to arrange it so that we have a sequence of Ricci flat manifolds $(M_i, q_i, \bar{g}_i)$ with

$$\lim_{r \to \infty} \frac{\text{vol}B\,(q_i, r)}{\omega_n r^n} \geq (1 - i^{-1}),$$

$$\|(M_i, \bar{g}_i)\|_{C^{1,\alpha}, 1}^{harm} \leq 1,$$

$$\|q_i \in (M_i, \bar{g}_i)\|_{C^{1,\alpha}, 1}^{harm} \in [0.5, 1].$$

From what we already know, we can then extract a subsequence that converges in the $C^{m,\alpha}$ topology to a Ricci flat manifold (M, q, g). In particular, we must have that metric balls of a given radius converge and that the volume forms converge. Thus, the limit space must satisfy

$$\lim_{r \to \infty} \frac{\text{vol}B\,(q, r)}{\omega_n r^n} = 1.$$

This means that we have maximal possible volume for all metric balls, and thus the manifold must be Euclidean. This, however, violates the continuity of the norm in the $C^{1,\alpha}$ topology, as the norm for the limit space would then have to be zero. $\square$

COROLLARY 41. *Let* $n \geq 2$, $-\infty < \lambda \leq \Lambda < \infty$, *and* D, $i_0 \in (0, \infty)$ *be given.* *There is a* $\delta = \delta\left(n, \lambda \cdot i_0^2\right)$ *such that the class of closed Riemannian n-manifolds satisfying*

$$(n-1)\Lambda \geq \text{Ric} \geq (n-1)\lambda,$$
$$\text{diam} \leq D,$$
$$\text{vol}B(p, i_0) \geq (1-\delta)v(n, \lambda, i_0)$$

is precompact in the $C^{1,\alpha}$ *topology for any* $\alpha \in (0, 1)$ *and in particular contains only finitely many diffeomorphism types.*

PROOF. We use the same techniques as when we had an injectivity radius bound. Observe that if we have a sequence $(M_i, p_i, \bar{g}_i)$ where $\bar{g}_i = k_i^2 g_i$, $k_i \to \infty$, and the (M_i, g_i) lie in the above class, then the volume condition now reads

$$\text{vol}B_{\bar{g}_i}(p_i, i_0 \cdot k_i) = k_i^n \text{vol}B_{g_i}(p_i, i_0)$$
$$\geq k_i^n(1-\delta)v(n, \lambda, i_0)$$
$$= (1-\delta)v\left(n, \lambda \cdot k_i^{-2}, i_0 \cdot k_i\right).$$

From relative volume comparison we can then conclude that for $r \leq i_0 \cdot k_i$ and very large i,

$$\text{vol}B_{\bar{g}_i}(p_i, r) \geq (1-\delta)v\left(n, \lambda \cdot k_i^{-2}, r\right) \sim (1-\delta)\omega_n r^n.$$

In the limit space we must therefore have

$$\text{vol}B(p, r) \geq (1-\delta)\omega_n r^n \text{ for all } r.$$

This limit space is also Ricci flat and is therefore Euclidean space. The rest of the proof goes as before, by getting a contradiction with the continuity of the norms. □

5.4. Curvature Pinching. Let us now turn our attention to some applications of these compactness theorems. One natural subject to explore is that of *pinching* results. Recall that we showed earlier that complete constant curvature manifolds have a uniquely defined universal covering. It is natural to ask whether one can in some topological sense still expect this to be true when one has close to constant curvature. Now, any Riemannian manifold (M, g) has curvature close to zero if we multiply the metric by a large scalar. Thus, some additional assumptions must come into play.

We start out with the simpler problem of considering Ricci pinching and then use this in the context of curvature pinching below. The results are very simple consequences of the convergence theorem we have already presented.

THEOREM 77. *Given* $n \geq 2$, i, $D \in (0, \infty)$, *and* $\lambda \in \mathbb{R}$, *there is an* $\varepsilon = \varepsilon(n, \lambda, D, i) > 0$ *such that any closed Riemannian n-manifold* (M, g) *with*

$$\text{diam} \leq D,$$
$$\text{inj} \geq i,$$
$$|\text{Ric} - \lambda g| \leq \varepsilon$$

is $C^{1,\alpha}$ *close to an Einstein metric with Einstein constant* λ.

PROOF. We already know that this class is precompact in the $C^{1,\alpha}$ topology no matter what ε we choose. If the result were not true, we could therefore find a sequence $(M_i, g_i) \to (M, g)$ that converges in the $C^{1,\alpha}$ topology to a closed Riemannian manifold of class $C^{1,\alpha}$, where in addition, $|\text{Ric}_{g_i} - \lambda g_i| \to 0$. Using harmonic coordinates as usual we can therefore conclude that the metric on the limit space must be a weak solution to

$$\frac{1}{2}\Delta g + Q\left(g, \partial g\right) = -\lambda g.$$

But this means that the limit space is actually Einstein, with Einstein constant λ, thus, contradicting that the spaces (M_i, g_i) were not close to such Einstein metrics. $\square$

Using the compactness theorem for manifolds with almost maximal volumes we see that the injectivity radius condition could have been replaced with an almost maximal volume condition. Now let us see what happens with sectional curvature.

THEOREM 78. *Given* $n \geq 2$, $v, D \in (0, \infty)$, *and* $\lambda \in \mathbb{R}$, *there is an* $\varepsilon = \varepsilon\left(n, \lambda, D, i\right) > 0$ *such that any closed Riemannian* n-*manifold* (M, g) *with*

$$\text{diam} \quad \leq \quad D,$$
$$\text{vol} \quad \geq \quad v,$$
$$|\sec - \lambda| \quad \leq \quad \varepsilon$$

is $C^{1,\alpha}$ *close to a metric of constant curvature* λ.

PROOF. In this case we first observe that Cheeger's lemma gives us a lower bound for the injectivity radius. The previous theorem then shows that such metrics must be close to Einstein metrics. We now have to check that if $(M_i, g_i) \to (M, g)$, where $|\sec_{g_i} - \lambda| \to 0$ and $\mathrm{Ric}_g = (n - 1)\lambda g$, then in fact (M, g) has constant curvature λ. To see this, it is perhaps easiest to observe that if

$$M_i \ni p_i \to p \in M,$$

then we can use polar coordinates around these points to write $g_i = dr^2 + g_{r,i}$ and $g = dr^2 + g_r$. Since the metrics converge in $C^{1,\alpha}$, we certainly have that $g_{r,i}$ converge to g_r. Using the curvature pinching, we conclude from chapter 6 that

$$\mathrm{sn}_{\lambda + \varepsilon_i}^2(r)\, ds_{n-1}^2 \leq g_{r,i} \leq \mathrm{sn}_{\lambda - \varepsilon_i}^2(r)\, ds_{n-1}^2,$$

where $\varepsilon_i \to 0$. In the limit we therefore have

$$\mathrm{sn}_\lambda^2(r)\, ds_{n-1}^2 \leq g_r \leq \mathrm{sn}_\lambda^2(r)\, ds_{n-1}^2.$$

This implies that the limit metric has constant curvature λ. $\square$

It is interesting that we had to go back and use the more geometric estimates for distance functions in order to prove the curvature pinching, while the Ricci pinching could be handled more easily with analytic techniques using harmonic coordinates. One can actually prove the curvature result with purely analytic techniques, but this requires that we study convergence in a more general setting where one uses L^p norms and estimates. This has been developed rigorously and can be used to improve the above results to situations were one has only L^p curvature pinching rather than the L^∞ pinching we use here (see [**79**], [**80**], and [**32**]).

When the curvature λ is positive, some of the assumptions in the above theorems are in fact not necessary. For instance, Myers' estimate for the diameter makes the diameter hypothesis superfluous. For the Einstein case this seems to be as far as we can go. In the positive curvature case we can do much better. In even dimensions, we already know from chapter 6, that manifolds with positive curvature have both bounded diameter and lower bounds for the injectivity radius, provided that there is an upper curvature bound. We can therefore show

COROLLARY 42. *Given* $2n \geq 2$, *and* $\lambda > 0$, *there is an* $\varepsilon = \varepsilon(n, \lambda) > 0$ *such that any closed Riemannian* $2n$*-manifold* (M, g) *with*

$$|\sec - \lambda| \leq \varepsilon$$

is $C^{1,\alpha}$ *close to a metric of constant curvature* λ.

This corollary is, in fact, also true in odd dimensions. This was proved by Grove-Karcher-Ruh in [**49**]. Notice that convergence techniques are not immediately applicable because there are no lower bounds for the injectivity radius. Their pinching constant is also independent of the dimension.

Also recall the quarter pinching results in positive curvature than we proved in chapter 6. There the conclusions were much weaker and purely topological. In a similar vein there is a nice result of Micaleff-Moore in [**66**]stating that any manifold with positive isotropic curvature has a universal cover that is homeomorphic to the sphere. However, this doesn't generalize the above theorem, for it is not necessarily true that two manifolds with identical fundamental groups and universal covers are homotopy equivalent.

In negative curvature some special things also happen. Namely, Heintze has proved that any complete manifold with $-1 \leq \sec < 0$ has a lower volume bound when the dimension ≥ 4 (see also [**46**] for a more general statement). The lower volume bound is therefore an extraneous condition when doing pinching in negative curvature. Unlike the situation in positive curvature, the upper diameter bound is, however, crucial. See, e.g., [**48**] and [**38**] for counterexamples.

This leaves us with pinching around 0. As any compact Riemannian manifold can be scaled to have curvature in $[-\varepsilon, \varepsilon]$ for any ε, we do need the diameter bound. The volume condition is also necessary, as the Heisenberg group from the exercises to chapter 3 has a quotient where there are metrics with bounded diameter and arbitrarily pinched curvature. This quotient, however, does not admit a flat metric. Gromov was nevertheless able to classify all n-manifolds with

$$|\sec| \leq \varepsilon(n),$$
$$\text{diam} \leq 1$$

for some very small $\varepsilon(n) > 0$. More specifically, they all have a finite cover that is a quotient of a nilpotent Lie group by a discrete subgroup. For more on this and collapsing in general, the reader can start by reading [**39**].

6. Further Study

Cheeger first proved his finiteness theorem and put down the ideas of C^k convergence for manifolds in [**21**]. They later appeared in journal form [**22**], but not all ideas from the thesis were presented in this paper. Also the idea of general pinching theorems as described here are due to Cheeger [**23**]. For more generalities on convergence and their uses we recommend the surveys by Anderson, Fukaya, Petersen, and Yamaguchi in [**45**]. Also for more on norms and convergence theorems the survey by Petersen in [**50**] might prove useful. The text [**47**] should also be mentioned again. It was probably the original french version of this book that really spread the ideas of Gromov-Hausdorff distance and the stronger convergence theorems to a wider audience. Also, the convergence theorem of Riemannian geometry, as stated here, appeared for the first time in this book.

We should also mention that S. Peters in [**77**] obtained an explicit estimate for the number of diffeomorphism classes in Cheeger's finiteness theorem. This also seems to be the first place where the modern statement of Cheeger's finiteness theorem is proved.

7. Exercises

(1) Find a sequence of 1-dimensional metric spaces that Hausdorff converge to the unit cube $[0,1]^3$ endowed with the metric coming from the maximum norm on $\mathbb{R}^3$. Then find surfaces (jungle gyms) converging to the same space.

(2) C. Croke has shown that there is a universal constant $c(n)$ such that any n-manifold with inj $\geq i_0$ satisfies $\mathrm{vol}B(p,r) \geq c(n) \cdot r^n$ for $r \leq \frac{i_0}{2}$. Use this to show that the class of n-dimensional manifolds satisfying inj $\geq i_0$ and vol $\leq V$ is precompact in the Gromov-Hausdorff topology.

(3) Develop a Bochner formula for Hess $\left(\frac{1}{2}g(X,Y)\right)$ and $\Delta\frac{1}{2}g(X,Y)$, where X and Y are vector fields with symmetric ∇X and ∇Y. Discuss whether it is possible to devise coordinates where Hess (g_{ij}) are bounded in terms of the full curvature tensor. If this were possible we would be able to get $C^{1,1}$ bounds for manifolds with bounded curvature. It is still an open question whether this is possible.

(4) Show that in contrast with the elliptic estimates, it is not possible to find C^α bounds for a vector field X in terms of C^0 bounds on X and div$X$.

(5) Define $C^{m,\alpha}$ convergence for incomplete manifolds. On such manifolds define the boundary ∂ as the set of points that lie in the completion but not in the manifold itself. Show that the class of incomplete spaces with $|\mathrm{Ric}| \leq \Lambda$ and inj $(p) \geq \min\{i_0, i_0 \cdot d(p,\partial)\}$, $i_0 < 1$, is precompact in the $C^{1,\alpha}$ topology.

(6) Define a *weighted norm* concept. That is, fix a positive function $\rho(R)$, and assume that in a pointed manifold (M,p,g) the distance spheres $S(p,R)$ have norm $\leq \rho(R)$. Prove the corresponding fundamental theorem.

(7) Suppose we have a class that is compact in the $C^{m,\alpha}$ topology. Show that there is a function $f(r)$ depending on the class such that $\|(M,g)\|_{C^{m,\alpha},r} \leq f(r)$ for all elements in this class, and also, $f(r) \to 0$ as $r \to 0$.

(8) The *local models* for a class of Riemannian manifolds are the types of spaces one obtains by scaling the elements of the class by a constant $\to \infty$. For example, if we consider the class of manifolds with $|\sec| \leq K$ for some K, then upon rescaling the metrics by a factor of λ^2, we have the condition $|\sec| \leq \lambda^{-2}K$, as $\lambda \to \infty$, we therefore arrive at the condition $|\sec| = 0$. This means that the local models are all the flat manifolds. Notice that we don't worry about any type of convergence here. If, in this example, we additionally assume that the manifolds have inj $\geq i_0$, then upon rescaling and letting $\lambda \to \infty$ we get the extra condition inj $= \infty$. Thus, the local model is Euclidean space. It is natural to suppose that any class that has Euclidean space as it only local model must be compact in some topology.

Show that a class of spaces is compact in the $C^{m,\alpha}$ topology if when we rescale a sequence in this class by constants that $\to \infty$, the sequence subconverges in the $C^{m,\alpha}$ topology to Euclidean space.

(9) Consider the singular Riemannian metric $dt^2 + (at)^2 d\theta^2$, $a > 1$, on $\mathbb{R}^2$. Show that there is a sequence of rotationally symmetric metrics on $\mathbb{R}^2$ with sec ≤ 0 and inj $= \infty$ that converge to this metric in the Gromov-Hausdorff topology.

(10) Show that the class of spaces with inj $\geq i$ and $\left|\nabla^k \text{Ric}\right| \leq \Lambda$ for $k = 0, \ldots, m$ is compact in the $C^{m+1,\alpha}$ topology.

(11) (S.-h. Zhu) Consider the class of complete or compact n-dimensional Riemannian manifolds with

$$\begin{aligned} \text{conj.rad} &\geq r_0, \\ |\text{Ric}| &\leq \Lambda, \\ \text{vol}B(p,1) &\geq v. \end{aligned}$$

Using the techniques from Cheeger's lemma, show that this class has a lower bound for the injectivity radius. Conclude that it is compact in the $C^{1,\alpha}$ topology.

(12) Using the Eguchi-Hanson metrics from the exercises to chapter 3 show that one cannot in general expect a compactness result for the class

$$\begin{aligned} |\text{Ric}| &\leq \Lambda, \\ \text{vol}B(p,1) &\geq v. \end{aligned}$$

Thus, one must assume either that v is large as we did before or that there a lower bound for the conjugate radius.

(13) The *weak (harmonic) norm* $\|(M,g)\|_{C^{m,\alpha},r}^{weak}$ is defined in almost the same way as the norms we have already worked with, except that we only insist that the charts $\varphi_s : B(0,r) \to U_s$ are *immersions*. The inverse is therefore only locally defined, but it still makes sense to say that it is harmonic.

(a) Show that if (M,g) has bounded sectional curvature, then for all $Q > 0$ there is an $r > 0$ such that $\|(M,g)\|_{C^{1,\alpha},r}^{weak} \leq Q$. Thus, the weak norm can be thought of as a generalized curvature quantity.

(b) Show that the class of manifolds with bounded weak norm is precompact in the Gromov-Hausdorff topology.

(c) Show that (M,g) is flat iff the weak norm is zero on all scales.

CHAPTER 11

Sectional Curvature Comparison II

In the first section we explain how one can find generalized gradients for distance functions in situations where the function might not be smooth. This critical point technique is used in the proofs of all the big theorems in this chapter. The other important technique comes from Toponogov's theorem, which we prove in the next section. The first applications of these new ideas are to sphere theorems. We then prove the soul theorem of Cheeger and Gromoll. Next, we discuss Gromov's finiteness theorem for bounds on Betti numbers and generators for the fundamental group. Finally, we show that these techniques can be adapted to prove the Grove-Petersen homotopy finiteness theorem.

Toponogov's theorem is a very useful refinement of Gauss's early realization that curvature and angle excess of triangles are related. The fact that Toponogov's theorem can be used to get information about the topology of a space seems to originate with Berger's proof of the quarter pinched sphere theorem. Toponogov himself proved these theorems in order to establish the splitting theorem for manifolds with nonnegative sectional curvature and the maximal diameter theorem for manifolds with a positive lower bound for the sectional curvature. As we saw in chapter 9, these results now hold in the Ricci curvature setting. The next use of Toponogov was to the soul theorem of Cheeger-Gromoll-Meyer. However, Toponogov's theorem is not truly needed for any of the results mentioned so far. With little effort one can actually establish these theorems with more basic comparison techniques. Still, it is convenient to have a workhorse theorem of universal use. It wasn't until Grove and Shiohama developed critical point theory to prove their diameter sphere theorem that Toponogov's theorem was put to serious use. Shortly after that, Gromov put these two ideas to even more nontrivial use, with his Betti number estimate for manifolds with nonnegative sectional curvature. After that, it became clear that in working with manifolds that have lower sectional curvature bounds, the two key techniques are Toponogov's theorem and the critical point theory of Grove-Shiohama. These two very geometric techniques are still being used to prove many interesting and nontrivial results.

1. Critical Point Theory

In the particular generalized critical point theory developed here, the object is to define generalized gradients of continuous functions and then use these gradients to conclude that certain regions of a manifold have no topology. The motivating basic lemma is the following:

LEMMA 54. *Let (M, g) be a Riemannian manifold and $f : M \to \mathbb{R}$ a proper smooth function. If f has no critical values in the closed interval $[a, b]$, then the preimages $f^{-1}([-\infty, b])$ and $f^{-1}([-\infty, a])$ are diffeomorphic. Furthermore, there*

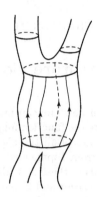

Figure 11.1

is a deformation retraction of $f^{-1}([-\infty, b])$ onto $f^{-1}([-\infty, a])$, in particular, the inclusion

$$f^{-1}([-\infty, a]) \hookrightarrow f^{-1}([-\infty, b])$$

is a homotopy equivalence.

PROOF. The idea is simply to move the level sets via the gradient of f. Since there are no critical points for f the gradient ∇f is nonzero everywhere on $f^{-1}([a, b])$ We then construct a bump function $\psi : M \to [0, 1]$ that is 1 on the compact set $f^{-1}([a, b])$ and zero outside some compact neighborhood of $f^{-1}([a, b])$. Finally consider the vector field

$$X = \psi \cdot \frac{\nabla f}{|\nabla f|^2}$$

This vector field has compact support and must therefore be complete (integral curves are defined for all time). Let F^t denote the flow for this vector field. (See Figure 11.1)

For fixed $q \in M$ consider the function $t \to f(F^t(q))$. The derivative of this function is $g(X, \nabla f)$, so as long as the integral curve $t \to F^t(q)$ remains in $f^{-1}([a, b])$, the function $t \to f(F^t(q))$ is linear with derivative 1. In particular, the diffeomorphism $F^{b-a} : M \to M$ must carry $f^{-1}([-\infty, a])$ diffeomorphically onto $f^{-1}([-\infty, b])$.

Moreover, by flowing backwards we can define the desired retraction:

$$r_t \quad : \quad f^{-1}([-\infty, b]) \to f^{-1}([-\infty, b]),$$

$$r_t(p) = \begin{cases} p & \text{if } f(p) \le a, \\ F^{t(a-f(p))}(p) & \text{if } a \le f(p) \le b. \end{cases}$$

Then $r_0 = id$, and r_1 maps $f^{-1}([-\infty, b])$ diffeomorphically onto $f^{-1}([-\infty, a])$. $\square$

Notice that we used in an essential way that the function is proper to conclude that the vector field is complete. In fact, if we delete a single point from the region $f^{-1}([a, b])$, then the function still won't have any critical values, but clearly the conclusion of the lemma is false.

We shall now try to generalize this lemma to functions that are not even C^1. To minimize technicalities we shall work exclusively with distance functions. Suppose

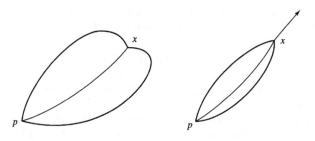

Figure 11.2

(M, g) is complete and $K \subset M$ a compact subset. Then the distance function

$$r(x) = d(x, K) = \min \{d(x, p) : p \in K\}$$

is proper. Wherever this function is smooth, we know that it has unit gradient and must therefore be noncritical at such points. However, it might also have local maxima, and at such points we certainly wouldn't want the function to be noncritical. To define the generalized gradient for such functions, let us list all the possible values it could have. Define $\Gamma(x, K)$, or simply $\Gamma(x)$, as the set of unit vectors in $T_x M$ that are tangent to a segment from K to x. That is, $v \in \Gamma(x, K) \subset T_x M$ if there is a unit speed segment $\sigma : [0, \ell] \to M$ such that $\sigma(0) \in K$, $\sigma(\ell) = x$, and $v = \dot{\sigma}(\ell)$. Note that σ is chosen such that no shorter curve from x to K exists. There might, however, be several such segments. In the case where r is smooth at x, we clearly have that $\{\nabla r\} = \Gamma(x, K)$. At other points, $\Gamma(x, K)$ might contain more vectors. We say that r is *regular, or noncritical,* at x if the set $\Gamma(x, K)$ is contained in an open hemisphere of the unit sphere in $T_x M$. The center of such a hemisphere is then a possible averaged direction for the gradient of r at x. Stated differently, we have that r is regular at x iff there is a vector $v \in T_x M$ such that the angles $\angle (v, w) < \pi/2$ for all $w \in \Gamma(x, K)$. If v is a unit vector, then it will be the center of the desired hemisphere. We can quantify being regular by saying that r is α-*regular* at x if there exist $v \in T_x M$ such that $\angle (v, w) < \alpha$ for all $w \in \Gamma(x, K)$. Thus, r is regular at x iff it is $\pi/2$-regular. The set of vectors v that can be used in the definition of α-regularity is denoted by $G_\alpha f(x)$, where G stands for *generalized gradient.*

Evidently, a point x is critical for $d(\cdot, p)$ if the segments from p to x spread out at x, while it is regular if they more or less point in the same direction. (See Figure 11.2) It was Berger who first realized and showed that a local maximum must be critical in the above sense. Berger's result is a consequence of the next proposition.

PROPOSITION 47. *Suppose (M, g) and $r = d(\cdot, K)$ are as above. Then:*

(1) $\Gamma(x, K)$ is closed and therefore compact for all x.

(2) The set of α-regular points is open in M.

(3) $G_\alpha r(x)$ is convex for all $\alpha \leq \frac{\pi}{2}$.

(4) If U is an open set of α-regular points for r, then there is a unit vector field X on U such that $X(x) \in G_\alpha r(x)$ for all $x \in U$. Furthermore, if γ is an integral curve for X and $s < t$, then

$$r(\gamma(t)) - r(\gamma(s)) > \cos(\alpha)(t - s).$$

PROOF. (1) Let $\sigma_i : [0, \ell] \to M$ be a sequence of unit speed segments from K to x with $\dot{\sigma}_i (\ell)$ converging to some unit vector $v \in T_x M$. Clearly,

$$\sigma(t) = \exp_x ((\ell - t) v)$$

is the limit of the segments σ_i and must therefore be a segment itself. Furthermore, since K is closed $\sigma(0) \in K$.

(2) Suppose $x_i \to x$, and x_i is not α-regular. We shall show that x is not α-regular. This means that for each $v \in T_x M$, we can find $w \in \Gamma(x, K)$ such that $\angle(v, w) \geq \alpha$. Now, for some fixed $v \in T_x M$, choose a sequence $v_i \in T_{x_i} M$ converging to v. For each i we can, by assumption, find $w_i \in \Gamma(x_i, K)$ with $\angle(v_i, w_i) \geq \alpha$. The sequence of unit vectors w_i must now subconverge to a vector $w \in T_x M$. Furthermore, the sequence of segments σ_i that generate w_i must also subconverge to a segment that is tangent to w. Thus, $w \in \Gamma(x, K)$.

(3) First observe that if $\alpha \leq \pi/2$, then for each $w \in T_x M$, the open cone

$$C_\alpha(w) = \{v \in T_x M : \angle(v, w) < \alpha\}$$

is convex. Then observe that $G_\alpha r(x)$ is the intersection of the cones $C_\alpha(w), w \in \Gamma(x, K)$, and is therefore itself convex.

(4) For each $p \in U$ we can find $v_p \in G_\alpha r(p)$. For each p, extend v_p to a vector field V_p. It now follows from the proof of (2) that $V_p(x) \in G_\alpha r(x)$ for x near p. We can then assume that V_p is defined on a neighborhood U_p on which it is a generalized gradient. We can now select a locally finite collection $\{U_i\}$ of U_p's and a corresponding partition of unity λ_i. Then property (3) tells us that the vector field

$$V = \sum \lambda_i V_i \in G_\alpha r.$$

In particular, it is nonzero and can therefore be normalized to a unit vector field.

The last property is clearly true at points where r is smooth, because in that case the derivative of $t \to r \circ \gamma$ is

$$g(X, \nabla r) = \cos \angle (X, \nabla r) > \cos \alpha.$$

Now observe that since r is Lipschitz continuous, this function is at least absolutely continuous. This implies that $r \circ \gamma$ is differentiable a.e. and is the integral of its derivative. It might, however, happen that $r \circ \gamma$ is differentiable at a point x where ∇r is not defined. To see what happens at such points we select a variation $\bar{\gamma}(s, t)$ such that $t \to \bar{\gamma}(0, t)$ is a segment from K to x, $\bar{\gamma}(s, 0) = \bar{\gamma}(0, 0)$, and $\bar{\gamma}(s, 1) = \gamma(s)$ is the integral curve for X through x. Thus

$$\frac{1}{2}(r \circ \gamma)^2 \leq \frac{1}{2}\left(\int_0^1 \left|\frac{\partial \gamma}{\partial t}\right| dt\right)^2$$

$$\leq \frac{1}{2}\int_0^1 \left|\frac{\partial \gamma}{\partial t}\right|^2 dt$$

$$= E(\gamma_s)$$

with equality holding for $s = 0$. Assuming that $r \circ \gamma$ is differentiable at $s = 0$ we get

$$
\begin{aligned}
r\left(\gamma\left(0\right)\right) \frac{dr \circ \gamma}{dt}\Big|_{s=0} &= \frac{dE}{ds}\Big|_{s=0} \\
&= g\left(\frac{\partial \bar{\gamma}}{\partial t}\left(0, b\right), \frac{\partial \bar{\gamma}}{\partial s}\left(0, b\right)\right) \\
&= g\left(\frac{\partial \bar{\gamma}}{\partial t}\left(0, b\right), X\right) \\
&= \left|\frac{\partial \bar{\gamma}}{\partial t}\right| \cos\left(\angle\left(X, \frac{\partial \bar{\gamma}}{\partial t}\right)\right) \\
&> r\left(\gamma\left(0\right)\right)\cos\alpha.
\end{aligned}
$$

This proves the desired property. $\qquad\square$

We can now generalize the above retraction lemma.

LEMMA 55. *Let (M, g) and $r = d\left(\cdot, K\right)$ be as above. Suppose that all points in $r^{-1}\left(\left[a, b\right]\right)$ are α-regular for $\alpha < \pi/2$. Then $r^{-1}\left(\left[-\infty, a\right]\right)$ is homeomorphic to $r^{-1}\left(\left[-\infty, b\right]\right)$, and $r^{-1}\left(\left[-\infty, b\right]\right)$ deformation retracts onto $r^{-1}\left(\left[-\infty, a\right]\right)$.*

PROOF. The construction is similar to the first lemma but a little more involved. We can construct a compactly supported vector field X such that the flow F^t for X satisfies

$$
r\left(F^t\left(p\right)\right) - r\left(p\right) > t \cdot \cos\left(\alpha\right), \ t \geq 0 \text{ if } p, F^t\left(p\right) \in r^{-1}\left(\left[a, b\right]\right).
$$

For each $p \in r^{-1}\left(b\right)$ we can therefore find a first time $t_p \leq \frac{b-a}{\cos\alpha}$ for which $F^{-t_p}\left(p\right) \in r^{-1}\left(a\right)$. The function $p \to t_p$ is continuous and thus we get the desired retraction

$$
r_t \ : \ r^{-1}\left(\left[-\infty, b\right]\right) \to r^{-1}\left(\left[-\infty, b\right]\right),
$$

$$
r_t\left(p\right) = \begin{cases} p & \text{if } r\left(p\right) \leq a \\ F^{-t \cdot t_p}\left(p\right) & \text{if } a \leq r\left(p\right) \leq b \end{cases}.
$$

$\qquad\square$

Note that as the level sets for r are not smooth, we can't expect to get diffeomorphic sublevels. It is now a question of how this can be used. As a very simple result let us mention

COROLLARY 43. *Suppose K is a compact submanifold of a complete Riemannian manifold (M, g) and suppose the distance function $r = d\left(\cdot, K\right)$ is regular everywhere on $M - K$. Then M is diffeomorphic to the normal bundle of K in M. In particular, if $K = \{p\}$, then M is diffeomorphic to $\mathbb{R}^n$.*

PROOF. We know that $M - K$ admits a vector field X, such that r is strictly increasing along the integral curves for X. Moreover, near K the distance function is smooth, and therefore X can be assumed to be equal to ∇r near K.

If

$$
\nu\left(K\right) = \{v \in T_p M : p \in K \text{ and } v \perp T_p K\},
$$

then we have the normal exponential map

$$
\exp : \nu\left(K\right) \to M.
$$

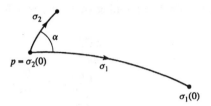

Figure 11.3

On a neighborhood of the zero section in $\nu(K)$ we know that this gives a diffeomorphism onto a neighborhood of K. Also, the curves $t \to \exp(tv)$ are, for small t, integral curves for X. In particular, we have for each $v \in \nu(K)$ a unique integral curve for X denoted $\gamma_v(t) : (0, \infty) \to M$ such that $\lim_{t \to 0} \dot{\gamma}_v(t) = v$. Now define our diffeomorphism $F : \nu(K) \to M$ by

$$F(0_p) = p \text{ for the origin in } \nu_p(K),$$
$$F(tv) = \gamma_v(t) \text{ where } |v| = 1.$$

This clearly defines a differentiable map. For small t this is just the exponential map. The map is one-to-one since integral curves for X can't intersect. It is onto, since r is proper, and therefore integral curves for X are defined for all time and must leave every compact set (since r is increasing along integral curves). Finally, as it is a diffeomorphism onto a neighborhood of K by the normal exponential map and the flow of a vector field always acts by local diffeomorphisms we see that it has nonsingular differential everywhere. $\qquad \square$

2. Distance Comparison

In this section we shall introduce the main results that will make it possible to conclude that various distance functions are noncritical. This obviously requires some sort of angle comparison. The most important step in this direction is supplied by the Toponogov theorem (or the *hinge version* of Toponogov's theorem; there are triangle and angle versions as well). The proof we present is probably the simplest available; and is based upon an idea by H. Karcher (see [28]).

Some preparations are necessary. Let (M, g) be a Riemannian manifold. We define two very natural geometric objects:

Hinge: A *hinge* consists of two segments σ_1 and σ_2 emanating from a common point p and forming an angle α. We shall always parametrize the geodesics by arc length and assume that

$$\sigma_1(\ell(\sigma_1)) = p = \sigma_2(0).$$

The angle α is then defined as

$$\alpha = \pi - \angle(\dot{\sigma}_1(\ell(\sigma_1)), \dot{\sigma}_2(0)).$$

Thus, the first segment ends at p, while the second begins there. The angle is the *interior* angle. See also Figure 11.3.

Triangle: A *triangle* consists of three segments that meet pairwise at three different points.

In both definitions one could use geodesics. It is then possible to have degenerate triangles where some vertices coincide without the joining geodesics being

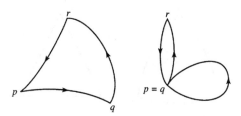

Figure 11.4

trivial. We shall use the more general hinges where σ_1 is a segment and σ_2 merely a geodesic in a few situations. In Figure 11.4 we have depicted a triangle consisting of segments, and a degenerate triangle where one of the sides is a geodesic loop and two of the vertices coincide.

Given a hinge (or a triangle), we can construct *comparison* hinges (or triangles) in the constant-curvature spaces S_k^n.

LEMMA 56. *Suppose (M,g) is complete and has $\sec \geq k$. Then for each hinge (or triangle) in M we can find a comparison hinge (or triangle) in S_k^n where the corresponding segments have the same length and the angle is the same (all corresponding segments have the same length).*

PROOF. Suppose we have three points $p, q, r \in M$. First, we know that in case $k > 0$, Myers' theorem implies

$$\operatorname{diam} M \leq \pi/\sqrt{k} = \operatorname{diam} S_k^n.$$

Thus, any segments between these three points have length $\leq \pi/\sqrt{k}$.

The hinge case. Here we have segments from p to q and from q to r forming an angle α at q. In the space form we can first choose $\bar{p}$ and $\bar{q}$ such that $d(\bar{p},\bar{q}) = d(p,q)$ and then join them by a segment. This is possible because $d(p,q) \leq \pi/\sqrt{k}$. At $\bar{q}$ we can then choose a direction that forms an angle α with the chosen segment. Then we take the unique geodesic going in this direction, and using the arc length parameter we go out distance $d(q,r)$ along this geodesic. This will now be a segment, as $d(q,r) \leq \pi/\sqrt{k}$. We have then found the desired hinge.

The triangle case is similar. First, pick $\bar{p}$ and $\bar{q}$ as above. Then, consider the two distance spheres $\partial B(\bar{p}, d(p,r))$ and $\partial B(\bar{q}, d(q,r))$. Since all possible triangle inequalities between p, q, r hold and $d(q,r), d(p,r) \leq \pi/\sqrt{k}$, these distance spheres are nonempty and intersect. Then, let $\bar{r}$ be any point in the intersection.

To be honest here, we must use Cheng's diameter theorem in case any of the distances is $\pi/\sqrt{k}$. In this case there is nothing to prove as $(M,g) = S_k^n$. □

We can now state the Toponogov comparison theorem.

THEOREM 79. (Toponogov, 1959) *Let (M,g) be a complete Riemannian manifold with $\sec \geq k$.*

Hinge Version: *Given any hinge with vertices $p, q, r \in M$ forming an angle α at q, it follows, that for any comparison hinge in S_k^n with vertices $\bar{p}, \bar{q}, \bar{r}$ we have:*
$d(p,r) \leq d(\bar{p},\bar{r})$.

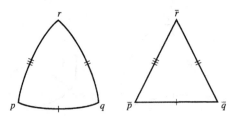

Figure 11.5

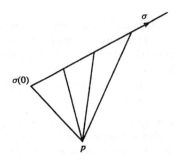

Figure 11.6

Triangle Version: *Given any triangle in M, it follows that the interior angles are larger than the corresponding interior angles for a comparison triangle in S_k^n. See also Figure 11.5*

The proof requires a little preparation. First, we claim that the hinge version implies the triangle version. This follows from the *law of cosines* in constant curvature. This law shows that if we have $p, q, r \in S_k^n$ and increase the distance $d(p, r)$ while keeping $d(p, q)$ and $d(q, r)$ fixed, then the angle at q increases as well. For simplicity, we shall only look at the cases where $k = 1, 0, -1$.

PROPOSITION 48. (Law of Cosines) *Let a triangle be given in S_k^n with side lengths a, b, c. If α denotes the angle opposite to a, then*

$$
\begin{aligned}
k = 0 \quad & a^2 = b^2 + c^2 - 2bc \cos \alpha. \\
k = -1 \quad & \cosh a = \cosh b \cosh c - \sinh b \sinh c \cos \alpha. \\
k = 1 \quad & \cos a = \cos b \cos c + \sin b \sin c \cos \alpha.
\end{aligned}
$$

PROOF. The general setup is the same in all cases. Namely, we suppose that a point $p \in S_k^n$ and a unit speed segment $\sigma : [0, c] \to S_k^n$ are given. We then investigate the restriction of the distance function from p to σ. If we denote $r(x) = d(p, x)$, then we are going to study $\varphi(t) = r \circ \sigma(t)$. See also Figure 11.6

Case $k = 0$: Note that $t \to d(p, \sigma(t))$ is not a very nice function, as it is the square root of a quadratic polynomial. This, however, indicates that the function will become more manageable if we square it. Thus, we consider

$$
\varphi(t) = \frac{1}{2}(r \circ \sigma(t))^2 = \frac{1}{2}|p - \sigma(t)|^2.
$$

We wish to compute the first and second derivatives of this function. This requires that we know the gradient and Hessian.

$$\nabla \frac{1}{2} r^2 \;=\; \nabla \frac{1}{2}\left(\left(x^1\right)^2 + \cdots + \left(x^n\right)^2\right)$$
$$=\; x^i \partial_i$$
$$=\; r \nabla r;$$

$$\mathrm{Hess}\,\frac{1}{2} r^2 \;=\; \nabla d\left(\frac{1}{2} r^2\right)$$
$$=\; \nabla \sum x^i dx^i$$
$$=\; \sum dx^i dx^i$$

As σ is a unit speed geodesic we get

$$\varphi'(t) \;=\; g\left(\dot\sigma, \nabla \frac{1}{2} r^2\right),$$
$$\varphi''(t) \;=\; \mathrm{Hess}\,\frac{1}{2} r^2\left(\dot\sigma, \dot\sigma\right) = 1.$$

So if we define $b = d\left(p, \sigma(0)\right)$ and let α be the interior angle between σ and the line joining p with $\sigma(0)$, then we have

$$\cos\left(\pi - \alpha\right) = -\cos\alpha = g\left(\dot\sigma(0), \nabla r\right).$$

After integration of $\varphi'' = 1$, we get

$$\varphi(t) \;=\; \varphi(0) + \varphi'(0) \cdot t + \frac{1}{2} t^2$$
$$=\; \frac{1}{2} b^2 - b \cdot \cos\alpha \cdot t + \frac{1}{2} t^2.$$

Now set $t = c$ and define $a = d\left(p, \sigma(c)\right)$, then

$$\frac{1}{2} a^2 = \frac{1}{2} b^2 - b \cdot c \cdot \cos\alpha + \frac{1}{2} c^2,$$

from which the law of cosines follows.

Case $k = -1$: This time we must modify the distance function in a different way. Namely, consider

$$\varphi(t) = \cosh\left(r \circ \sigma(t)\right) - 1.$$

Then

$$\varphi'(t) \;=\; \sinh\left(r \circ \sigma(t)\right) g\left(\nabla r, \dot\sigma\right),$$
$$\varphi''(t) \;=\; \cosh\left(r \circ \sigma(t)\right) = \varphi(t) + 1.$$

As before, we have $b = d\left(p, \sigma(0)\right)$, and the interior angle satisfies

$$\cos\left(\pi - \alpha\right) = -\cos\alpha = g\left(\dot\sigma(0), \nabla d\right).$$

Thus, we must solve the initial value problem

$$\varphi'' - \varphi \;=\; 1,$$
$$\varphi(0) \;=\; \cosh(b) - 1,$$
$$\varphi'(0) \;=\; -\sinh(b) \cos\alpha.$$

The general solution is

$$\varphi(t) = C_1 \cosh t + C_2 \sinh t - 1$$
$$= (\varphi(0) + 1) \cosh t + \varphi'(0) \sinh t - 1.$$

So if we let $t = c$ and $a = d(p, \sigma(c))$ as before, we arrive at

$$\cosh a - 1 = \cosh b \cosh c - \sinh b \sinh c \cos \alpha - 1,$$

which implies the law of cosines again.

Case $k = 1$: This case is completely analogous to the case $k = -1$. We set

$$\varphi = 1 - \cos(r \circ \sigma(t))$$

and arrive at the initial value problem

$$\varphi'' + \varphi = 1,$$
$$\varphi(0) = 1 - \cos(b),$$
$$\varphi'(0) = -\sin b \cos \alpha.$$

Then,

$$\varphi(t) = C_1 \cos t + C_2 \sin t + 1$$
$$= (\varphi(0) - 1) \cos t + \varphi'(0) \sin t + 1,$$

and consequently

$$1 - \cos a = -\cos b \cos c - \sin b \sin c \cos \alpha + 1,$$

which implies the law of cosines.   □

The proof of the law of cosines suggests that in working in space forms it is easier to work with a modified distance function, the main advantage being that the Hessian is much simpler. Something similar can be done in variable curvature.

LEMMA 57. *Let (M, g) be a complete Riemannian manifold, $p \in M$, and $r(x) = d(x, p)$. If $\sec M \geq k$, then the Hessian of r satisfies*

$k = 0$: *The function $r_0 = \frac{1}{2} r^2$ satisfies $\mathrm{Hess}\, r_0 \leq g$ in the support sense everywhere.*

$k = -1$: *The function $r_{-1} = \cosh r - 1$ satisfies $\mathrm{Hess}\, r_{-1} \leq (\cosh r) g = (r_{-1} + 1) g$ in the support sense everywhere.*

$k = 1$: *The function $r_1 = 1 - \cos r$ satisfies $\mathrm{Hess}\, r_1 \leq (\cos r) g = (-r_1 + 1) g$ in the support sense everywhere.*

PROOF. All three proofs are, of course, similar so we concentrate just on the first case. The comparison estimates from chapter 6 imply that whenever r is smooth and w is perpendicular to ∇r, then

$$\mathrm{Hess}\, r(w, w) \leq \frac{1}{r} g(w, w).$$

For such w one can therefore immediately see that

$$\mathrm{Hess}\, r_0(w, w) \leq g(w, w).$$

If instead, $w = \nabla r$, then it is trivial that this holds, whence we have established the Hessian estimate at points where r is smooth. At all other points we just use the same trick by which we obtained the Laplacian estimates with lower Ricci curvature bounds in chapter 9.   □

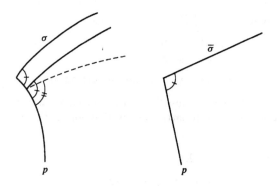

Figure 11.7

We are now ready to prove the hinge version of Toponogov's theorem. The proof is divided into the three cases: $k = 0, -1, 1$. But the setup is the same in all cases. We shall assume that a point $p \in M$ and a *geodesic* $\sigma : [0, \ell] \to M$ are given. Correspondingly, we assume that a point $\bar{p} \in S_k^n$ and segment $\bar{\sigma} : [0, \ell] \to S_k^n$ are given. Given the appropriate initial conditions, we claim that

$$d(p, \sigma(t)) \le d(\bar{p}, \bar{\sigma}(t)).$$

We shall for simplicity assume that $d(x, p)$ is smooth at $\sigma(0)$. Then the initial conditions are

$$d(p, \sigma(0)) \le d(\bar{p}, \bar{\sigma}(0)),$$
$$g(\nabla r, \dot{\sigma}(0)) \le g_k \left(\nabla \bar{r}, \frac{d}{dt} \bar{\sigma}(0) \right).$$

In case r is not smooth at $\sigma(0)$, we can just slide σ down along a segment joining p with $\sigma(0)$ and use a continuity argument. This also shows that we can use the stronger initial condition

$$d(p, \sigma(0)) < d(\bar{p}, \bar{\sigma}(0)).$$

In Figure 11.7 we have shown how σ can be changed by moving it down along a segment joining p and $\sigma(0)$. We have also shown how the angles can be slightly decreased. This will be important in the last part of the proof.

PROOF OF $k = 0$. We consider the modified functions

$$\varphi(t) = \frac{1}{2} (r \circ \sigma(t))^2,$$
$$\bar{\varphi}(t) = \frac{1}{2} (\bar{r} \circ \bar{\sigma}(t))^2.$$

For small t these functions are smooth and satisfy

$$\varphi(0) < \bar{\varphi}(0),$$
$$\varphi'(0) \le \bar{\varphi}'(0).$$

Moreover, for the second derivatives we have

$$\varphi'' \le 1 \text{ in the support sense,}$$
$$\bar{\varphi}'' = 1,$$

whence the difference $\psi(t) = \bar{\varphi}(t) - \varphi(t)$ satisfies

$$\psi(0) > 0,$$
$$\psi'(0) \geq 0,$$
$$\psi''(t) \geq 0 \text{ in the support sense.}$$

This shows that ψ is a convex function that is positive and increasing for small t, and hence increasing, and in particular positive, for all t. This proves the hinge version.   □

PROOF OF $k = -1$. Consider

$$\varphi(t) = \cosh r \circ \sigma(t) - 1,$$
$$\bar{\varphi}(t) = \cosh \bar{r} \circ \bar{\sigma}(t) - 1.$$

Then

$$\varphi(0) < \bar{\varphi}(0),$$
$$\varphi'(0) \leq \bar{\varphi}'(0),$$
$$\varphi'' \leq \varphi + 1 \text{ in the support sense,}$$
$$\bar{\varphi}'' = \bar{\varphi} + 1.$$

Then the difference $\psi = \bar{\varphi} - \varphi$ satisfies

$$\psi(0) > 0,$$
$$\psi'(0) \geq 0,$$
$$\psi''(t) \geq \psi(t) \text{ in the support sense.}$$

The first condition again implies that ψ is positive for small t. The last condition shows that as long as ψ is positive, it is also convex. The second condition then shows that ψ is increasing to begin with. It must now follow that ψ keeps increasing. Otherwise, there would be a positive maximum, and that violates convexity at points where ψ is positive.   □

PROOF OF $k = 1$. Case $k = 1$: This case is considerably harder. We begin as before by defining

$$\varphi(t) = 1 - \cos(r \circ \sigma(t)),$$
$$\bar{\varphi}(t) = 1 - \cos(\bar{r} \circ \bar{\sigma}(t))$$

and then observing that the difference $\psi = \bar{\varphi} - \varphi$ satisfies

$$\psi(0) > 0,$$
$$\psi'(0) \geq 0,$$
$$\psi''(t) \geq -\psi(t) \text{ in the support sense.}$$

That, however, doesn't look very promising. Even though the function starts out being positive, the last condition only gives a *negative* lower bound for the second derivative. At this point some people might recall that perhaps Sturm-Liouville theory could save us. But for that to work well it is best to assume $\psi'(0) > 0$. Thus, another little continuity argument is necessary as we need to perturb σ again to decrease the interior angle. If the interior angle is positive, this can clearly be done, and in the case where this angle is zero the hinge version is trivially true anyway. Now define $\zeta(t)$ by

$$
\begin{aligned}
\zeta'' &= -(1+\varepsilon)\zeta, \\
\zeta(0) &= \psi(0) = \alpha > 0, \\
\zeta'(0) &= \psi'(0) = \beta > 0,
\end{aligned}
$$

this means that

$$
\zeta(t) = \sqrt{\alpha^2 + \frac{\beta^2}{1+\varepsilon}} \cdot \sin\left(\sqrt{1+\varepsilon}\cdot t + \arctan\left(\frac{\alpha\cdot\sqrt{1+\varepsilon}}{\beta}\right)\right).
$$

For small t we have

$$
\begin{aligned}
(\psi(t) - \zeta(t))'' &\geq -\psi(t) + (1+\varepsilon)\zeta(t) \\
&= \zeta(t) - \psi(t) + \varepsilon\zeta(t) \\
&> 0.
\end{aligned}
$$

Thus we have $\psi(t) - \zeta(t) \geq 0$ for small t. We now wish to extend this to the interval where $\zeta(t)$ is positive, i.e., for

$$
t < \frac{\pi - \arctan\left(\frac{\alpha\cdot\sqrt{1+\varepsilon}}{\beta}\right)}{\sqrt{1+\varepsilon}},
$$

To get this to work, consider the quotient

$$
h = \frac{\psi}{\zeta}.
$$

So far, we know that this function satisfies

$$
\begin{aligned}
h(0) &= 1, \\
h(t) &\geq 1 \text{ for small } t.
\end{aligned}
$$

Should it therefore dip below 1 before reaching the end of the interval, then h would have a positive local maximum at some t_0. At this point we can use support functions ψ_δ for ψ from below, and conclude that also $\frac{\psi_\delta}{\zeta}$ has a local maximum at t_0. Thus, we have

$$
\begin{aligned}
0 &\geq \frac{d^2}{dt^2}\left(\frac{\psi_\delta}{\zeta}\right)(t_0) \\
&= \frac{\psi_\delta''(t_0)}{\zeta(t_0)} - 2\frac{\zeta'(t_0)}{\zeta(t_0)}\cdot\frac{d}{dt}\left(\frac{\psi_\delta}{\zeta}\right)_{t=t_0} - \frac{\psi_\delta(t_0)}{\zeta^2(t_0)}\zeta''(t_0) \\
&\geq \frac{-\psi_\delta(t_0) - \delta}{\zeta(t_0)} + \frac{\psi_\delta(t_0)}{\zeta(t_0)}(1+\varepsilon) \\
&= \frac{\varepsilon\cdot\psi_\delta(t_0) - \delta}{\zeta(t_0)}.
\end{aligned}
$$

But this becomes positive as $\delta \to 0$, since we assumed $\psi_\delta(t_0) > 0$, and so we have a contradiction. Next, we can let $\varepsilon \to 0$ and finally, let $\alpha \to 0$ to get the desired estimate for all $t \leq \pi$ using continuity. $\qquad\square$

Note that we never really use in the proof that we work with segments. The only thing that must hold is that the geodesics in the space form are segments. For $k \leq 0$ this is of course always true, but when $k = 1$ this means that the geodesic must have length $\leq \pi$. This was precisely the important condition in the last part of the proof.

3. Sphere Theorems

Our first applications of the Toponogov theorem are to the case of positively curved manifolds. Using scaling, we shall assume throughout this section that we work with a closed Riemannian n-manifold (M, g) with sec ≥ 1. For such spaces we have established

(1) diam $(M, g) \leq \pi$, with equality holding only if $M = S^n(1)$.
(2) If n is odd, then M is orientable.
(3) If n is even and M is orientable, then M is simply connected and inj $(M) \geq \pi / \sqrt{\max \sec}$.
(4) If n is even and max sec is close to 1, then (M, g) is close to a constant curvature metric. In particular, M must be a sphere when it is simply connected.
(5) It has also been mentioned that Klingenberg has shown that if M is simply connected and max sec < 4, then inj $(M) \geq \pi / \sqrt{\max \sec}$.
(6) If M is simply connected and max sec < 4, then M is homotopy equivalent to a sphere.

The penultimate result is quite subtle and is beyond what we can prove here. Gromov (see [**36**]) has a proof of this that in spirit goes as follows: One considers $p \in M$. If the upper curvature bound is $4 - \delta$, then we know that if we pull the metric back to the tangent bundle, then there are no conjugate points on the disc $B\left(0, \pi / \sqrt{4 - \delta}\right)$. Consider the modified distance r_1 to the origin in $T_p M$. This function is smooth on $B\left(0, \pi / \sqrt{4 - \delta}\right)$ and satisfies

$$\text{Hess} r_1 \leq (1 - r_1) g = (\cos r) g.$$

On the region

$$B\left(0, \frac{\pi}{\sqrt{4 - \delta}}\right) - \bar{B}(0, \pi/2)$$

this function will therefore have strictly negative Hessian. In particular, the level sets for r or r_1 that lie in that region are strictly concave. Now map these level sets down into M via the exponential map. As this map is nonsingular they will be mapped to strictly concave, possibly immersed, hypersurfaces in M. In the case where M is simply connected, one can prove an analogue to the Hadamard theorem for immersed convex hypersurfaces, namely, that they must be embedded spheres (this also uses that M has nonnegative curvature). However, if these hypersurfaces are embedded, then the exponential map must be an embedding on $B\left(0, \pi / \sqrt{4 - \delta}\right)$, and in particular, we obtain the desired injectivity radius estimate.

We can now prove the celebrated Rauch-Berger-Klingenberg sphere theorem, also known as the quarter pinched sphere theorem. Note that the conclusion is stronger than Berger's result mentioned in chapter 6. The part of the proof presented below is also due the Berger.

THEOREM 80. (1951-1961) *If M is a simply connected closed Riemannian manifold with $1 \leq \sec \leq 4 - \delta$, then M is homeomorphic to a sphere.*

PROOF. We gave a different proof of this in chapter 6 that used index estimation.

We have shown that the injectivity radius is $\geq \pi/\sqrt{4 - \delta}$. Thus, we have large discs around every point in M. Now select two points $p, q \in M$ such that $d(p, q) =$ diam$M$. Note that

$$\text{diam} M \geq \text{inj} M > \frac{\pi}{2}.$$

We now claim that every point $x \in M$ lies in one of the two balls $B\left(p, \pi/\sqrt{4 - \delta}\right)$, or $B\left(q, \pi/\sqrt{4 - \delta}\right)$, and thus M is covered by two discs. This certainly makes M look like a sphere as it is the union of two discs. Below we shall construct an explicit homeomorphism to the sphere in a more general setting.

Now take $x \in M$. Let $d = \text{diam}M = d(p, q)$, $a = d(p, x)$, and $b = d(x, q)$. If, for instance, $b > \pi/2$, then we claim that $a < \pi/2$. First, observe that since q is at maximal distance from p, it must follow that q cannot be a regular point for the distance function to p. Therefore, if we select any segment σ_1 from x to q, then we can find a segment σ_2 from p to q that forms an angle $\alpha \leq \pi/2$ with σ_1 at q. Then we can consider the hinge σ_1, σ_2 with angle α. The hinge version of Toponogov's theorem implies

$$\begin{aligned} \cos a \quad &\geq \quad \cos b \cos d + \sin b \sin d \cos \alpha \\ &\geq \quad \cos b \cos d. \end{aligned}$$

Now, both $b, d > \pi/2$, so the left hand side is positive. This implies that $a < \pi/2$, as desired. $\square$

Recall from the last chapter that Micaleff and Moore proved a similar theorem for manifolds that only have positive isotropic curvature.

Note that the theorem does not say anything about the non-simply connected situation. Thus we cannot conclude that such spaces are homeomorphic to spaces of constant curvature. Only that the universal covering is a sphere.

The above proof suggests, perhaps, that the conclusion of the theorem should hold as long as the manifold has large diameter. This is the content of the next theorem. This theorem was first proved by Berger for simply connected manifolds with a different proof and a slightly weaker conclusion. The present version is known as the Grove-Shiohama diameter sphere theorem. It was for the purpose of proving this theorem that Grove and Shiohama introduced critical point theory.

THEOREM 81. (Berger, 1962 and Grove-Shiohama, 1977) *If (M, g) is a closed Riemannian manifold with $\sec \geq 1$ and diam $> \pi/2$, then M is homeomorphic to a sphere.*

PROOF. We first give Berger's index estimation proof that follows his index proof of the quarter pinched sphere theorem. The goal is to find $p \in M$ such that all geodesic loops at p have length $> \pi$. The proof from chapter 6, then carries over verbatim. To this end select $p, q \in M$ such that

$$d(p, q) = \text{diam}M = d > \pi/2.$$

We claim that p has the desired property. Supposing otherwise we get a geodesic loop $\gamma : [0, 1] \to M$ based at p of length $\leq \pi$. As p is at maximal distance from q,

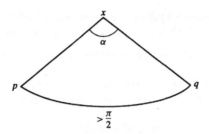

Figure 11.8

we can find a segment σ from q to p such that the hinge spanned by σ and γ has angle $\leq \pi/2$. While γ is not a segment it is sufficiently short that the hinge version of Toponogov's theorem still holds. Thus we must have

$$
\begin{aligned}
0 > \quad & \cos\left(d\left(p,q\right)\right) \\
= \quad & \cos\left(d\left(\sigma\left(0\right), \gamma\left(1\right)\right)\right) \\
\geq \quad & \cos\left(d\left(\sigma\left(0\right), \sigma\left(1\right)\right)\right)\cos\left(\ell\left(\gamma\right)\right) \\
= \quad & \cos\left(d\left(p,q\right)\right)\cos\left(\ell\left(\gamma\right)\right).
\end{aligned}
$$

This is clearly not possible unless $\ell\left(\gamma\right) = 0$.

We now give the Grove-Shiohama proof. Fix $p, q \in M$ with

$$
d\left(p,q\right) = \mathrm{diam}M = d > \pi/2.
$$

The claim is that the distance function from p has only q as a critical point. To see this, let $x \in M - \{p,q\}$ and let α be the angle between any two geodesics from x to p and q. If we suppose that $\alpha \leq \pi/2$ and set $b = d\left(p,x\right)$ and $c = d\left(x,q\right)$, then the hinge version of Toponogov's theorem implies

$$
\begin{aligned}
0 > \quad & \cos d \geq \cos b \cos c + \sin b \sin c \cos\alpha \\
\geq \quad & \cos b \cos c.
\end{aligned}
$$

But then $\cos b$ and $\cos c$ have opposite signs. If, for example, $\cos b \in (0,1)$, then we have $\cos d > \cos c$, which implies $c > d = \mathrm{diam}M$. Thus we have arrived at a contradiction, and hence we must have $\alpha > \pi/2$. See also Figure 11.8

We can now construct a vector field X that is the gradient field for $x \to d\left(x,p\right)$ near p and the negative of the gradient field for $x \to d\left(x,q\right)$ near q. Furthermore, the distance to p increases along integral curves for X. For each $x \in M - \{p,q\}$ there is a unique integral curve $\gamma_x\left(t\right)$ for X through x. Suppose that x varies over a small distance sphere $\partial B\left(p,\varepsilon\right)$ that is diffeomorphic to S^{n-1}. After time t_x this integral curve will hit the distance sphere $\partial B\left(q,\varepsilon\right)$ which can also be assumed to be diffeomorphic to S^{n-1}. The function $x \to t_x$ is continuous and in fact smooth as both distance spheres are smooth submanifolds. Thus we have a diffeomorphism defined by

$$
\begin{aligned}
\partial B\left(p,\varepsilon\right) \times [0,1] \quad &\to \quad M - \left(B\left(p,\varepsilon\right) \cup B\left(q,\varepsilon\right)\right), \\
\left(x,t\right) \quad &\to \quad \gamma_x\left(t \cdot t_x\right)
\end{aligned}
$$

Gluing this map together with the two discs $B\left(p,\varepsilon\right)$ and $B\left(q,\varepsilon\right)$ then yields a continuous bijection $M \to S^n$. Note that the construction does not guarantee smoothness of this map on $\partial B\left(p,\varepsilon\right)$ and $\partial B\left(q,\varepsilon\right)$. □

Aside from the fact that the conclusions in the above theorems could possibly be strengthened to diffeomorphism, we have optimal results. Complex projective space has curvatures in $[1, 4]$ and diameter $\pi/2$ and the real projective space has constant curvature 1 and diameter $\pi/2$. If one relaxes the conditions slightly, it is, however, still possible to say something.

THEOREM 82. *Suppose (M, g) is simply connected of dimension n with $1 \leq \sec \leq 4 + \varepsilon$.*

(1) (Berger, 1983) If n is even, then there is $\varepsilon(n) > 0$ such that M must be homeomorphic to a sphere or diffeomorphic to one of the spaces $\mathbb{C}P^{n/2}$, $\mathbb{H}P^{n/4}$, $\mathbb{O}P^2$.

(2) (Abresch-Meyer, 1994) If n is odd, then there is an $\varepsilon > 0$, which can be chosen independently of n, such that M is homeomorphic to a sphere.

The spaces $\mathbb{C}P^{n/2}$, $\mathbb{H}P^{n/4}$, or $\mathbb{O}P^2$ are known as the compact rank 1 symmetric spaces (CROSS). The complex projective space has already been studied in chapters 3 and 8. The quaternionic projective space is $\mathbb{H}P^n = S^{4n+3}/S^3$, but the octonion plane is a bit more exotic: $F_4/Spin(9) = \mathbb{O}P^2$ (see also chapter 8 for more on these spaces). The proof of (1) uses convergence theory. First, it is shown that if $\varepsilon = 0$, then M is either homeomorphic to a sphere or isometric to one of the CROSSs. Then using the injectivity radius estimate in even dimensions, we can apply the convergence machinery.

For the diameter situation we have

THEOREM 83. *(Grove-Gromoll, 1987 and Wilking, 2001) Suppose (M, g) is closed and satisfies $\sec \geq 1$, $\operatorname{diam} \geq \frac{\pi}{2}$ Then one of the following cases holds:*

(1) M is homeomorphic to a sphere.

(2) M is isometric to a finite quotient $S^n(1)/\Gamma$, where the action of Γ is reducible (has an invariant subspace).

(3) M is isometric to one of $\mathbb{C}P^{n/2}$, $\mathbb{H}P^{n/4}$, $\mathbb{C}P^{n/2}/\mathbb{Z}_2$ for $n = 2$ mod 4.

(4) M is isometric to $\mathbb{O}P^2$.

Grove and Gromoll settled all but part (4), where they only showed that M had to have the cohomology ring of $\mathbb{O}P^2$. It was Wilking who finally settled this last case (see [**94**]).

4. The Soul Theorem

Let us commence by stating the theorem we are aiming to prove and then slowly work our way through the rather intricate and technical proof.

THEOREM 84. *(Cheeger-Gromoll-Meyer, 1969, 1972) If (M, g) is a complete non-compact Riemannian manifold with $\sec \geq 0$, then M contains a soul $S \subset M$, which is a closed totally convex submanifold, such that M is diffeomorphic to the normal bundle over S. Moreover, when $\sec > 0$, the soul is a point and M is diffeomorphic to $\mathbb{R}^n$.*

The history is briefly that Gromoll-Meyer first showed that if $\sec > 0$, then M is diffeomorphic to $\mathbb{R}^n$. Soon after Cheeger-Gromoll established the full theorem. The Gromoll-Meyer theorem is in itself rather remarkable.

We shall use critical point theory to establish this theorem. The problem lies in finding the soul. When this is done, it will be easy to see that the distance function to the soul has only regular points, and then we can use the results from the first section.

Before embarking on the proof, it might be instructive to show the following less ambitious result, whose proof will be used in the next section.

LEMMA 58. (Gromov's critical point estimate, 1981) *If (M, g) is a complete open manifold of nonnegative sectional curvature, then for every $p \in M$ the distance function $d(\cdot, p)$ has no critical points outside some ball $B(p, R)$. In particular, M must have the topology of a compact manifold with boundary.*

PROOF. We shall use a contradiction argument. So suppose we have a sequence p_k of critical points for $d(\cdot, p)$, where $d(p_k, p) \to \infty$. After passing to a subsequence we can without loss of generality assume that

$$d(p_{k+1}, p) \geq 2d(p_k, p).$$

Now select segments σ_k from p to p_k. The above inequality implies that the angle at p between any two segments is $\geq 1/6$. To see this, suppose σ_k and σ_{k+l} form an angle $< 1/6$ at p. The hinge version of Toponogov's theorem then implies

$$
\begin{aligned}
(d(p_k, p_{k+l}))^2 \quad &< \quad (d(p, p_{k+l}))^2 + (d(p_k, p))^2 - 2d(p, p_{k+l}) d(p_k, p) \cos \frac{1}{6} \\
&\leq \quad \left(d(p, p_{k+l}) - \frac{3}{4} d(p_k, p) \right)^2.
\end{aligned}
$$

Now use that p_k is critical for p to conclude that there are segments from p to p_k and p_{k+l} to p_k that from an angle $\leq \pi/2$ at p_k. Then use the hinge version again to conclude

$$
\begin{aligned}
(d(p, p_{k+l}))^2 \quad &\leq \quad (d(p_k, p))^2 + (d(p_k, p_{k+l}))^2 \\
&\leq \quad (d(p_k, p))^2 + \left(d(p, p_{k+l}) - \frac{3}{4} d(p_k, p) \right)^2 \\
&= \quad \frac{25}{16} (d(p_k, p))^2 + (d(p, p_{k+l}))^2 - \frac{3}{2} d(p, p_{k+l}) d(p_k, p),
\end{aligned}
$$

which implies

$$d(p, p_{k+l}) \leq \frac{25}{24} d(p_k, p).$$

But this contradicts our assumption that

$$d(p, p_{k+l}) \geq d(p_{k+1}, p) \geq 2d(p_k, p).$$

Now that all the unit vectors $\dot{\sigma}_k(0)$ form angles of at least $1/6$ with each other, we can conclude that there can't be infinitely many such vectors. Hence, there cannot be critical points infinitely far away from p.

Observe that the vectors $\dot{\sigma}_k(0)$ lie on the unit sphere in $T_p M$ and are distance $1/6$ away form each other. Thus, the balls $B(\dot{\sigma}_k(0), 1/12)$ are disjoint in the unit sphere and hence there are at most

$$\frac{v(n-1, 1, \pi)}{v(n-1, 1, \frac{1}{12})} \leq 100^n$$

such points.   □

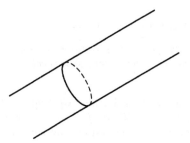

Figure 11.9

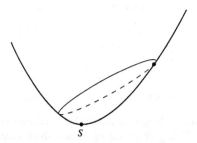

Figure 11.10

We now have to explain what it means for a submanifold, or more generally a subset, to be totally convex. The notion is similar to being totally geodesic. A subset $A \subset M$ of a Riemannian manifold is said to be *totally convex* if any geodesic in M joining two points in A actually lies in A. There are in fact several different kinds of convexity, but as they are not important for any other developments here, we shall confine ourselves to total convexity. The first observation is that this definition agrees with the usual definition for convexity in Euclidean space. Other than that, it is not clear that any totally convex sets exist at all. For example, if $A = \{p\}$, then A is totally convex only if there are no geodesic loops based at p. This means that points will almost never be totally convex. In fact, if M is closed, then M is the only totally convex subset. This is not completely trivial, but using the energy functional as in chapter 6 we note that if $A \subset M$ is totally convex, then $A \subset M$ is k-connected for any k. It is however, not possible for a closed n-manifold to have n-connected nontrivial subsets as this would violate Poincaré duality. On complete manifolds it is sometimes possible to find totally convex sets.

EXAMPLE 59. *Let (M, g) be the flat cylinder $\mathbb{R} \times S^1$. All of the circles $\{p\} \times S^1$ are geodesics and totally convex. This also means that no point in M can be totally convex. In fact, all of those circles are souls. See also Figure 11.9*

EXAMPLE 60. *Let (M, g) be a smooth rotationally symmetric metric on $\mathbb{R}^2$ of the form $dr^2 + \varphi^2(r) d\theta^2$, where $\varphi'' < 0$. Thus, (M, g) looks like a parabola of revolution. The radial symmetry implies that all geodesics emanating from the origin $r = 0$ are rays going to infinity. Thus the origin is a soul and totally convex. Most other points, however, will have geodesic loops based there. See also Figure 11.10.*

The way to find totally convex sets is via

LEMMA 59. *If $f : (M, g) \to \mathbb{R}$ is concave, in the sense that the Hessian is weakly nonpositive everywhere, then every superlevel set $A = \{x \in M : f(x) \geq a\}$ is totally convex.*

PROOF. Given a geodesic γ in M, we have that the function $f \circ \gamma$ has nonpositive weak second derivative. Thus, $f \circ \gamma$ is concave as a function on $\mathbb{R}$. In particular, the minimum of this function on any compact interval is obtained at one of the endpoints. This finishes the proof.   □

We are now left with the problem of the existence of proper concave functions on complete manifolds with nonnegative sectional curvature.

LEMMA 60. *Suppose (M, g) is as in the theorem and that $p \in M$. If we take all rays $\{\gamma_\alpha\}$ emanating from p and construct*

$$f = \inf_\alpha b_{\gamma_\alpha},$$

where b_γ denotes the Busemann function, then f is both proper and concave.

PROOF. First we show that in nonnegative sectional curvature all Busemann functions are concave. Using that, we can then show that the given function is concave and proper.

Recall that in nonnegative Ricci curvature Busemann functions are superharmonic. The proof of concavity is almost identical. Instead of the Laplacian estimate for distance functions, we must use a similar Hessian estimate. If $r = d(\cdot, p)$, then we know that Hess$r$ vanishes on radial directions $\partial_r = \nabla r$ and satisfies

$$\text{Hess} r \leq \frac{1}{r} g$$

on vectors perpendicular to the radial direction. In particular, Hess$r \leq \frac{1}{r} g$ at all smooth points. We can then extend this estimate to the points where r isn't smooth as we did for modified distance functions. We can now proceed as in the Ricci curvature case to show that Busemann functions have nonpositive Hessians in the weak sense and are therefore concave.

The infimum of a collection of concave functions is clearly concave. So we must now show that the superlevel sets for f are compact. Suppose, on the contrary, that some superlevel set $A = \{x \in M : f(x) \geq a\}$ is noncompact. If $a > 0$, then also $A = \{x \in M : f(x) \geq 0\}$ is noncompact. So we can assume that $a \leq 0$. As all of the Busemann functions b_{γ_α} are zero at p also $f(p) = 0$. In particular, $p \in A$. Using noncompactness select a sequence $p_n \in A$ that goes to infinity. Then join p_n to p by a segment, and as in the construction of rays, choose a subsequence of these segments converging to a ray emanating from p. As A is totally convex, all of these segments lie in A. Since A is closed the ray must also lie in A and therefore be one of the rays γ_α. But

$$f(\gamma_\alpha(t)) \leq b_{\gamma_\alpha}(\gamma_\alpha(t)) = -t \to -\infty,$$

so we have a contradiction.   □

We now need to establish a few fundamental properties of totally convex sets.

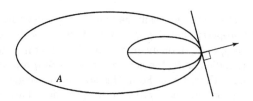

Figure 11.11

LEMMA 61. *If $A \subset (M, g)$ is totally convex, then A has an interior, denoted by int$A$, and a boundary ∂A. The interior is a totally convex submanifold of M, and the boundary has the property that for each $\dot{x} \in \partial A$ there is an inward pointing vector $w \in T_x M$ with the property: If $\gamma(t) : [0, a] \to A$ is a geodesic with $\gamma(0) = x$ and $\gamma(a) \in$ int$A$, then $\angle(w, \dot{\gamma}(0)) < \frac{\pi}{2}$.*

Some comments are in order before the proof. The words *interior* and *boundary*, while describing fairly accurately what the sets look like, are not meant in the topological sense. Most convex sets will, of course, not have any topological interior at all. The property about the boundary is what is often called the *supporting hyperplane property*. Namely, the interior of the convex set is supposed to lie on one side of a hyperplane at any of the boundary points. The vector w is the normal to this hyperplane and can be taken to be tangent to some geodesic that goes into the interior. It is important to note that the supporting hyperplane property shows that the distance function to a subset of int$A$ cannot have any critical points on ∂A. See also Figure 11.11.

PROOF. The convexity radius estimate from chapter 6 will be used in many places. Specifically we shall use that there is a positive function $\varepsilon(p) : M \to (0, \infty)$ such that the distance function $r_p(x) = d(x, p)$ is smooth and strictly convex on $B(p, \varepsilon(p))$.

First, let us identify points in the interior and on the boundary. To make the identifications simpler we assume that A is closed.

Find the maximal integer k such that A contains a k-dimensional submanifold of M. If $k = 0$, then A must be a point. For if A contains two points, then A also contains a segment joining these points and therefore a 1-dimensional submanifold. Now define $N \subset A$ as being the union of all k-dimensional submanifolds in M that are contained in A. We claim that N is a k-dimensional totally convex submanifold whose closure is A. We shall thus identify int$A$ with N and ∂A with $A - N$.

To see that it is a submanifold, pick $p \in N$ and let $N_p \subset A$ be a k-dimensional submanifold of M containing p. By shrinking N_p if necessary, we also assume that it is embedded. We can therefore find $\delta \in (0, \varepsilon(p))$ so that $B(p, \delta) \cap N_p = N_p$. We now claim that also $B(p, \delta) \cap A = N_p$. If this were not true, then we could find

$$q \in A \cap B(p, \delta) - N_p.$$

Now assume that δ is so small that also $\delta < \text{inj}_q$. Then we can join each point in $B(p, \delta) \cap N_p$ to q by a unique segment. The union of these segments will, away from q, form a cone that is a $(k + 1)$-dimensional submanifold which is contained in A (see Figure 11.12), thus contradicting maximality of k. In particular, N must be an embedded submanifold as we have $B(p, \delta) \cap N = N_p$.

What we have just proved can easily be modified to show that for points $p \in N$ and $q \in A$ with the property that $d(p,q) < \text{inj}_q$ there is a k-dimensional submanifold $N_p \subset N$ such that $q \in \bar{N}_p$, namely, just take a $(k-1)$-dimensional submanifold through p in N perpendicular to the segment from p to q and consider the cone over this submanifold with vertex q. From this statement we get the property that if $\gamma : [0,a] \to A$ is a geodesic, then $\gamma(0,a) \subset N$ provided that, say, $\gamma(0) \in N$. In particular, N is dense in A.

Having identified the interior and boundary, we now have to establish the supporting hyperplane property. First we note that since N is totally geodesic its tangent spaces $T_q N$ are preserved by parallel translation along curves in N. For $p \in \partial A$ we therefore have a well-defined k-dimensional tangent space $T_p A \subset T_p M$ coming from parallel translating the tangent spaces to N along curves in N that end at p. Next define the tangent cone at $p \in \partial A$

$$C_p A = \left\{ v \in T_p M : \exp_p(tv) \in N \text{ for some } t > 0 \right\}.$$

Note that in fact $\exp_p(tv) \in N$ for all small $t > 0$. This shows that $C_p A$ is a cone. Clearly $C_p A \subset T_p A$ and in fact spans it as we can easily find k linearly independent vectors in $C_p A$. Finally, we see that $C_p A$ is an open subset of $T_p A$.

For $\varepsilon > 0$ small, suppose we can select

$$q \in A_\varepsilon = \{ x \in A : d(x, \partial A) \geq \varepsilon \}$$

such that $d(q,p) = \varepsilon$. The set of such points is clearly 2ε-dense in ∂A. So the set of points $p \in \partial A$ for which we can find an $\varepsilon > 0$ and $q \in A_\varepsilon$ such that $d(q,p) = \varepsilon$ is dense in ∂A. As the supporting plane property is an open property (this follows from critical point theory), it suffices to prove it for such p. We can also suppose ε is so small that $r_q = d(\cdot, q)$ is smooth and convex on a neighborhood containing p. The claim is that $\angle(-\nabla r_q, v) < \frac{\pi}{2}$ for all $v \in C_p A$. To see this, observe that we have a convex set

$$A' = A \cap \bar{B}(q, \varepsilon),$$

with interior

$$N' = A \cap B(q, \varepsilon) \subset N$$

and $p \in \partial A'$. Thus $C_p A' \subset C_p A$. In addition we see that $T_p A = T_p A'$. The tangent cone of $\bar{B}(q, \varepsilon)$ is given by

$$C_p \bar{B}(q, \varepsilon) = \left\{ v \in T_p M : \angle(v, -\nabla r_q) < \frac{\pi}{2} \right\}$$

as r is smooth at p, thus

$$C_p A' = \left\{ v \in T_p A : \angle(v, -\nabla r_q) < \frac{\pi}{2} \right\}$$

If now

$$C_p A' \subsetneqq C_p A,$$

then openness of $C_p A$ in $T_p A$ implies that there must be a $v \in C_p A$ such that also $-v \in C_p A$. But this implies that $p \in N$, as it becomes a point on a geodesic whose endpoints lie in N. (See Figure 11.13.) $\square$

The last lemma we need is

LEMMA 62. Let (M,g) have $\sec \geq 0$. If $A \subset M$ is totally convex, then the distance function $r : A \to \mathbb{R}$ defined by $r(x) = d(x, \partial A)$ is concave on A, and strictly concave if $\sec > 0$.

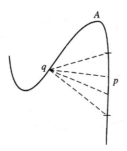

Figure 11.12

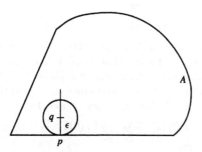

Figure 11.13

PROOF. We shall show that the Hessian is nonpositive in the support sense. Fix $q \in \mathrm{int}A$, and find $p \in \partial A$ so that $d(p, q) = d(q, \partial A)$. Then select $\sigma : [0, a] \to A$ to be a segment from p to q. Using exponential coordinates at p we create a hypersurface H which is the image of the hyperplane perpendicular to $\dot{\sigma}(0)$. This hypersurface is perpendicular to $\dot{\sigma}(0)$, the second fundamental form for H at p is zero, and $H \cap \mathrm{int}A = \emptyset$. (See Figure 11.14.) We have that $f(x) = d(x, H)$ is a support function from above for $d(\cdot, \partial A)$ at $\sigma(t)$ for all $t \in [0, a]$. Moreover f is smooth at $\sigma(t)$ for all $t < a$.

We start by showing that the support function f is concave at $\sigma(t)$ as long as f is smooth at $\sigma(t)$. Note that σ is an integral curve for ∇f. Evaluating the fundamental equation on a parallel field along σ that starts out being tangent to H, i.e., perpendicular to σ, therefore yields:

$$\frac{d}{dt}\mathrm{Hess}f(E, E) = -g(R(E, \dot{\sigma})\dot{\sigma}, E) - \mathrm{Hess}^2 f(E, E)$$
$$\leq 0.$$

Since $\mathrm{Hess}f(E, E) = 0$ at $t = 0$ we see that $\mathrm{Hess}f(E, E) \leq 0$ along σ (and < 0 if sec > 0). This shows that we have a smooth support function for $d(\cdot, \partial A)$ on an open and dense subset in A.

If f is not smooth at $\sigma(a)$ we can for $t < a$ find a hypersurface H_t as above that is perpendicular to $\dot{\sigma}(t)$ at $\sigma(t)$ and has vanishing second fundamental form at $\sigma(t)$. For t close to a we have that $f_t = d(\cdot, H_t)$ is smooth at q and therefore also has nonpositive (negative) Hessian at q. In this case we claim that $t + f_t$ is a support function for $d(\cdot, \partial A)$. Clearly, the functions are equal at q. If x is close to

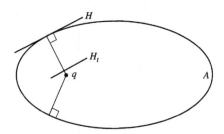

Figure 11.14

q, then we can select $z \in H_t$ so that

$$d(x, \partial A) = d(x, z) + f_t(z).$$

Thus we are reduced to showing that $d(z, \partial A) \leq t$ for each $z \in H_t$.

As f is smooth at $\sigma(t)$ it follows that $d(\cdot, \partial A)$ is concave in a neighborhood of $\sigma(t)$. Now select a short geodesic $\gamma(s)$ from $z \in H_t$ to $\sigma(t)$. By the construction of H_t we can assume that this geodesic is contained in H_t and therefore perpendicular to $\sigma(t)$. Concavity of $s \to d(\gamma(s), \partial A)$ then shows that

$$d(\gamma(s), \partial A) \leq d(\gamma(0), \partial A) = t.$$

This establishes our claim.   □

We are now ready to prove the soul theorem. Start with the proper concave function f constructed from the Busemann functions. The maximum level set

$$C_1 = \{x \in M : f(x) = \max f\}$$

is nonempty and convex since f is proper and concave. Moreover, it follows from the previous lemma that C_1 is a point if sec > 0. This is because the superlevel sets

$$A = \{x \in M : f(x) \geq a\}$$

are convex with $\partial A = f^{-1}(a)$, so $f = d(\cdot, \partial A)$ on A. Now, a strictly concave function (Hessian in support sense is negative) must have a unique maximum or no maximum, thus showing that C_1 is a point. If C_1 is a submanifold, then we are also done. In this case $d(\cdot, C_1)$ has no critical points, as any point lies on the boundary of a convex superlevel set. Otherwise, C_1 is a convex set with nonempty boundary. But then $d(\cdot, \partial C_1)$ is concave. The maximum set C_2 is again nonempty, since C_1 is compact and convex. If it is a submanifold, then we again claim that we are done. For the distance function $d(\cdot, C_2)$ has no critical points, as any point lies on the boundary for a superlevel set for either f or $d(\cdot, \partial C_1)$. We can now iterate to get a sequence of convex sets

$$C_1 \supset C_2 \supset \cdots \supset C_k.$$

We claim that in at most $n = \dim M$ steps we arrive at a point or submanifold S, that we call the soul (see Figure 11.15). This is because $\dim C_i > \dim C_{i+1}$. To see this suppose $\dim C_i = \dim C_{i+1}$, then $\operatorname{int} C_{i+1}$ will be an open subset of $\operatorname{int} C_i$. So if $p \in \operatorname{int} C_{i+1}$, then we can find δ such that

$$B(p, \delta) \cap \operatorname{int} C_{i+1} = B(p, \delta) \cap \operatorname{int} C_i.$$

Now choose a segment σ from p to ∂C_i. Clearly $d(\cdot, \partial C_i)$ is strictly increasing along this curve. This curve, however, runs through $B(p, \delta) \cap \operatorname{int} C_i$, thus showing that $d(\cdot, \partial C_i)$ must be constant on the part of the curve close to p.

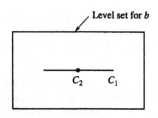

Figure 11.15

Much more can be said about complete manifolds with nonnegative sectional curvature. A rather complete account can be found in Greene's survey in [**50**]. We briefly mention two important results:

THEOREM 85. *Let S be a soul of a complete Riemannian manifold with sec ≥ 0, arriving from the above construction.*

(1) (Sharafudtinov, 1978) There is a distance-nonincreasing map $sh : M \to S$ such that $sh|_S = id$. In particular, all souls must be isometric to each other.

(2) (Perel'man, 1993) The map $sh : M \to S$ is a submetry. From this it follows that S must be a point if all sectional curvatures based at just one point are positive.

Having reduced all complete nonnegatively curved manifolds to bundles over closed nonnegatively curved manifolds, it is natural to ask the converse question: Given a closed manifold S with non-negative curvature, which bundles over S admit complete metrics with sec ≥ 0? Clearly, the trivial bundles do. When $S = T^2$ Özaydın-Walschap in [**75**] have shown that this is the only 2-dimensional vector bundle that admits such a metric. Still, there doesn't seem to be a satisfactory general answer. If, for instance, we let $S = S^2$, then any 2-dimensional bundle is of the form $\left(S^3 \times \mathbb{C}\right)/S^1$, where S^1 is the Hopf action on S^3 and acts by rotations on $\mathbb{C}$ in the following way: $\omega \times z = \omega^k z$ for some integer k. This integer is the Euler number of the bundle. As we have a complete metric of nonnegative curvature on $S^3 \times \mathbb{C}$, the O'Neill formula from chapter 3 shows that these bundles admit metrics with sec ≥ 0.

There are some interesting examples of manifolds with positive and zero Ricci curvature that show how badly the soul theorem fails for such manifolds. In 1978, Gibbons-Hawking in [**43**] constructed Ricci flat metrics on quotients of $\mathbb{C}^2$ blown up at any finite number of points. Thus, one gets a Ricci flat manifold with arbitrarily large second Betti number. About ten years later Sha-Yang showed that the infinite connected sum

$$\left(S^2 \times S^2\right) \sharp \left(S^2 \times S^2\right) \sharp \cdots \sharp \left(S^2 \times S^2\right) \sharp \cdots$$

admits a metric with positive Ricci curvature, thus putting to rest any hopes for general theorems in this direction. Sha-Yang have a very nice survey in [**45**] describing these and other examples. The construction uses doubly warped product metrics on $I \times S^2 \times S^1$ as described in chapter 3.

5. Finiteness of Betti Numbers

The theorem we wish to prove is

THEOREM 86. (Gromov, 1978, 1981) *There is a constant $C(n)$ such that any complete manifold (M, g) with sec ≥ 0 satisfies*

(1) $\pi_1(M)$ can be generated by $\leq C(n)$ generators.

(2) For any field F of coefficients the Betti numbers are bounded:

$$\sum_{i=0}^{n} b_i(M, F) = \sum_{i=0}^{n} \dim H_i(M, F) \leq C(n).$$

Part (2) of this result is considered one of the deepest and most beautiful results in Riemannian geometry. Before embarking on the proof, let us put it in context. First, we should note that the Gibbons-Hawking and Sha-Yang examples show that a similar result cannot hold for manifolds with nonnegative Ricci curvature. Sha-Yang also exhibited metrics with positive Ricci curvature on the connected sums

$$\underbrace{(S^2 \times S^2) \sharp (S^2 \times S^2) \sharp \cdots \sharp (S^2 \times S^2)}_{k \text{ times}}.$$

For large k, the Betti number bound shows that these connected sums cannot have a metric with nonnegative sectional curvature. Thus, we have simply connected manifolds that admit positive Ricci curvature but not nonnegative sectional curvature. The reader should also consult our discussion of manifolds with nonnegative curvature operator at the end of chapters 7 and 8 to get an appreciation for how rigid manifolds with nonnegative curvature operator are. Let us list the open problems that were posed there and settled for manifolds with nonnegative curvature operator:

(i) (H. Hopf) Does $S^2 \times S^2$ admit a metric with positive sectional curvature?

(ii) (H. Hopf) If M is even-dimensional, does sec ≥ 0 (> 0) imply $\chi(M) \geq 0$ (> 0)?

(iii) (Gromov) If sec ≥ 0, is $\sum_{i=0}^{n} b_i(M, F) \leq 2^n$?

Recall that these questions were also discussed in chapter 7 under additional assumptions about the isometry group.

First we establish part (1) of Gromov's theorem. The proof resembles that of the critical point estimate lemma from the previous section.

PROOF OF (1). We shall construct what is called a *short set* of generators for $\pi_1(M)$. We consider $\pi_1(M)$ as acting by deck transformations on the universal covering $\tilde{M}$ and fix $p \in \tilde{M}$. We then inductively select a generating set $\{g_1, g_2, \ldots\}$ such that

(a) $d(p, g_1(p)) \leq d(p, g(p))$ for all $g \in \pi_1(M) - \{e\}$.

(b) $d(p, g_k(p)) \leq d(p, g(p))$ for all $g \in \pi_1(M) - \langle g_1, \ldots, g_{k-1} \rangle$.

Now join p and $g_k(p)$ by segments σ_k (see Figure 11.16). We claim that the angle between any two such segments is $\geq \pi/3$.

Otherwise, the hinge version of Toponogov's theorem would imply

$$
\begin{aligned}
(d(g_{k+l}(p), g_k(p)))^2 &< (d(p, g_k(p)))^2 + (d(p, g_{k+l}(p)))^2 \\
&\quad - d(p, g_k(p)) d(p, g_{k+l}(p)) \\
&\leq (d(p, g_{k+l}(p)))^2.
\end{aligned}
$$

But then

$$d(g_{k+l}^{-1} \circ g_k(p), p) < d(p, g_{k+l}(p)),$$

which contradicts our choice of g_{k+l}.

It now follows that there can be at most

$$\frac{v(n-1, 1, \pi)}{v(n-1, 1, \frac{\pi}{6})}$$

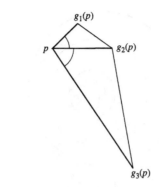

Figure 11.16

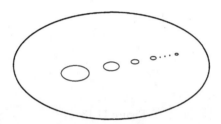

Figure 11.17

elements in the set $\{g_1, g_2, \ldots\}$. We have therefore produced a generating set with a bounded number of elements. $\square$

The proof of the Betti number estimate is established through several lemmas. First, we need to make three definitions for metric balls. Throughout, we fix a Riemannian n-manifold M with sec ≥ 0 and a field F of coefficients for our homology theory

$$H_* (\cdot, F) = H_* (\cdot) = H_0 (\cdot) \oplus \cdots \oplus H_n (\cdot).$$

The field will be suppressed throughout the proof.

Content: The *content* of a metric ball $B (p, r) \subset M$ is

$$\text{cont} B (p, r) = \text{rank} \left(H_* \left(B \left(p, \frac{1}{5} r \right) \right) \to H_* (B (p, r)) \right).$$

The reason for working with content, rather that just the rank of $H_* (B (p, r))$ itself, is that metric balls might not have infinitely generated homology. However, if $O_1 \subset M$ is any bounded subset of a manifold and $\bar{O}_1 \subset O_2 \subset M$, then the image of $H_* (O_1)$ in $H_* (O_2)$ is finitely generated. In Figure 11.17 we have taken a planar domain and extracted infinitely many discs of smaller and smaller size. This yields a compact set with infinite topology. Nevertheless, this set has finitely generated topology when mapped into any neighborhood of itself, as that has the effect of canceling all of the smallest holes.

Corank: The *corank* of a set $A \subset M$ is defined as the largest integer k such that we can find k metric balls $B (p_1, r_1), \ldots, B (p_k, r_k)$ with the properties

(a) There is a critical point x_i for p_i with $d (p_i, x_i) = 10 r_i$.

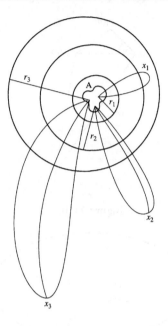

Figure 11.18

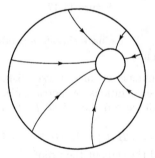

Figure 11.19

(b) $r_i \geq 3r_{i-1}$ for $i = 2, \ldots, k$.

(c) $A \subset \bigcap_{i=1}^{k} B(p_i, r_i)$.

In Figure 11.18 we have a picture of how the set A and the larger circles might be situated relative to each other.

Compressibility: We say that a ball $B(p, r)$ is *compressible* if it contains a ball $B(q, r') \subset B(p, r)$ such that

(a) $r' \leq \frac{r}{2}$.

(b) $\operatorname{cont} B(q, r') \geq \operatorname{cont} B(p, r)$.

If a ball is not compressible we call it *incompressible*. Note that any ball with content > 1, can be successively compressed to an incompressible ball. Figure 11.19 gives a schematic picture of a ball that can be compressed to a smaller ball.

We shall now tie these three concepts together through some lemmas that will ultimately lead us to the proof of the Betti number estimate. Observe that for large r, the ball $B(p, r)$ contains all the topology of M, so

$$\text{cont}\, B(p, r) = \sum_i b_i(M).$$

Also, the corank of such a ball must be zero, as there can't be any critical points outside this ball. The idea is now to compress this ball until it becomes incompressible and then estimate its content in terms of balls that have corank 1. We shall in this way successively be able to estimate the content of balls of fixed corank in terms of the content of balls with one higher corank. The proof is then finished first, by showing that the corank of a ball is uniformly bounded by 100^n, second, by observing that balls of maximal corank must be contractible and therefore have content 1 (otherwise they would contain critical points for the center, and the center would have larger corank).

LEMMA 63. *The corank of any set $A \subset M$ is bounded by 100^n.*

PROOF. Suppose that A has corank larger than 100^n. Select balls $B(p_1, r_1)$, $\ldots$, $B(p_k, r_k)$ with corresponding critical points $x_1, \ldots, x_k$, where $k > 100^n$. Now choose $z \in A$ and join z to x_i by segments σ_i. As in the critical point estimate lemma from the previous section, we can then find two of these segments σ_i and σ_j that form an angle $< 1/6$ at z.

For simplicity, suppose $i < j$ and define

$$
\begin{aligned}
a_i &= \ell(\sigma_i) = d(z, x_i), \\
a_j &= \ell(\sigma_j) = d(z, x_j), \\
l &= d(x_i, x_j),
\end{aligned}
$$

and observe that

$$
\begin{aligned}
b_i &= d(z, p_i) \leq r_i, \\
b_j &= d(z, p_j) \leq r_j.
\end{aligned}
$$

Figure 11.20 gives two pictures explaining the notation in the proof.

The triangle inequality implies

$$
\begin{aligned}
a_i &\leq 10r_i + b_i \leq 11r_i, \\
a_j &\geq 10r_j - r_j \geq 9r_j.
\end{aligned}
$$

Also, $r_j \geq 3r_i$, so we see that $a_j > a_i$. As in the critical point estimate lemma, we can conclude that

$$l \leq a_j - \frac{3}{4}a_i.$$

Now use the triangle inequality to conclude

$$
\begin{aligned}
c = d(p_i, x_j) &\geq a_j - b_i \\
&\geq 10r_j - b_j - b_i \\
&\geq 8r_j \\
&\geq 24r_i \\
&\geq 20r_i = 2d(p_i, x_i).
\end{aligned}
$$

Yet another application of the triangle inequality will then imply

$$l \geq d(x_i, p_i).$$

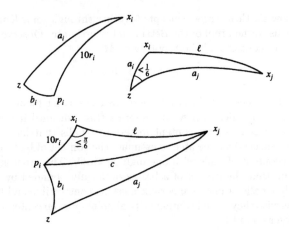

Figure 11.20

Since x_i is critical for p_i, we can now use the hinge version of Toponogov's theorem to conclude

$$c^2 \leq (d(p_i, x_i))^2 + l^2$$
$$\leq \left(l + \frac{1}{2}d(p_i, x_i)\right)^2.$$

Thus,

$$c \leq l + \frac{1}{2}d(p_i, x_i)$$
$$\leq l + 5r_i.$$

The triangle inequality then implies

$$a_j \leq c + b_i \leq c + r_i \leq l + 6r_i.$$

However, we also have

$$a_i \geq 10r_i - b_i \geq 9r_i,$$

which together with

$$l \leq a_j - \frac{3}{4}a_i$$

implies

$$l \leq a_j - \frac{27}{4}r_i.$$

Thus, we have a contradiction:

$$l + \frac{27}{4}r_i \leq a_j \leq l + 6r_i.$$

$\square$

Having established a bound on the corank, we can now try to check how the topology changes when we pass from balls of lower corank to balls of higher corank. Let $\mathcal{C}(k)$ denote the set of balls in M of corank $\geq k$, and $\mathcal{B}(k)$ the largest content of any ball in $\mathcal{C}(k)$.

LEMMA 64. *There is a constant $C(n)$ depending only on dimension such that*

$$\mathcal{B}(k) \le C(n) \mathcal{B}(k+1).$$

PROOF. The number $\mathcal{B}(k)$ is, of course, realized by some incompressible ball $B(p,R)$. Now consider a ball $B(x,r)$ where $x \in B(p,R/4)$ and $r \le R/20$. We claim that this ball lies in $\mathcal{C}(k+1)$. To see this, consider the ball

$$B(x, R/2) \subset B(p, R) \subset B(x, 2R).$$

Since $B(p,R)$ is assumed to be incompressible, there must be a critical point for x in the annulus $B(x, 2R) - B(x, R/2)$. For otherwise we could deform $B(p,R)$ to $B(x, R/2)$ inside $B(x, 2R)$. This would imply that $\text{cont} B(p,R) \le \text{cont} B(x, R/2)$ and thus contradict incompressibility of $B(p,R)$. We can now show that $B(x,r) \in \mathcal{C}(k+1)$. Using that $B(p,R) \in \mathcal{C}(k)$, select $B(p_1, r_1), \ldots, B(p_l, r_l), l \ge k$, as in the definition of corank. Then pick a critical point y for x in $B(x, 5R) - B(x, R/2)$ and consider the ball $B(x, d(x,y)/10)$. Then the balls $B(p_1, r_1), \ldots, B(p_l, r_l)$, $B(x, d(x,y)/10)$ can be used to show that $B(x,r)$ has corank $\ge l+1 > k$.

Now cover $B(p, R/5)$ by balls $B(p_i, R/100), i = 1, \ldots, m$. If we suppose that the balls $B(p_i, R/200)$ are pairwise disjoint, then we must have:

$$m \le \frac{v(n, 0, 2R)}{v\left(n, 0, \frac{1}{200}R\right)} = 400^n.$$

Next consider the sets

$$B\left(p_i, \frac{1}{2}R\right) \subset B(p, R).$$

First, we claim that

$$\text{cont} B(p, R) \le \text{rank}\left(H_*\left(\bigcup_{i=1}^{m} B\left(p_i, \frac{1}{100}R\right)\right) \to H_*\left(\bigcup_{i=1}^{m} B\left(p_i, \frac{1}{2}R\right)\right)\right)$$

This follows from the simple observation that if $A \subset B \subset C \subset D$, then

$$\text{rank}(H_*(A) \to H_*(D)) \le \text{rank}(H_*(B) \to H_*(C))$$

To estimate the right-hand side of the above inequality, it is natural to suppose that we can use a Mayer-Vietoris argument, together with induction on m, to show

$$\text{rank}\left(H_*\left(\bigcup_{i=1}^{m} B\left(p_i, \frac{1}{100}R\right)\right) \to H_*\left(\bigcup_{i=1}^{m} B\left(p_i, \frac{1}{2}R\right)\right)\right)$$

$$\le \sum_{\substack{i_1 < \cdots < i_s \\ 1 \le s \le m}} \text{rank}\left(H_*\left(\bigcap_{t=1}^{s} B\left(p_{i_t}, \frac{1}{100}R\right)\right) \to H_*\left(\bigcap_{t=1}^{s} B\left(p_{i_t}, \frac{1}{2}R\right)\right)\right).$$

We then observe that if

$$\bigcap_{t=1}^{s} B\left(p_{i_t}, \frac{1}{100}R\right) \ne \emptyset,$$

then the triangle inequality implies (see also below)

$$\bigcap_{t=1}^{s} B\left(p_{i_t}, \frac{1}{100}R\right) \subset B\left(p_{i_1}, \frac{1}{100}R\right) \subset B\left(p_{i_1}, \frac{1}{20}R\right) \subset \bigcap_{t=1}^{s} B\left(p_{i_t}, \frac{1}{2}R\right).$$

As each of the balls $B(p_i, R/10) \in \mathcal{C}(k+1)$, and there can be at most 2^m nonempty intersections, we then arrive at the estimate

$$\mathrm{cont}\, B(p, R) = \mathcal{B}(k) \leq 2^{400^n} \cdot \mathcal{B}(k+1).$$

This is the desired inequality. □

We now claim that

$$\mathrm{cont}\, M \leq 2^{40000^n},$$

which will, of course, prove the theorem. The above lemma clearly yields that

$$
\begin{aligned}
\mathrm{cont}\, M &= \mathcal{B}(0) \\
&\leq \mathcal{B}(k) \cdot \left(2^{400^n}\right)^k \\
&= \mathcal{B}(k) \cdot 2^{k \cdot 400^n} \\
&\leq \mathcal{B}(k) \cdot 2^{40000^n},
\end{aligned}
$$

where $k \leq 100^n$ is the largest possible corank in M. It then remains to check that $\mathcal{B}(k) = 1$. However, it follows from the above that if $\mathcal{C}(k)$ contains an incompressible ball, then $\mathcal{C}(k+1) \neq \emptyset$. Thus, all balls in $\mathcal{C}(k)$ are compressible, but then they must have minimal content 1.

The above estimate on the rank of the inclusion

$$H_* \left(\bigcup_{i=1}^m B\left(p_i, \frac{R}{100}\right)\right) \to H_* \left(\bigcup_{i=1}^m B\left(p_i, \frac{R}{2}\right)\right),$$

in terms of the ranks of all the intersections, is in fact not quite right. One actually needs to consider the doubly indexed family $B\left(p_i, 10^{-j-1}R\right)$, $j = 1, \ldots, n+2$, where we assume that for each fixed j the family covers $B\left(p, \frac{1}{5}R\right)$. The correct estimate is then that the rank of the inclusion

$$H_* \left(\bigcup_{i=1}^m B\left(p_i, \frac{R}{10^{n+2}}\right)\right) \to H_* \left(\bigcup_{i=1}^m B\left(p_i, \frac{R}{2}\right)\right)$$

is bounded by the rank of all of the possible intersections

$$H_* \left(\bigcap_{t=1}^s B\left(p_{i_t}, \frac{R}{10^{j+1}}\right)\right) \to H_* \left(\bigcap_{t=1}^s B\left(p_{i_t}, \frac{R}{2 \cdot 10^{j-1}}\right)\right)$$

Whenever such an intersection

$$\bigcap_{t=1}^s B\left(p_{i_t}, 10^{-j-1}R\right) \neq \emptyset,$$

we still have the inclusions

$$
\begin{aligned}
\bigcap_{t=1}^s B\left(p_{i_t}, \frac{R}{10^{j+1}}\right) &\subset B\left(p_{i_1}, \frac{R}{10^{j+1}}\right) \\
&\subset B\left(p_{i_1}, \frac{R}{2 \cdot 10^j}\right) \\
&\subset \bigcap_{t=1}^s B\left(p_{i_t}, \frac{R}{2 \cdot 10^{j-1}}\right).
\end{aligned}
$$

So we can still estimate those ranks by the content of balls in $\mathcal{C}(k+1)$. We have, however, more intersections and also more balls, as this time the smaller balls

$B\left(p_i, 10^{-n-1}R\right)$ have to cover. One can easily compute the correct Betti number estimate with these modifications. The reader should consult the survey by Cheeger in [**24**] for the complete story.

The Betti number theorem can easily be proved in the more general context of manifolds with lower sectional curvature bounds, but one must then also assume an upper diameter bound. Otherwise, the ball covering arguments, and also the estimates using Toponogov's theorem, won't work. Thus, there is a constant $C\left(n, D, k\right)$ such that any closed Riemannian n-manifold (M, g) with sec $\geq k$ and diam $\leq D$ has the properties that

(1) $\pi_1\left(M\right)$ can be generated by $\leq C\left(n, k, D\right)$ elements,
(2) $\sum_{i=0}^n b_i\left(M, F\right) \leq C\left(n, D, k\right)$.

6. Homotopy Finiteness

This section is devoted to a result that interpolates between Cheeger's finiteness theorem and Gromov's Betti number estimate. We know that in Gromov's theorem the class under investigation contains infinitely many homotopy types, while if we have a lower volume bound and an upper curvature bound as well, Cheeger's result says that we have finiteness of diffeomorphism types.

THEOREM 87. (Grove-Petersen, 1988) *Given an integer $n > 1$ and numbers $v, D, k \in (0, \infty)$, the class of Riemannian n-manifolds with*

$$
\begin{aligned}
\text{diam} &\leq D, \\
\text{vol} &\geq v, \\
\text{sec} &\geq -k^2
\end{aligned}
$$

contains only finitely many homotopy types.

As with the other proofs in this chapter we need to proceed in stages. First, we present the main technical result.

LEMMA 65. *For a manifold as in the above theorem, we can find $\alpha = \alpha\left(n, D, v, k\right) \in \left(0, \frac{\pi}{2}\right)$ and $\delta = \delta\left(n, D, v, k\right) > 0$ such that if $p, q \in M$ satisfy $d\left(p, q\right) \leq \delta$, then either p is α-regular for q or q is α-regular for p.*

PROOF. The proof is by contradiction and based on a suggestion by Cheeger. Assume there is a pair of points $p, q \in M$ that are not α-regular with respect to each other, and set $l = d\left(p, q\right) \leq \delta$. Let $\Gamma\left(p, q\right)$ denote the set of unit speed segments from p to q, and define

$$
\begin{aligned}
\dot{\Gamma}_{pq} &= \left\{v \in T_p M : v = \dot{\sigma}\left(0\right), \sigma \in \Gamma\left(p, q\right)\right\}, \\
\dot{\Gamma}_{qp} &= \left\{-v \in T_q M : v = \dot{\sigma}\left(r\right), \sigma \in \Gamma\left(p, q\right)\right\}.
\end{aligned}
$$

Then the two sets $\dot{\Gamma}_{pq}$ and $\dot{\Gamma}_{qp}$ of unit vectors are by assumption $(\pi - \alpha)$-dense in the unit sphere. It is a simple exercise to show that if $A \subset S^{n-1}$, then the function

$$
t \to \frac{\text{vol}B\left(A, t\right)}{v\left(n - 1, 1, t\right)}
$$

is nonincreasing (see also exercises to chapter 9). In particular, for any $(\pi - \alpha)$-dense set $A \subset S^{n-1}$

$$
\begin{aligned}
\text{vol}\left(S^{n-1} - B\left(A, \alpha\right)\right) &= \text{vol} S^{n-1} - \text{vol} B\left(A, \alpha\right) \\
&\leq \text{vol} S^{n-1} - \text{vol} S^{n-1} \cdot \frac{v\left(n-1, 1, \alpha\right)}{v\left(n-1, 1, \pi - \alpha\right)} \\
&= \text{vol} S^{n-1} \cdot \frac{v\left(n-1, 1, \pi - \alpha\right) - v\left(n-1, 1, \alpha\right)}{v\left(n-1, 1, \pi - \alpha\right)}.
\end{aligned}
$$

Now choose $\alpha < \frac{\pi}{2}$ such that

$$
\text{vol} S^{n-1} \cdot \frac{v\left(n-1, 1, \pi - \alpha\right) - v\left(n-1, 1, \alpha\right)}{v\left(n-1, 1, \pi - \alpha\right)} \cdot \int_0^D \left(\text{sn}_k\left(t\right)\right)^{n-1} dt = \frac{v}{6}.
$$

Thus, the two cones (see exercises to chapter 9) satisfy

$$
\begin{aligned}
\text{vol} B^{S^{n-1} - B\left(\dot{\Gamma}_{pq}, \alpha\right)}\left(p, D\right) &\leq \frac{v}{6}, \\
\text{vol} B^{S^{n-1} - B\left(\dot{\Gamma}_{qp}, \alpha\right)}\left(q, D\right) &\leq \frac{v}{6}.
\end{aligned}
$$

We now use Toponogov's theorem to choose δ such that any point in M that does not lie in one of these two cones must be close to either p or q (Figure 11.21 shows how a small δ will force the other leg in the triangle to be smaller than r). To this end, pick $r > 0$ such that

$$
v\left(n, -k^2, r\right) = \frac{v}{6}.
$$

We now claim that if δ is sufficiently small, then

$$
M = B\left(p, r\right) \cup B\left(q, r\right) \cup B^{S^{n-1} - B\left(\dot{\Gamma}_{pq}, \alpha\right)}\left(p, D\right) \cup B^{S^{n-1} - B\left(\dot{\Gamma}_{qp}, \alpha\right)}\left(q, D\right).
$$

This will, of course, lead to a contradiction, as we would then have

$$
\begin{aligned}
v &\leq \text{vol} M \\
&\leq \text{vol}\left(B\left(p, r\right) \cup B\left(q, r\right) \cup B^{S^{n-1} - B\left(\dot{\Gamma}_{pq}, \alpha\right)}\left(p, D\right) \cup B^{S^{n-1} - B\left(\dot{\Gamma}_{qp}, \alpha\right)}\left(q, D\right)\right) \\
&\leq 4 \cdot \frac{v}{6} < v.
\end{aligned}
$$

To see that these sets cover M, observe that if

$$
x \notin B^{S^{n-1} - B\left(\dot{\Gamma}_{pq}, \alpha\right)}\left(p, D\right),
$$

then there is a segment from x to p and a segment from p to q that form an angle $\leq \alpha$. (See Figure 11.22.)

Thus, we have from Toponogov's theorem that

$$
\cosh d\left(x, q\right) \leq \cosh l \cosh d\left(x, p\right) - \sinh l \sinh d\left(x, p\right) \cos\left(\alpha\right).
$$

If also

$$
x \notin B^{S^{n-1} - B\left(\dot{\Gamma}_{qp}, \alpha\right)}\left(q, D\right),
$$

we have in addition,

$$
\cosh d\left(x, p\right) \leq \cosh l \cosh d\left(x, q\right) - \sinh l \sinh d\left(x, q\right) \cos\left(\alpha\right).
$$

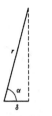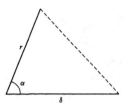

Figure 11.21

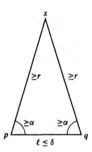

Figure 11.22

If in addition $d(x,p) > r$ and $d(x,q) > r$, we get

$$
\begin{aligned}
\cosh d(x,q) \quad &\leq \quad \cosh l \cosh d(x,p) - \sinh l \sinh d(x,p) \cos(\alpha) \\
&\leq \quad \cosh d(x,p) \\
&\quad + (\cosh l - 1) \cosh D - \sinh l \sinh r \cos(\alpha)
\end{aligned}
$$

and

$$
\begin{aligned}
\cosh d(x,p) \quad &\leq \quad \cosh d(x,q) \\
&\quad + (\cosh l - 1) \cosh D - \sinh l \sinh r \cos(\alpha).
\end{aligned}
$$

However, as $l \to 0$, we see that the quantity

$$
\begin{aligned}
f(l) \quad &= \quad (\cosh l - 1) \cosh D - \sinh l \sinh r \cos(\alpha) \\
&= \quad (-\sinh r \cos \alpha) l + O(l^2)
\end{aligned}
$$

becomes negative. Thus, we can find $\delta(D,r,\alpha) > 0$ such that for $l \leq \delta$ we have

$$
(\cosh l - 1) \cosh D - \sinh l \sinh r \cos(\alpha) < 0.
$$

We have then arrived at another contradiction, as this would imply

$$
\cosh d(x,q) < \cosh d(x,p)
$$

and

$$
\cosh d(x,p) < \cosh d(x,q)
$$

at the same time. Thus, the sets cover as we claimed. As this covering is also impossible, we are lead to the conclusion that under the assumption that $d(p,q) \leq \delta$, we must have that either p is α-regular for q or q is α-regular for p. □

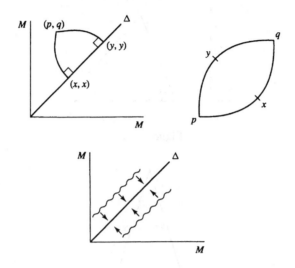

Figure 11.23

As it stands, this lemma seems rather strange and unmotivated. A little analysis will, however, enable us to draw some very useful conclusions from it.

Consider the product $M \times M$ with the product metric. Geodesics in this space are of the form (γ_1, γ_2), where both γ_1, γ_2 are geodesics in M. In $M \times M$ we have the diagonal $\Delta = \{(x, x) : x \in M\}$, which is a compact submanifold. Note that

$$T_{(p,p)}\Delta = \{(v, v) : v \in T_p M\},$$

and consequently, the normal bundle is

$$\nu(\Delta) = \{(v, -v) : v \in T_p M\}.$$

Therefore, if

$$(\sigma_1, \sigma_2) : [a, b] \to M \times M$$

is a segment from (p, q) to Δ, then we must have that $\dot{\sigma}_1(b) = -\dot{\sigma}_2(b)$. Thus these two segments can be joined at the common point $\sigma_1(b) = \sigma_2(b)$ to form a geodesic from p to q in M. This geodesic is, in fact, a segment, for otherwise, we could find a shorter curve from p to q. Dividing this curve in half would then produce a shorter curve from (p, q) to Δ. Thus, we have a bijective correspondence between segments from p to q and segments from (p, q) to Δ. Moreover,

$$\sqrt{2} \cdot d((p, q), \Delta) = d(p, q).$$

The above lemma now implies

COROLLARY 44. *Any point within distance $\delta/\sqrt{2}$ of Δ is α-regular for Δ.*

Figure 11.23 shows how the contraction onto the diagonal works and also how segments to the diagonal are related to segments in M.

Thus, we can find a curve of length $\leq \frac{1}{\cos \alpha} \cdot d((p, q), \Delta)$ from any point in this neighborhood to Δ. Moreover, this curve depends continuously on (p, q). We can translate this back into M. Namely, if $d(p, q) < \delta$, then p and q are joined by a curve $t \to H(p, q, t)$, $0 \leq t \leq 1$, whose length is $\leq \frac{\sqrt{2}}{\cos \alpha} \cdot d(p, q)$. Furthermore, the map $(p, q, t) \to H(p, q, t)$ is continuous. For simplicity, we let $C = \frac{\sqrt{2}}{\cos \alpha}$ in the constructions below.

We now have the first ingredient in our proof.

COROLLARY 45. *If $f_0, f_1 : X \to M$ are two continuous maps such that*

$$d(f_0(x), f_1(x)) < \delta$$

for all $x \in X$, then f_0 and f_1 are homotopy equivalent.

For the next construction, recall that a k-simplex Δ^k can be thought of as the set of affine linear combinations of all the basis vectors in $\mathbb{R}^{k+1}$, i.e.,

$$\Delta^k = \left\{ (x^0, \ldots, x^k) : x^0 + \cdots + x^k = 1 \text{ and } x^0, \ldots, x^k \in [0, 1] \right\}.$$

The basis vectors $e_i = \left(\delta_i^1, \ldots, \delta_i^k \right)$ are called the vertices of the simplex.

LEMMA 66. *Suppose we have $k + 1$ points $p_0, \ldots, p_k \in B(p, r) \subset M$. If*

$$2r \frac{C^k - 1}{C - 1} < \delta,$$

then we can find a continuous map

$$f : \Delta^k \to B\left(p, r + 2r \cdot C \cdot \frac{C^k - 1}{C - 1} \right),$$

where $f(e_i) = p_i$.

PROOF. Figure 11.24 gives the essential idea of the proof. The proof goes by induction on k. For $k = 0$ there is nothing to show.

Suppose now that the statement holds for k and that we have $k + 2$ points $p_0, \ldots, p_{k+1} \in B(p, r)$. First, we find a map

$$f : \Delta^k \to B\left(p, 2r \cdot C \cdot \frac{C^k - 1}{C - 1} + r \right)$$

with $f(e_i) = p_i$ for $i = p_0, \ldots, p_k$. We then define

$$\bar{f} \quad : \quad \Delta^{k+1} \to B\left(p, r + 2r \cdot C \cdot \frac{C^{k+1} - 1}{C - 1} \right),$$

$$\bar{f}\left(x^0, \ldots, x^k, x^{k+1} \right) \quad = \quad H\left(f\left(\frac{x^0}{\sum_{i=1}^k x^i}, \ldots, \frac{x^k}{\sum_{i=1}^k x^i} \right), p_{k+1}, x^{k+1} \right).$$

This clearly gives a well-defined continuous map as long as

$$d\left(f\left(\frac{x^0}{\sum_{i=1}^k x^i}, \ldots, \frac{x^k}{\sum_{i=1}^k x^i} \right), p_{k+1} \right)$$

$$\leq \quad d\left(f\left(\frac{x^0}{\sum_{i=1}^k x^i}, \ldots, \frac{x^k}{\sum_{i=1}^k x^i} \right), p \right) + d(p, p_{k+1})$$

$$\leq \quad \left(2r \cdot C \cdot \frac{C^k - 1}{C - 1} + r \right) + r$$

$$= \quad 2r \cdot \frac{C^{k+1} - 1}{C - 1}$$

$$< \quad \delta.$$

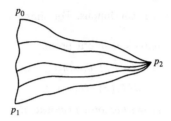

$$\text{Figure 11.24}$$

Moreover, it has the property that

$$d\left(p, \bar{f}\left(\cdot\right)\right) \;\leq\; d\left(p, p_{k+1}\right) + d\left(p_{k+1}, \bar{f}\left(\cdot\right)\right)$$

$$\leq\; r + 2r \cdot C \cdot \frac{C^{k+1} - 1}{C - 1}.$$

This concludes the induction step.   □

Note that if we select a face spanned by, say, $(e_1, \ldots, e_k)$ of the simplex Δ^k, then we could, of course, construct a map in the above way by mapping e_i to p_i. The resulting map will, however, be the same as if we constructed the map on the entire simplex and restricted it to the selected face.

We can now prove finiteness of homotopy types. Observe that the class we work with is precompact in the Gromov-Hausdorff distance as we have an upper diameter bound and a lower bound for the Ricci curvature. Thus it suffices to prove

LEMMA 67. *There is an* $\varepsilon = \varepsilon\left(n, k, v, D\right) > 0$ *such that if two Riemannian n-manifolds* (M, g_1) *and* (N, g_2) *satisfy*

$$\text{diam} \;\leq\; D,$$
$$\text{vol} \;\geq\; v,$$
$$\text{sec} \;\geq\; -k^2,$$

and

$$d_{G-H}\left(M, N\right) < \varepsilon,$$

then they are homotopy equivalent.

PROOF. Suppose M and N are given as in the lemma, together with a metric d on $M \amalg N$, inside which the two spaces are ε Hausdorff close. The size of ε will be found through the construction.

First, triangulate both manifolds in such a way that any simplex of the triangulation lies in a ball of radius ε. Using the triangulation on M, we can now construct a continuous map $f : M \to N$ as follows. First we use the Hausdorff approximation to map all the vertices $\{p_\alpha\} \subset M$ of the triangulation to points $\{q_\alpha\} \subset N$ such that $d\left(p_\alpha, q_\alpha\right) < \varepsilon$. If now $(p_{\alpha_0}, \ldots, p_{\alpha_n})$ forms a simplex in the triangulation of M, then we constructed the triangulation such that

$$\left(p_{\alpha_0}, \ldots, p_{\alpha_n}\right) \subset B\left(x, \varepsilon\right)$$

for some $x \in M$. Thus

$$\left(q_{\alpha_0}, \ldots, q_{\alpha_n}\right) \subset B\left(q_{\alpha_0}, 4\varepsilon\right).$$

Therefore, if

$$8\varepsilon \frac{C^n - 1}{C - 1} < \delta,$$

then we can use the above lemma to define f on the simplex spanned by $(p_{\alpha_0}, \ldots, p_{\alpha_n})$. In this way we get a map $f : M \to N$ by constructing it on each simplex as just described. To see that it is continuous, we must check that the construction agrees on common faces of simplices. But this follows, as the construction is natural with respect to restriction to faces of simplices. We now need to estimate how good a Hausdorff approximation f is. To this end, select $x \in M$ and suppose that it lies in the face spanned by the vertices $(p_{\alpha_0}, \ldots, p_{\alpha_n})$. Then we have

$$
\begin{aligned}
d(x, f(x)) &\leq d(x, p_{\alpha_0}) + d(p_{\alpha_0}, f(x)) \\
&\leq 2\varepsilon + \varepsilon + d(q_{\alpha_0}, f(x)) \\
&\leq 3\varepsilon + 4\varepsilon + 8\varepsilon \cdot C \cdot \frac{C^n - 1}{C - 1} \\
&= 7\varepsilon + 8\varepsilon \cdot C \cdot \frac{C^n - 1}{C - 1}.
\end{aligned}
$$

We can now construct $g : N \to M$ in the same manner. This map will, of course, also satisfy

$$d(y, g(y)) \leq 7\varepsilon + 8\varepsilon \cdot C \cdot \frac{C^n - 1}{C - 1}.$$

It is now possible to estimate how close the compositions $f \circ g$ and $g \circ f$ are to the identity maps on N and M, respectively, as follows:

$$
\begin{aligned}
d(y, f \circ g(y)) &\leq d(y, g(y)) + d(g(y), f \circ g(y)) \\
&\leq 14\varepsilon + 16\varepsilon \cdot C \cdot \frac{C^n - 1}{C - 1}; \\
d(x, g \circ f(x)) &\leq 14\varepsilon + 16\varepsilon \cdot C \cdot \frac{C^n - 1}{C - 1}.
\end{aligned}
$$

As long as

$$14\varepsilon + 16\varepsilon \cdot C \cdot \frac{C^n - 1}{C - 1} < \delta,$$

we can then conclude that these compositions are homotopy equivalent to the respective identity maps. In particular, the two spaces are homotopy equivalent. $\square$

Note that as long as

$$16\varepsilon \cdot \frac{C^{n+1} - 1}{C - 1} < \delta,$$

the two spaces are homotopy equivalent. Thus, ε depends in an explicit way on $C = \frac{\sqrt{2}}{\cos \alpha}$ and δ. It is possible, in turn, to estimate α and δ from $n, k, v,$ and D. We can therefore get an explicit estimate for how close spaces must be to ensure that they are homotopy equivalent. Given this explicit ε, it is then possible, using our work from the section on Gromov-Hausdorff distance, to find an explicit estimate for the number of homotopy types.

To conclude, let us compare the three finiteness theorems by Cheeger, Gromov, and Grove-Petersen. We have inclusions of classes of closed Riemannian n-manifolds

$$
\left\{ \begin{array}{ll} \text{diam} & \leq D \\ \text{sec} & \geq -k^2 \end{array} \right\} \supset \left\{ \begin{array}{ll} \text{diam} & \leq D \\ \text{vol} & \geq v \\ \text{sec} & \geq -k^2 \end{array} \right\} \supset \left\{ \begin{array}{ll} \text{diam} & \leq D \\ \text{vol} & \geq v \\ |\text{sec}| & \leq k^2 \end{array} \right\}
$$

with strengthening of conclusions from bounded Betti numbers to finitely many homotopy types to compactness in the $C^{1,\alpha}$ topology. In the special case of non-negative curvature Gromov's estimate actually doesn't depend on the diameter, thus yielding obstructions to the existence of such metrics on manifolds with complicated topology. For the other two results the diameter bound is still necessary. Consider for instance the family of lens spaces $\{S^3/\mathbb{Z}_p\}$ with curvature $= 1$. Now rescale these metrics so that they all have the same volume. Then we get a class which contains infinitely many homotopy types and also satisfies

$$\text{vol} = v,$$
$$1 \geq \sec > 0.$$

The family of lens spaces $\{S^3/\mathbb{Z}_p\}$ with curvature $= 1$ also shows that the lower volume bound is necessary in both of these theorems.

Some further improvements are possible in the conclusion of the homotopy finiteness result. Namely, one can strengthen the conclusion to state that the class contains finitely many homeomorphism types. This was proved for $n \neq 3$ in [51] and in a more general case in [76]. One can also prove many of the above results for manifolds with certain types of integral curvature bounds, see for instance [79] and [80]. The volume [50] also contains complete discussions of generalizations to the case where one has merely Ricci curvature bounds.

7. Further Study

There are many texts that partially cover or expand the material in this chapter. We wish to attract attention to the surveys by Grove in [45], by Abresch-Meyer, Colding, Greene, and Zhu in [50], by Cheeger in [24], and by Karcher in [28]. The most glaring omission from this chapter is probably that of the Abresch-Gromoll theorem and other uses of the excess function. The above-mentioned articles by Zhu and Cheeger cover this material quite well.

8. Exercises

(1) Let (M, g) be a closed simply connected positively curved manifold. Show that if M contains a totally geodesic closed hypersurface (i.e., the shape operator is zero), then M is homeomorphic to a sphere. (Hint: first show that the hypersurface is orientable, and then show that the signed distance function to this hypersurface has only two critical points - a maximum and a minimum. This also shows that it suffices to assume that $H^1(M, \mathbb{Z}_2) = 0$.)

(2) Show that the converse of Toponogov's theorem is also true. In other words, if for some k the conclusion to Toponogov's theorem holds when hinges (or triangles) are compared to the same objects in S_k^2, then $\sec \geq k$.

(3) (Heintze-Karcher) Let $\gamma \subset (M, g)$ be a geodesic in a Riemannian n-manifold with $\sec \geq -k^2$. Let $T(\gamma, R)$ be the normal tube around γ of radius R, i.e., the set of points in M that can be joined to γ by a segment of length $\leq R$ that is perpendicular to γ. The last condition is superfluous when γ is a closed geodesic, but if it is a loop or a segment, then not all points in M within distance R of γ will belong to this tube. On this tube introduce coordinates (r, s, θ), where r denotes the distance to γ, s is the arc-length parameter on γ, and $\theta = (\theta^1, \ldots, \theta^{n-2})$ are spherical

coordinates normal to γ. These give adapted coordinates for the distance r to γ. Show that as $r \to 0$ the metric looks like

$$g(r) = \begin{pmatrix} 1 & 0 & 0 & \cdots & 0 \\ 0 & 1 & 0 & \cdots & 0 \\ 0 & 0 & 0 & \cdots & 0 \\ \vdots & \vdots & \vdots & \ddots & \vdots \\ 0 & 0 & 0 & \cdots & 0 \end{pmatrix} + \begin{pmatrix} 0 & 0 & 0 & \cdots & 0 \\ 0 & 0 & 0 & \cdots & 0 \\ 0 & 0 & 1 & \cdots & 0 \\ \vdots & \vdots & \vdots & \ddots & \vdots \\ 0 & 0 & 0 & \cdots & 1 \end{pmatrix} \cdot r^2 + O\left(r^3\right)$$

Using the lower sectional curvature bound, find an upper bound for the volume density on this tube. Conclude that

$$\mathrm{vol}T\left(\gamma, R\right) \le f\left(n, k, R, \ell\left(\gamma\right)\right),$$

for some continuous function f depending on dimension, lower curvature bound, radius, and length of γ. Moreover, as $\ell\left(\gamma\right) \to 0$, $f \to 0$. Use this estimate to prove Cheeger's lemma from Chapter 10 and the main lemma on mutually critical points from the homotopy finiteness theorem. This shows that Toponogov's theorem is not needed for the latter result.

(4) Show that any vector bundle over a 2-sphere admits a complete metric of nonnegative sectional curvature. Hint: You need to know something about the classification of vector bundles over spheres. In this case k-dimensional vector bundles are classified by homotopy classes of maps from S^1, the equator of the 2-sphere, into $SO(k)$. This is the same as $\pi_1\left(SO\left(k\right)\right)$, so there is only one 1-dimensional bundle, the 2-dimensional bundles are parametrized by $\mathbb{Z}$, and 2 higher dimensional bundles.

(5) Use Toponogov's theorem to show that b_γ is convex when $\sec \ge 0$.

De Rham Cohomology

We shall in this appendix explain the main ideas surrounding de Rham co-homology. This is done as a service to the reader who has learned about tensors and algebraic topology but had only sporadic contact with Stokes' theorem. First we give an introduction to Lie derivatives on manifolds. We then give a digest of forms and important operators on forms. Then we explain how one integrates forms and prove Stokes' theorem for manifolds without boundary. Finally, we define de Rham cohomology and show how the Poincaré lemma and the Meyer-Vietoris lemma together imply that de Rham cohomology is simply standard cohomology. The cohomology theory that comes closest to de Rham cohomology is Čech cohomology. As this cohomology theory often is not covered in standard courses on algebraic topology, we define it here and point out that it is easily seen to satisfy the same properties as de Rham cohomology.

1. Lie Derivatives

Let X be a vector field and F^t the corresponding locally defined flow on a smooth manifold M. Thus $F^t(p)$ is defined for small t and the curve $t \to F^t(p)$ is the integral curve for X that goes through p at $t = 0$. The Lie derivative of a tensor in the direction of X is defined as the first order term in a suitable Taylor expansion of the tensor when it is moved by the flow of X.

Let us start with a function $f : M \to \mathbb{R}$. Then

$$f\left(F^t(p)\right) = f(p) + t\left(L_X f\right)(p) + o(t),$$

where the Lie derivative $L_X f$ is just the directional derivative $D_X f = df(X)$. We can also write this as

$$
\begin{aligned}
f \circ F^t &= f + t L_X f + o(t), \\
L_X f &= D_X f = df(X).
\end{aligned}
$$

When we have a vector field Y things get a little more complicated. We wish to consider $Y|_{F^t}$, but this can't be directly compared to Y as the vectors live in different tangent spaces. Thus we look at the curve $t \to DF^{-t}\left(Y|_{F^t(p)}\right)$ that lies in T_pM. Then we expand for t near 0 and get

$$DF^{-t}\left(Y|_{F^t(p)}\right) = Y|_p + t\left(L_X Y\right)|_p + o(t)$$

for some vector $(L_X Y)|_p \in T_p M$. This Lie derivative also has an alternate definition.

PROPOSITION 49. *For vector fields X, Y on M we have*

$$L_X Y = [X, Y].$$

PROOF. We see that the Lie derivative satisfies

$$DF^{-t}(Y|_{F^t}) = Y + tL_X Y + o(t)$$

or equivalently

$$Y|_{F^t} = DF^t(Y) + tDF^t(L_X Y) + o(t).$$

It is therefore natural to consider the directional derivative of a function f in the direction of $Y|_{F^t} - DF^t(Y)$.

$$
\begin{aligned}
D_{Y|_{F^t} - DF^t(Y)} f &= D_{Y|_{F^t}} f - D_{DF^t(Y)} f \\
&= (D_Y f) \circ F^t - D_Y (f \circ F^t) \\
&= D_Y f + tD_X D_Y f + o(t) \\
&\quad - D_Y (f + tD_X f + o(t)) \\
&= t(D_X D_Y f - D_Y D_X f) + o(t) \\
&= tD_{[X,Y]} f + o(t).
\end{aligned}
$$

This shows that

$$
\begin{aligned}
L_X Y &= \lim_{t \to 0} \frac{Y|_{F^t} - DF^t(Y)}{t} \\
&= [X, Y].
\end{aligned}
$$

$\square$

We are now ready to define the Lie derivative of a $(0, p)$-tensor T and also give an algebraic formula for this derivative. We define

$$(F^t)^* T = T + t(L_X T) + o(t)$$

or more precisely

$$
\begin{aligned}
\left((F^t)^* T\right)(Y_1, ..., Y_p) &= T(DF^t(Y_1), ..., DF^t(Y_p)) \\
&= T(Y_1, ..., Y_p) + t(L_X T)(Y_1, ..., Y_p) + o(t).
\end{aligned}
$$

PROPOSITION 50. *If X is a vector field and T a $(0, p)$-tensor on M, then*

$$(L_X T)(Y_1, ..., Y_p) = D_X (T(Y_1, ..., Y_p)) - \sum_{i=1}^{p} T(Y_1, ..., L_X Y_i, ..., Y_p)$$

PROOF. We restrict attention to the case where $p = 1$. The general case is similar but requires more notation. Using that

$$Y|_{F^t} = DF^t(Y) + tDF^t(L_X Y) + o(t)$$

we get

$$
\begin{aligned}
\left((F^t)^* T\right)(Y) &= T(DF^t(Y)) \\
&= T(Y|_{F^t} - tDF^t(L_X Y)) + o(t) \\
&= T(Y) \circ F^t - tT(DF^t(L_X Y)) + o(t) \\
&= T(Y) + tD_X(T(Y)) - tT(DF^t(L_X Y)) + o(t).
\end{aligned}
$$

Thus

$$
\begin{aligned}
(L_X T)(Y) &= \lim_{t \to 0} \frac{((F^t)^* T)(Y) - T(Y)}{t} \\
&= \lim_{t \to 0} \left(D_X(T(Y)) - T(DF^t(L_X Y)) \right) \\
&= D_X(T(Y)) - T(L_X Y).
\end{aligned}
$$

$\square$

Finally we have that Lie derivatives satisfy all possible product rules. From the above propositions this is already obvious when multiplying functions with vector fields or $(0, p)$-tensors. However, it is less clear when multiplying tensors.

PROPOSITION 51. *Let T_1 and T_2 be $(0, p_i)$-tensors, then*

$$
L_X(T_1 \cdot T_2) = (L_X T_1) \cdot T_2 + T_1 \cdot (L_X T_2).
$$

PROOF. Recall that for 1-forms and more general $(0, p)$-tensors we define the product as

$$
T_1 \cdot T_2 (X_1, ..., X_{p_1}, Y_1, ..., Y_{p_2}) = T_1(X_1, ..., X_{p_1}) \cdot T_2(Y_1, ..., Y_{p_2}).
$$

The proposition is then a simple consequence of the previous proposition and the product rule for derivatives of functions. $\square$

PROPOSITION 52. *Let T be a $(0, p)$-tensor and $f : M \to \mathbb{R}$ a function, then*

$$
L_{fX} T(Y_1, ..., Y_p) = f L_X T(Y_1, ..., Y_p) + df(Y_i) \sum_{i=1}^{p} T(Y_1, ..., X, ..., Y_p).
$$

PROOF. We have that

$$
\begin{aligned}
L_{fX} T(Y_1, ..., Y_p) &= D_{fX}(T(Y_1, ..., Y_p)) - \sum_{i=1}^{p} T(Y_1, ..., L_{fX} Y_i, ..., Y_p) \\
&= f D_X(T(Y_1, ..., Y_p)) - \sum_{i=1}^{p} T(Y_1, ..., [fX, Y_i], ..., Y_p) \\
&= f D_X(T(Y_1, ..., Y_p)) - f \sum_{i=1}^{p} T(Y_1, ..., [X, Y_i], ..., Y_p) \\
&\quad + df(Y_i) \sum_{i=1}^{p} T(Y_1, ..., X, ..., Y_p)
\end{aligned}
$$

$\square$

The case where $X|_p = 0$ is of special interest when computing Lie derivatives. We note that $F^t(p) = p$ for all t. Thus $DF^t : T_p M \to T_p M$ and

$$
\begin{aligned}
L_X Y|_p &= \lim_{t \to 0} \frac{DF^{-t}(Y|_p) - Y|_p}{t} \\
&= \frac{d}{dt}(DF^{-t})|_{t=0}(Y|_p).
\end{aligned}
$$

This shows that $L_X = \frac{d}{dt}(DF^{-t})|_{t=0}$ when $X|_p = 0$. From this we see that if θ is a 1-form then $L_X \theta = -\theta \circ L_X$ at points p where $X|_p = 0$. This is related to the following interesting result.

LEMMA 68. *If a function $f : M \to \mathbb{R}$ has a critical point at p then the Hessian of f at p does not depend on the metric.*

PROOF. Assume that $X = \nabla f$ and $X|_p = 0$. Next select coordinates x^i around p such that the metric coefficients satisfy $g_{ij}|_p = \delta_{ij}$. Then we see that

$$
\begin{aligned}
L_X \left(g_{ij} dx^i dx^j \right)|_p &= L_X \left(g_{ij} \right)|_p + \delta_{ij} L_X \left(dx^i \right) dx^j + \delta_{ij} dx^i L_X \left(dx^j \right) \\
&= \delta_{ij} L_X \left(dx^i \right) dx^j + \delta_{ij} dx^i L_X \left(dx^j \right) \\
&= L_X \left(\delta_{ij} dx^i dx^j \right)|_p.
\end{aligned}
$$

Thus $\mathrm{Hess} f|_p$ is the same if we compute it using g and the Euclidean metric in the fixed coordinate system. □

Lie derivatives also come in handy when working with Lie groups. For a Lie group G we have the inner automorphism $\mathrm{Ad}_h : x \to hxh^{-1}$ and its differential at $x = e$ denoted by the same letters

$$\mathrm{Ad}_h : \mathfrak{g} \to \mathfrak{g}.$$

LEMMA 69. *The differential of $h \to \mathrm{Ad}_h$ is given by $U \to \mathrm{ad}_U (X) = [U, X]$*

PROOF. If we write $\mathrm{Ad}_h (x) = R_{h^{-1}} L_h (x)$, then its differential at $x = e$ is given by $\mathrm{Ad}_h = DR_{h^{-1}} DL_h$. Now let F^t be the flow for U. Then $F^t (g) = gF^t (e) = L_g (F^t (e))$ as both curves go through g at $t = 0$ and have U as tangent everywhere since U is a left-invariant vector field. This also shows that $DF^t = DR_{F^t(e)}$. Thus

$$
\begin{aligned}
\mathrm{ad}_U (X)|_e &= \frac{d}{dt} DR_{F^{-t}(e)} DL_{F^t(e)} \left(X|_e \right)|_{t=0} \\
&= \frac{d}{dt} DR_{F^{-t}(e)} \left(X|_{F^t(e)} \right)|_{t=0} \\
&= \frac{d}{dt} DF^{-t} \left(X|_{F^t(e)} \right)|_{t=0} \\
&= L_U X = [U, X].
\end{aligned}
$$

□

This is used in the next Lemma.

LEMMA 70. *Let $G = Gl(V)$ be the Lie group of invertible matrices on V. The Lie bracket structure on the Lie algebra $\mathfrak{gl}(V)$ of left invariant vector fields on $Gl(V)$ is given by commutation of linear maps. i.e., if $X, Y \in T_I Gl(V)$, then*

$$[X, Y]|_I = XY - YX.$$

PROOF. Since $x \to hxh^{-1}$ is a linear map on the space $\mathrm{hom}(V, V)$ we see that $\mathrm{Ad}_h (X) = hXh^{-1}$. The flow of U is given by $F^t (g) = g (I + tU + o(t))$ so we have

$$
\begin{aligned}
[U, X] &= \frac{d}{dt} \left(F^t (I) X F^{-t} (I) \right)|_{t=0} \\
&= \frac{d}{dt} ((I + tU + o(t)) X (I - tU + o(t)))|_{t=0} \\
&= \frac{d}{dt} (X + tUX - tXU + o(t))|_{t=0} \\
&= UX - XU.
\end{aligned}
$$

□

2. Elementary Properties

Given p 1-forms $\omega_i \in \Omega^1(M)$ on a manifold M we define

$$(\omega_1 \wedge \cdots \wedge \omega_p)(v_1, ..., v_p) = \det([\omega_i(v_j)])$$

where $[\omega_i(v_j)]$ is the matrix with entries $\omega_i(v_j)$. We can then extend the wedge product to all forms using linearity and associativity. This gives the *wedge product* operation

$$\Omega^p(M) \times \Omega^q(M) \quad \rightarrow \quad \Omega^{p+q}(M),$$
$$(\omega, \psi) \quad \rightarrow \quad \omega \wedge \psi.$$

This operation is bilinear and antisymmetric in the sense that:

$$\omega \wedge \psi = (-1)^{pq} \psi \wedge \omega.$$

The wedge product of a function and a form is simply standard multiplication.

There are three other important operations defined on forms: the *exterior derivative*

$$d : \Omega^p(M) \rightarrow \Omega^{p+1}(M),$$

the *Lie derivative*

$$L_X : \Omega^p(M) \rightarrow \Omega^p(M),$$

and the *interior product*

$$i_X : \Omega^p(M) \rightarrow \Omega^{p-1}(M).$$

The exterior derivative of a function is simply its usual differential, while if we are given a form

$$\omega = f_0 df_1 \wedge \cdots \wedge df_p,$$

then we declare that

$$d\omega = df_0 \wedge df_1 \wedge \cdots \wedge df_p.$$

The Lie derivative was defined in the previous section and the interior product is just evaluation in the first variable

$$(i_X \omega)(Y_1, \ldots, Y_{p-1}) = \omega(X, Y_1, \ldots, Y_{p-1}).$$

These operators satisfy the derivation properties:

$$\begin{aligned}
d(\omega \wedge \psi) &= (d\omega) \wedge \psi + (-1)^p \omega \wedge (d\psi), \\
i_X(\omega \wedge \psi) &= (i_X\omega) \wedge \psi + (-1)^p \omega \wedge (i_X\psi), \\
L_X(\omega \wedge \psi) &= (L_X\omega) \wedge \psi + \omega \wedge (L_X\psi),
\end{aligned}$$

and the composition properties

$$\begin{aligned}
d \circ d &= 0, \\
i_X \circ i_X &= 0, \\
L_X &= d \circ i_X + i_X \circ d, \\
L_X \circ d &= d \circ L_X, \\
i_X \circ L_X &= L_X \circ i_X.
\end{aligned}$$

The third property $L_X = d \circ i_X + i_X \circ d$ is also known a H. Cartan's formula (son of the geometer E. Cartan). It can be used to give an inductive definition of the exterior derivative via

$$i_X \circ d = L_X - d \circ i_X.$$

On a Riemannian manifold we can use the covariant derivative to define an exterior derivative:

$$d^\nabla \omega \left(X_0, ..., X_p \right) = \sum_{i=0}^{p} (-1)^i \left(\nabla_{X_i} \omega \right) \left(X_0, ..., \hat{X}_i, ..., X_n \right).$$

If $\omega = f$, then $d^\nabla f = df$ and if $\omega = df$, then

$$
\begin{aligned}
\left(d^\nabla df \right) (X, Y) &= \left(\nabla_X df \right) (Y) - \left(\nabla_Y df \right) (X) \\
&= \operatorname{Hess} f (X, Y) - \operatorname{Hess} f (X, Y) \\
&= 0.
\end{aligned}
$$

Using that this derivative, like the exterior derivative, has the derivation property

$$d^\nabla \left(\omega \wedge \psi \right) = \left(d^\nabla \omega \right) \wedge \psi + (-1)^p \, \omega \wedge \left(d^\nabla \psi \right)$$

we see that

$$d^\nabla \left(f_0 df_1 \wedge \cdots \wedge df_p \right) = d \left(f_0 df_1 \wedge \cdots \wedge df_p \right).$$

Thus we have a nice metric dependent formula for the exterior derivative.

3. Integration of Forms

We shall assume that M is an oriented n-manifold. Thus, M comes with a covering of charts

$$\varphi_\alpha = \left(x_\alpha^1, ..., x_\alpha^n \right) : U_\alpha \longleftrightarrow B(0, 1) \subset \mathbb{R}^n$$

such that the transition functions $\varphi_\alpha \circ \varphi_\beta^{-1}$ preserve the usual orientation on Euclidean space, i.e., $\det \left(D \left(\varphi_\alpha \circ \varphi_\beta^{-1} \right) \right) > 0$. In addition, we shall also assume that a partition of unity with respect to this covering is given. In other words, we have smooth functions $\lambda_\alpha : M \to [0, 1]$ such that $\lambda_\alpha = 0$ on $M - U_\alpha$ and $\sum_\alpha \lambda_\alpha = 1$. For the last condition to make sense, it is obviously necessary that the covering also be locally finite.

Given an n-form ω on M we wish to define:

$$\int_M \omega.$$

When M is not compact, it might be necessary to assume that the form has compact support, i.e., it vanishes outside some compact subset of M.

In each chart we can write

$$\omega = f_\alpha dx_\alpha^1 \wedge \cdots \wedge dx_\alpha^n.$$

Using the partition of unity, we then obtain

$$
\begin{aligned}
\omega &= \sum_\alpha \lambda_\alpha \omega \\
&= \sum_\alpha \lambda_\alpha f_\alpha dx_\alpha^1 \wedge \cdots \wedge dx_\alpha^n,
\end{aligned}
$$

where each of the forms

$$\lambda_\alpha f_\alpha dx_\alpha^1 \wedge \cdots \wedge dx_\alpha^n$$

has compact support in U_α. Since U_α is identified with $B(0,1)$, we simply declare that

$$\int_{U_\alpha} \lambda_\alpha f_\alpha dx_\alpha^1 \wedge \cdots \wedge dx_\alpha^n = \int_{B(0,1)} \lambda_\alpha f_\alpha dx^1 \cdots dx^n,$$

where the right-hand side is the integral of the function $\phi_\alpha f_\alpha$ viewed as a function on $B(0,1)$. Then define

$$\int_M \omega = \sum_\alpha \int_{U_\alpha} \lambda_\alpha f_\alpha dx_\alpha^1 \wedge \cdots \wedge dx_\alpha^n$$

whenever this sum converges. Using the standard change of variables formula for integration on Euclidean space, we see that this definition is indeed independent of the choice of coordinates.

We can now state and prove Stokes' theorem for manifolds without boundary.

THEOREM 88. *For any $\omega \in \Omega^{n-1}(M)$ with compact support we have*

$$\int_M d\omega = 0.$$

PROOF. If we use the trick

$$d\omega = \sum_\alpha d(\phi_\alpha \omega),$$

then we see that it suffices to prove the theorem in the case $M = B(0,1) \subset \mathbb{R}^n$ and ω has compact support on $B(0,1)$. Then write

$$\omega = \sum_{i=1}^n f_i dx^1 \wedge \cdots \wedge \widehat{dx^i} \wedge \cdots \wedge dx^n,$$

where the functions f_i are zero near the boundary of $B(0,1)$. The differential of ω is now easily computed:

$$\begin{aligned}
d\omega &= \sum_{i=1}^n (df_i) \wedge dx^1 \wedge \cdots \wedge \widehat{dx^i} \wedge \cdots \wedge dx^n \\
&= \sum_{i=1}^n \left(\frac{\partial f_i}{\partial x^i}\right) dx^i \wedge dx^1 \wedge \cdots \wedge \widehat{dx^i} \wedge \cdots \wedge dx^n \\
&= \sum_{i=1}^n (-1)^{i-1} \left(\frac{\partial f_i}{\partial x^i}\right) dx^1 \wedge \cdots \wedge dx^i \wedge \cdots \wedge dx^n.
\end{aligned}$$

Thus,

$$\begin{aligned}
\int_{B(0,1)} d\omega &= \int_{B(0,1)} \sum_{i=1}^n (-1)^{i-1} \left(\frac{\partial f_i}{\partial x^i}\right) dx^1 \cdots dx^n \\
&= \sum_{i=1}^n (-1)^{i-1} \int_{B(0,1)} \left(\frac{\partial f_i}{\partial x^i}\right) dx^1 \cdots dx^n \\
&= \sum_{i=1}^n (-1)^{i-1} \int \left(\int \left(\frac{\partial f_i}{\partial x^i}\right) dx^i\right) dx^1 \cdots \widehat{dx^i} \cdots dx^n.
\end{aligned}$$

The fundamental theorem of calculus tells us that

$$\int \left(\frac{\partial f_i}{\partial x^i}\right) dx^i = 0,$$

as f_i is zero near the boundary of the range of x^i. In particular, the entire integral must be zero. $\square$

Stokes' theorem leads to some important formulae on Riemannian manifolds.

COROLLARY 46. (The Divergence Theorem) *If X is a vector field on (M, g) with compact support, then*

$$\int_M \operatorname{div} X \cdot d\mathrm{vol} = 0.$$

PROOF. Just observe

$$\begin{aligned}
\operatorname{div} X \cdot d\mathrm{vol} &= L_X d\mathrm{vol} \\
&= i_X d\,(d\mathrm{vol}) + d\,(i_X d\mathrm{vol}) \\
&= d\,(i_X d\mathrm{vol})
\end{aligned}$$

and use Stokes' theorem. $\square$

COROLLARY 47. (Green's Formulae) *If f_1, f_2 are two compactly supported functions on (M, g), then*

$$\int_M (\Delta f_1) \cdot f_2 \cdot d\mathrm{vol} = -\int_M g\,(\nabla f_1, \nabla f_2) = \int_M f_1 \cdot (\Delta f_2) \cdot d\mathrm{vol}.$$

PROOF. Just use that

$$\operatorname{div}(f_1 \cdot \nabla f_2) = g\,(\nabla f_1, \nabla f_2) + f_1 \cdot \Delta f_2,$$

and apply the divergence theorem to get the desired result. $\square$

COROLLARY 48. (Integration by Parts) *If S, T are two $(1, p)$ tensors with compact support on (M, g), then*

$$\int_M g\left(S^{\flat}, \nabla \operatorname{div} T\right) \cdot d\mathrm{vol} = -\int_M g\,(\operatorname{div} S, \operatorname{div} T) \cdot d\mathrm{vol},$$

where $S^{\flat}$ denotes the $(0, p+1)$-tensor defined by

$$S^{\flat}(X, Y, Z, \ldots) = g\,(X, S\,(Y, Z, \ldots)).$$

PROOF. For simplicity, first assume that S and T are vector fields X and Y. Then the formula can be interpreted as

$$\int_M g\,(X, \nabla \operatorname{div} Y) \cdot d\mathrm{vol} = -\int_M \operatorname{div} X \cdot \operatorname{div} Y \cdot d\mathrm{vol}.$$

We can then use that

$$\operatorname{div}(f \cdot X) = g\,(\nabla f, X) + f \cdot \operatorname{div} X.$$

Therefore, if we define $f = \operatorname{div} Y$ and use the divergence theorem, we get the desired formula.

In general, choose an orthonormal frame E_i, and observe that we can define a vector field by

$$X = \sum_{i_1, \ldots, i_p} S\left(E_{i_1}, \ldots, E_{i_p}\right) \operatorname{div} T\left(E_{i_1}, \ldots, E_{i_p}\right).$$

In other words, if we think of $g\,(V, S\,(X_1, \ldots, X_p))$ as a $(0, p)$-tensor, then X is implicitly defined by

$$g\,(X, V) = g\,(g\,(V, S), \operatorname{div} T).$$

Then we have

$$\mathrm{div}X = g\left(\mathrm{div}S, \mathrm{div}T\right) + g\left(S^\flat, \nabla\mathrm{div}T\right),$$

and the formula is established as before. □

It is worthwhile pointing out that usually

$$\int_M g\left(S^\flat, \mathrm{div}\nabla T\right) \neq -\int_M g\left(\nabla S, \nabla T\right),$$

even when the tensors are vector fields. On Euclidean space, for example, simply define $S = T = x^1\partial_1$. Then

$$\begin{aligned}
\nabla\left(x^1\partial_1\right) &= dx^1\partial_1, \\
\left|dx^1\partial_1\right| &= 1, \\
\mathrm{div}\left(dx^1\partial_1\right) &= 0.
\end{aligned}$$

Of course, the tensors in this example do not have compact support, but that can easily be fixed by multiplying with a compactly supported function.

4. Čech Cohomology

Before defining de Rham cohomology, we shall briefly mention how Čech cohomology is defined. This is the cohomology theory that seems most natural from a geometric point of view. Also, it is the cohomology that is most naturally associated with de Rham cohomology

For a manifold M, suppose that we have a covering of contractible open sets U_α such that all possible nonempty intersections $U_{\alpha_1} \cap \cdots \cap U_{\alpha_k}$ are also contractible. Such a covering is called a *good cover*. Now let I^k be the set of ordered indices that create nontrivial intersections

$$I^k = \{(\alpha_0, \ldots, \alpha_k) : U_{\alpha_0} \cap \cdots \cap U_{\alpha_k} \neq 0\}.$$

Čech cycles with values in a ring R are defined as a space of alternating maps

$$\check{Z}^k = \left\{f : I^k \to R : f \circ \tau = -f \text{ where } \tau \text{ is a transposition of two indices}\right\}.$$

The differential, or coboundary operator, is now defined by

$$d \quad : \quad \check{Z}^k \to \check{Z}^{k+1},$$

$$df\left(\alpha_0, \ldots, \alpha_{k+1}\right) = \sum_{i=0}^{k} (-1)^i f\left(\alpha_0, \ldots, \hat{\alpha}_i, \ldots, \alpha_{k+1}\right).$$

Čech cohomology is then defined as

$$H^k(M, R) = \frac{\ker\left(d : \check{Z}^k \to \check{Z}^{k+1}\right)}{\mathrm{im}\left(d : \check{Z}^{k-1} \to \check{Z}^{k+1}\right)}.$$

The standard arguments with refinements of covers can be used to show that this cohomology theory is independent of the choice of good cover. Below, we shall define de Rham cohomology for forms and prove several properties for that cohomology theory. At each stage one can easily see that Čech cohomology satisfies those same properties. Note that Čech cohomology seems almost purely combinatorial. This feature makes it very natural to work with in many situations.

5. De Rham Cohomology

Throughout we let M be an n-manifold. Using that $d \circ d = 0$, we trivially get that the exact forms

$$B^p(M) = d\left(\Omega^{p-1}(M)\right)$$

are a subset of the closed forms

$$Z^p(M) = \{\omega \in \Omega^p(M) : d\omega = 0\}.$$

The de Rham cohomology is then defined as

$$H^p(M) = \frac{Z^p(M)}{B^p(M)}.$$

Given a closed form ψ, we let $[\psi]$ denote the corresponding cohomology class.

The first simple property comes from the fact that any function with zero differential must be locally constant. On a connected manifold we therefore have

$$H^0(M) = \mathbb{R}.$$

Given a smooth map $f : M \to N$, we get an induced map in cohomology:

$$H^p(N) \quad \to \quad H^p(M),$$
$$f^*([\psi]) \quad = \quad [f^*\psi].$$

This definition is independent of the choice of ψ, since the pullback f^* commutes with d.

The two key results that are needed for a deeper understanding of de Rham cohomology are the Meyer-Vietoris sequence and the Poincaré lemma.

LEMMA 71. (The Mayer-Vietoris Sequence) *If $M = A \cup B$ for open sets $A, B \subset M$, then there is a long exact sequence*

$$\cdots \to H^p(M) \to H^p(A) \oplus H^p(B) \to H^p(A \cap B) \to H^{p+1}(M) \to \cdots.$$

PROOF. The proof is given in outline, as it is exactly the same as the corresponding proof in algebraic topology.

First, we need to define the maps. We clearly have inclusions

$$H^p(M) \quad \to \quad H^p(A),$$
$$H^p(M) \quad \to \quad H^p(B),$$
$$H^p(A) \quad \to \quad H^p(A \cap B),$$
$$H^p(B) \quad \to \quad H^p(A \cap B).$$

By adding the first two, we get

$$H^p(M) \quad \to \quad H^p(A) \oplus H^p(B),$$
$$[\psi] \quad \to \quad ([\psi|_A], [\psi|_B]).$$

Subtraction of the last two, yields

$$H^p(A) \oplus H^p(B) \quad \to \quad H^p(A \cap B),$$
$$([\omega], [\psi]) \quad \to \quad [\omega|_{A \cap B}] - [\psi|_{A \cap B}].$$

With these definitions it is not hard to see that the sequence is exact at $H^p(A) \oplus H^p(B)$.

The coboundary operator $H^p(A \cap B) \to H^{p+1}(M)$ is as usual defined by considering the exact diagram

$$\begin{array}{ccccccccc}
0 & \to & \Omega^{p+1}(M) & \to & \Omega^{p+1}(A) \oplus \Omega^{p+1}(B) & \to & \Omega^{p+1}(A \cap B) & \to & 0 \\
& & \uparrow d & & \uparrow d & & \uparrow d & & \\
0 & \to & \Omega^p(M) & \to & \Omega^p(A) \oplus \Omega^p(B) & \to & \Omega^p(A \cap B) & \to & 0
\end{array}$$

If we take a closed form $\omega \in Z^p(A \cap B)$, then we have $\psi \in \Omega^p(A) \oplus \Omega^p(B)$ that is mapped onto ω. Then $d\psi \in \Omega^{p+1}(A) \oplus \Omega^{p+1}(B)$ is zero when mapped to $\Omega^{p+1}(A \cap B)$, as we assumed that $d\omega = 0$. But then exactness tells us that $d\psi$ must come from an element in $\Omega^{p+1}(M)$. It is now easy to see that in cohomology, this element is well defined and gives us a linear map

$$H^p(A \cap B) \to H^{p+1}(M)$$

that makes the Meyer-Vietoris sequence exact.   □

LEMMA 72. (The Poincaré Lemma) *The cohomology of the open unit disk* $B(0,1) \subset \mathbb{R}^n$ *is:*

$$\begin{aligned}
H^0(B(0,1)) &= \mathbb{R}, \\
H^p(B(0,1)) &= \{0\} \text{ for } p > 0.
\end{aligned}$$

PROOF. Evidently, the proof hinges on showing that any closed p-form ω is exact when $p > 0$. Using that the form is closed, we see that for any vector field

$$L_X \omega = d i_X \omega.$$

We shall use the radial field $X = \sum x^i \partial_i$ to construct a map

$$H : \Omega^p \to \Omega^p$$

that satisfies

$$\begin{aligned}
H \circ L_X &= id, \\
d \circ H &= H \circ d.
\end{aligned}$$

This is clearly enough, as we would then have

$$\omega = d(H(i_X \omega)).$$

Since L_X is differentiation in the direction of the radial field, the map H should be integration in the same direction. Motivated by this, define

$$H\left(f dx^{i_1} \wedge \cdots \wedge dx^{i_p}\right) = \left(\int_0^1 t^{p-1} f(tx) \, dt\right) dx^{i_1} \wedge \cdots \wedge dx^{i_p}$$

and extend it to all forms using linearity. We now need to check the two desired properties. This is done by direct calculations:

$$\begin{aligned}
& H \circ L_X \left(f dx^{i_1} \wedge \cdots \wedge dx^{i_p}\right) \\
=\; & H\left(x^i \partial_i f dx^{i_1} \wedge \cdots \wedge dx^{i_p} + f L_X\left(dx^{i_1} \wedge \cdots \wedge dx^{i_p}\right)\right) \\
=\; & H\left(x^i \partial_i f dx^{i_1} \wedge \cdots \wedge dx^{i_p} + p f dx^{i_1} \wedge \cdots \wedge dx^{i_p}\right) \\
=\; & \left(\left(\int_0^1 t^{p-1}\left(tx^i\right) \partial_i f(tx) \, dt\right) + p\left(\int_0^1 p t^{p-1} f(tx) \, dt\right)\right) \cdot dx^{i_1} \wedge \cdots \wedge dx^{i_p} \\
=\; & \left(\int_0^1 \frac{d}{dt}\left(t^p \cdot f(tx)\right) dt\right) dx^{i_1} \wedge \cdots \wedge dx^{i_p} \\
=\; & f(x) \, dx^{i_1} \wedge \cdots \wedge dx^{i_p};
\end{aligned}$$

$$
\begin{aligned}
H \circ d \left(f dx^{i_1} \wedge \cdots \wedge dx^{i_p} \right) &= H \left(\partial_i f \cdot dx^i \wedge dx^{i_1} \wedge \cdots \wedge dx^{i_p} \right) \\
&= \left(\left(\int_0^1 t^p \partial_i f\,(tx)\,dt \right) dx^i \right) \wedge dx^{i_1} \wedge \cdots \wedge dx^{i_p} \\
&= d \left(\int_0^1 t^{p-1} f\,(tx)\,dt \right) \wedge dx^{i_1} \wedge \cdots \wedge dx^{i_p} \\
&= d \left(\left(\int_0^1 t^{p-1} f\,(tx)\,dt \right) \wedge dx^{i_1} \wedge \cdots \wedge dx^{i_p} \right) \\
&= d \circ H \left(f dx^{i_1} \wedge \cdots \wedge dx^{i_p} \right).
\end{aligned}
$$

This finishes the proof. □

We can now prove de Rham's theorem.

THEOREM 89. (de Rham, 1931) *If M is a closed manifold, then the de Rham cohomology groups $H^p(M)$ are the same as the Čech, or singular, cohomology groups $H^p(M, \mathbb{R})$ with real coefficients. In particular, all the cohomology groups are finitely generated.*

PROOF. We first observe that both theories have natural Meyer-Vietoris sequences. Therefore, if M has a finite covering by open sets U_α with the property that

$$
H^p \left(U_{\alpha_1} \cap \cdots \cap U_{\alpha_k} \right) = H^p \left(U_{\alpha_1} \cap \cdots \cap U_{\alpha_k}, \mathbb{R} \right)
$$

for all p and intersections $U_{\alpha_1} \cap \cdots \cap U_{\alpha_k}$, then using induction on the number of elements in the covering, we see that the two cohomologies of M are the same.

To find such a covering, take a Riemannian metric on M. Then find a covering of convex balls $B(p_\alpha, r)$. The intersections of convex balls are clearly diffeomorphic to the unit ball. Thus, the Poincaré lemma ensures that the two cohomology theories are the same on all intersections. In case the covering is infinite we also need to make sure that it is countable and locally finite. This is clearly possible.

It also follows from this proof that the cohomology groups of a compact space are finitely generated. □

Suppose now we have two manifolds M and N with good coverings $\{U_\alpha\}$ and $\{V_\beta\}$. A map $f : M \to N$ is said to preserve these coverings if for each β we can find $\alpha(\beta)$ such that

$$
\bar{U}_{\alpha(\beta)} \subset f^{-1}(V_\beta).
$$

Given a good cover of N and a map $f : M \to N$, we can clearly always find a good covering of M such that f preserves these covers. The induced map: $f^* : H^p(N) \to H^p(M)$ is now completely determined by the combinatorics of the map $\beta \to \alpha(\beta)$. This makes it possible to define f^* for all continuous maps. Moreover, since the set of maps f that satisfy

$$
\bar{U}_{\alpha(\beta)} \subset f^{-1}(V_\beta)
$$

is open, we see that any map close to f induces the same map in cohomology. Consequently, homotopic maps must induce the same map in cohomology. This gives a very important result.

THEOREM 90. *If two manifolds, possibly of different dimension, are homotopy equivalent, then they have the same cohomology.*

6. Poincaré Duality

The last piece of information we need to understand is how the wedge product acts on cohomology. It is easy to see that we have a map

$$H^p(M) \times H^q(M) \quad \to \quad H^{p+q}(M),$$
$$([\psi], [\omega]) \quad \to \quad [\psi \wedge \omega].$$

We are interested in understanding what happens in case $p + q = n$. This requires a surprising amount of preparatory work. First we have

THEOREM 91. *If M is an oriented closed n-manifold, then we have a well-defined isomorphism*

$$H^n(M) \quad \to \quad \mathbb{R},$$
$$[\omega] \quad \to \quad \int_M \omega.$$

PROOF. That the map is well-defined follows from Stokes' theorem. It is also onto, since any form with the property that it is positive when evaluated on a positively oriented frame is integrated to a positive number. Thus, we must show that any form with $\int_M \omega = 0$ is exact. This is not easy to show, and in fact, it is more natural to show this in a more general context: If M is an oriented n-manifold that can be covered by finitely many charts, then any compactly supported n-form ω with $\int_M \omega = 0$ is exact.

The proof of this result is by induction on the number of charts it takes to cover M. But before we can start the inductive procedure, we must establish the result for the n-sphere.

Case 1: $M = S^n$. Cover M by two open discs whose intersection is homotopy equivalent to S^{n-1}. Then use induction on n together with the Meyer-Vietoris sequence to show that for each $n > 0$,

$$H^p(S^n) = \begin{cases} 0, & p \neq 0, n, \\ \mathbb{R}, & p = 0, n. \end{cases}$$

The induction apparently starts at $n = 0$ and S^0 consists of two points and therefore has $H^0(S^0) = \mathbb{R} \oplus \mathbb{R}$. Having shown that $H^n(S^n) = \mathbb{R}$, it is then clear that the map $\int : H^n(S^n) \to \mathbb{R}$ is an isomorphism.

Case 2: $M = B(0,1)$. We can think of M as being an open hemisphere of S^n. Any compactly supported form ω on M therefore yields a form on S^n. Given that $\int_M \omega = 0$, we therefore also get that $\int_{S^n} \omega = 0$. Thus, ω must be exact on S^n. Let $\psi \in \Omega^{n-1}(S^n)$ be chosen such that $d\psi = \omega$. Use again that ω is compactly supported to find an open disc N such that ω vanishes on N and $N \cup M = S^n$. Then ψ is clearly closed on N and must by the Poincaré lemma be exact. Thus, we can find $\theta \in \Omega^{n-2}(N)$ with $d\theta = \psi$ on N. Now observe that $\psi - d\theta$ is actually defined on all of S^n, as it vanishes on N. But then we have found a form $\psi - d\theta$ with support in M whose differential is ω.

Case 3: $M = A \cup B$ where the result holds on A and B. Select a partition of unity $\lambda_A + \lambda_B = 1$ subordinate to the cover $\{A, B\}$. Given an n-form ω with $\int_M \omega = 0$, we get two forms $\lambda_A \cdot \omega$ and $\lambda_B \cdot \omega$ with support in A and B, respectively.

Using our assumptions, we see that

$$0 = \int_M \omega$$

$$= \int_A \lambda_A \cdot \omega + \int_B \lambda_B \cdot \omega.$$

On $A \cap B$ we can select an n-form ψ with compact support inside $A \cap B$ such that

$$\int_{A \cap B} \tilde{\omega} = \int_A \lambda_A \cdot \omega.$$

Using $\tilde{\omega}$ we can create two forms,

$$\lambda_A \cdot \omega - \tilde{\omega},$$

$$\lambda_B \cdot \omega + \tilde{\omega},$$

with support in A and B, respectively. From our constructions it follows that they both have integral zero. Thus, we can by assumption find ψ_A and ψ_B with support in A and B, respectively, such that

$$d\psi_A = \lambda_A \cdot \omega - \tilde{\omega},$$

$$d\psi_B = \lambda_B \cdot \omega + \tilde{\omega}.$$

Then we get a globally defined form $\psi = \psi_A + \psi_B$ with

$$d\psi = \lambda_A \cdot \omega - \tilde{\omega} + \lambda_B \cdot \omega + \tilde{\omega}$$

$$= (\lambda_A + \lambda_B) \cdot \omega$$

$$= \omega.$$

The theorem now follows by using induction on the number of charts it takes to cover M.   □

The above proof indicates that it might be more convenient to work with compactly supported forms. This leads us to *compactly supported cohomology*, which is defined as follows: Let $\Omega_c^p(M)$ denote the compactly supported p-forms. With this we have the compactly supported exact and closed forms $B_c^p(M) \subset Z_c^p(M)$ (note that $d : \Omega_c^p(M) \to \Omega_c^{p+1}(M)$). Then define

$$H_c^p(M) = \frac{Z_c^p(M)}{B_c^p(M)}.$$

Needless to say, for closed manifolds the two cohomology theories are identical. For open manifolds, on the other hand, we have that the closed 0-forms must be zero, as they also have to have compact support. Thus $H_c^0(M) = \{0\}$ if M is not closed.

Note that only proper maps $f : M \to N$ have the property that they map $f^* : \Omega_c^p(N) \to \Omega_c^p(M)$. In particular, if $A \subset M$ is open, we do not have a map $H_c^p(M) \to H_c^p(A)$. Instead we observe that there is a natural inclusion $\Omega_c^p(A) \to \Omega_c^p(M)$, which induces

$$H_c^p(A) \to H_c^p(M).$$

The above proof, stated in our new terminology, says that

$$H_c^n(M) \to \mathbb{R},$$

$$[\omega] \to \int_M \omega$$

is an isomorphism for oriented n-manifolds. Moreover, using that $B(0,1) \subset S^n$, we can easily prove the following version of the Poincaré lemma:

$$H_c^p(B(0,1)) = \begin{cases} 0, & p \neq n, \\ \mathbb{R}, & p = n. \end{cases}$$

In order to carry out induction proofs with this cohomology theory, we also need a Meyer-Vietoris sequence:

$$\cdots \leftarrow H_c^p(M) \leftarrow H_c^p(A) \oplus H_c^p(B) \leftarrow H_c^p(A \cap B) \leftarrow H_c^{p+1}(M) \leftarrow \cdots.$$

This is established in the same way as before using the diagram

$$\begin{array}{ccccccccc}
0 & \leftarrow & \Omega_c^{p+1}(M) & \leftarrow & \Omega_c^{p+1}(A) \oplus \Omega_c^{p+1}(B) & \leftarrow & \Omega_c^{p+1}(A \cap B) & \leftarrow & 0 \\
& & \uparrow d & & \uparrow d & & \uparrow d & & \\
0 & \leftarrow & \Omega_c^p(M) & \leftarrow & \Omega_c^p(A) \oplus \Omega_c^p(B) & \leftarrow & \Omega_c^p(A \cap B) & \leftarrow & 0.
\end{array}$$

THEOREM 92. *Let M be an oriented n-manifold that can be covered by finitely many charts. The pairing*

$$\begin{array}{rcl}
H^p(M) \times H_c^{n-p}(M) & \to & \mathbb{R}, \\
([\omega], [\psi]) & \to & \int_M \omega \wedge \psi
\end{array}$$

is well-defined and nondegenerate. In particular, the two cohomology groups $H^p(M)$ and $H_c^{n-p}(M)$ are dual to each other and therefore have the same dimension as finite-dimensional vector spaces.

PROOF. We proceed by induction on the number of charts it takes to cover M. For the case $M = B(0,1)$, this theorem follows from the two versions of the Poincaré lemma. In general suppose $M = A \cup B$, where the theorem is true for A, B, and $A \cap B$. Note that the pairing gives a natural map

$$H^p(N) \to \left(H_c^{n-p}(N)\right)^* = \text{Hom}\left(H_c^{n-p}(N), \mathbb{R}\right)$$

for any manifold N. We apparently assume that this map is an isomorphism for $N = A, B, A \cap B$. Using that taking duals reverses arrows, we obtain a diagram where the left- and right most columns have been eliminated

$$\begin{array}{ccccccc}
\to & H^p(A \cap B) & \to & H^{p+1}(M) & \to & H^{p+1}(A) \oplus H^p(B) & \to \\
& \downarrow & & \downarrow & & \downarrow & \\
\to & (H_c^p(A \cap B))^* & \to & \left(H_c^{p+1}(M)\right)^* & \to & \left(H^{p+1}(A)\right)^* \oplus (H^p(B))^* & \to.
\end{array}$$

Each square in this diagram is either commutative or anticommutative (i.e., commutes with a minus sign.) As all vertical arrows, except for the middle one, are assumed to be isomorphisms, we see by a simple diagram chase (the five lemma) that the middle arrow is also an isomorphism. $\square$

COROLLARY 49. *On a closed oriented n-manifold M we have that $H^p(M)$ and $H^{n-p}(M)$ are isomorphic.*

7. Degree Theory

Given the simple nature of the top cohomology class of a manifold, we see that maps between manifolds of the same dimension can act only by multiplication on the top cohomology class. We shall see that this multiplicative factor is in fact an integer, called the *degree* of the map.

To be precise, suppose we have two oriented n-manifolds M and N and also a proper map $f : M \to N$. Then we get a diagram

$$
\begin{array}{ccc}
H_c^n(N) & \xrightarrow{f^*} & H_c^n(M) \\
\downarrow \int & & \downarrow \int \\
\mathbb{R} & \xrightarrow{d} & \mathbb{R}.
\end{array}
$$

Since the vertical arrows are isomorphisms, the induced map f^* yields a unique map $d : \mathbb{R} \to \mathbb{R}$. This map must be multiplication by some number, which we call the degree of f, denoted by $\deg f$. Clearly, the degree is defined by the property

$$
\int_M f^* \omega = \deg f \cdot \int_N \omega.
$$

LEMMA 73. *If $f : M \to N$ is a diffeomorphism between oriented n-manifolds, then $\deg f = \pm 1$, depending on whether f preserves or reverses orientation.*

PROOF. Note that our definition of integration of forms is independent of co-ordinate changes. It relies only on a choice of orientation. If this choice is changed then the integral changes by a sign. This clearly establishes the lemma. □

THEOREM 93. *If $f : M \to N$ is a proper map between oriented n-manifolds, then $\deg f$ is an integer.*

PROOF. The proof will also give a recipe for computing the degree. First, we must appeal to Sard's theorem. This theorem ensures that we can find $y \in N$ such that for each $x \in f^{-1}(y)$ the differential $Df : T_x M \to T_y N$ is an isomorphism. The inverse function theorem then tells us that f must be a diffeomorphism in a neighborhood of each such x. In particular, the preimage $f^{-1}(y)$ must be a discrete set. As we also assumed the map to be proper, we can conclude that the preimage is finite: $\{x_1, \ldots, x_k\} = f^{-1}(y)$. We can then find a neighborhood U of y in N, and neighborhoods U_i of x_i in M, such that $f : U_i \to U$ is a diffeomorphism for each i. Now select $\omega \in \Omega_c^n(U)$ with $\int \omega = 1$. Then we can write

$$
f^* \omega = \sum_{i=1}^{k} f^* \omega|_{U_i},
$$

where each $f^* \omega|_{U_i}$ has support in U_i. The above lemma now tells us that

$$
\int_{U_i} f^* \omega|_{U_i} = \pm 1.
$$

Hence,

$$
\begin{aligned}
\deg f &= \deg f \cdot \int_N \omega \\
&= \deg f \cdot \int_U \omega \\
&= \int_M f^* \omega \\
&= \sum_{i=1}^{k} \int_{U_i} f^* \omega|_{U_i}
\end{aligned}
$$

is an integer.   □

Note that $\int_{U_i} f^* \omega|_{U_i}$ is ± 1, depending simply on whether f preserves or reverses the orientations at x_i. Thus, the degree simply counts the number of preimages for regular values with sign. In particular, a finite covering map has degree equal to the number of sheets in the covering.

On an oriented Riemannian manifold (M, g) we always have a canonical volume form denoted by $d\text{vol}_g$. Using this form, we see that the degree of a map between closed Riemannian manifolds $f : (M, g) \to (N, h)$ can be computed as

$$\deg f = \frac{\int_M f^* (d\text{vol}_h)}{\text{vol}(N)}.$$

In case f is locally a Riemannian isometry, we must have that:

$$f^* (d\text{vol}_h) = \pm d\text{vol}_g.$$

Hence,

$$\deg f = \pm \frac{\text{vol}M}{\text{vol}N}.$$

This gives the well-known formula for the relationship between the volumes of Riemannian manifolds that are related by a finite covering map.

8. Further Study

There are several texts that expand on the material covered here. The book by Warner [**92**] is more than sufficient for most purposes. There is also a very nice book by Bott and Tu [**16**] that in addition covers characteristic classes. This book only has the small defect that it doesn't mention how one can compute characteristic classes using curvature forms. This can, however, be found in [**87**, vol. V]. The more recent book [**63**] is also an excellent book that is easy to read.

Bibliography

[1] C.B. Allendoerfer and A. Weil, *The Gauss-Bonnet theorem for Riemannian polyhedra*, TAMS 53 (1943) 101-129.

[2] M. Anderson, *Short Geodesics and Gravitational Instantons*, J. Diff. Geo. 31 (1990) 265-275.

[3] M. Anderson, *Metrics of positive Ricci curvature with large diameter*, Manu. Math. 68 (1990) 405-415.

[4] C. S. Aravinda and F. T. Farrell, *Nonpositivity: curvature vs. curvature operator*. Proc. Amer. Math. Soc. 133 (2005), no. 1, 191–192

[5] V.I. Arnol'd, *Mathematical methods of classical mechanics*, New York: Springer-Verlag, 1989.

[6] W. Ballmann, *Non-positively curved manifolds of higher rank*, Ann. Math. 122 (1985) 597-609.

[7] W. Ballmann, *Spaces of non-positive curvature*, Basel: Birkhäuser, 1995.

[8] W. Ballmann and P. Eberlein, *Fundamental groups of manifolds of non-positive curvature*, J. Diff. Geo. 25 (1987) 1-22.

[9] W. Ballmann, V. Schroeder and M. Gromov, *Manifolds of non-positive curvature*, Boston: Birkhäuser, 1985.

[10] M. Berger, *Riemannian geometry during the second half of the twentieth century*, University Lecture series, AMS 17.

[11] A.L. Besse, *Einstein Manifolds*, Berlin-Heidelberg: Springer-Verlag, 1978.

[12] A.L. Besse, *Manifolds all of whose geodesics are closed*, Berlin-Heidelberg: Springer-Verlag, 1987.

[13] R.L. Bishop and R.J. Crittenden, *Geometry of Manifolds*, New York: Academic Press, 1964.

[14] R.L. Bishop and S.I. Goldberg, *Tensor analysis on manifolds*, Dover, 1980.

[15] A. Borel, *Seminar on Transformation Groups*, Ann. Math. Studies 46. Princeton: Princeton University Press 1960.

[16] R. Bott and L.W. Tu, *Differential forms in algebraic topology*, New York: Springer-Verlag, 1982.

[17] K. Burns and R. Spatzier, *On topological Tits buildings and their classification*, IHES Publ. Math. 65 (1987) 5-34.

[18] M.P. do Carmo, *Differential forms and applications*, Berlin-Heidelberg: Springer Verlag, 1994.

[19] M.P. do Carmo, *Riemannian Geometry*, Boston: Birkhäuser, 1993.

[20] I. Chavel, *Riemannian Geometry, A Modern Introduction*, New York: Cambridge University Press, 1995.

[21] J. Cheeger, *Comparison and finiteness theorems for Riemannian manifolds*, Ph. D. thesis, Princeton University.

[22] J. Cheeger, *Finiteness theorems for Riemannian manifolds*, Am. J. Math. 92 (1970), 61-75.

[23] J. Cheeger, *Pinching theorems for a certain class of Riemannian manifolds*, Am. J. Math. 92 (1970), 807-834.

[24] J. Cheeger et al., *Geometric Topology: Recent developments*, LNM 1504, Berlin-Heidelberg: Springer-Verlag, 1991.

[25] J. Cheeger and T. H. Colding, *On the structure of space with Ricci curvature bounded below*, J. Diff. Geo. 46 (1997), 406-480.

[26] J. Cheeger and D. G. Ebin, *Comparison Theorems in Riemannian Geometry*, New York: North-Holland/Elsevier, 1975.

[27] J. Cheeger and D. Gromoll, *The splitting theorem for manifolds of non-negative Ricci curvature*, J. Diff. Geo. 6 (1971) 119-128.

[28] S.S. Chern, ed., *Global Geometry and Analysis*, 2nd edition, MAA Studies 27, Washington: Mathematical Assocoation of America, 1989.

[29] T. H. Colding, *Ricci curvature and volume convergence*, Ann. Math. 145 (1997) 477-501.

[30] R. Courant and D. Hilbert, *Methods of Mathematical Physics*, vol. II, New York: Wiley Interscience, 1962.

[31] B. Chow and D. Yang, *Rigidity of non-negatively curved compact quaternionic-Kähler manifolds*, J. Diff. Geo 29 (1989) 361-372.

[32] X. Dai, P. Petersen and G. Wei, *Integral pinching theorems*, Manuscripta Math. 101 (2000), no. 2, 143-152.

[33] D. DeTurk and J. Kazdan, *Some regularity theorems in Riemannian geometry*, Ann. scient. Éc. Norm. Sup. 14 (1981) 249-260.

[34] P. Eberlein, *Geometry of Nonpositively Curved Manifolds*, Chicago: The University of Chicago Press, 1996.

[35] R. Edwards and R. Kirby, *Deformations of spaces of embeddings*, Ann. of Math. 93 (1971), 63-88.

[36] J.-H. Eschenburg, *Local convexity and non-negative curvature- Gromov's proof of the sphere theorem*, Invt. Math. 84 (1986) 507-522.

[37] J.-H. Eschenburg and E. Heintze, *An elementary proof of the Cheeger-Gromoll splitting theorem*, Ann. Glob. Ana. and Geo. 2 (1984) 141-151.

[38] T. Farrell and L. Jones, *Negatively curved manifolds with exotic smooth structures*, J. AMS 2 (1989) 899-908.

[39] K. Fukaya, *Hausdorff convergence of Riemannian manifolds and its applications*, Advanced Studies in Pure Math. 18-I (1990) Recent topics in Differential and Analytic Geometry pp143-238.

[40] S. Gallot, *Isoperimetric inequalities based on integral norms of Ricci curvature*, Astérisque, 157-158, (1988) pp191-216.

[41] S. Gallot, D. Hulin and J. Lafontaine, *Riemannian Geometry*, Berlin-Heidelberg: Springer-Verlag, 1987.

[42] S. Gallot and D. Meyer, *Opérateur de courbure et laplacien des formes différentielles d'une variété riemannianne*, J. Math. Pures. Appl. 54 (1975), 259-284.

[43] G. Gibbons and S. Hawking, *Gravitational multi-instantons*, Phys. Lett. B 78 (1978) 430-432.

[44] D. Gilbarg and N.S. Trudinger, *Elliptic Partial Differential Equations of Second Order*, 2nd edition, Berlin-Heidelberg: Springer-Verlag, 1983.

[45] R. Greene and S.T. Yau, eds., Proc. Symp. Pure Math. 54 vol 3 (1994).

[46] M. Gromov, *Manifolds of negative curvature*, J. Diff. Geo. 12 (1978) 223-230.

[47] M. Gromov, *Metric Structures for Riemannian and Non-Riemannian Spaces*, Boston: Birkhäuser, 1999.

[48] M. Gromov and W. Thurston, *Pinching constants for hyperbolic manifolds*, Invt. Math. 89 (1987) 1-12.

[49] K. Grove, H. Karcher, and E. Ruh, *Group actions and curvature*, Invt. Math. 23 (1974), 31-48.

[50] K. Grove and P. Petersen, eds., *Comparison Geometry*, MSRI publications vol. 30, New York: Cambridge University Press, 1997.

[51] K. Grove, P. Petersen, and J.-Y. Wu, *Geometric finiteness theorems via controlled topology*, Invt. Math. 99 (1990) 205-213, *Erratum* Invt. Math. 104 (1991) 221-222.

[52] R.S. Hamilton, *The formation of singularities in the Ricci flow*, Surveys in Diff. Geo. vol. 2, International Press (1995) 7-136.

[53] S. Helgason, *Differential Geometry, Lie Groups and Symmetric spaces*, New York-London: Academic Press, 1962.

[54] W.-Y. Hsiang and B. Kleiner, *On the topology of positively curved 4-manifolds with symmetry*, J. Diff. Geo. 29 (1989), 615-621.

[55] M. Ise and M. Takeuchi, *Lie Groups I, II*, Translations of Mathematical Monographs vol 85. Providence: AMS, 1991.

[56] J. Jost, *Riemannian Geometry and Geometric Analysis*, Berlin-Heidelberg: Springer-Verlag, 1995.

[57] J. Jost and H. Karcher, *Geometrische Methoden zur Gewinnung von a-priori-Schranken für harmonische Abbildungen*, Manu. Math. 19 (1982) 27-77.

[58] W. Klingenberg, *Riemannian geometry*. Second edition. Berlin: Walter de Gruyter & Co., 1995.

[59] S. Kobayashi, *Transformation Groups in Differential Geometry*, Berlin-Heidelberg: Springer-Verlag, 1972.

[60] S. Kobayashi and K. Nomizu, *Foundations of Differential Geometry*, vols. I, II, New York: Wiley-Interscience, 1963.

[61] H.B. Lawson Jr. and M.-L. Michelsohn, *Spin Geometry*, Princeton: Princeton University Press, 1989.

[62] J.M. Lee and T.H. Parker, *The Yamabe problem*, Bull. AMS 17 (1987), 37-91.

[63] J.M. Lee, *Introduction to Smooth Manifolds*, New York: Springer Verlag, 2003.

[64] A. Lichnerowicz, *Géométrie des groupes de transformations*, Paris: Dunod, 1958.

[65] A. Lichnerowicz, *Propagateurs et Commutateurs en relativité générale*, Publ. Math. IHES 10 (1961) 293-343.

[66] M.J. Micaleff and J.D. Moore, *Minimal 2-spheres and the topology of manifolds with positive curvature on totally isotropic 2-planes*, Ann. of Math. 127 (1988), 199-227.

[67] R.K. Miller and A.N. Michel, *Ordinary differential equations*, New York-London: Academic Press, 1982.

[68] J.W. Milnor, *Morse Theory*, Princeton: Princeton University Press, 1963.

[69] J.W. Milnor and J.D. Stasheff, *Characteristic Classes*, Princeton: Princeton University Press, 1974.

[70] C.W. Misner, K.S. Thorne and J.A. Wheeler, *Gravitation*, New York: Freeman, 1973.

[71] N. Mok, *The uniformization theorem for compact Kähler manifolds of non-negative holomorphic bi-sectional curvature*, J. Diff. Geo. 27 (1988), 179-214.

[72] J.W. Morgan, *The Seiberg-Witten equations and applications to the topology of smooth four-manifolds*, Princeton: Princeton Univ. Press, 1996.

[73] B. O'Neill, *Semi-Riemannian Geometry*, New York-London: Academic Press, 1983.

[74] Y. Otsu, *On manifolds of positive Ricci curvature with large diameters*, Math. Z. 206 (1991) 255-264.

[75] M. Özaydın and G. Walschap, *Vector bundles with no soul*, PAMS 120 (1994) 565-567.

[76] G. Perel'man, *Alexandrov's spaces with curvatures bounded from below II*, preprint.

[77] S. Peters, *Cheeger's finiteness theorem for diffeomorphism classes of manifolds*, J. Reine Angew. Math. 349 (1984) 77-82.

[78] P. Petersen, *Aspects of Global Riemannian Geometry*, Bull AMS 36 (1999), 297-344.

[79] P. Petersen, S. Shteingold and G. Wei, *Comparison geometry with integral curvature bounds*, Geom Func Anal 7 (1997), 1011-1030.

[80] P. Petersen and G. Wei, *Relative volume comparison with integral Ricci curvature bounds*, Geom Func Anal 7 (1997), 1031-1045.

[81] G. de Rham, *Differentiable Manifolds*, Berlin-Heidelberg: Springer-Verlag, 1984.

[82] X. Rong, *The almost cyclicity of the fundamental groups of positively curved manifolds*, Invt. Math. 126 (1996) 47-64. 126 (1996) 47-64. And *Positive curvature, local and global symmetry, and fundamental groups*, preprint, Rutgers, New Brunswick.

[83] X. Rong, *A Bochner Theorem and Applications*, Duke Math. J. 91 (1998), 381-392.

[84] X. Rong and X. Su, *The Hopf Conjecture for Manifolds with Abelian Group Actions*, Comm. Contemp. Math. 1 (2005) 1-16.

[85] Y. Shen and S.-h. Zhu, *A sphere theorem for 3-manifolds with positive Ricci curvature and large diameter*, preprint, Dartmouth.

[86] E.H. Spanier, *Algebraic Topology*, New York-Berlin-Heidelberg: Springer-Verlag, 1966.

[87] M. Spivak, *A Comprehensive Introduction to Differential Geometry*, vols. I-V, Wilmington: Publish or Perish, 1979.

[88] J. J. Stoker, *Differential Geometry*, New York: Wiley-Interscience, 1989.

[89] S. Tachibana, *A theorem on Riemannian manifolds of positive curvature operator*, Proc. Japan Acad. 50 (1974), 301-302.

[90] M.E. Taylor, *Partial differential equations*, vols. I-III, New York: Springer Verlag, 1996.

[91] T.Y. Thomas, *Riemann spaces of class one and their characterization*, Acta Math. 67 (1936) 169-211.

[92] F.W. Warner, *Foundations of Differentiable Manifolds and Lie Groups*, New York: Springer-Verlag, 1983.

[93] H. Weyl, *The classical groups*, Princeton: Princeton University Press, 1966.

[94] B. Wilking, *Index parity of closed geodesics and rigidity of Hopf fibrations.* Invent. Math. 144 (2001), no. 2, 281–295.

[95] B. Wilking, *Torus actions on manifolds of positive curvature,* Acta Math. 191 (2003), 259–297.

[96] J. Wolf, *Spaces of Constant Curvature,* Wilmington: Publish or Perish, 1984.

[97] H.H. Wu, *The Bochner Technique in Differential Geometry,* Mathematical Reports vol. 3, part 2, London: Harwood Academic Publishers, 1988.

Index

Graduate Texts in Mathematics

(continued from page ii)

《国外数学名著系列》(影印版)

(按初版出版时间排序)